The Neurolab Spacelab Mission: Neuroscience Research in Space

Results from the STS-90, Neurolab Spacelab Mission

NASA SP-2003-535

Edited by:
Jay C. Buckey, Jr., M.D.
Jerry L. Homick, Ph.D.

National Aeronautics and
Space Administration

Lyndon B. Johnson Space Center
Houston, Texas

Library of Congress Cataloging-in-Publication Data

The Neurolab Spacelab mission : neuroscience research in space : results
from the STS-90, Neurolab Spacelab mission / edited by Jay C. Buckey
Jr., Jerry L. Homick.
     p. ; cm.
     "NASA SP-2003-535."
     Includes bibliographical references and index.
     1. Central nervous system—Effect of space flight on. 2. Space
flight—Physiological effect. 3. Spacelab Program.
[DNLM: 1. Spacelab Program. 2. Nervous System Physiology.
3. Weightlessness. 4. Adaptation, Physiological. 5. Neurosciences.
6. Research Design. 7. Space Flight. WL 102 N494513 2003] I. Buckey,
Jay C. II. Homick, Jerry L., 1942-
RC1151.C45N48 2003
616.9'80214—dc21

                                     2003000685

Foreword

Neurolab (STS-90) represents a major scientific achievement that built upon the knowledge and capabilities developed during the preceding 15 successful Spacelab module missions. NASA proposed a dedicated neuroscience research flight in response to a Presidential declaration that the 1990's be the Decade of the Brain. Criteria were established for selecting research proposals in partnership with the National Institutes of Health (NIH), the National Science Foundation, the Department of Defense, and a number of the International Space Agencies. The resulting Announcement of Opportunity for Neurolab in 1993 resulted in 172 proposals from scientists worldwide. After an NIH-managed peer review, NASA ultimately selected 26 proposals for flight on the Neurolab mission.

The mission presented an unprecedented challenge of technical complexity with sophisticated experiments representing a wide range of neuroscience research. These experiments represented a number of historic "firsts" in space life science research. The successful demonstration of microneurography, extracellular multiple unit recording, general anesthesia, and complex microsurgical procedures have set the stage for transition to the International Space Station as a research platform. The data from Neurolab will have far-reaching implications in the neuroscience community while helping answer terrestrial research questions and furthering the goals of human space exploration.

Educational outreach achievements were as impressive as the outstanding scientific productivity of the mission. Educational activities ranged from the inclusion of histological specimens from the collection of Santiago Ramón y Cajal, 1906 Nobel Laureate in Medicine, to the Morehouse School of Medicine educational curriculum, televised documentary productions, interactive websites, and the Professor in Space program. This mission captured the minds of both the young and old in their quest to understand neuroscience.

Finally, on the behalf of the STS-90 crew, it was an honor and pleasure to work with the many world-renowned scientists on the investigator teams. The success of the Neurolab mission was the result of a highly coordinated team effort. There were scores of individuals—managers, support scientists, engineers, and technicians—in the United States, Europe, Canada, and Japan whose dedicated efforts contributed significantly to the ultimate success of the mission. Not all can be named here but several stand out as leaders in their respective areas of responsibility.

Neurolab was the brainchild of three people: Frank Sulzman from NASA Headquarters, Rodolfo Llinás from NYU School of Medicine, and J. Wally Wolf of Universities Space Research Association. They provided the visionary inspiration for the creation of the mission. Dr. Mary Ann Frey, the Neurolab Program Scientist, ensured that the overall scientific integrity of the mission was maintained. Ms. LeLe Newkirk, the Neurolab Mission Manager, enabled the overall integration of the experiment hardware systems into the Shuttle and Spacelab and ensured that the scientific goals of the mission were not compromised as the experiments moved from the laboratory to flight. Her advocacy for the science on Neurolab made a tremendous difference.

Dr. Mel Buderer and Ms. Suzanne McCollum at the Johnson Space Center led the implementation of the human experiments on the mission, while Dr. Louis Ostrach and Mr. Chris Maese at the Ames Research Center led the implementation of the animal experiments. Dr. Smith Johnston, the Neurolab Flight Surgeon, assured the health and safety of the flight crew. Dr. Joe Bielitski, the NASA Chief Veterinarian, provided oversight for the housing, handling, and welfare of the various animal species flown on the mission. The international Neurolab Steering Committee provided invaluable recommendations regarding the selection of experiments for the mission and assisted in the resolution of experiment-related issues as they arose. U.S members of the committee included Dr. Joel Davis from the Office of Naval Research; Dr. Kathy Olsen and Dr. Christopher Platt, both from the National Science Foundation; and Dr. Anthony Demsey, Dr. William Heetderks, Dr. James Kiley, Dr. Andrew Monjan, and Dr. Daniel Sklare, all representing the National Institutes of Health. International members of the Steering Committee were Dr. Antonio Guell of the French Space Agency; Dr. Alan Mortimer of the Canadian Space Agency; Dr. Shunji Nagaoka of the Japanese Space Agency; Dr. Heinz Oser of the European Space Agency; and Dr. Gunter Ruyters of the German Space Agency.

We would like to thank all of those who helped put the mission together and all of the many trainers who ensured that we were up to the task of completing it successfully.

Dave Williams MD FCFP FRCP
Mission Specialist 3 (MS3) STS-90 Neurolab

Director of the Space and Life Sciences Directorate
Johnson Space Center
August 2002

This book is dedicated to
Rodolfo Llinás, Frank Sulzman, and J. Wally Wolf,
the initiators of the Neurolab mission

The Neurolab mission carried slides made by the Nobel prize winning neuroscientist Santiago Ramón y Cajal. This poster commemorates his work and shows the growth in neuroscience from his ground-breaking experiments at the beginning of the 20th century, to the ability to fly complex neuroscience experiments in Earth orbit at the century's end.

Preface

Neurolab was an extraordinary space mission. The 16-day flight was dedicated to one area of research: the brain and nervous system. The mission brought together an international complement of scientists, all of whom completed a demanding, peer-reviewed science program. The Neurolab crew, which included competitively selected scientific payload specialists, was directly involved in developing the experiments for flight, as well as executing them in space. The resulting experiments fully demonstrated the capabilities of the Spacelab program. The researchers' findings will have long-term influence, both on biomedical research here on Earth and on future spaceflights.

Neurolab's focus on brain and nervous system research allowed for in-depth studies and provided a series of complementary results. Even though performing science experiments in weightlessness presents significant logistical and operational challenges, the guiding philosophy on Neurolab was to surmount operational challenges to meet the science needs, rather than alter the science to meet the demands of spaceflight. As a result, in most cases, the facilities in space on Neurolab were the equal of Earth-based laboratories. Where they weren't equivalent, every effort was made to provide comparable results.

Because of this effort, the Neurolab mission provides a remarkable scientific story. This story offers a valuable lesson, both because of the experimental results and the methods used to get them. The lessons learned on Neurolab can help guide future space research. Immediately after the mission, however, it was not clear how the Neurolab story would be told. The results from previous life science Spacelab missions were scattered throughout the scientific literature. For these flights, there was no single book where readers could find out about the goals of a mission and review all its accomplishments. In most medical libraries, the only book on the shelf with individual spaceflight mission results is the *Biomedical Results from Skylab*, published in 1975. This book does an excellent job of describing not only the scientific results, but also the equipment and facilities that were developed. It has offered a valuable guide for those planning long-duration spaceflights and showed just what kind of work is possible in space. It also served as the inspiration for this book on the Neurolab mission.

The Neurolab Spacelab Mission is designed to provide a general scientific reader with an overview of Neurolab. The book is divided into three sections. The first contains scientific reports from the individual investigator teams. The investigators describe their experiments and present their results. The reports are divided into five major research areas: the balance system; sensory integration and navigation; nervous system development in weightlessness; blood pressure control; and circadian rhythms, sleep, and respiration.

The book's second section provides technical reports. These are descriptions of particularly interesting procedures or pieces of equipment that were developed for the flight. As those who have been involved in space research know well, developing hardware or procedures for spaceflight is a demanding and challenging task. The fact that the equipment and procedures described here functioned as advertised is a tribute to the hard work of the people who developed them. This is not a complete presentation of all the experiment hardware developed for Neurolab, and we apologize for any items that have been omitted. All those who worked to develop solid scientific hardware—while also meeting the demands of space qualification—deserve recognition.

The last section in the book offers perspectives on the mission. Crewmembers describe what they felt were some of the most significant aspects of the flight. This section also includes discussion of problems that arose inflight and how those problems might be corrected in the future.

The aim of this book is to enable both generalists and experts to understand the science aboard Neurolab and get the information they need. Those who want just a general overview will find the section introductions helpful. In these introductions, illustrations provide an overview of the area and a summary describes the main findings. Readers who are interested in a particular area, but may not have an extensive background in it, can turn to the abstracts and introductions of the individual reports. These are written to provide a general scientific reader with a summary of the project and why it was done. Finally, readers who are knowledgeable about a particular area will find that individual reports have details comparable to what would appear in scientific literature. Also, reference lists will guide readers to the published papers from the experiments.

To make the book easy to use, each report is written to stand on its own. Interested readers can go directly to an individual report to find out about a given experiment; they do not have to review the section introductions or other reports to get background information. This approach has introduced some redundancy. For example, a description of the physiology underlying blood pressure control appears at the beginning of several reports. Similarly, the physiology and anatomy of the inner ear is discussed in various areas. Each of these descriptions, however, is slightly different and relevant to a particular experiment. We felt that the repetition was a reasonable price to pay for flexibility.

The book would not have been possible without the help and encouragement of many people. Sid Jones, from Indyne, at the NASA-Johnson Space Center, established the overall graphic design for the book and did the layout for all of the chapters. His dedication, professionalism, and artistic flair made a tremendous difference. Sharon Hecht, also from Indyne, provided invaluable help with editing the text, correcting errors, and maintaining a consistent style. William Scavone from Kestrel Illustrations created beautiful illustrations for the section introductions. A special thanks is due to Sarah Masters Buckey for her expert and thoughtful help with editing, even when the time was short and the hour was late. Dee Dee Thomas, from Lockheed Martin, was very helpful in finding pictures and information from the mission to supplement the reports. Thanks are due to Joan Austin for helping to keep things organized. We are also grateful to Judy Robinson, at NASA-Johnson Space Center, for her consistent support

Although the Neurolab mission was the last of the Spacelab series, we hope that the lessons from Neurolab can be passed on with this book.

Jay C. Buckey, Jr.
Hanover, New Hampshire

Jerry L. Homick
Houston, Texas

2002

Contents

Scientific Reports

Section 1 **The Balance System**

Background *How spaceflight affects balance*

INTRODUCTION

After returning from space, astronauts often find that even small head movements give an exaggerated sense of body movement. Inclining the head slightly, for example, produces a sensation of tumbling forward; a small sideways tilt feels like a significant lean. Everyday activities, like walking or climbing stairs, require concentration.

These sensations reflect changes in the balance system that occur in orbit. While crewmembers are weightless, their balance systems lose the up and down cues that are usually given by the inner ear. As a result, the balance system adapts to make the best use of the remaining senses. The crewmembers don't notice these changes until gravity is reintroduced during landing, and the balance system once again receives information about gravity.

The adaptation to weightlessness and the later re-adaptation to gravity are complex processes involving different brain areas, nervous pathways, and gravity sensors. One of the main goals of the Neurolab mission was to understand how the nervous system adapts to weightlessness and where in the brain the changes occur. This spaceflight research provides information that also can be used to understand balance disorders on Earth.

THE PHYSIOLOGY OF BALANCE

Plates 1 and 2 illustrate some of the key portions of the balance system. Plate 1 shows the main gravity sensor in the body—the inner ear. Within the inner ear apparatus are two critical gravity-sensing areas: the saccule (oriented vertically) and the utricle (oriented horizontally). These two sensors provide information important for determining whether the body is upright, tilting sideways, or pitching front-to-back. The utricle and saccule both contain a layer of nerve cells called a macula. The macula (see bottom left diagram in Plate 1) constitutes the core of the gravity sensor. Within the macula, the main gravity-sensing cells are called hair cells, because of their hairlike projections.

These hairs contact the otolithic membrane, a structure containing small crystals called otoliths. As shown on the right bottom of Plate 1, when gravity acts on the crystals in the membrane, this bends the hairs. The hair cells sense this bending and transmit the information to the brain. In weightlessness, when the membrane is no longer loaded by gravity, the hairs are not bent, and the brain loses this information about up and down.

The hairs also will be bent by inertia. Imagine a small stone sitting on the dashboard of a car. If the car accelerates forward quickly, the stone will move backward on the dashboard. Similarly, head accelerations (moving side-to-side, front-to-back, or tilting) will cause the otoliths to move on the hair cells. This information can be integrated with the information the brain receives from other senses (such as vision and position sense) to produce an integrated sense of body motion.

Plate 2 shows the pathways taken by information from the inner ear. Signals from the inner ear go to the brain stem, where they enter the vestibular nuclei. The vestibular nuclei, in turn, connect to the cerebellum—the part of the brain involved with the planning and execution of movements as well as with maintaining a stable posture. Other parts of the brain, such as the inferior olive, the locus coeruleus, and the areas controlling eye movements (the oculomotor, trochlear, and abducens nuclei) also receive information from the inner ear. These areas are important for sharing information on gravity with other body systems, such as the cardiovascular centers (to help maintain blood pressure when standing) and the eye movement areas (to help keep the eyes fixed on a target even when the body is moving).

As Plate 2 shows, eye movements are closely integrated with vestibular information. If a cameraman made a video of the goalpost while running down a football field, the result would be a bouncing and jerking image that would be difficult to view. Nevertheless, a player can run down the same field and maintain a clear, stable image of the goal. This example demonstrates the power of the reflexes connecting the vestibular system and the eyes (the vestibulo-ocular reflex). The vestibular system senses motion and quickly adjusts the eyes to maintain a stable view of the world.

1

This close connection between the vestibular system and the eyes can also be used in reverse to study the vestibular system. Since eye movements can be controlled, at least in part, by the vestibular system, measuring eye movements when the vestibular system is stimulated can reveal information about how the vestibular system interprets that stimulation.

For example, when a motorcycle rider enters a tight turn, the otoliths sense the combined effects of the turn (inertia) and gravity. The body, head, and eyes are all oriented towards the sum of these two forces, and the rider tilts into the turn. The forces the motorcycle rider experiences can be simulated with a centrifuge, as shown on the bottom left in Plate 2. As the crewmember rotates while seated in the chair, a sideways centripetal force is created. Gravity provides a downward force. Just like the motorcycle rider, the crewmember in the chair feels tilted when the rotator is spinning. An examination of eye movements during rotation reveals how the vestibular system is interpreting the inputs from centripetal acceleration and gravity.

THE NEUROLAB BALANCE EXPERIMENTS

Two of the Neurolab experiments looked at the gravity sensors themselves. Dr. Muriel Ross and her team extended their previous experiments on the connections among hair cells in the macula of the utricle and saccule. Her previous experiments and the Neurolab work show that in the absence of gravity, more connections are made among the hair cells, which might serve to change the sensitivity of the macula (i.e., the output from the gravity sensor for a given gravitational input). The experiment by Dr. Steven Highstein and his colleagues determined how the sensitivity of the utricle changes as a consequence of weightlessness exposure. Using novel electrode technology, they were able to show that the

output from the utricle in the oyster toadfish (*Opsanus tau*) was increased significantly after spaceflight.

A pair of experiments looked further into the nervous system. Dr. Ottavio Pompeiano and his group examined the parts of the rat brain where vestibular information travels (vestibular nuclei, inferior olive, locus coeruleus) and found marked changes in the nerve cells there both after entering weightlessness and upon reentering Earth's gravity. Dr. Gay Holstein and her laboratory uncovered changes within the parts of the rat cerebellum that receive input from the gravity-sensing organs of the inner ear. These findings show there are structural changes in the nervous system that accompany the adaptation to weightlessness.

Dr. Cohen and his team showed that in contrast to previous studies, the rolling response of the eyes to the accelerations received in the chair was not altered by spaceflight. This raised the intriguing possibility that the intermittent centrifugation received by the crew during the flight may have prevented the anticipated changes. The team from Dr. Clément's laboratory showed that the tilt sensations produced in the chair increased greatly during the flight, as might be expected when the gravitational acceleration is removed. Interestingly, this change did not occur right away in space (it took several days to develop); and in contrast to what had been predicted in the past, no crewmember ever felt translation (sideways movement) instead of tilt in the chair.

Taken together, these Neurolab experiments provided a detailed look at the vestibular system's adaptation to weightlessness—ranging from the sensations experienced by the crewmembers, to the changes that occur within the brain itself. These findings have practical benefits for the space program. They clearly show that, after weightlessness, astronauts reenter Earth's gravity with a nervous system that has changed significantly in space. Their responses to tilting and rolling motions will likely be altered, which should be taken into account when performing tasks that involve sensing motion, like flying or landing the Shuttle.

Plate 1: The Balance System: The Inner Ear

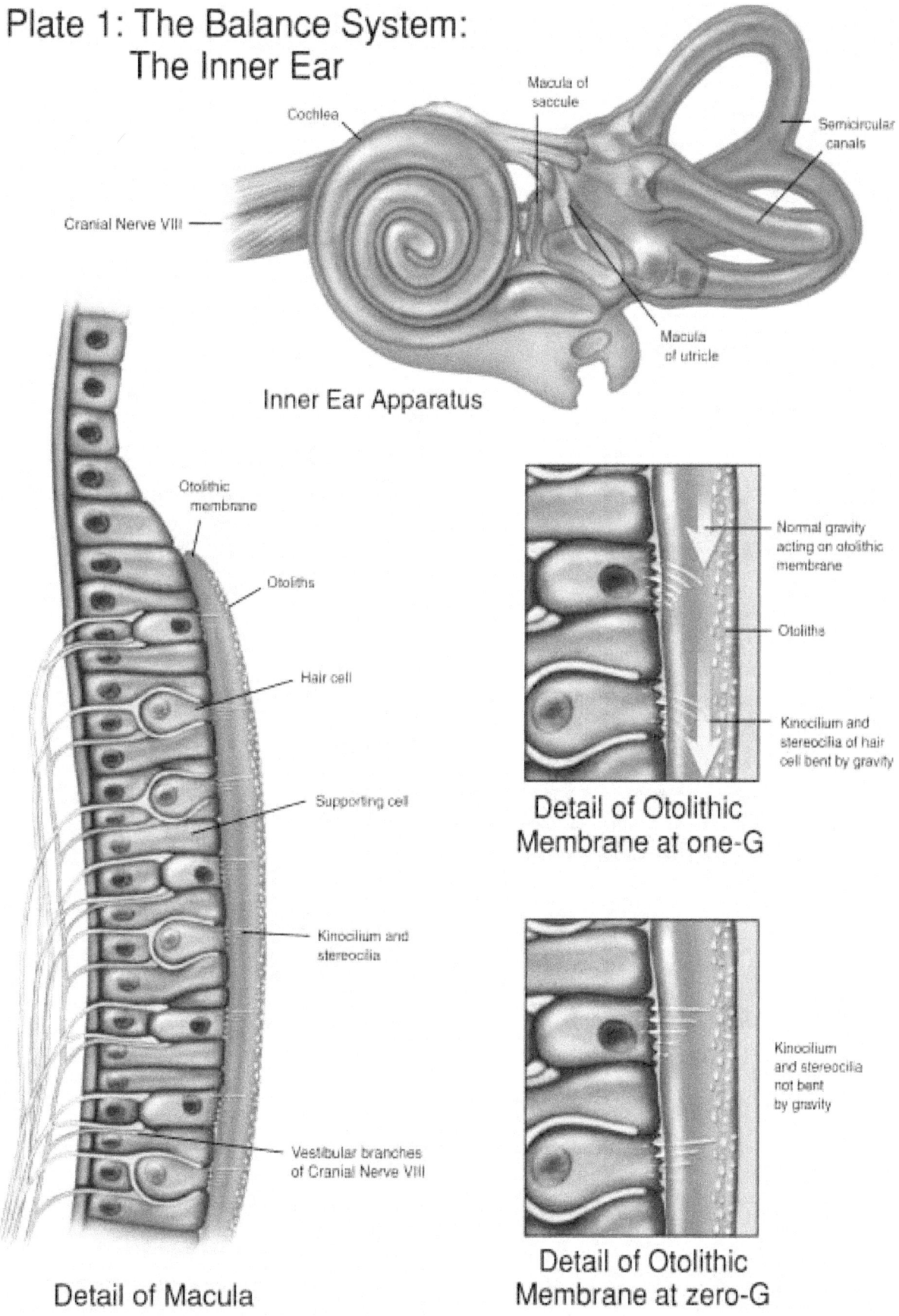

Cochlea

Cranial Nerve VIII

Macula of saccule

Semicircular canals

Macula of utricle

Inner Ear Apparatus

Otolithic membrane

Otoliths

Hair cell

Supporting cell

Kinocilium and stereocilia

Vestibular branches of Cranial Nerve VIII

Detail of Macula

Normal gravity acting on otolithic membrane

Otoliths

Kinocilium and stereocilia of hair cell bent by gravity

Detail of Otolithic Membrane at one-G

Kinocilium and stereocilia not bent by gravity

Detail of Otolithic Membrane at zero-G

Plate 2: The Balance System: Vestibular Ocular Reflex

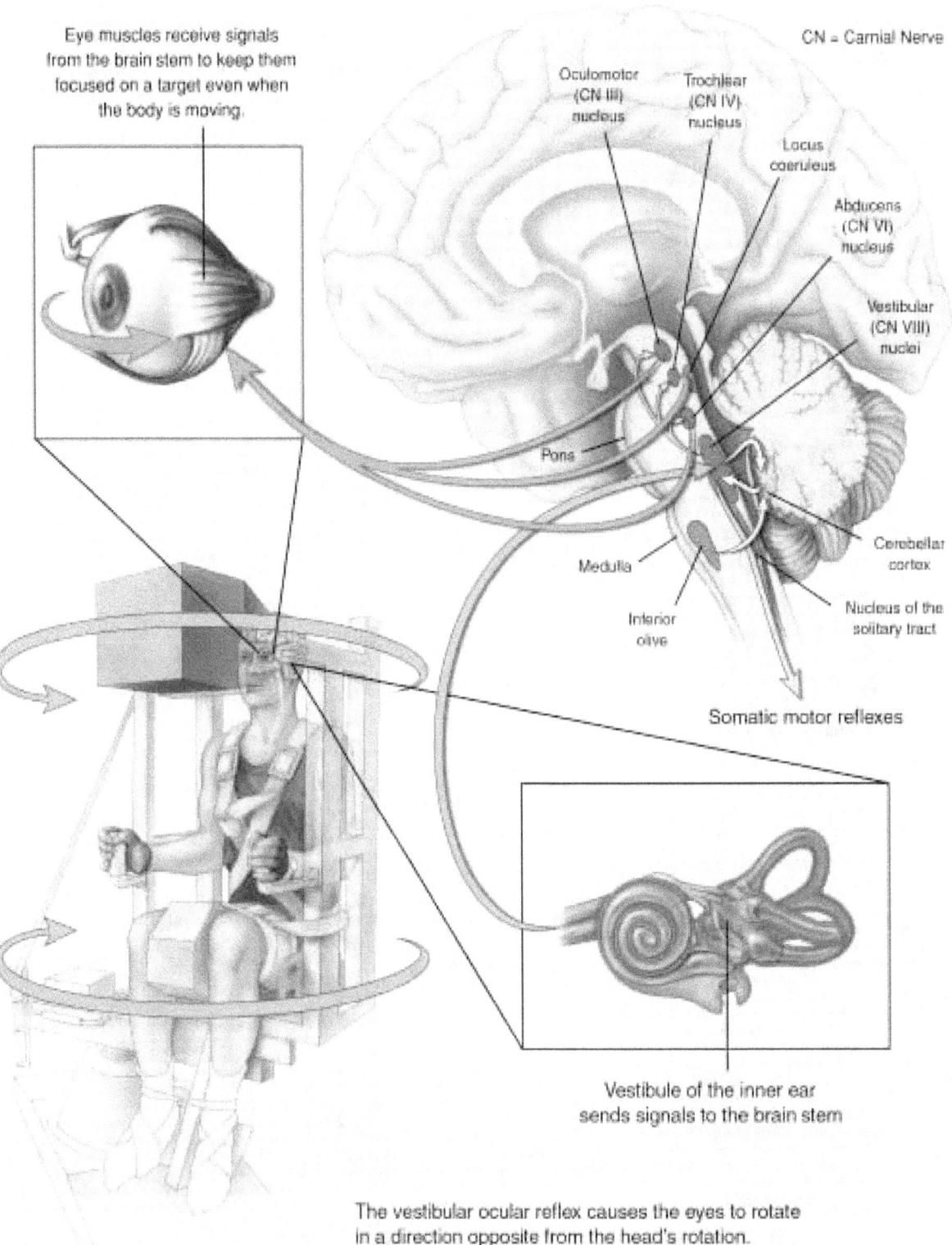

Eye muscles receive signals from the brain stem to keep them focused on a target even when the body is moving.

CN = Carnial Nerve

Oculomotor (CN III) nucleus

Trochlear (CN IV) nucleus

Locus coeruleus

Abducens (CN VI) nucleus

Vestibular (CN VIII) nuclei

Pons

Medulla

Inferior olive

Cerebellar cortex

Nucleus of the solitary tract

Somatic motor reflexes

Vestibule of the inner ear sends signals to the brain stem

The vestibular ocular reflex causes the eyes to rotate in a direction opposite from the head's rotation.

Perception of the Spatial Vertical During Centrifugation and Static Tilt

Authors

Gilles Clément, Alain Berthoz, Bernard Cohen,
Steven Moore, Ian Curthoys, Mingjia Dai,
Izumi Koizuka, Takeshi Kubo, Theodore Raphan

Experiment Team

Principal Investigator: **Gilles Clément[1]**

Co-Investigators: **Alain Berthoz[2]**
Bernard Cohen[3]
Steven Moore[3]
Ian Curthoys[4]
Mingjia Dai[3]
Izumi Koizuka[5]
Takeshi Kubo[6]
Theodore Raphan[7]

[1]Centre National de la Recherche Scientifique, Paris, France
[2]Collège de France, Paris, France
[3]Mount Sinai School of Medicine, New York City, USA
[4]University of Sydney, Sydney, Australia
[5]St. Marianna University School of Medicine, Kawasaki, Japan
[6]Osaka University, Osaka, Japan
[7]Brooklyn College, New York City, USA

ABSTRACT

The otolith organs in the inner ear utilize small crystals to sense both gravity and body movements. They determine whether an individual is upright or tilted by sensing gravity. They also sense whether an individual is translating; i.e., moving side-to-side, up-and-down, or from front-to-back. Often the inner ear receives a combination of inputs. For example, when a motorcycle rider turns a sharp corner, the rider's otoliths sense the combined effects of turning and gravity. In spaceflight, the inner ear no longer senses gravity because the astronauts are weightless. However, inputs from side-to-side, up-down, and front-back translation still persist. Since tilt with regard to gravity is essentially meaningless in space, it has been hypothesized that, as a result of the process of adaptation to weightlessness, the brain begins to reinterpret all otolith signals to indicate primarily translation, not tilt. This experiment tested that hypothesis using a human-rated centrifuge. This centrifuge generated constant centrifugal forces, similar to those experienced by the motorcycle rider in a tight turn. Astronauts riding in the centrifuge during orbital flight were tested to evaluate whether they perceived the centrifugal accelerations as tilt or translation. Results showed that all subjects perceived tilt, but that the magnitude of the perceived tilt changed throughout the mission. Only after two weeks in space did the perceived tilt match the actual direction of centripetal acceleration. The results indicate that orientation to gravitoinertial acceleration is maintained in space and the tilt-translation hypothesis was not supported. The findings also provided valuable experience for the use of artificial gravity during long-term or planetary exploration missions.

INTRODUCTION

On Earth, when a ball drops it accelerates in a straight line toward the ground. This is an example of a linear acceleration, and these kinds of accelerations are very common in everyday life. On the Earth's surface, two major sources of linear acceleration exist. One is related to the Earth's gravity. Gravity significantly affects most of our movement (motor) behavior (it has been estimated that about 60% of our musculature is devoted to opposing gravity), and it provides a constant reference for up and down. It is present under all conditions on Earth, and it forms one of the major pillars of spatial orientation (Howard, 1982; Schöne, 1984). Other sources of linear acceleration arise in the side-to-side, up-and-down, or front-to-back translations that commonly occur during walking or running, and from the centrifugal force that we feel when going around turns or corners. As Einstein noted, all linear acceleration is equivalent, whether it is produced by gravity or motion; and when we are in motion, the linear accelerations sum. The body responds to the resultant, and we tend to align our long body axis with the resultant linear acceleration vector, called the gravitoinertial acceleration (GIA) vector (Figure 1). For example, when a person is either walking or running around a turn, the inward linear acceleration is added to the upward gravitational acceleration to form a GIA that is tilted in toward the center of the turn. Unconsciously, the head, body, and eyes are oriented so that they tend to align with the GIA. The angle of tilt of these body parts depends on the speed of turning (Imai, 2001). Put in simple terms, people align with gravity when standing upright and tilt into the direction of the turn when in motion. If they don't, they lose balance and fall. Perhaps the most graphic illustration of the importance of body tilt toward the resultant linear acceleration vector is provided by motorcycle drivers, who tilt their machines 30-45 degrees into the direction of the turn to maintain their balance. People do the same when walking or running around a curve.

In spaceflight, gravitational force is no longer sensed because the spaceflight crews are experiencing the effects of microgravity. Also, since there is little locomotion in space, the exposure to centripetal forces is reduced; but the linear accelerations due to side-to-side, up-and-down, and front-to-back motions (translations) persist. Since tilt is meaningless in space (there is no vertical reference from gravity), it has been hypothesized that, during adaptation to weightlessness, the brain would reinterpret all otolith signals to indicate primarily translation, not tilt (Parker, 1985; Young, 1984). It was postulated that this adaptation within the brain underlies the amelioration of space motion sickness symptoms over time. This otolith tilt-translation reinterpretation (OTTR) hypothesis has received some support from perceptual studies done after spaceflight, but it had never been tested during spaceflight.

In this experiment, astronauts were rotated in a centrifuge. When the centrifuge started, they felt rotation, but this feeling of rotation disappeared after 30–45 seconds of rotation. The net effect of the centrifugation on Earth was that, when seated, the crewmembers felt as if they were tilted either 25 degrees to the side (at 0.5-G acceleration) or 45 degrees to the side (at one-G acceleration). When lying down, the crewmembers felt as if they were tilted backwards approximately 15 degrees (0.5-G) or 30 degrees (one-G). In space, when weightless, the input to the inner ear from gravity is gone, and the inner ear will only sense the accelerations due to chair rotation. When seated in the chair, this could mean that instead of feeling tilted 30 or 45 degrees, crewmembers would feel as if they were tilted 90 degrees (i.e., as if they were lying on their side). According to the OTTR hypothesis, however, during centrifugation in space crewmembers should not perceive themselves as being tilted 90 degrees relative to their perceived upright, but instead should feel as if they are being translated (moving to one side). The purpose of this study was to determine whether in space the astronauts felt a sense of tilt or translation during constant-velocity centrifugation, as compared to their original position before the centrifuge started (Figure 1).

METHODS

Seven subjects participated in this study. Four astronauts were tested before, during, and after the STS-90 Neurolab mission. During the spaceflight, the astronauts were tested on flight days (FDs) 2, 7, 10, 11, 12, and 16. Baseline data were collected on the same subjects 90 days (L–90), 60 days (L–60), 30 days (L–30), and 15 days (L–15) days prior to flight using a replicate of the flight centrifuge. The same tests were repeated 24 hours following landing (R+1) and on subsequent days (R+2, R+4, and R+9) for studying re-adaptation of the response to the presence of Earth's gravitational acceleration. Two other subjects were tested during preflight testing only (L–90, L–60, L–30, and L–15 days). Finally, one astronaut was tested at L–30 and L–15 days before the nine-day STS-86 space mission, and again on R+3 and R+4 days after the landing. The astronauts' reports were also collected during all training sessions using the centrifuge in order to evaluate a possible training effect.

The equipment used for this experiment is described in detail in a technical report (see technical report by Cohen et al. in this publication). The subjects were rotated on a short-arm centrifuge, either in darkness or during the presentation of visual stimuli. The results presented in this report concern only those recorded when the subjects were in complete darkness. The centrifuge was accelerated in complete darkness and after 40 seconds; i.e., after the perception of rotation ceased (and the associated eye movements, called nystagmus, had also stopped). The subjects were then prompted by an operator to verbally report whether they had a perceived sensation of tilt or motion during steady-state centrifugation. A typical trial consisted of clockwise (CW) and counterclockwise (CCW) rotation with the left-ear-out (LEO) orientation, CCW and CW rotation with the right-ear-out (REO) orientation, and CW rotation with the lying-on-back (LOB) orientation. The basis for the perceptual measurements on Earth was whether subjects perceived tilt of their body vertical relative to their perception of the spatial vertical. For the LEO/REO orientations, subjects used a scale of 0–90 degrees to represent their roll tilt perception, where zero degree indicated that the subjects felt upright and 90 degrees meant that they felt as if they were "lying-on-side." A similar scale was used in the LOB configuration. If the subject felt horizontal, this would be reported as zero tilt, and a tilt of –90 degrees indicated that the subjects felt "upside-down." Inflight, subjects had no perception of tilt when the centrifuge was stationary. When exposed to the 0.5-G or one-G centripetal acceleration during rotation, if subjects felt as if they were tilted, they used the same criteria as on Earth to report their tilt relative to a perceived spatial vertical. That is, if the subjects reported a roll tilt of 90 degrees in LEO/REO orientations, this indicated that they felt as if they would have been "lying-on-side" on Earth. A –90-degree pitch tilt when in the LOB position indicated that the subjects would have been "upside-down" on Earth. Subjects were also asked to report any sense of linear motion using a simple estimate of magnitude (in meters/second).

In addition to the centrifugation study, the astronauts' perception during static full-body roll tilt was studied during

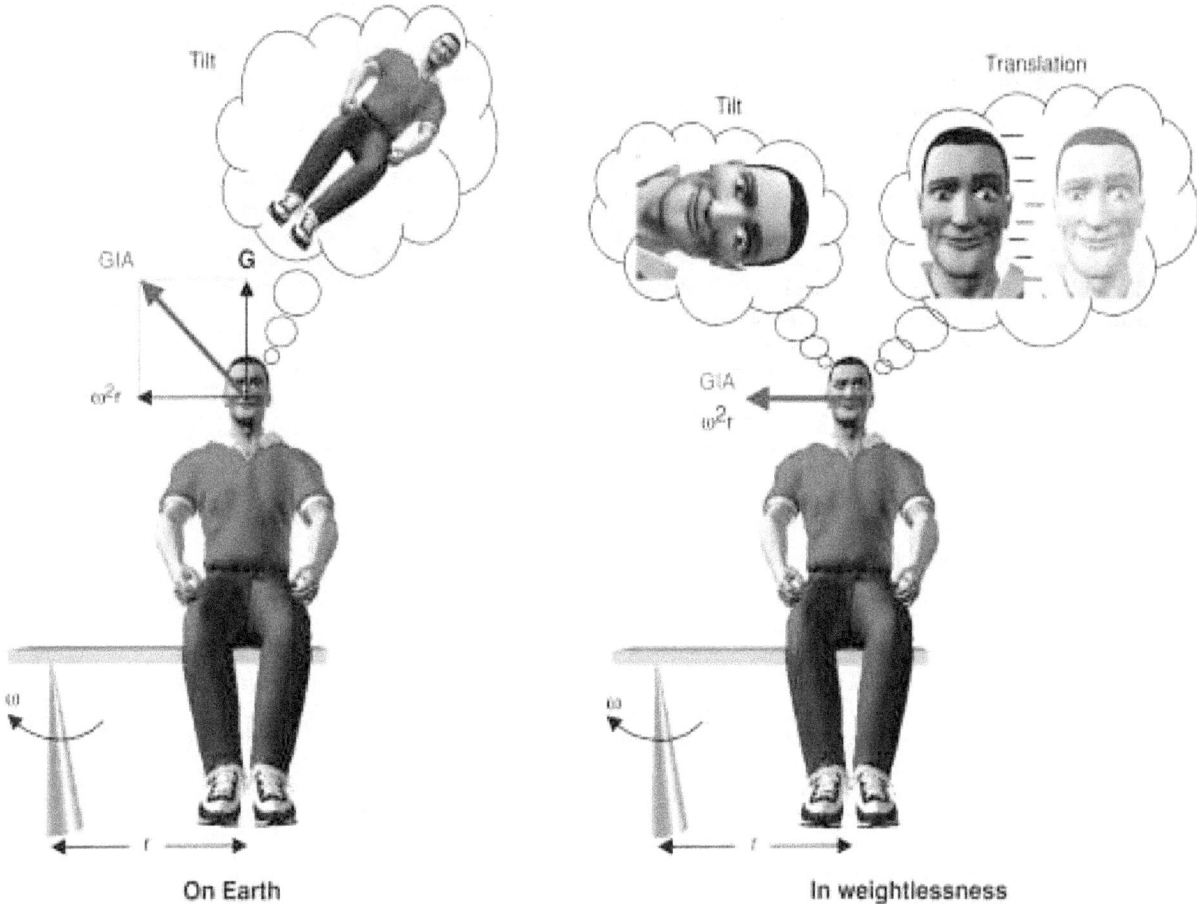

On Earth   In weightlessness

Figure 1. On Earth, during centrifugation, oriented with the LEO, the one-G centripetal acceleration (w2r) sums with the acceleration of gravity (G) to tilt the gravitoinertial acceleration (GIA) vector 45 degrees with respect to the head. Subjects perceive the GIA as being the spatial vertical and, therefore, feel tilted in roll. In weightlessness, the GIA is equivalent to the centripetal acceleration and is directed along the ear-to-ear axis. The question was whether the astronauts perceived this centripetal acceleration as tilt or translation while in space.

preflight and postflight testing. Postflight testing started as soon as two hours after landing (R+0). Subjects sat in a tilt chair in darkness and their bodies were passively tilted left-ear-down from the upright position around an axis located under their feet. The chair was tilted in increments of 15 degrees up to 90 degrees. Subjects stayed tilted at each angle for about 40 seconds before reporting their perceived angle of tilt. Three astronauts of the STS-78 Life and Microgravity Spacelab (LMS) mission were also tested during static roll tilt of 30 degrees on L–30, R+0, R+1, R+4 and R+7.

RESULTS

On Earth when sitting upright on a centrifuge and facing into the direction of motion, subjects at first sense that they are in a steep turn and then feel that they are tilted outward. They perceive a one-G centripetal acceleration as a tilt of about 45 degrees, although they are upright. (The one-G acceleration of

gravity adds to the one-G centripetal acceleration to cause the 45-degree tilt). Similarly, they perceive a 0.5-G centripetal acceleration as a tilt of about 25 degrees. When they are centrifuged while lying on their back, they perceive a body tilt toward the head-down position. These effects are called somatogravic illusions and are present in every person with an intact vestibular system (Gillingham, 1985).

In our test subjects, the perceived body tilt was larger during the first 4–5 runs on the centrifuge (Figure 2), but fairly constant afterward. This result shows that there is a training effect for the somatogravic illusion, which might explain the different values across studies depending on whether the test subjects were naive or not. After several trials, the mean perceived body tilt during one-G and 0.5-G centrifugation in our subjects was smaller (35 degrees and 20 degrees, respectively) than that of the GIA (45 degrees and 27 degrees) because the subjects were asked to report their angle of body tilt only 40 seconds after constant-velocity rotation. Ground-based studies have shown that it takes about 80 seconds for the sensation of body tilt to reach its full magnitude after the centrifuge has reached its final velocity (Clark, 1966).

At no point during or after the mission did the subjects perceive translation during constant-velocity centrifugation. Instead during spaceflight, the perceived body tilt increased from about 45 degrees on FD1 to nearly 90 degrees on FD16 during centrifugation at one-G (Figure 3). Inflight tilt perception during 0.5-G centrifugation on FD7 and FD12 was approximately half that reported during one-G centrifugation. The error in perceived tilt during all phases of testing by our subjects is shown in Figure 4. For comparison, the GIA was actually tilted by 45 degrees and 27 degrees on Earth and by 90 degrees in space relative to the spatial vertical during one-G and 0.5-G centrifugation respectively. The error in perceived tilt was very large early inflight and early postflight (Figure 4). A similar error was seen postflight during static full-body tilt (Figure 5). Prior to flight, the perceived tilt angle was close to the actual tilt angle, and subjects underestimated or overestimated the extent of body tilt by only 3–5 degrees for tilts larger than 60 degrees, similar to the precision of setting a visual line to the vertical (Wade, 1970). On R+0 and R+1, the extent of body tilt was overestimated by about 15 degrees. Four days after landing, the astronauts' perception of tilt during static body tilt had returned to preflight values (Figure 5). Thus, tilt was underestimated at the beginning of spaceflight during centrifugation and overestimated on return to Earth during both centrifugation and actual body tilt. The time course of return of subjects' estimates to preflight values was similar in both conditions.

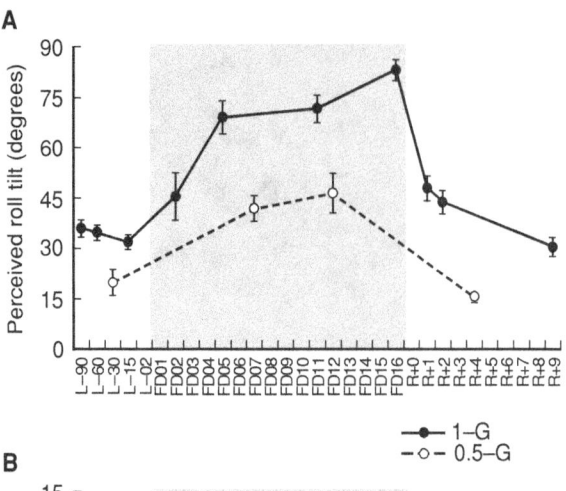

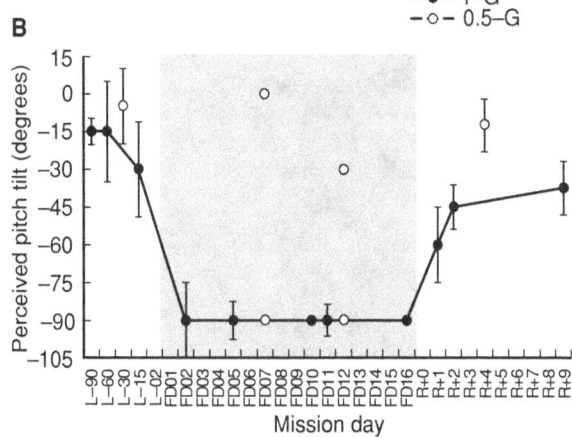

Figure 3. (A) Perceived body tilt in roll (mean and SE) during centrifugation along the interaural axis (LEO/REO orientations) in four subjects for centripetal linear accelerations of one-G and 0.5-G. Roll tilt perception during one-G inflight centrifugation approached 90 degrees as the mission progressed. Tilt perception during 0.5-G inflight centrifugation was approximately half that during one-G centrifugation. (B) Perceived body tilt (median and interquartile range) in pitch during centrifugation along the longitudinal body axis (LOB orientation) in two subjects for one-G and 0.5-G centripetal acceleration. Subjects felt as if they were "upside down" during inflight one-G centrifugation. During inflight 0.5-G centrifugation, one subject felt "upside down," whereas the other crewmember reported little perception of tilt.

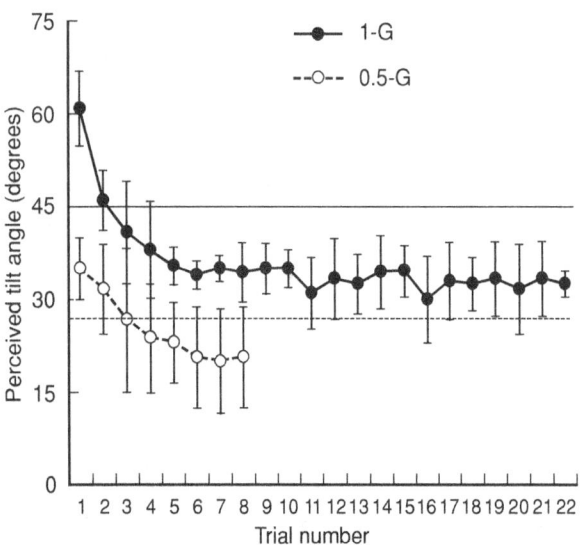

Figure 2. Perceived body tilt (mean and standard error (SE)) in roll during centrifugation in seven subjects for centripetal linear accelerations of one-G and 0.5-G along an axis connecting the ears (LEO/REO orientations). The actual tilt of the gravitoinertial acceleration vector was 45 degrees during one-G centrifugation (solid horizontal line) and 27 degrees during 0.5-G centrifugation (dotted horizontal line).

DISCUSSION

Before the Neurolab mission, astronauts had experienced sustained centripetal acceleration in space only on rare occasions. During the Gemini XI flight, in 1966, the manned spacecraft was tethered to an Agena target vehicle by a long Dacron line. This caused the two vehicles to spin slowly around each other for several minutes. According to the Gemini commander, a television camera fell "down" in the direction of the centrifugal force, but the crew on board Gemini did not sense the centripetal acceleration. Subjects sitting on a linear sled flown during the Spacelab-D1 mission in 1985 (Arrott, 1986) perceived linear acceleration but not tilt. Similarly, during off-axis rotation on the

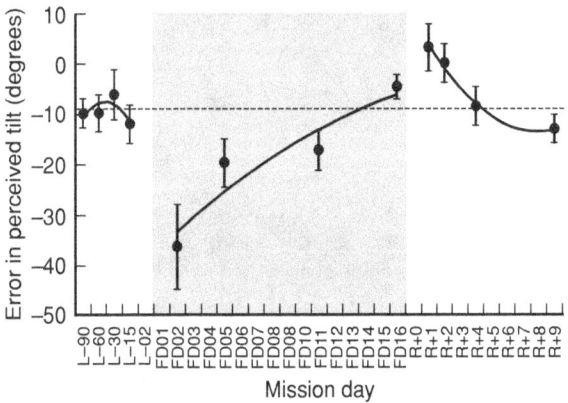

Figure 4. Error in perceived roll and pitch tilt (mean and SE) in four subjects during centrifugation at one-G and 0.5-G, calculated as the difference between the perceived tilt and the actual direction of the gravitational acceleration vector. For example, during centrifugation at one-G preflight, the GIA was tilted 45 degrees relative to the spatial vertical and the subjects perceived themselves as being tilted about 35 degrees; hence an error of approximately –10 degrees (the horizontal dotted line shows the mean preflight value). Inflight, the GIA was only due to centripetal acceleration; so the subjects should have felt tilted by 90 degrees relative to their original position before centrifugation. However, early inflight subjects felt tilted by only 45 degrees (hence an error of –45 degrees) during one-G centrifugation. Late inflight, in the same conditions, they indeed felt a tilt close to 90 degrees (error zero degrees).

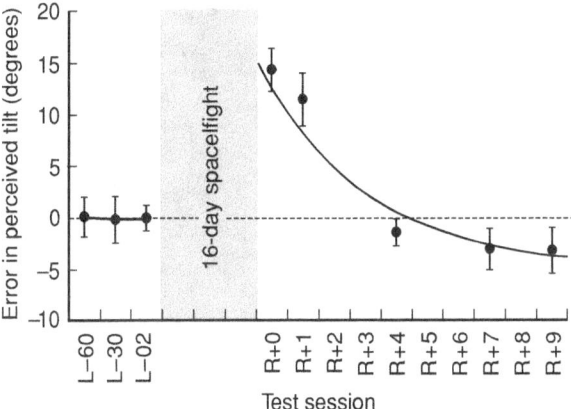

Figure 5. Error in perceived roll tilt in seven subjects during static, full-body, left-ear-down tilt preflight and postflight, calculated as the difference between perceived tilt and actual tilt. Each value represents the mean and SE of errors measured for actual tilt angles of 15, 30, 45, 60, 75, and 90 degrees. Results demonstrate an exaggerated sense of roll tilt after landing compared to preflight values, which is comparable to the exaggerated sense of tilt during centrifugation postflight (see Figure 3). Recovery of the sense of tilt in both conditions is also similar.

Spacelab IML-1 mission, subjects perceived only rotation, not tilt (Benson, 1997). In both experiments, however, linear accelerations were below 0.22-G, which ground-based studies have shown to be insufficiently strong to yield a perception of tilt

(Mittelstaedt, 1992). Humans had never been exposed to steady-state linear acceleration of 0.5-G and one-G in space before Neurolab. The results showed that the astronauts perceived a body tilt relative to a perceived spatial vertical when exposed to 0.5-G and one-G, and that the magnitude of this perception adapted throughout the mission.

After two weeks in space, the subjects perceived an almost 90-degree tilt when they received a one-G sideways linear acceleration in space, and about half of this when they received a 0.5-G acceleration. Although they had never encountered this stimulus before, their perception was essentially veridical in that it represented the actual levels of linear acceleration experienced by the graviceptors. It suggests that the otoliths are operating normally in space when exposed to 0.5-G and one-G steady-state linear acceleration, after the initial period of adaptation. The reduced response to the 0.5-G stimulus, whether it was directed along the interaural axis or the longitudinal axis, shows that not only the direction of GIA but also its magnitude is taken into account by the brain. This result could not have been obtained on the Earth's surface in a one-G environment.

The finding that none of the astronauts felt translation instead of tilt in response to the 0.5-G or one-G constant linear accelerations in space indicates that the OTTR hypothesis is incorrect. Tilt is perceived as tilt, regardless of whether the subjects are in microgravity or the one-G environment of Earth, and is not sensed as translation. A model, which references perceptions of tilt with regard to a weighted sum of all linear acceleration and body vertical (idiotropic vector) (Mittelstaedt, 1992) as the perceived spatial vertical, could explain these results (Clément, 2001). The underestimation of tilt at the beginning of the flight suggests that the subjects continued to weight their internal estimate of body vertical to compute the direction of the GIA. However, as the flight progressed, the weight of this internal estimate of body vertical gradually decreased and the subjects finally adopted the centripetal acceleration as the new spatial vertical. On return to Earth, perceived body tilt was larger than preflight. This overestimation of body tilt can be interpreted as the result of the continued small weighting of the internal representation of body vertical in estimating the spatial vertical, after adaptation to the weightless environment. Eye movements during both centrifugation in darkness and horizontal optokinetic stimulation shifted toward the GIA in space, consistent with the perceptual data. Thus, the underestimation of tilt on entry into microgravity, and the exaggerated sense of tilt on return, could both be due to the lag in readjusting the weight of the sense of body vertical in determining the perceived spatial vertical reference. Eye movement recordings during these studies also showed that the vector of eye velocity in darkness and of horizontal optokinetic nystagmus during centrifugation continued to shift toward the GIA in space as on Earth. Therefore, both the eye movement data and perceptual findings are consistent and do not support the OTTR hypothesis.

Information from this research could be used to develop countermeasures to overcome lags in adaptation or changes in gaze and balance that occur after return from space. Such information and countermeasures are critical in the long-duration spaceflights planned for planetary exploration. When astronauts go to Mars, for example, they may have to fend for themselves immediately after landing on a planet with a significant gravitational force (0.38-G), although they will have been in a microgravity environment for months. Anything that could hasten their re-adaptation to a gravitational environment would be valuable and important to them in overcoming difficulties with gaze, posture, walking, and running. One consequence of our findings is that if low-frequency linear acceleration is always perceived as tilt—whether subjects are in weightlessness or on Earth—long-duration missions can proceed with the expectation that the astronauts will respond normally to artificial gravity or to the gravitational fields of other planets.

There are also substantial clinical implications from these experiments. We have little understanding of why there is imbalance when the vestibular system is damaged. We also do not understand why older people are so prone to falling. Alignment of the body axis to the GIA during walking or turning is likely to be an important source of this imbalance. The evaluation of the perceived tilt during centrifugation might prove to be a useful test of the capability for the brain to evaluate the direction of the GIA in a dynamic situation.

REFERENCES

FACTORS CONTRIBUTING TO THE DELAY IN THE PERCEPTION OF THE OCULOGRAVIC ILLUSION. B. Clark and A. Graybiel. *Am. J. Psychol.*, Vol. 79, pages 377–388; 1966.

THE EFFECT OF DIFFERENT PSYCHOPHYSICAL METHODS ON VISUAL ORIENTATION DURING TILT. N.J Wade. *Psycholog. Sci.*, Vol. 19, pages 201–212; 1970.

HUMAN VISUAL ORIENTATION. I.P. Howard. John Wiley & Sons, 1982.

SPATIAL ORIENTATION IN WEIGHTLESSNESS AND READAPTATION TO EARTH'S GRAVITY. L.R. Young, C.M. Oman, D.G.D. Watt, K.E. Money, and B.K. Lichtenberg. *Science,* Vol. 225, pages 205–208; 1984.

SPATIAL ORIENTATION: THE SPATIAL CONTROL OF BEHAVIOR IN ANIMALS AND MAN. H. Schöne: Princeton University Press, 1984.

SPATIAL ORIENTATION IN FLIGHT. K.K. Gillingham and J.W. Wolfe in *Fundamentals of Aerospace Medicine*, edited by R.L. Dehart Lea & Febiger Inc., pages 299–381; 1985.

OTOLITH TILT-TRANSLATION REINTERPRETATION FOLLOWING PROLONGED WEIGHTLESSNESS: IMPLICATIONS FOR PRE-FLIGHT TRAINING. D.E. Parker, M.F. Reschke, A.P. Arrott, J.L. Homick, and B.K. Lichtenberg. *Aviat. Space Environ. Med.,* Vol. 56, pages 601–606; 1985.

PERCEPTION OF LINEAR ACCELERATION IN WEIGHTLESSNESS. A.P. Arrott, L.R. Young and D.M. Merfeld. *Aviat. Space Environ. Med.,* Vol. 61, pages 319–326; 1986.

SOMATIC VERSUS VESTIBULAR GRAVITY PERCEPTION IN MAN. H.H. Mittelstaedt. *Annals New York Academy of Sciences*, edited by B. Cohen, D. Tomko, and F. Guedry, Vol. 656, pages 124–139; 1992.

MICROGRAVITY VESTIBULAR INVESTIGATIONS: PERCEPTION OF SELF-ORIENTATION AND SELF-MOTION. A.J. Benson, F.E. Guedry, D.E. Parker, and M.F. Reschke. *J. Vestib. Res.,* Vol. 7, pages 453–457; 1997.

INTERACTION OF THE BODY, HEAD AND EYES DURING WALKING AND TURNING. T. Imai, S.T. Moore, B. Cohen, and T. Raphan *Exp. Brain Res.*, Vol. 136, pages 1–18, 2001.

PERCEPTION OF TILT (SOMATOGRAVIC ILLUSION) IN RESPONSE TO SUSTAINED LINEAR ACCELERATION DURING SPACE FLIGHT. G. Clément, S.T. Moore, T. Raphan, and B. Cohen. *Exp. Brain Res.,* Vol. 138, pages 410–418, 2001.

Ocular Counter-Rolling During Centrifugation and Static Tilt

Authors

Steven Moore, Bernard Cohen, Gilles Clément,
Ian Curthoys, Mingjia Dai, Izumi Koizuka,
Takeshi Kubo, Theodore Raphan

Experiment Team

Principal Investigator: **Bernard Cohen[1]**

Co-Investigators: **Gilles Clément[2]**
Steven Moore[1]
Ian Curthoys[3]
Mingjia Dai[1]
Izumi Koizuka[4]
Takeshi Kubo[5]
Theodore Raphan[6]

[1]Mount Sinai School of Medicine, New York City, USA
[2]Centre National de la Recherche Scientifique, Paris, France
[3]University of Sydney, Sydney, Australia
[4]St. Marianna University School of Medicine, Kawasaki, Japan
[5]Osaka University, Osaka, Japan
[6]Brooklyn College, New York City, USA

ABSTRACT

Activation of the gravity sensors in the inner ear—the otoliths—generates reflexes that act to maintain posture and gaze. Ocular counter-rolling (OCR) is an example of such a reflex. When the head is tilted to the side, the eyes rotate around the line of sight in the opposite direction (i.e., counter-rolling). While turning corners, undergoing centrifugation, or making side-to-side tilting head movements, the OCR reflex orients the eyes towards the sum of the accelerations from body movements and gravity. Deconditioning of otolith-mediated reflexes following adaptation to microgravity has been proposed as the basis of many of the postural, locomotor, and gaze control problems experienced by returning astronauts. Evidence suggests that OCR is reduced postflight in about 75% of astronauts tested; but the data are sparse, primarily due to difficulties in recording rotational eye movements.

During the Neurolab mission, a short-arm human centrifuge was flown that generated sustained sideways accelerations of 0.5-G and one-G to the head and upper body. This produces OCR; and so for the first time, the responses to sustained centrifugation could be studied without the influence of Earth's gravity on the results. This allowed us to determine the relative importance of sideways and vertical acceleration in the generation of OCR. This also provided the first test of the effects of exposure to artificial gravity in space on postflight otolith-ocular reflexes.

There was little difference between the responses to centrifugation in microgravity and on Earth. In both conditions, the induced OCR was roughly proportional to the applied acceleration, with the OCR magnitude during 0.5-G centrifugation approximately 60% of that generated during one-G centrifugation. The overall mean OCR from the four payload crewmembers in response to one-G of sideways acceleration was 5.7±1.1 degree (mean and SD) on Earth. Inflight one-G centrifugation generated 5.1±0.9 degree of OCR, which was a small but significant decrease in OCR magnitude. The postflight OCR was 5.9±1.4 degree, which was not significantly different from preflight values. During both 0.5-G and one-G centrifugation in microgravity, where the head vertical gravitational component was absent, the OCR magnitude was not significantly different from that produced by an equivalent acceleration during static tilt on Earth. This suggests that the larger OCR magnitude observed during centrifugation on Earth was due to the larger body vertical linear acceleration component, which may have activated either the otoliths or the body tilt receptors. In contrast to previous studies, there was no decrease in OCR gain postflight. Our findings raise the possibility that inflight exposure to artificial gravity, in the form of intermittent one-G and 0.5-G centripetal acceleration, may have been a countermeasure to deconditioning of otolith-based orientation reflexes.

INTRODUCTION

The otolith organs of the inner ear, the utricle and saccule, are the primary gravity sensors of the body. Activation of the otoliths by linear acceleration (including that of gravity) generates various spinal and ocular reflexes that act to maintain posture and gaze. Ocular counter-rolling (OCR) is one example of an otolith-ocular reflex in response to activation of the otoliths. When the head is tilted laterally, the eyes rotate around the line of sight. Termed OCR, this torsion is an orienting reflex that tends to align the eyes with the spatial vertical. During centrifugation, during side-to-side head movements, and while turning corners, the OCR reflex orients the eyes towards the sum of the imposed accelerations from body movements and gravity. For example, when walking or driving around a bend, there is an inward linear (centripetal) acceleration that sums with gravity to create a vector that is tilted into the turn (Figure 1). This is called the gravitoinertial acceleration (GIA) vector. Recent work has shown that rolling the head and eyes towards alignment with this tilted GIA vector plays a role in maintaining balance and gaze when walking around sharp turns.

The semicircular canals in the inner ear, which sense angular acceleration of the head, also induce torsional eye movements during rapid head movements; but these responses are transient. In contrast, otolith-induced OCR responses are sustained during static tilts of the head or the GIA vector, with a gain (amount of ocular torsion/head tilt angle) of approximately 0.1. The magnitude of OCR is related to the angle of head tilt.

Deconditioning[1] of otolith-mediated spinal and ocular reflexes following adaptation to microgravity has been proposed as the basis of many of the postural, locomotor, and gaze control problems experienced by returning astronauts. Consequently, OCR has been used in many postflight studies to gauge the effect of microgravity exposure on otolith function. There is evidence that OCR is reduced postflight in about 75% of astronauts tested; but the data are sparse, primarily due to difficulties in recording torsional eye movements. OCR was reduced in two cosmonauts for 14 days after landing (Yakovleva, 1982). Following the 10-day Spacelab-1 mission, OCR to leftward roll tilts was reduced by 28–56% in three subjects and was unchanged in one subject (Vogel, 1986). Asymmetries in the OCR response to left and right static roll tilts were also observed. OCR was reduced by 57% in one astronaut for five days after the 1992 Russian Mir mission (Hofstetter-Degen, 1993). OCR was also reduced in two subjects during postflight side-to-side oscillations at 0.4 and 0.8 Hz (Arrott, 1986). OCR gain was depressed in four subjects following the two-week SLS-2 mission (Young, 1998). In addition, asymmetries in OCR to left/right roll tilt were observed in all subjects studied on SLS-2. The development of video-oculography (see Moore, 1996, for a review) has led to significant improvements in OCR measurements in humans,

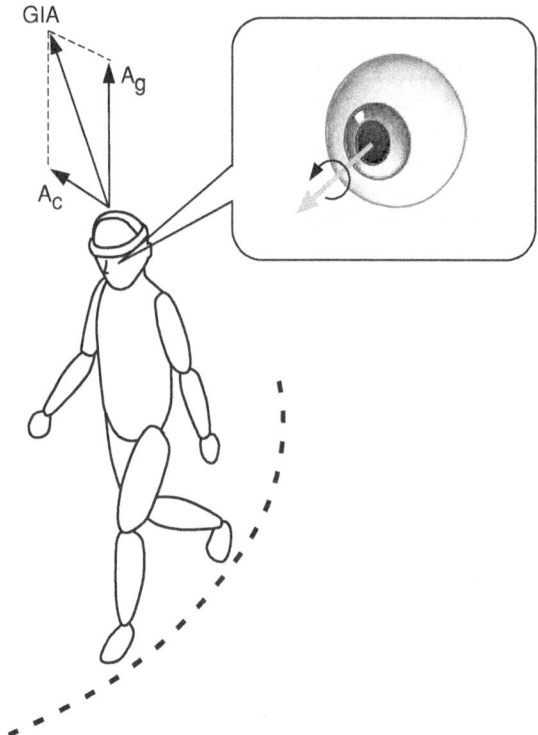

Figure 1. When walking around a turn, there is an inward centripetal linear acceleration, Ac, that sums with gravity, Ag, to form the GIA vector. To maintain balance and gaze during turns, the head, body, and eyes (inset) tend to roll towards alignment with the GIA. Rolling of the eyes around the line of sight towards a GIA vector that is tilted with regard to the head is termed OCR.

compared to the techniques used in the results cited above. OCR gain, measured using video-oculography following a 30-day Mir mission, decreased in one astronaut but increased in two other astronauts who had been in space for 180 days (Diamond, 1998).

Strong evidence for deficits in postflight otolith function was obtained from two monkeys following a 14-day COSMOS mission (Dai, 1994). Torsional eye position was measured postflight using a robust and accurate measure of ocular torsion (search coils). The eye movements were measured both during static roll tilt and during off-vertical axis rotation (OVAR). OVAR presents a sinusoidal linear acceleration stimulus to the otoliths suitable for averaging. There was a highly significant (70%) reduction (>2 SD) in OCR gain, which persisted over the 11 days of postflight testing. In addition, vergence of the eyes, an otolith-mediated response to front-to-back linear acceleration, was also reduced during this 11-day period. Thus, although the data are not entirely consistent, the majority of subjects tested have exhibited a decrease in their OCR response following short-duration missions. In this paper, we present a direct comparison of the OCR responses during preflight, inflight, and postflight centrifugation, as well as during pre- and postflight static tilt.

---

[1]Defined as a decrease in gain of otolith-mediated reflexes.

METHODS

Centrifugation

Four payload crewmembers of the Neurolab STS-90 Space Shuttle mission served as subjects and as inflight operators for this experiment. Over the 16-day mission, the astronauts were exposed to inflight one-G centrifugation on flight days (FDs) 2, 5, 10, 11, and 16. Two crewmembers were also exposed to 0.5-G centrifugation on FDs 7 and 12. Approximately 80% of the exposure to sustained one-G linear acceleration was in the form of centripetal acceleration directed along a line connecting the ears (Figure 2). In addition, subjects underwent centrifugation where the centripetal acceleration was directed along the spine (see technical report by Cohen et al. in this publication). Cumulative flight exposure times to one-G centripetal acceleration were 50 minutes, 62 minutes, 37 minutes, and 62 minutes for the four subjects. Three subjects had their first exposure to centrifugation on FD2, less than 24 hours into the mission. One subject did not experience inflight centrifugation until FD5. In addition, due to mission operational constraints, this subject's exposure time was limited to approximately half that of the other payload crew's.

Baseline one-G data using the same paradigms as in flight were collected at Johnson Space Center in Houston 90, 60, and 15 days prior to launch (L–90, L–60, and L–15) on a centrifuge that was a replicate of the flight centrifuge (see technical report by Cohen et al. in this publication). The same tests were repeated in Houston 24 hours after return (R+1) and on subsequent days (R+2 and R+9). Baseline 0.5-G data were obtained from all four subjects on L–30 and R+4.

In each run, subjects were accelerated at 26 degrees/second^2 in darkness to a constant angular velocity of 254 degrees/second or 179 degrees/second, which generated a one-G or 0.5-G centripetal acceleration along a line connecting the ears. Subjects were oriented with their left ear facing away from the center of rotation (left-ear-out (LEO)), as in Figure 2A) or right-ear-out (REO). After 65 seconds at constant velocity in darkness, subjects were presented with a centering display dot for 9.5 seconds and instructed to fixate the dot. Eye movement data recorded during this period were used to calculate OCR. Subjects were then decelerated at 26 degrees/second^2 to rest in darkness either immediately following the center display, or after optokinetic and smooth pursuit stimuli were displayed (not considered in this report). A typical trial consisted of clockwise (CW) LEO centrifugation (facing-motion), counterclockwise (CCW) LEO (back-to-motion), and CCW and CW REO (facing- and back-to-motion). The video eye monitors were calibrated by having the subject fixate on 25 points at known gaze angles prior to the first LEO and REO runs. For a complete description of the methodology, refer to the technical report by Cohen et al. in this publication.

Static Tilt

Full-body static roll tilt was performed in Houston using the tilt mode of the ground centrifuge, and at Kennedy Space Center in Cape Canaveral using a static tilt chair developed at Mount Sinai Medical Center. After subjects were positioned in the tilt chair and the video system was calibrated, subjects were roll-tilted from the upright (zero degree) to 60 degrees left-ear-down in 15-degree increments. The chair was locked in place at each tilt angle, and after 60-second video images were recorded for approximately 10 seconds while the subject viewed a centering dot on the visual display. This segment was used to measure OCR. Preflight baseline data collection was carried out 60 and 30 days prior to launch (L–60 and L–30) in Houston. Postflight data were obtained on the day of return (R+0) two to four hours after landing at Kennedy Space Center, and in Houston on R+1, R+4, and R+9.

RESULTS

OCR during centrifugation

During constant angular velocity sideways centrifugation (Gy centrifugation), there is a radial inward linear (centripetal) acceleration, Ac, regardless of the direction of rotation (Figure 2A) along a line connecting the ears (the interaural axis). On Earth, the equivalent acceleration of gravity, Ag, is aligned with the head vertical axis. The sum of Ag and Ac, known as the GIA vector, is tilted with respect to the head vertical axis as shown in Figure 2A. Ground-based centrifugation with one-G of centripetal acceleration generated a GIA vector with a magnitude of 1.4-G tilted 45 degrees with respect to the head. As a result, the subjects felt like they were tilted 45 degrees while rotating at this speed on Earth. Gy centrifugation at 0.5-G on Earth generated a GIA magnitude and tilt of 1.1-G and 27 degrees, respectively. In microgravity, however, the gravitational component was negligible, and the GIA was equivalent to the centripetal acceleration; i.e., it was directed along the interaural axis with a magnitude of either one-G or 0.5-G.

Robust torsional movements of the eyes were induced during Gy centrifugation both on Earth and inflight. The OCR was characterized by dynamic and static components (Figure 2C). The dynamic component, shown in Figure 2C, decayed at the onset of constant velocity and was dependent on the direction of rotation (Figure 2D). For example, a LEO CCW angular acceleration (back-to-motion) generated CW torsional eye movements in which the upper pole of the eye rolled to the subject's right. CW rotation (facing-motion) initially generated CCW ocular torsion. Thus, the dynamic component added to the static component of OCR when moving back-to-motion and subtracted from it when facing-motion.

When back-to-motion, ocular torsion developed rapidly during angular acceleration, reaching a maximum of 10 degrees at the onset of constant-velocity rotation before decaying to a steady-state value of approximately six degrees after 10 seconds at constant velocity. During facing-motion centrifugation, the eye initially torted in the CCW direction then rolled back in the CW direction, reaching a plateau of approximately six degrees following 10 seconds of constant velocity rotation—as it did for back-to-motion centrifugation. Since the OCR had the same polarity for facing- and back-to-motion, the dynamic

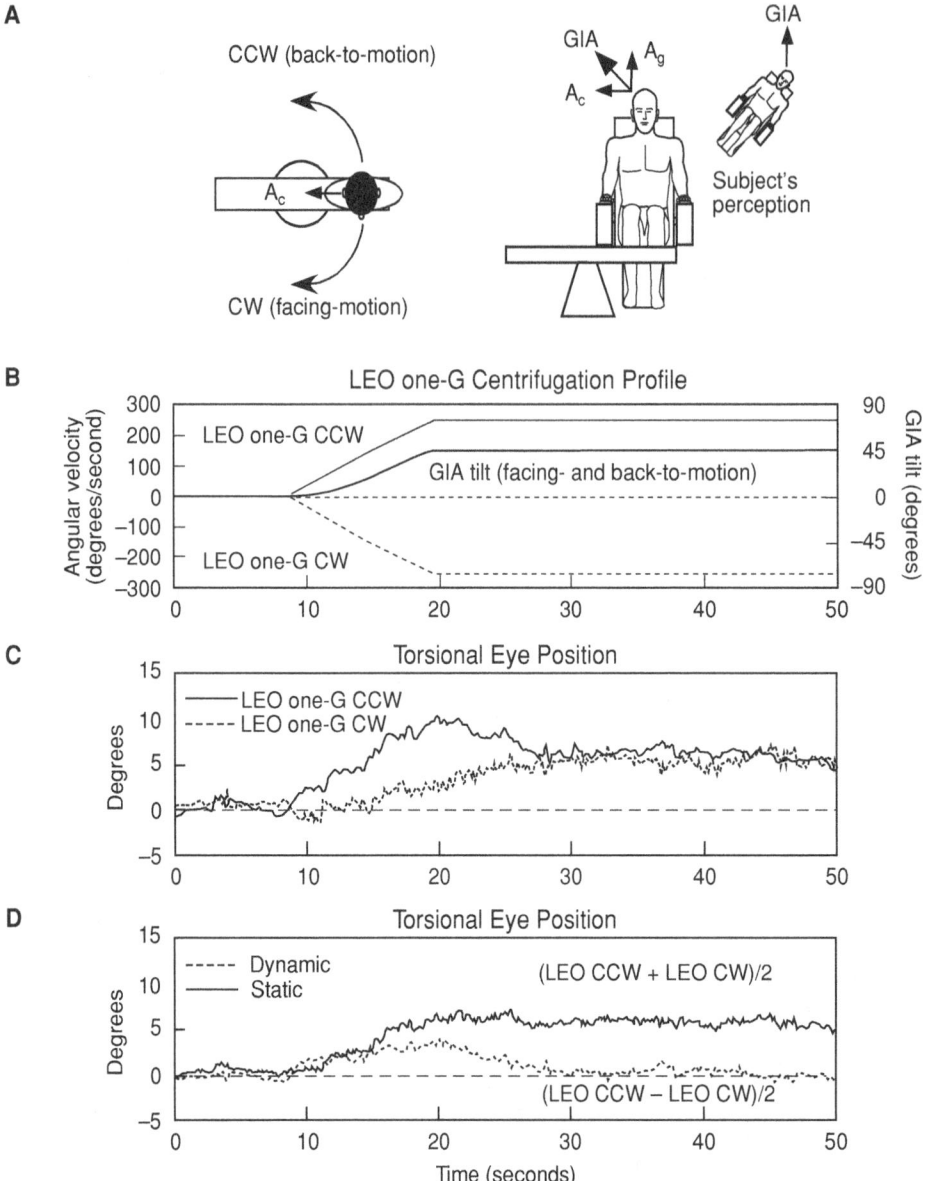

Figure 2. (A) Positioning of the subject for LEO centrifugation, which directs the centripetal linear acceleration along a line connecting the ears (also know as the interaural axis). This sideways acceleration is termed Gy centrifugation. LEO constant velocity centrifugation generates a centripetal acceleration, Ac, that sums with the equivalent acceleration of gravity, Ag, to tilt the GIA vector in the roll plane relative to the subject's head. Subjects tend to perceive the GIA as the spatial vertical, and feel tilted in the roll plane away from the rotation axis (inset). (B) Rotation velocity profiles for LEO back-to-motion (CCW) and facing-motion (CW) centrifugation that generated a one-G centripetal acceleration at steady state and a 45 degree roll-tilt of the GIA. The direction of GIA tilt was CW during LEO centrifugation (from the subject's point of view), regardless of the direction of rotation. During REO centrifugation (not shown), the roll-tilt of the GIA was CCW. (C) Torsional right eye position data from one subject during LEO one-G centrifugation on Earth. During angular acceleration, there was a dynamic ocular torsion component whose direction was dependent on the direction of rotation. Upon reaching constant angular velocity, this dynamic component decayed, and static OCR was generated by the otoliths in response to the tilted GIA. We sampled the OCR magnitude approximately 65 seconds into rotation, long after the dynamic torsional eye movements had ceased. (D) The dynamic and static components of the ocular torsion response could be isolated by superposing the torsional eye position records during facing- and back-to-motion centrifugation. The dynamic torsional response to angular acceleration (dashed trace) reached a maximum at onset of steady state and decayed over the following 10 seconds. The static OCR component (solid trace) rose in a linear fashion with the GIA tilt, reaching a plateau of approximately six degrees at onset of constant velocity. (From Moore, 2001, with permission; reproduced from *Experimental Brain Research*.)

component could be isolated by subtracting the two torsional eye position traces and halving the result (Figure 2D). Dynamic torsional eye position reached a peak of approximately four degrees at the onset of constant-velocity centrifugation and then decayed with a time constant of approximately six seconds. The directional dependence of the dynamic ocular torsion response has previously been observed during on-center rotation and is likely a semicircular canal response.

The static OCR component was generated by the otoliths in response to the tilt of the GIA with regard to the head, reaching a plateau during constant-velocity centrifugation after 10 seconds. In contrast to the dynamic component, static OCR was in the same direction (towards the GIA vector) for a given subject orientation (LEO or REO) regardless of the direction of rotation. This was because the centripetal acceleration, and therefore the GIA tilt, was in the same direction during facing- and back-to-motion centrifugation. The static OCR response was extracted by averaging the facing- and back-to-motion traces, which cancelled the oppositely directed dynamic components. The static OCR response followed the tilt of the GIA, reaching a maximum of approximately six degrees where the GIA tilt reached a plateau of 45 degrees at onset of constant velocity. We obtained our measures of OCR magnitude after approximately one minute of constant velocity rotation, when the dynamic contribution had ceased.

Torsional eye position showed little difference between the responses in microgravity and on Earth (Figure 3). The OCR response was roughly proportional to the applied interaural linear acceleration, with OCR magnitude during 0.5-G centrifugation approximately 60% of that generated during one-G centrifugation. The overall mean OCR response was determined by combining LEO and REO OCR data from the four payload crewmembers (Figure 3). On Earth, one-G Gy centrifugation elicited an OCR response of 5.7±1.1 degrees (mean and SD). During inflight one-G centrifugation there was a small but significant (p=0.0025) 10% decrease in OCR magnitude to 5.1±0.9

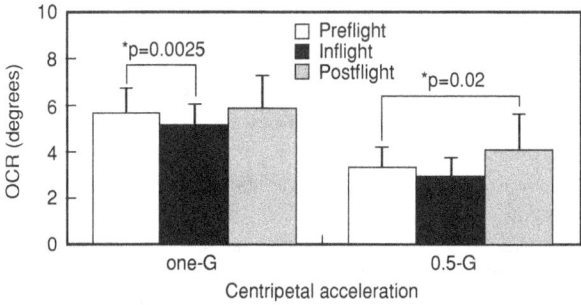

Figure 3. (A) OCR (mean and SD; combined LEO and REO data) of all four subjects. Relative to preflight OCR, there was a significant 10% decrease in inflight OCR but not in postflight OCR during one-G Gy centrifugation. (B) There was a similar inflight decrease in mean OCR during 0.5-G Gy centrifugation, but this did not reach significance due to the small number of inflight subjects (N=2). Postflight OCR was slightly but significantly increased during 0.5-G centrifugation from preflight values. (From Moore, 2001, with permission; reproduced from *Experimental Brain Research*.)

degree. The magnitude of OCR during postflight one-G centrifugation was 5.9±1.4 degrees, which was not significantly different from preflight values. A similar trend was observed during 0.5-G Gy centrifugation. Preflight centrifugation generated 3.3±0.9 degree of OCR. There was an 11% decrease observed in OCR during inflight 0.5-G centrifugation to 3.0±0.8 degree (mean and SD of two subjects only), but this was not statistically significant. Postflight 0.5-G centrifugation generated a weak but significant (p=0.02) increase in postflight OCR to 4.1±1.5 degrees as compared to preflight values.

It is interesting to note that, in contrast with the other three subjects whose responses were symmetrical, one subject developed a marked OCR asymmetry in response to left/right tilts of the GIA during inflight centrifugation. This subject exhibited a significant (p=0.0002) 26% decrement in mean OCR inflight relative to preflight values during one-G REO centrifugation, but had only a 7.5% decrease during LEO centrifugation. This asymmetry was maintained after landing. In response to postflight one-G REO centrifugation, OCR magnitude returned to preflight values, but there was a highly significant 28.9% increase relative to preflight (p=0.0001) during one-G LEO centrifugation. The asymmetry was also apparent during postflight 0.5-G centrifugation, where the OCR was also larger when in the LEO orientation.

OCR during static tilt

The OCR response was further investigated by tilting the body left-ear-down (LED) in a chair during pre- and postflight testing. Consistent with the centrifugation results, there was no significant change (p>0.05) in OCR two to four hours after landing (R+0) and on subsequent postflight test days as compared to preflight values (Figure 4A). It is interesting to note that the magnitude of OCR generated by 45-degree LED static tilt (three to four degrees) was significantly less than that induced by a 45-degree tilt of the GIA during preflight one-G Gy centrifugation (5.7 degrees). Previous studies have suggested that OCR is linearly related to the magnitude of interaural linear acceleration. Our OCR data exhibited a linear relationship with interaural linear acceleration during static tilt (mean of all pre- and postflight data), with a slope of 5.04 degrees/G (R=0.995) (Figure 4B). The magnitude of OCR during preflight centrifugation followed this linear relationship, but still had a significantly larger magnitude than for static tilt at an equivalent interaural linear acceleration (0.5-G centrifugation: p=0.03; one-G centrifugation: p=0.01). During both 0.5-G and one-G centrifugation in microgravity, where the head dorsoventral gravitational component was absent, the magnitude of OCR was not significantly different from that induced by static tilts on Earth with equivalent interaural linear acceleration.

DISCUSSION

The OCR reflex in response to sustained interaural linear acceleration induced by Gy centrifugation was essentially maintained in microgravity. In addition, the OCR magnitude was proportional to the magnitude of the applied interaural linear

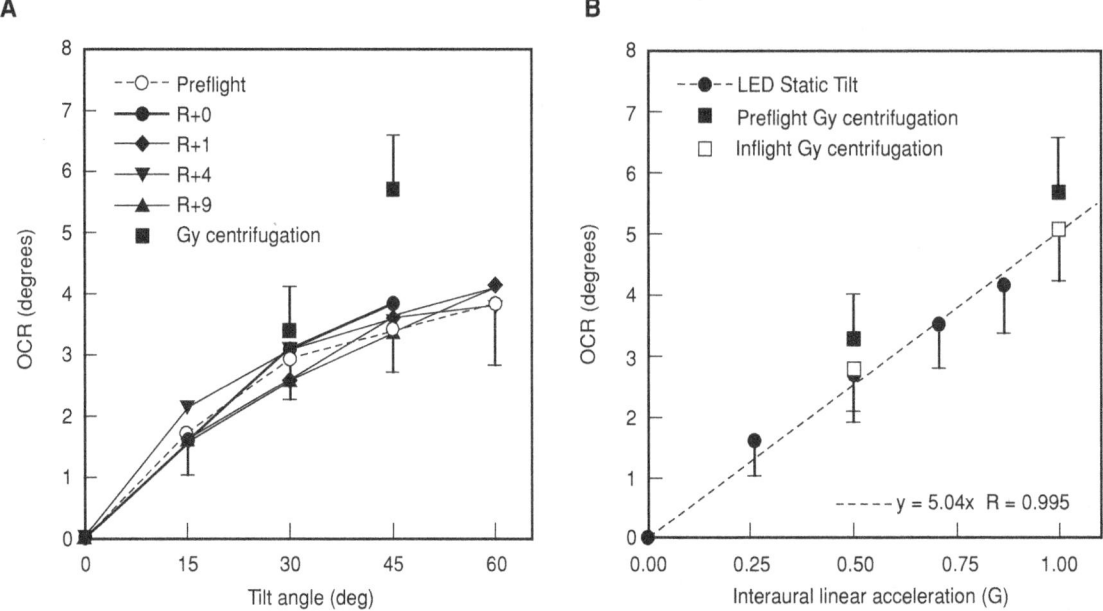

Figure 4. (A) Magnitude of OCR in response to static left-ear-down (LED) tilts during pre- and postflight testing. Each point is mean of the OCR of the right eye of the four payload crewmembers. Consistent with the results of centrifugation, there were no significant postflight changes in OCR on the day of landing (R+0) and on subsequent postflight test days (R+1, R+4, and R+9). The preflight data are the mean of OCR measures obtained 60 and 30 days prior to launch. For clarity, the SD is shown only for the preflight data. Superimposed on the plot is OCR measured during preflight Gy centrifugation (filled squares; mean and SD). OCR measured at a 45-degree static tilt angle was significantly less than that generated by a 45-degree GIA tilt during centrifugation. (B) OCR during LED static tilt (mean and SD of all test sessions) exhibited a linear relationship with interaural linear acceleration (filled circles – dashed line). OCR generated by preflight Gy centrifugation was significantly larger than that induced by a static tilt with equivalent interaural linear acceleration (filled squares). In contrast, there was no difference between OCR produced by inflight Gy centrifugation and by static tilt with equivalent interaural linear acceleration (open squares). This suggests that the larger vertical acceleration component during centrifugation on Earth (i.e., gravity) contributed to the OCR response. (From Moore, 2001, with permission; reproduced from *Experimental Brain Research*.)

acceleration. Mean values of inflight OCR were slightly lower (10%) than preflight values. In contrast to previous studies, there was no significant difference in OCR generated during pre- and postflight testing, suggesting that in the Neurolab astronauts' adaptation to microgravity and re-adaptation to gravity did not alter the gain of OCR. Consistent with this, there was no change in the OCR induced by static tilt before and after flight.

Our finding that preflight OCR during static tilts and Gy centrifugation were approximately linear functions of interaural acceleration up to one-G are in accord with previous studies. Our data are also consistent with findings that Gy centrifugation on Earth generates significantly greater OCR than during static roll tilt with an equivalent interaural acceleration. This difference has been attributed to the larger head vertical linear acceleration (one-G) during centrifugation. Centrifugation in space gave us a unique opportunity to test this hypothesis in that it generated the same interaural linear acceleration as on Earth, but with no head vertical gravitational component. During both 0.5-G and one-G Gy centrifugation in microgravity, OCR was generated with the same magnitude as that induced by static roll tilts on Earth with an equivalent interaural acceleration. These data are consistent with the hypothesis that OCR is pri-

marily generated in response to interaural linear acceleration. The increased OCR during terrestrial centrifugation was likely due to the larger vertical linear acceleration, which contributed approximately 10% of the total OCR magnitude. This could be attributed to activation of either the saccules, which lie approximately orthogonal to the utricles, or to activation of the body tilt receptors in response to the larger GIA magnitude during centrifugation.

Although there is some variability, in previous studies approximately 75% of subjects tested have exhibited a decrease in postflight OCR (Arrott, 1986; Diamond, 1998; Hofstetter-Degen, 1993; Vogel, 1986; Yakovleva, 1982; Young, 1998). Our finding—that there was no reduction in postflight OCR magnitude compared to preflight values in all four Neurolab payload crewmembers—raises the possibility that inflight exposure to artificial gravity, in the form of intermittent one-G and 0.5-G centripetal acceleration, may have been a countermeasure to oppose the deconditioning of otolith-based orientation reflexes. The only subject to exhibit signs of alteration in otolith responses —i.e., a substantial asymmetry in OCR to right and left tilts of the GIA—was not centrifuged until five days into the mission, and was exposed to significantly less artificial gravity than the other payload crew. This subject's OCR asymmetry developed in space and persisted throughout the nine days of postflight testing. Asymmetries in low-frequency otolith sensitivity to roll-tilts of the GIA have previously been observed in astronauts postflight (Vogel, 1986; Young, 1998) and may have a significant impact on postural control, especially when turning corners. If intermittent inflight centrifugation did in fact act to prevent otolith-ocular deconditioning, the results suggest that any countermeasure effect may be reliant on early and/or cumulative exposure to artificial gravity.

Acknowledgements

We gratefully acknowledge the efforts of the payload crew, Drs. Rick Linnehan, Dave Williams, Jay Buckey, and Jim Pawelczyk, for serving as subjects and operators during the mission; the alternate payload specialists, Drs. Alex Dunlap and Chiaki Mukai; Dr. Mel Buderer, Dan Harfe, Gwenn Sandoz, and Nasser Ayub (JSC); Jacqui Van Tweste (KSC); Mike Cork and Dr. Thierry Dewandre (ESA/ESTEC); and Frederic Bellossi (Aerospatiale). Supported by NASA Contract NAS 9-19441 (Drs. Cohen, Moore and Raphan) and the Centre National d'Etudes Spatiales (Dr. Clément).

REFERENCES

RESULTS OF VESTIBULAR FUNCTION AND SPATIAL PERCEPTION OF THE COSMONAUTS FOR THE 1ST AND 2ND EXPLOITATION ON STATION OF SALUT 6. I.Y. Yakovleva, L.N. Kornilova, G.D. Serix, I.K. Tarasov, and V.N. Alekseev. *Space Biol.* (Russia), Vol. 1, pages 19–22; 1982.

EUROPEAN VESTIBULAR EXPERIMENTS ON THE SPACELAB-1 MISSION: 7. OCULAR COUNTERROLLING MEASUREMENTS PRE- AND POST-FLIGHT. H. Vogel and J.R. Kass. *Exp. Brain Res.,* Vol. 64, pages 284–290; 1986.

M.I.T./CANADIAN VESTIBULAR EXPERIMENTS ON THE SPACELAB-1 MISSION: 6. VESTIBULAR REACTIONS TO LATERAL ACCELERATION FOLLOWING 10 DAYS OF WEIGHTLESSNESS. A.P. Arrott and L. R. Young. *Exp. Brain Res.,* Vol. 64, pages 347–357; 1986.

OCULOVESTIBULAR INTERACTIONS UNDER MICROGRAVITY. K. Hofstetter-Degen, J. Weizig, and R. Von Baumgarten. *Clin. Investigations,* Vol. 10, pages 749–756; 1993.

EFFECTS OF SPACEFLIGHT ON OCULAR COUNTERROLLING AND THE SPATIAL ORIENTATION OF THE VESTIBULAR SYSTEM. M. Dai, L. Mcgarvie, I.B. Kozlovskaya, T. Raphan, and B. Cohen. *Exp. Brain Res.,* Vol. 102, pages 45–56; 1994.

A GEOMETRIC BASIS FOR MEASUREMENT OF THREE-DIMENSIONAL EYE POSITION USING IMAGE PROCESSING. S.T. Moore, T. Haslwanter, I.S. Curthoys, and S.T. Smith. *Vision Res.,* Vol. 36, pages 445–459; 1996.

SPACEFLIGHT INFLUENCES ON OCULAR COUNTERROLLING AND OTHER NEUROVESTIBULAR REACTIONS. L.R. Young and P. Sinha. *Otolaryngology—Head And Neck Surgery,* Vol. 118, pages S31–S34; 1998.

THE EFFECT OF SPACE MISSIONS ON GRAVITY-RESPONSIVE TORSIONAL EYE MOVEMENTS. S.G. Diamond and C.H. Markham. *J. Vestib. Res.,* Vol. 8, pages 217–231; 1998.

OCULAR COUNTERROLLING INDUCED BY CENTRIFUGATION DURING ORBITAL SPACE FLIGHT. S.T. Moore, G. Clément, T. Raphan, and B. Cohen. *Exp. Brain Res.,* Vol. 137, pages 323–335; 2001.

The Effect of Spaceflight on the Ultrastructure of the Cerebellum

Authors

Gay R. Holstein, Giorgio P. Martinelli

Experiment Team

Principal Investigator: **Gay R. Holstein**

Co-Investigator: **Giorgio P. Martinelli**

Postdoctoral Fellow: **Ewa Kuielka**

Technical Assistants: **Rosemary Lang, E. Douglas MacDonald II**

Mount Sinai School of Medicine, New York City, USA

ABSTRACT

In weightlessness, astronauts and cosmonauts may experience postural illusions as well as motion sickness symptoms known as the "space adaptation syndrome." Upon return to Earth, they have irregularities in posture and balance. The adaptation to microgravity and subsequent re-adaptation to Earth occurs over several days. At the cellular level, a process called neuronal plasticity may mediate this adaptation. The term plasticity refers to the flexibility and modifiability in the architecture and functions of the nervous system. In fact, plastic changes are thought to underlie not just behavioral adaptation, but also the more generalized phenomena of learning and memory. The goal of this experiment was to identify some of the structural alterations that occur in the rat brain during the sensory and motor adaptation to microgravity.

One brain region where plasticity has been studied extensively is the cerebellar cortex—a structure thought to be critical for motor control, coordination, the timing of movements, and, most relevant to the present experiment, motor learning. Also, there are direct as well as indirect connections between projections from the gravity-sensing otolith organs and several subregions of the cerebellum. We tested the hypothesis that alterations in the ultrastructural (the structure within the cell) architecture of rat cerebellar cortex occur during the early period of adaptation to microgravity, as the cerebellum adapts to the absence of the usual gravitational inputs. The results show ultrastructural evidence for neuronal plasticity in the central nervous system of adult rats after 24 hours of spaceflight.

Qualitative studies conducted on tissue from the cerebellar cortex (specifically, the nodulus of the cerebellum) indicate that ultrastructural signs of plasticity are present in the cerebellar zones that receive input from the gravity-sensing organs in the inner ear (the otoliths). These changes are not observed in this region in cage-matched ground control animals. The specific changes include the formation of lamellar bodies, profoundly enlarged Purkinje cell mitochondria, the presence of inter-neuronal cellular protrusions in the molecular layer, and signs of degeneration in the distal dendrites of the Purkinje cells. Since these morphologic signs are not apparent in the control animals, they are not likely to be due to caging or tissue processing effects. The particular nature of the structural alterations in the nodulus, most notably the formation of lamellar bodies and the presence of degeneration, further suggests that excitotoxicity (damaging overstimulation of neurons) may play a role in the short-term neural response to spaceflight.

These findings suggest a structural basis for the neuronal and synaptic plasticity accompanying the central nervous system response to altered gravity and help identify the cellular bases underlying the vestibular abnormalities experienced by astronauts during periods of adaptation and re-adaptation to different gravitational forces. Also, since the short- and long-term changes in neural structure occurring during such periods of adaptation resemble the neuronal alterations that occur in some neurologic disorders such as stroke, these findings may offer guidance in the development of strategies for rehabilitation and treatment of such disorders.

INTRODUCTION

Upon entering weightlessness, astronauts and cosmonauts can experience postural illusions and symptoms of motion sickness (a constellation of symptoms referred to as the "space adaptation syndrome"). After the flight, crewmembers may have irregularities in posture and balance, and motion sickness symptoms may return as they re-adapt to Earth's gravity. Behavioral adaptation to the microgravity environment (and re-adaptation to Earth) usually occurs within several days (Nicogossian, 1989).

At the cellular level, this behavioral adaptation is thought to be mediated by a process called neuronal plasticity. The term plasticity conveys the flexibility and modifiability in the architecture and functions of the nervous system. The nervous system can make new connections or use existing connections in new ways. Plastic changes in the brain are thought to underlie not just behavioral adaptation, but also the more general phenomena of learning and memory. The goal of this Neurolab experiment was to identify the structural alterations that occur in the rat brain during sensory and motor adaptation to microgravity.

One brain region in which plasticity has been studied extensively is the cerebellar cortex—a structure thought to be critical for motor control, coordination, the timing of movements, and, most relevant to the present experiment, motor learning. The anatomical and functional organization of the cerebellar cortex is well documented (Ito, 1984). The cerebellar cortex has three layers (molecular, ganglion/Purkinje, granular) and five main cell types (basket, stellate, Purkinje, granule, Golgi). Interneuronal basket and stellate cells are found in the outermost molecular layer; Purkinje cells are the only neuronal cell bodies of the ganglion or Purkinje cell layer; and granule and Golgi cells are present in the cell-dense granular layer.

The cerebellar cortex receives two main functional inputs. Climbing fibers take origin from cells in the inferior olivary complex and terminate directly on Purkinje cells. Mossy fibers, in contrast, take origin from a wide variety of brain structures and innervate cerebellar granule cells. The granule cells, in turn, give rise to long axons called parallel fibers. These axons course transversely through the molecular layer, perpendicular to the Purkinje cell dendritic trees, and provide innervation to those dendrites. The ultrastructure features of cerebellar cells have been thoroughly characterized. The Purkinje cell cytoplasm has a whorl-like arrangement of organelles surrounding the nucleus. Both granular and agranular endoplasmic reticula form a loose filigree throughout the perikaryon, including an extensive system of cisterns called hypolemmal cisternae that are located just inside the plasma membrane. The Purkinje cell mitochondria are highly variable in shape and pervasive in the cytoplasm, except in the dendritic thorns. Microtubules invade the entire dendritic tree, again except in the thorns, which have their own cytoplasmic features.

Classically, the cerebellum is subdivided into anterior, posterior, and flocculo-nodular lobes; each is composed of transversely oriented lobules. The flocculo-nodular lobe, or vestibular cerebellum, is comprised of two lobules, the midline or vermal nodulus and the laterally placed hemispheric flocculus. Gravity information sensed by receptors in the otolith organs of the ear is conveyed to particular zones in the nodulus. These zones were the focus for the present report. In addition, the semilunar lobule (hemisphere lobule VIII) was selected as an internal tissue control, since this region does not participate directly in vestibular, postural, or balance functions.

We tested the hypothesis that alterations in the ultrastructural anatomy of rat cerebellar cortex occur during the early period of adaptation to the microgravity environment. We chose to test the question of spaceflight-induced neuronal plasticity in the cerebellar cortex for three reasons: (1) various forms and manifestations of neuronal and synaptic plasticity have been clearly demonstrated in the adult rat cerebellar cortex; (2) the cerebellar cortex has been widely implicated in motor learning; and (3) there are direct as well as indirect projections from the gravity-sensing otolith organs to several regions of the cerebellar cortex.

METHODS

Hindbrain tissue was obtained from rats flown on the Neurolab mission (STS-90). Tissue for the present report was obtained from four adult male Fisher 344 rats on orbit during flight day 2 (FD2), 24 hours after launch, and from equal numbers of vivarium control rats and control rats housed in flight-type cages maintained on Earth (cage controls). These control rats were studied 48 and 96 hours after the flight dissections, respectively. The flight tissues could not be preserved on the Space Shuttle using vascular perfusion, which is the preferred method for this type of research. Instead, ground-based studies were conducted to establish the optimal conditions for immersion fixation of the cerebellar tissue. A technical report included in this volume provides detailed information regarding the development and verification of this fixation paradigm (see technical report by Holstein et al., in this publication). All experiments were performed in accordance with the *Principles of Laboratory Animal Care Guide* (National Institutes of Health Pub. 85-23) and were Animal Care and Use Committee approved by both NASA and the Mount Sinai School of Medicine.

Hindbrains were immersion-fixed for 45 minutes in 4% paraformaldehyde/0.1% glutaraldehyde in 0.1M phosphate buffer (pH 7.3), and were then transferred to a 4% paraformaldehyde solution in 0.1M phosphate buffer for 18 days at 4°C. After this fixation period, each cerebellum was photographed (Figure 1), dissected away from the ventral portion of the brain stem, and then re-photographed (Figure 2). All of the tissue collection and processing protocols were identical for the flight and control specimens.

After the brain stem was dissected away, the paraflocculi were separated from the cerebellum and a small notch was carved into the dorsal aspect of each paraflocculus to aid in orientating the structure in the future. Each cerebellum was then sectioned in the midline, and both halves were mounted

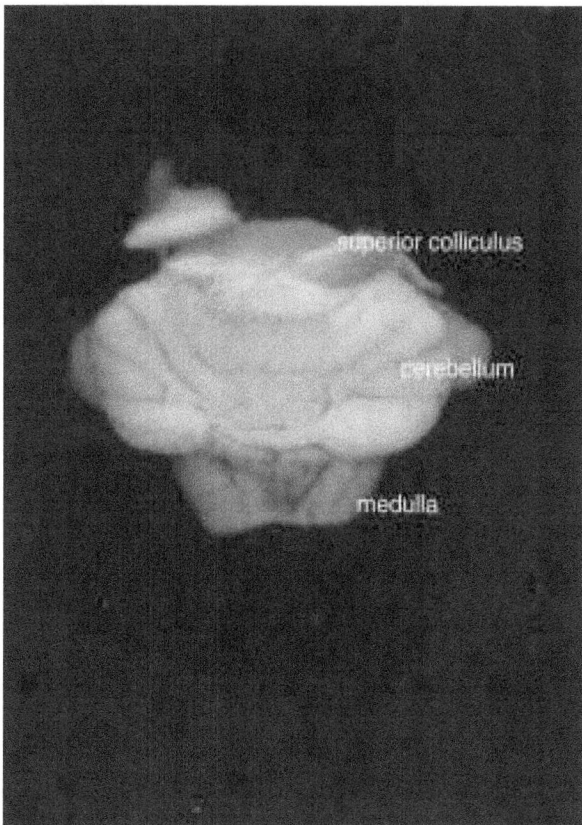

Figure 1. The brain stem of a flight rat after 18 days of immersion fixation. The hindbrain is resting in a petri dish and immersed in buffer just prior to separation of the cerebellum. At the top, the superior colliculus connects to higher brain structures, and the medulla oblongata at the bottom connects to the spinal cord.

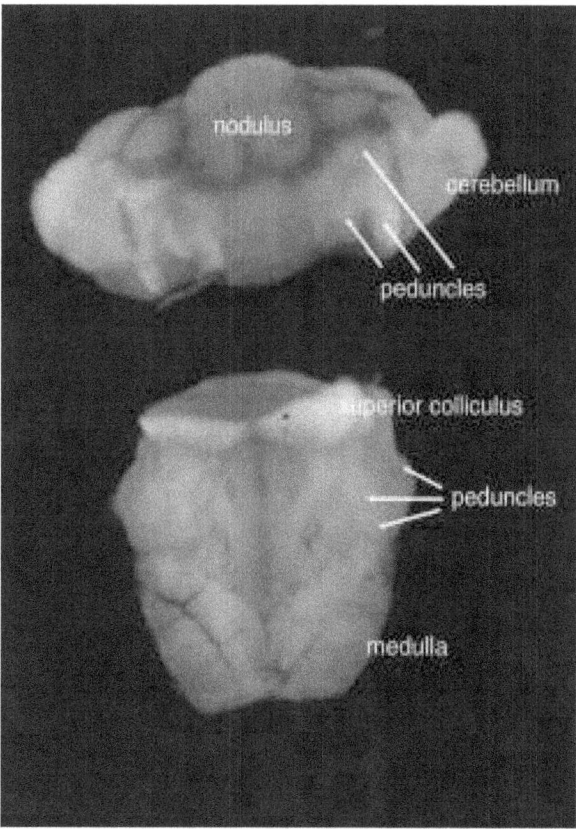

Figure 2. The same brain stem seen in Figure 1, immediately following the sectioning of the three pairs of cerebellar peduncles that form the attachment between the cerebellum and the brain stem. The nerve traffic from the gravity-sensing organs travels through these peduncles to the cerebellum. The cerebellum has been rotated forward, exposing the nodulus and the cut fibers of the peduncles.

together on a glass stage of a vibrating microtome. These specimens were then cut into 100-µm sections. The sections were collected serially and processed for electron microscopy by osmication (1% OsO₄ in deionized water for 30 minutes), dehydration in a graded series of methanol solutions, en bloc staining with uranyl acetate at the 70% methanol stage, infiltration with Epon-Araldite resin, and embedment in resin as tissue wafers between plastic coverslips.

The ultrastructure of the otolith-recipient zones of the cerebellar nodulus was analyzed in tissue from the FD2 flight and cage-control ground-based rats. To obtain the appropriate tissue specimens for this portion of the study, the entire series of wafers from each FD2 rat was examined to identify those sections containing the nodulus. All such wafers were photographed and then traced using a Trisimplex inverted projector. These tracings were used to reconstruct the nodulus and to determine the tissue wafer containing the middle section for each subject. Using this calculated midline as the zero

point, the wafers from two zones—1300–1400 µm and 1600–1700 µm bilaterally from that midline—were identified for each rat. These zones were selected because they receive indirect gravity-related otolith input via the inferior olivary complex (Voogd, 1996) and, therefore, were most likely to be affected by exposure to an altered gravitational environment. These tracings were also used to obtain estimates of the relative volume (section thickness was assumed to be 100 µm) of the nodulus in flight and control rats, as well as the relative volume of the molecular layer of the nodulus as a partial volume fraction (Holstein, 1999).

Specimens from the otolith-recipient zones of the nodulus were dissected from these wafers and mounted on blank resin blocks. Trapezoidal blockfaces were carved from the wafers (Figure 3), thin-sectioned (70-nm thick) by ultramicrotome onto Formvar-coated copper slot grids, and examined using a Hitachi 7500 transmission electron microscope. No post-microtomy staining was performed.

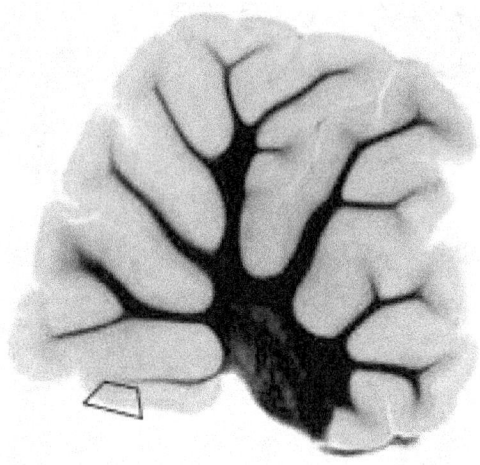

Figure 3. A 100-μm thick Vibratome section embedded in plastic resin and sandwiched between plastic coverslips. The section is 1300–1400 μm lateral to the midline, and it was obtained from the cerebellum of a flight rat. The trapezoid indicates the region thin-sectioned for electron microscopic observations of the nodulus from this and all other flight and control animals.

RESULTS

Qualitative ultrastructural comparisons were conducted using tissue from the zones of the cerebellar nodulus that receive gravity-related inputs from the otoliths. Both the FD2 flight and ground-based, cage-control rats were studied. In addition, sections of the semilunar lobule (hemispheric lobule VIII) from the same FD2 flight rats were examined because this tissue provided an internal tissue control. This lobule does not receive direct vestibular or other inputs that could be affected by microgravity exposure. However, this internal cerebellar control tissue from the flight rats was exposed to the same launch- and Space Shuttle-related stimuli (including noise, vibration, and radiation) that the otolith-recipient cerebellar regions experienced.

The comparisons completed to date suggest that several architectural alterations occur both in the Purkinje cell cytoplasm and in the molecular layer of the nodulus of rats exposed to 24 hours of spaceflight. These structural alterations have not been apparent in the nodulus from the FD2 cage-control animals (Figure 4), and have not been observed thus far in the semilunar lobule of the FD2 flight rats.

The most dramatic alteration observed in the nodular tissue from flight animals was seen in the organelles of the Purkinje cells. These cells normally contain a system of cisterns of smooth endoplasmic reticulum. In Purkinje cells of the otolith-recipient zones of the nodulus in the FD2 flight rats, such cisterns were substantially enlarged and more complex. The increased complexity of the cisterns resulted in the formation of long, stacked lamellar bodies. These were observable throughout entire Purkinje cells, including their somata, dendrites,

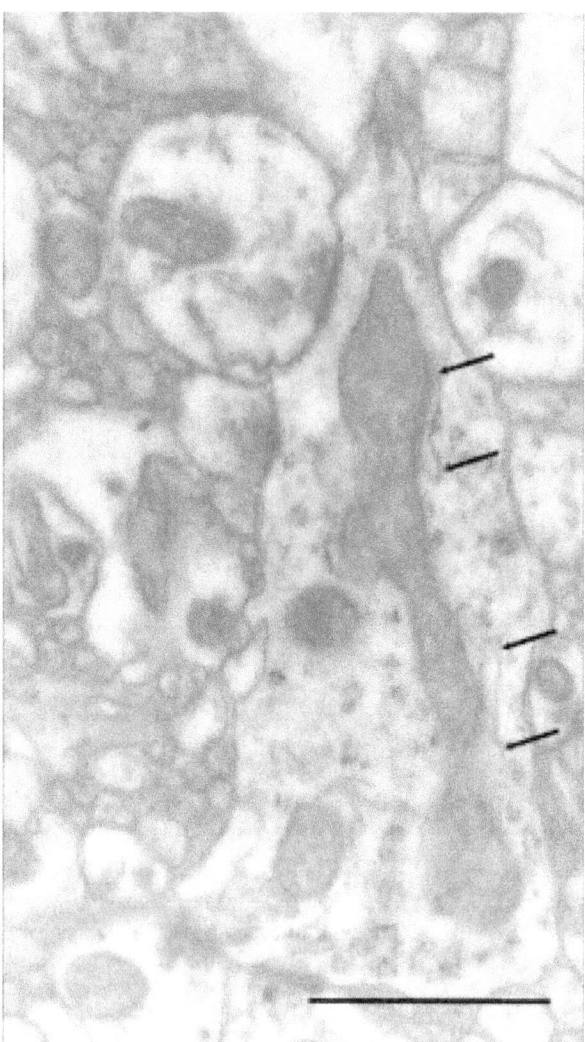

Figure 4. An electron micrograph of the molecular layer of the nodulus from a cage-control rat that was treated identically to the flight rats but was not exposed to spaceflight. The arrows point to examples of subsurface cisterns in the Purkinje cell dendrite. No lamellar bodies are present. Scale bar: 1 μm.

spines, thorns, and axon terminals (Figure 5). In general, these organelles were closely associated with adjacent mitochondria and the nearby plasma membrane.

In addition, occasional enormous mitochondria, some more than 1 μm in cross-sectional diameter, were present in the Purkinje cells of flight rats (Figure 6). Alterations in the molecular layer of the nodulus included frequent and sometimes large protrusions of neuronal elements into neighboring structures (Figure 7). Lastly, ultrastructural signs of electron-dense degeneration were apparent in the dendrites of Purkinje cells from the nodulus of the flight rats. Such structures contained increased numbers of lysosomes and degenerated mitochondria, but they maintained apparently healthy synaptic contacts with parallel fiber terminals that, themselves, appeared normal (Figure 8).

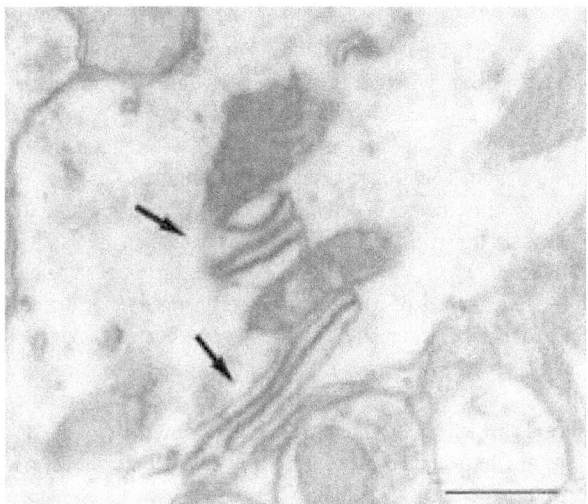

Figure 5. An electron micrograph through a Purkinje cell dendrite from an otolith-recipient zone of the nodulus from an FD2 flight rat. The arrows indicate long, stacked lamellar bodies that were observed throughout entire Purkinje cells, including the somata, dendrites, and thorns of flight rats. The lamellar bodies are closely associated with the adjacent mitochondria and nearby plasma membrane. Scale bar: 0.5 μm.

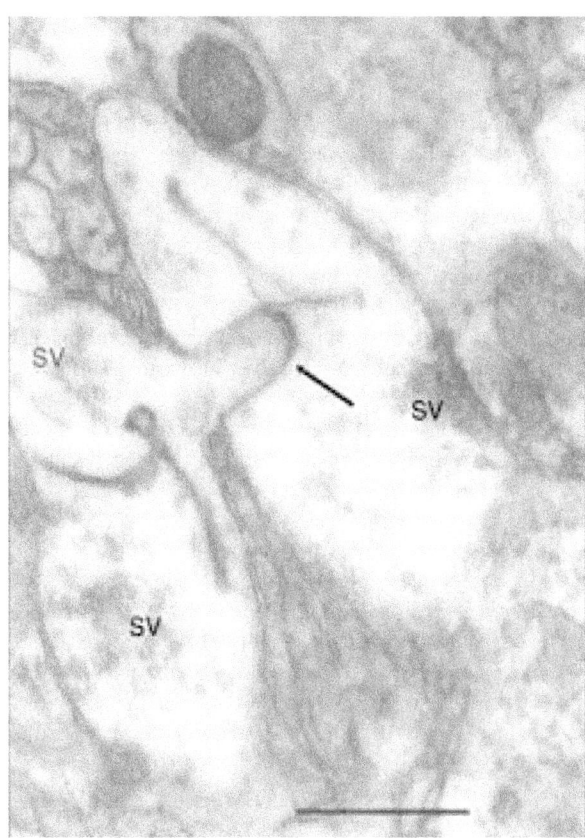

Figure 7. Electron micrograph of the molecular layer of the nodulus from an FD2 flight rat. The arrow indicates an example of a cellular protrusion between adjacent structures. Both structures are axon terminals, as indicated by the presence of clusters of synaptic vesicles (sv). Scale bar: 0.5 μm.

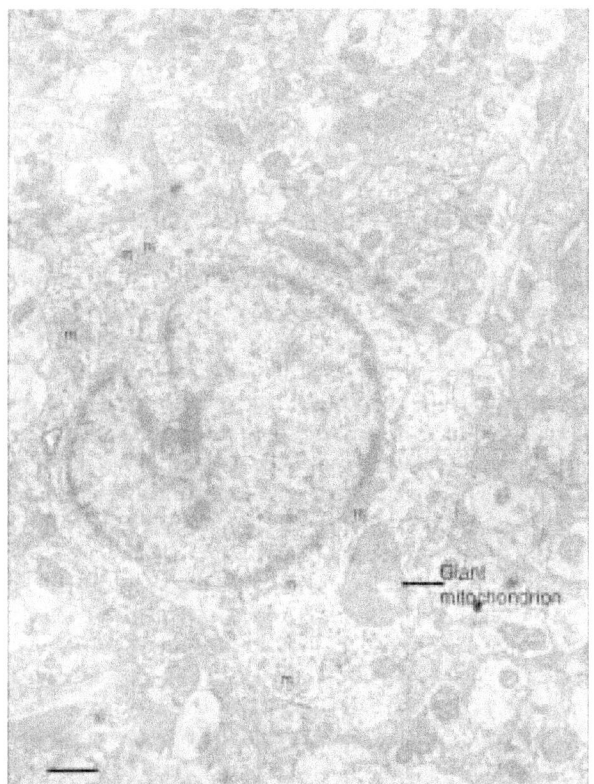

Figure 6. Electron micrograph through a Purkinje cell body from an otolith-recipient zone of the nodulus from an FD2 flight rat. An example of an extremely large mitochondrion is indicated, and normal mitochondria in the same cell body are labeled (m) for comparison. Scale bar: 1 μm.

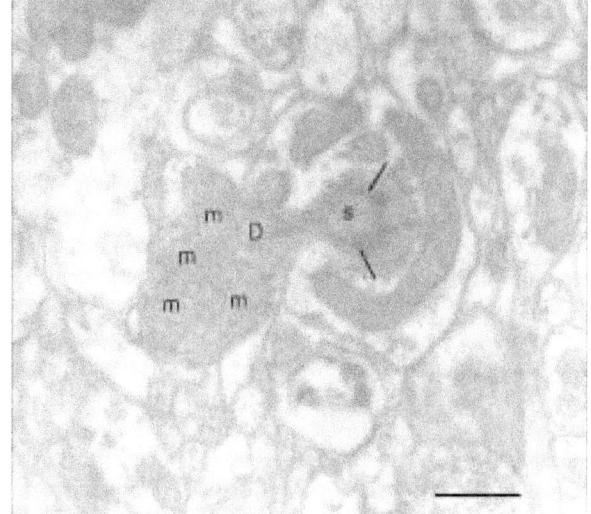

Figure 8. Electron-dense degeneration is present in a distal Purkinje cell dendrite (D) and spine (s) in the molecular layer of the nodulus from an FD2 flight rat. Degenerated mitochondria (m) are also present in the dendrite. Black arrows indicate synaptic contacts with the degenerated cell. Scale bar: 0.5 μm.

DISCUSSION

Qualitative studies conducted after 24 hours of spaceflight on the cerebellar nodulus show ultrastructural signs of plasticity in the otolith-recipient zones of the nodulus, which are not observed in this region in cage-matched ground-control rats. Such alterations include the formation of lamellar bodies and profoundly enlarged Purkinje cell mitochondria, the presence of inter-neuronal cellular protrusions in the molecular layer, and signs of degeneration in the distal dendrites of the Purkinje cells. Since these morphologic signs are not apparent in the control rats, they are not likely to be due to caging or tissue processing effects.

In the flight tissue, the subsurface cisterns of smooth endoplasmic reticulum, which are common in the cytoplasm of normal Purkinje cells, develop into complex lamellar bodies. These organelles are closely associated with adjacent mitochondria and the nearby plasmalemma. In the initial descriptions of these organelles, the lamellar bodies were shown to consist of stacks of several (between four and 12) flattened, parallel, regularly spaced (300–400 A) tubular cisterns alternating with plates of dense granular material. The stacks were often observed to be subjacent to the plasma membrane in close proximity to the smooth endoplasmic reticulum, and closely apposed to the outer membranes of mitochondria. They were sometimes, but not always, associated with the rough endoplasmic reticulum as well. More recently, some investigators have suggested that these organelles may be artifacts of fixation or lack of oxygen before death. However, most studies to date view the cisternal stacks as distinct neuronal structures (Yamamoto, 1991). In fact, lamellar bodies have been observed in Purkinje cells using a variety of fixation protocols and embedding and sectioning methods (Yamamoto, 1991).

It has been suggested that markedly increased lamellar body numbers or complexity result from experimental manipulations and/or pathological conditions (Bestetti, 1980; Hansson, 1981). For example, lamellar body formation in Purkinje cell dendrites can be induced by a brief (five-minute) period without oxygen, or by the administration of L-glutamate into the brain. Although L-glutamate is normally found in the brain and participates in the transmission impulses between some neurons, too much glutamate can overstimulate neurons and kill them. This process is called excitotoxicity. In one study, the enhanced formation of cisternal stacks was inhibited by co-administration of glutamate and a glutamate receptor antagonist. Since the antagonist prevents the formation of the stacks, this suggests that glutamate excitotoxicity may play a role in cisternal stack formation. Finding these stacks (lamellae) in the brain (as occurred in our study) may be an indication of excitotoxicity. It is also possible that the entire network of smooth endoplasmic reticulum in Purkinje cells has the relatively unique property to form cisternal aggregates. The number of cisternal stacks present in situ under a given set of experimental conditions may reflect a dynamic equilibrium between the tendencies for aggregation and dispersion of the aggregates.

Regarding the other major findings of the present study, the presence of gigantic mitochondria in Purkinje cells has been suggested by other investigators to serve as an ultrastructural sign of early cell degeneration reflecting an underlying process of synaptic remodeling. Moreover, the presence of inter-neuronal cellular protrusions suggests enhanced membrane fluidity, also possibly reflecting an underlying process of neuronal plasticity. Lastly, electron-dense degeneration was apparent in the Purkinje cell dendrites of the flight rats. Such structures contained increased numbers of lysosomes and degenerated mitochondria, but maintained apparently healthy synaptic contacts with normal-appearing axon terminals.

To date, there are few studies on the impact of exposure to altered gravitational environments on the central nervous system structure. In the granular layer of the nodulus of rats raised in a two-G environment for 60 days, 80% of the glomeruli have been reported to show modifications in synaptic morphology. These alterations include changes in the density of pre- and postsynaptic membranes, increased thickness of the postsynaptic density, enlargement of the synaptic cleft, increased packing density of synaptic vesicles, enlarged mitochondria, and an increase in the number of microtubules (Krasnov, 1986; Krasnov, 1991). Two days after such animals return to a one-G environment, the ultrastructure of the nodulus is reported to resemble that of control rats. The synaptic vesicle packing density is diminished, and the number of microtubules is decreased, suggesting that such gravity-induced effects are reversible (Krasnov, 1991). Complementary morphologic changes have been reported to occur in the nodulus of rats following spaceflight (Krasnov, 1986; Krasnov, 1990). In these rats, ultrastructural changes in Purkinje cell dendritic synaptology have been reported.

In the peripheral vestibular system, light microscopic observations of the rat sacculus have been reported following development in 2.3-G and 4.15-G (Lim, 1974). In 60-day-old rats raised pre- and postnatally in a two-G environment induced by continuous centrifugation, the large-sized otoconia (stones in the otoliths) present in normal control rats do not develop. Type I receptor cells exhibit abnormal ultrastructural features, including increased chromatin, increased perinuclear space, and increased intercellular space (Krasnov, 1987; Krasnov, 1991). An increase in otoconial size has been reported in the utricle of rats after seven days of exposure to microgravity (Ross, 1985), suggesting that the otoconia are capable of plastic changes in response to altered gravity conditions. Ultrastructural studies have been conducted on the utricle and sacculus of adult rats after 30 days of exposure to two-G and on the utricular hair cells of adult rats flown on Space Shuttle missions. In these studies, Type II hair cells, and to a lesser extent Type I hair cells, showed evidence of neuronal plasticity. Increases in the mean number of synaptic ribbons were observed in such cells from rats studied on FD13 of a 14-day mission as well as in those sacrificed immediately following Space Shuttle landing, and others sacrificed after a mission-length, one-G re-adaptation period. Taken together, these data support the notion that the adult utricle retains the potential for plastic morphologic reorganization, and that the Type II hair cells are more directly involved in these adaptations to altered gravity conditions (Ross, 1997).

While there has been evidence for plasticity in the peripheral nervous system, the present results provide ultrastructural evidence for neuronal plasticity in the central nervous system of adult rats after 24 hours of exposure to spaceflight. The particular nature of the structural alterations in the nodulus, including the formation of lamellar bodies and the presence of degeneration, further suggests that excitotoxity may play a role in the short-term neural response to spaceflight. The specificity of the alterations for particular organelles of the Purkinje cell, the overall acceptability of the ultrastructural tissue preservation, and the paucity of similar observations in control tissues all suggest the soundness of the observations. Nevertheless, it remains critical to demonstrate the extent of these morphological alterations in perfusion-fixed tissue as well.

Our findings suggest a structural basis for the neuronal and synaptic plasticity accompanying the central nervous system response to altered gravity. Specifically, the studies of neuronal degeneration provide new insight into the immediate and short-term cellular responses to varying gravitational fields. Since the short- and long-term changes in neural structure occurring during such periods of adaptation resemble the neuronal alterations that occur in some neurologic disorders such as stroke, these findings may offer guidance in the developing of strategies for rehabilitation and treatment of such disorders. Lastly, since habituation of the time constant of the vestibulo-ocular reflex critically involves the cerebellar nodulus, it seems clear that the vestibular portion of the cerebellar cortex is critical for mediating plastic changes compatible with sensory-motor integration. In that light, the present results help to identify the cellular bases underlying the vestibular abnormalities experienced by astronauts during periods of adaptation and re-adaptation to different gravitational forces.

Acknowledgements

The authors are grateful to Dr. Louis Ostrach and Ms. Lisa Baer of the Ames Research Center, NASA, for tireless support of this project; and Dr. Ewa Kukielka, Ms. Rosemary Lang, and Mr. E. Douglas MacDonald II at Mount Sinai School of Medicine for invaluable assistance with all aspects of the research. This work was aided by NASA grant NAG2-946 and NIH grant DC02451 from the National Institute for Deafness and Other Communication Disorders.

REFERENCES

OBSERVATIONS ON SACCULUS OF RATS EXPOSED TO LONG-TERM HYPERGRAVITY. D. J. Lim, J. A. Stith, and J. Oyama. *Aerospace Med.,* Vol. 45(7), pages 705–710; 1974.

THE OCCURRENCE OF CYTOPLASMIC LAMELLAR BODIES IN NORMAL AND PATHOLOGIC CONDITIONS. G. Bestetti, and G. L. Rossi. *Acta Neuropathol.,* Vol. 49, pages 75–78; 1980.

LAMELLAR BODIES IN PURKINJE NERVE CELLS EXPERIMENTALLY INDUCED BY ELECTRIC FIELD. H. A. Hansson. *Brain Res.,* Vol. 216, pages 187–191; 1981.

THE CEREBELLUM AND NEURAL CONTROL. M. Ito, New York, Raven Press; 1984.

OTOCONIAL MORPHOLOGY IN SPACE-FLOWN RATS. M. D. Ross, K. Donovan, and O. Chee. *The Physiologist*, Vol. 28 (Suppl. 1), pages 219–220; 1985.

ULTRASTRUCTURE OF THE CORTEX OF THE CEREBELLAR NODULUS IN RATS AFTER A FLIGHT ON THE BIOSATELLITE KOSMOS 1514. I. B. Krasnov and L. N. Dyachkova, *Kosm. Biol. Aviak. Med.,* Vol. 20(5), pages 45–48; 1986.

INCREASE IN THE SENSIBILITY OF OTOLITH APPARATUS IN WEIGHTLESSNESS. MORPHOLOGICAL EVIDENCE. I. B. Krasnov, Proceedings of the Intercosmos Space Biology Medical Group Meeting, Vol. 10, page 127; 1987.

OVERALL PHYSIOLOGICAL RESPONSE TO SPACE FLIGHT. A. E. Nicogossian, In: *Space Physiology and Medicine.* A. E. Nicogossian, C. Leach-Huntoon and S. L. Pool. eds. Philadelphia, Lea and Febiger, pages 139–153; 1989.

THE EFFECT OF SPACE FLIGHT ON THE ULTRASTRUCTURE OF THE RAT CEREBELLAR AND HEMISPHERE CORTEX. I. B. Krasnov, and L. N. Dyachkova. *The Physiologist*, Vol. 33 (Suppl. 1), pages 29–30; 1990.

STACKS OF FLATTENED SMOOTH ENDOPLASMIC RETICULUM HIGHLY ENRICHED IN INOSITOL [1,4,5]-TRIPHOSPHATE (INSP3) RECEPTOR IN MOUSE CEREBELLAR PURKINJE CELLS. A. Yamamoto, H. Otsu, T. Yoshimori, N. Maeda, K. Mikoshiba, and Y. Tashiro, *Cell Struct. and Funct.,* Vol. 16, pages 419–432;1991.

THE OTOLITH APPARATUS AND CEREBELLAR NODULUS IN RATS DEVELOPED UNDER 2-G GRAVITY. I. B. Krasnov, *The Physiologist*, Vol. 34 (Suppl. 1.), pages S206–S207; 1991.

ORGANIZATION OF THE VESTIBULOCEREBELLUM. J. N. Voogd, M. Gerrits, et al., In: *New Directions in Vestibular Research.* S. M. Highstein, B. Cohen and J. A. Büttner-Ennever eds., New York, New York Academy of Sciences, Vol. 781, pages 553–579; 1996.

ADAPTIVE RESPONSES OF RAT VESTIBULAR MACULAR HAIR CELLS TO MICROGRAVITY, ISOLATION, AND OTHER STRESS. M. D. Ross. *Proceedings of the Association for Research in Otolaryngology Midwinter Research Meeting*, Vol. 38; 1997.

ANATOMICAL OBSERVATIONS OF THE RAT CEREBELLAR NODULUS AFTER 24 HR OF SPACEFLIGHT. G. R. Holstein, E. Kukielka, *J. Gravit. Physiol.*; 1999.

Gene Expression in the Rat Brain During Spaceflight

Author

Ottavio Pompeiano

Experiment Team

Principal Investigator: **Ottavio Pompeiano**

Co-Investigators: **Paola d'Ascanio, Claudia Centini, Maria Pompeiano, Chiara Cirelli, Giulio Tononi**

Università di Pisa, Pisa, Italy

ABSTRACT

The vestibular system senses the position and movement of the head in space, controls the activity of postural and eye muscles, influences the cardiovascular and respiratory systems, and monitors body orientation. During spaceflight, astronauts show: (a) changes in balance and eye movements; (b) alterations in the control of cardiovascular and respiratory activities; (c) changes in body orientation and perception; and (d) sleep disturbances. After flight, balance is changed markedly. These deficits are compensated over time.

The compensation for these changes involves plasticity—the ability of nerves to make new connections or to use existing connections in different ways. One way to determine which nervous structures are involved in plasticity is to measure changes in immediate early genes (IEGs). Changes in neural activity produce changes in the expression of IEGs, such as *c-fos* and *FRAs* (fos-related antigens). The products from these IEGs can be detected within minutes after stimulation, while the corresponding proteins produced by activating these genes persist for hours (Fos) or days (FRAs). Fos and FRAs can also affect the expression of other genes. These, in turn, may convert the nerve activation induced during spaceflight into short-term or long-term biochemical responses. Such molecular changes may thus contribute to the plastic events responsible for adaptation to zero-G and the re-adaptation to one-G.

We measured Fos and FRAs to identify the cellular and molecular changes occurring during spaceflight. The entire brain (except for the cerebellum) of 24 adult male albino rats (Fisher 344 rats) was studied at different time points of the spaceflight and compared to the entire brain of 48 control rats.

Changes in Fos and/or FRAs expression were observed in the following brain structures:

(a) the vestibular nuclei, which control posture and eye movements either directly or through other brain structures, such as the inferior olive and the lateral reticular nucleus;

(b) the nucleus of the tractus solitarius (NTS) (the nucleus of the solitary tract) and other areas involved in the spontaneous and reflex regulation of cardiovascular and respiratory functions;

(c) the area postrema (AP), which plays a role in motion sickness;

(d) the central nucleus of the amygdala, a structure that is functionally related to both the NTS and the AP;

(e) cortical and subcortical areas involved in body orientation and perception;

(f) hypothalamic and brain stem nuclei, such as the locus coeruleus, involved in the regulation of the sleep-wake cycle.

The results identified the brain regions most affected by the adaptation to weightlessness and the re-adaptation to Earth's gravity.

INTRODUCTION

Vestibular receptors in the inner ear detect head movements and provide these signals to the brain. Vestibular stimulation plays a prominent role in controlling eye, trunk, and limb muscles—thus stabilizing eye and body position during movements (Wilson, 1979). The vestibular system helps maintain stable blood pressure and respiration during changes in posture (Yates, 1998), and it is essential for eliciting nausea and vomiting during motion sickness (Miller, 1994). Vestibular signals may also contribute to body orientation and perception. Finally, changes in vestibular activity may modify the sleep-wake cycle. During and after spaceflight, crews show some impairment in the functions mentioned above. These deficits, however, are compensated over time.

The brain can adapt and respond to perturbations such as weightlessness. Adaptation occurs both during exposure to microgravity and after reentry to the Earth's atmosphere. However, the specific brain structures affected by changes in gravity, as well as the cellular and molecular mechanisms responsible for the adaptation to zero-G and re-adaptation to one-G, are largely unknown. Understanding the cellular and molecular basis of these adaptations is one of the greatest challenges of modern neuroscience. If the brain areas involved in the adaptation were identified and the biochemistry of these nerves were known, scientists would be offered the possibility to intervene in the adaptation where it might be beneficial.

Since some of these adaptations persist after the perturbation ceases, we postulated that regulation of gene expression represents one of the important mechanisms involved. One way to measure gene expression is to use immediate early genes (IEGs). IEGs are genes that rapidly produce proteins after the cell is stimulated. The pattern of IEG expression can show which areas in an organ, such as the brain, have been stimulated. We studied the expression of the IEGs *c-fos* and *FRAs* (fos-related antigens). The IEGs *c-fos* and *FRAs* in turn produce the proteins Fos and FRAs, respectively. These proteins bind to DNA regulatory regions and control the expression of several other target genes (Herdegen, 1998). The induction of *c-fos* is transient, since Fos levels peak within two to four hours from the original stimulus and return to baseline values within six to eight hours. FRA levels, on the other hand, peak later and then persist in the brain for days or weeks after the initial stimulus. Therefore, while Fos induction mediates short-lived changes in gene expression, FRAs mediate long-term molecular changes that can lead to long-lasting regulation of brain functions.

Both of these markers have been used in our laboratory to identify the brain regions that undergo changes during: (1) the normal sleep-wake cycle; (2) different periods of sleep deprivation; (3) unilateral labyrinthectomy (i.e., when the balance organ is removed on one side) (Cirelli, 1996); and (4) activation of the reflex arc involved in vomiting (Miller, 1994). These studies showed marked changed in the vestibular nuclei (VN), the inferior olive (IO), and the lateral reticular nucleus (LRt)—all regions involved with balance, posture, and movement. We hypothesized that these areas would also be affected by spaceflight. Other areas of interest are the medullary structures involved in the vestibular regulation of cardiovascular or respiratory functions (see Figure 1) such as the nucleus of the tractus solitarius (Yates, 1998). The area postrema (AP), a structure involved in producing space motion sickness, and the amygdala, a structure that controls several cardiovascular and respiratory functions, could also be affected by spaceflight. The locus coeruleus (LC), a central noradrenergic structure that projects to broad areas of the brain (including those involved in the response to gravity signals), could be affected (Figure 1). The LC intervenes not only in the regulation of the sleep-wake cycle (Barnes, 1991), but also in the adaptation and compensation of vestibular syndromes (Pompeiano, 1994). An activation of the LC during different flight conditions would suggest a prominent role for this area in mediating the adaptation to microgravity and/or the re-adaptation to the terrestrial environment.

In this study, we used Fos and FRAs immunocytochemistry to identify which neuronal systems underwent genetic activation during the NASA Neurolab Space Shuttle mission (STS-90). We compared the pattern of induction of these transcription factors in numerous brain structures of rats exposed to different levels of gravity to matched controls on Earth.

METHODS

Experimental design

Experimental design – The experiments were performed on 72 adult male albino Fisher 344 rats divided into three groups of 24 rats. The first group was selected to support the inflight investigation (flight [FLT] group). Flight rats were housed in Research Animal Holding Facility (RAHF) cages and loaded on board the Space Shuttle 33 hours prelaunch. With a delay of two to three days with respect to FLT rats, two comparable groups of ground-based control rats were individually housed either in cages similar to those used for the flight rats (4×4.25×10 inches: simulated RAHF or asynchronous ground control [AGC] group) or in standard vivarium (VIV) cages (18.5×10.25×8.5 inches: VIV group). Control rats were maintained at one-G under the same temperature (23±1°C) and lighting conditions as the FLT rats. All animals were fed on SLO Foodbars (flight diet) and their body weight underwent only slight modifications. Animal care was provided by a veterinarian crewmember in flight and by specialized personnel on ground postflight. All animal procedures complied with the National Institutes of Health (NIH) *Guide for the Care and Use of Laboratory Animals*.

The rats used in this study were also the rats used in the experiment of Fuller (see science report by Fuller et al. in this publication), which involved the administration of a light pulse (LP). Half of the rats studied were exposed to an LP of 300 lux for 60 minutes (LP rats) before collecting the brain tissue, while the other rats were not exposed to an LP (non-light pulse (NLP) rats) before collecting the brain tissue. Also, 18 rats were submitted to constant dim light (light intensity <30 lux) at time point R+13. The remaining rats were submitted to a 12 hours on, 12 hours off lighting cycle (light intensity=30 lux).

Rats were studied at two time points during flight and at two symmetrical time points after landing: FD2 (flight day 2)= one day (24 hours) after launch (n=4); FD14=13 days after launch (n=9); R+1 (recovery+1)=one day after the reentry (n=5); R+13 = 13 days after landing (n=6). During launch, G forces increased from one to two-Gs during the first two minutes, then fell back to one-G and gradually increased to three-Gs over six minutes. The launch lasted eight minutes. During reentry, the G-forces increased gradually from zero-G to 1.5-Gs over 30 minutes, before stabilizing at one-G.

Utilization of the brain tissue – We examined the entire brain of rats for IEG expression with two exceptions: other groups studied the forebrain and the cerebellum.

Limitations – The superior vestibular (SuVe) nucleus, the dorsal part of lateral vestibular (LVe) nuclei, the LC, and the parabrachial nuclei were partially damaged during dissection. The other tissues were intact.

Methods – Adjacent brain sections from all rats were alternatively stained with Fos or FRA antibodies, according to standard immunocytochemical protocols (Cirelli, 1996). A commercial polyclonal antibody was used for Fos (Oncogene Research Products, Cambridge, MA), while the antibody used to detect FRAs was a generous gift of Dr. M. J. Iadarola (National Institute of Dental Research (NIDR), NIH, Bethesda, MD). The different brain structures were identified according to the Atlas of Paxinos and Watson (Paxinos, 1998). The number of labeled cells observed in all of the examined structures for all rats was evaluated by two observers who were blind to the origin of the sections. Dr. E. Balaban performed a statistical analysis of results using a Scheirer-Ray-Hare nonparametric two-way analysis of variance with a Bonferroni correction for multiple comparisons.

RESULTS

The effects of spaceflight were evaluated by comparing the number of labeled cells in a given structure of the FLT group with that of the corresponding AGC group. The reason for choosing the AGC group as the control group was that AGC rats were housed in cages of the same size as those used for the FLT rats (see Methods). AGC rats showed a stronger Fos labeling than did the VIV rats, probably because they were constrained in cages much smaller than the cages that were used for the VIV rats. Although the number of Fos-positive cells was usually higher in rats that received the LP compared to those that did not, the general patterns of Fos expression were comparable. Unless specified, the following description will refer to the results obtained in both LP and NLP rats.

Fos protein expression in forebrain structures implicated in the regulation of sleep and waking.

Ground-based experiments performed in Wistar and Fisher 344 rats showed that the expression of *c-fos* was negligible or absent in most areas of the neocortex, hypothalamus, lateral septum, aspecific (but not specific) nuclei of the thalamus,

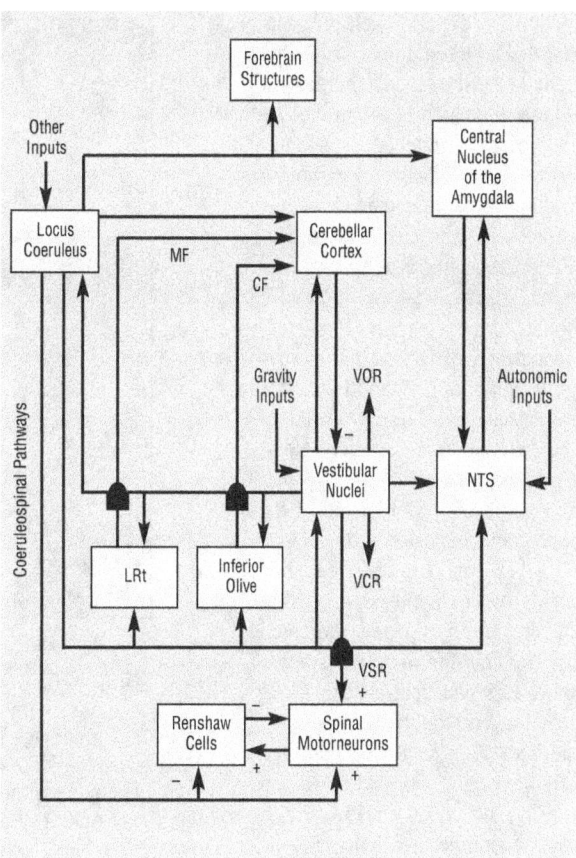

Figure 1. General organization of connections made by the LC. The LC connects to the vestibular nuclei, and the vestibular nuclei send gravity information to the LC. The vestibular nuclei, in turn, contribute to the reflexes that connect the inner ear to the eyes (the vestibulo-ocular reflex (VOR)), the neck (the vestibulocollic reflex (VCR)), and the spinal cord (the vestibulospinal reflex (VSR)). These reflexes are important in maintaining proper eye fixation and body posture during movement. The LC also projects to the cerebellum and precerebellar nuclei (the inferior olive (IO) and lateral reticular nucleus (LRt)), which are structures involved in the control of posture and movement. Lastly, the LC connects to structures involved in vegetative functions (cardiovascular, respiratory, and gastrointestinal functions) such as the nucleus of the tractus solitarius (NTS) and the central nucleus of the amygdala. CF, climbing fibers; MF, mossy fibers.

hippocampus, and several brain stem structures, including the LC (see next section) after a few hours of sleep. Expression was high if the rat had been either spontaneously awake or sleep-deprived for a few hours. In contrast to these findings, long-term (4–14 days) sleep-deprived rats showed only a small amount of Fos expression in most cortical areas, the hippocampal formation, and the septum. When long-term sleep-deprived rats were allowed to sleep, they showed a significant rebound of rapid eye movement (REM) sleep. During this recovery sleep, *c-fos* expression was induced in several forebrain regions, including the central nucleus of the amygdala, as well as in several brain stem regions involved in the generation of REM sleep.

In our experiments, FD2 rats showed a large number of labeled cells in most of the brain areas in which Fos had been previously identified as induced by spontaneous or forced (for a few hours) waking. A similar pattern of Fos expression was also observed in FD14 rats with only minor variations in labeling, which was in some cases stronger than that occurring in FD2 rats. This suggests that both FD2 and FD14 rats might have been mostly awake in the hours before sacrifice. However, no difference was found between AGC and FLT rats at the time points indicated above.

R+1 rats showed very low levels of Fos expression in most of the brain regions indicated above. This finding suggests that these rats had been asleep for at least several hours before being studied. This would be expected after a period of forced waking during the reentry. However, a higher number of labeled cells was observed in the central nucleus of the amygdala of FLT with respect to AGC rats. Control rats actually showed a waking pattern of Fos expression characterized by the presence of labeled cells in many cortical and subcortical regions. Finally, R+13 rats showed either low levels or high levels of Fos expression in cortical and subcortical structures comparable to levels occurring during sleep or waking, respectively. However, none of these rats showed significant staining in the central nucleus of the amygdala. Finally, most AGC rats studied at R+13 showed many Fos-positive cells in cortical and subcortical areas, as in FD2 and FD14 rats.

Fos and FRA protein expression in the locus coeruleus and its relationship to gravity signals

The noradrenergic LC nucleus is a structure with diffuse projections (Figure 1) that has neurons that discharge during waking but not during sleep.

In our experiments, Fos protein expression in the LC was on the average higher in FLT than in AGC rats, corresponding to 7.5 and 3.9 labeled cells/section/side, respectively, as evaluated over the four time points of the spaceflight in NLP rats. These values, however, increased on the average to 24.0 and 10.8 labeled cells/section/side, respectively, in LP rats. The higher number of Fos-positive cells in LP with respect to NLP rats can be due to the fact that LC neurons are particularly activated by novel stimuli rather than by familiar stimuli (Barnes, 1991). There were no consistent differences in Fos

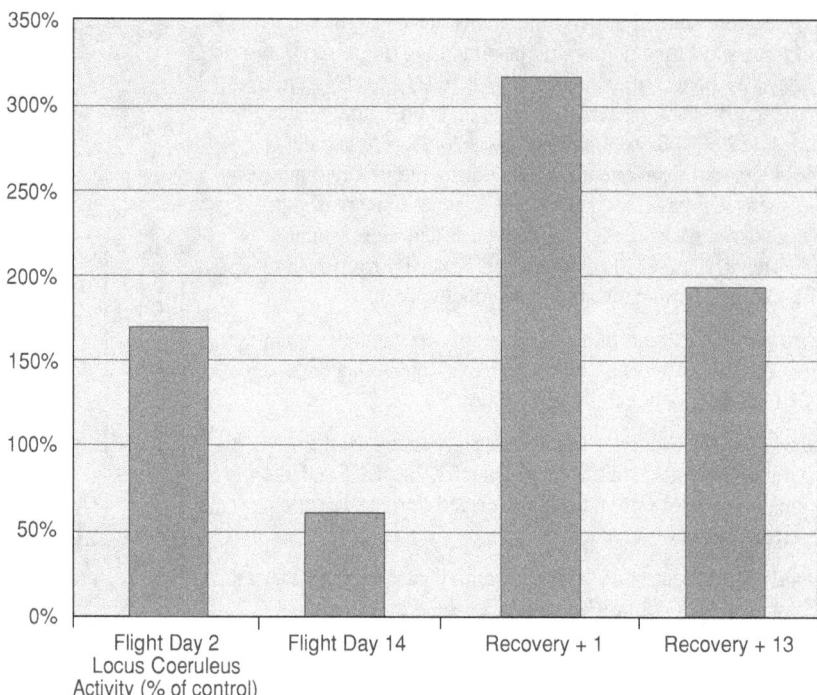

Figure 2. Percentage change in FRA expression compared to the ground controls (AGC). FRA expression increased on FD2, but was below the levels in the ground controls at FD14. FRA expression rose markedly one day after reentry (R+1), and was still somewhat higher than AGC levels 13 days after landing.

expression at the different time points. To help interpret these data, we decided to study the expression in the LC of FRA proteins, which have a more prolonged expression with respect to Fos.

The FRA results are summarized in Figure 2. In FLT-FD2 rats, the number of labeled cells increased on average to 169.9% of the AGC controls. This effect could be attributed to the 10-minute period of the acceleration stimulus that brought the Space Shuttle from one-G to three-Gs, before reaching orbit (~zero-G). In addition, this effect was possibly even attenuated by the suppression of gravity, which occurred during most of the 24 hours following launch. This hypothesis is supported by the fact that the number of FRA-positive cells observed in FLT-FD14 rats decreased on average to 61.0% of the corresponding AGC control. Most remarkably, the number of FRA-positive cells increased to 316.5% of the AGC controls in FLT-R+1 rats, when the peak gravity force increased from zero-G to 1.5-G to 1.8-G before reaching one-G after landing (Figure 2). Finally, the number of FRA-positive cells decreased on average to 193.1% of the control in FLT-R+13 rats; i.e., after re-adaptation to the terrestrial environment (one-G).

Fos protein expression in brain areas concerned with balance: the vestibular nuclei (VN)

The VN, the connections of which are schematically drawn in Figure 1, showed changes in Fos expression during different

Table 1. Data reported in this table refer to 32 rats studied at the indicated time of the spaceflight. In particular, four rats (two LP and two NLP) were sacrificed for each time point of the spaceflight (FLT) (FD2, FD14, R+1, R+13), and the results were compared to the corresponding controls (AGC). For each examined region, mean value ±SD represents the average number of Fos-positive cells observed per side per section for each rat (average of the mean values obtained from each single rat), while the total number of observations performed on each structure, on one or both sides of individual sections, is indicated in parentheses. The structures examined correspond to: (A) the VN, which controls directly the activity of neck and eye neurons; (B) the vegetative areas of the medulla, which control cardiovascular and respiratory activities; and (C&D) the precerebellar structures; i.e., the IO and the LRt, which connect to the cerebellum. Results obtained in 16 FLT rats can be compared with those obtained in a corresponding number of control (AGC) rats. Occasionally, two (^) or three rats (#) only were examined instead of four rats per group.

	FD2		FD14		R+1		R+13	
	AGC	FLT	AGC	FLT	AGC	FLT	AGC	FLT
A. Vestibular nuclei								
MVeMC	3.48±2.06 (39)	4.02±3.23 (43)	2.19±0.93 (53)	1.14±0.20 (49)	2.28±1.43 (49)	4.19±1.96 (50)	1.25±0.99 (47)	1.51±1.31 (46)
MVePC	7.28±4.11 (39)	6.61±1.55 (43)	4.71±2.47 (53)	1.81±0.59 (49)	3.19±0.51 (49)	9.84±1.75 (50)*	3.46±2.60 (47)	3.33±1.40 (46)
MVe	9.63±5.66 (14)	18.52±6.43 (20)	10.59±5.63 (17)	4.27±3.11 (18)	7.56±5.14 (16)	17.08±9.92 (22)	3.02±2.97 (15)	4.85±4.31 (20)
SpVe+FVe	5.94±2.19(40)	9.95±5.24 (44)	2.94±0.97 (48)	2.85±0.54 (50)	3.49±1.71 (47)	26.15±8.77 (51)*	1.47±0.91 (43)	3.53±3.74 (46)
Pr	2.67±1.05 (36)	3.73±0.75 (46)	1.73±0.40 (46)	1.99±0.49 (45)	1.73±0.73 (48)	1.74±0.41 (38)	1.13±0.73 (34)	1.31±0.22 (38)
B. Vegetative areas of medulla								
NTS	39.61±11.52 (66)	22.77±14.65 (72)	30.99±10.84 (84)	20.84±9.34 (79)	22.76±9.09 (87)	50.38±10.85 (86)*	16.06±8.76 (68)	19.94±15.12 (74)
AP	9.25±5.81 (7)	3.13±2.84 (7)	^5.50±2.12 (4)	#7.78±5.92 (7)	#13.0±5.57 (7)	42.03±15.25 (14)*	5.67±2.0 (11)	14.63±20.33(11)
PCRt	18.77±14.05 (44)	14.71±16.72 (52)	17.83±11.55 (50)	12.56±5.05 (54)	14.53±7.33 (60)	13.28±3.10 (50)	9.10±3.55 (58)	6.0±2.46 (54)
MdD	12.23±4.23 (37)	10.07±9.65 (40)	12.23±5.26 (56)	11.56±5.89 (46)	10.73±4.34 (47)	13.65±3.51 (51)	8.77±5.53 (38)	6.65±2.26 (42)
IRt	8.83±6.09 (66)	10.03±11.88 (72)	8.66±4.44 (83)	8.51±2.36 (78)	6.51±3.59 (88)	6.10±3.58 (80)	4.81±3.71 (68)	1.88±1.28 (75)
cardiovasc.and ventral resp.areas	5.95±1.58 (50)	5.94±1.48 (73)	5.37±2.27 (86)	5.34±2.29 (72)	4.14±1.04 (79)	7.89±2.01 (68)*	3.70±2.81 (80)	1.67±0.80 (74)
C. Inferior olive								
IOA/(IOA+B+C)	4.78±3.77 (53)	17.31±5.29 (60)*	2.64±0.99 (55)	12.39±6.81 (62)*	3.87±2.36 (63)	1.27±0.96 (66)	3.55±2.76 (61)	3.61±3.22 (63)
IOBe	1.26±1.62 (18)	2.10±1.53 (22)	0.10±0.12 (20)	1.69±0.60 (28)	1.37±1.44 (28)	0±0 (22)	0.03±0.07 (26)	0±0 (30)
IOK	0±0 (12)	0.78±1.01 (16)	0±0 (14)	0±0 (12)	0.13±0.25 (16)	0.17±0.34 (18)	#0±0 (14)	0.50±1.0 (18)
IOD	6.87±6.34 (40)	5.12±3.02 (48)	3.89±3.88 (37)	2.62±2.44 (48)	5.81±4.65 (52)	2.32±1.64 (48)	3.90±4.92 (46)	1.45±1.32 (46)
IOPr	0.63±0.72 (24)	0.98±0.97 (26)	0.58±0.96 (26)	0.73±0.67 (28)	0.40±0.26 (28)	0.28±0.36 (26)	0.72±0.41 (28)	1.15±1.07 (25)
D. Lateral reticular nucleus								
LRt	#8.31±3.05 (25)	14.22±5.47 (31)*	7.90±2.71 (34)	16.20±5.70 (35)*	6.65±4.30 (38)	10.98±6.13 (35)	14.04±12.70 (34)	7.72±3.34 (41)
LRtPC	#12.93±1.49 (25)	18.85±9.38 (31)	14.63±6.09 (34)	21.72±7.45 (35)	20.01±17.25 (38)	14.86±9.28 (35)	22.82±10.75 (34)	17.60±3.86 (41)

Within the indicated structures (left column), + refers to the sum of the labeled cells, and / refers to the average number of cells labeled in the neighbouring structures.
* Values indicated by asterisks in FLT rats are significantly different from those obtained in the corresponding AGC rats (p < 0.05).
For other symbols, see legend of this Table.

times of spaceflight (Table 1, A). In particular, FLT-FD2 rats showed a slight increase in Fos protein expression with respect to the AGC controls in the caudal parts of the medial vestibular nuclei (MVe) and spinal vestibular nuclei (SpVe). Surprisingly, no Fos protein expression was observed in the LVe, even though this area—similar to the MVe and the SpVe—receives signals from the macula (one of the gravity-sensing organs in the inner ear).

FLT-FD14 rats showed a prominent decrease in Fos protein expression in parts of the MVe and SpVe with respect to controls, possibly due to adaptation to zero-G. The most relevant effect, however, occurred in FLT-R+1 rats, which showed prominent increases in Fos expression as compared to the AGC controls in several areas of the VN. No Fos-positive cells were found in the LVe. Fos protein expression in FLT-R+1 rats was more prominent than that observed in FLT-FD2 rats. Only low levels of Fos expression were present in FLT-R+13 rats; i.e.,

after adaptation to the terrestrial environment. Most of these findings were also confirmed by using the marker for FRA proteins. The results obtained in LP rats were comparable to those obtained in NLP rats. The results are summarized graphically in Figure 3.

Results similar to those observed in the MVe and SpVe were also observed in the SuVe during different spaceflight conditions. However, only qualitative rather than quantitative data were obtained due to the damage that sometimes affected this structure during dissection of the cerebellar peduncles.

Fos protein expression in areas of the brain connecting to cardiovascular, gastrointestinal, and respiratory functions: nucleus of the tractus solitarius (NTS), area postrema (AP), and amygdala.

Both the NTS and the AP receive inputs from gravity receptors (Figure 4). The NTS projects, in turn, to areas that affect the

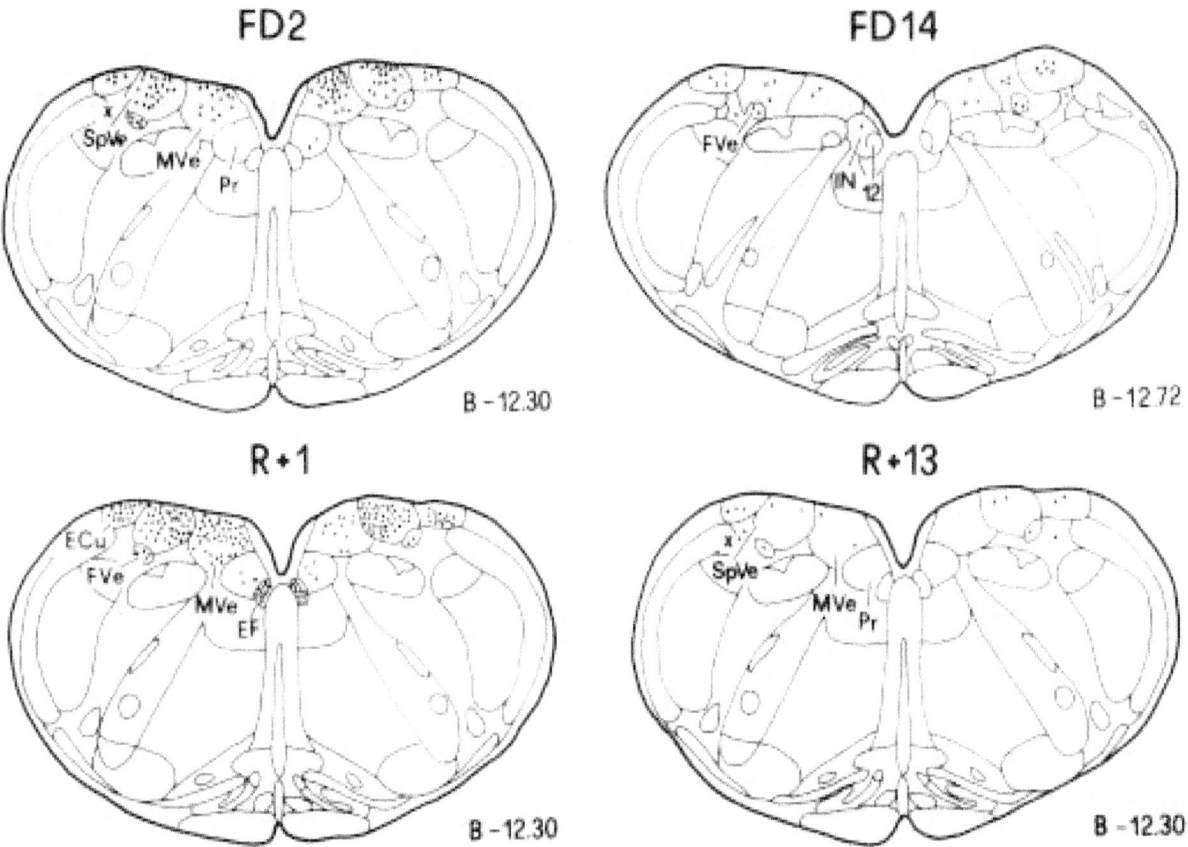

Figure 3. Visualization of Fos protein in the vestibular nuclei studied at the second flight day (FD2), the fourteenth flight day (FD14), one day after landing (R+1), and 13 days after landing (R+13). Expression of Fos protein is markedly increased on FD2 and on the first day after landing. Each dot corresponds to the nucleus of a Fos-positive cell. In this and the following figures, the number following the B corresponds to the level of the coronal section of the brain stem relative to the bregma (B), as indicated in the Paxinos and Watson (Paxinos, 1998) atlas. 12, hypoglossal nucleus; ECu, external cuneate nucleus; EF, epifascicular nucleus; FVe, F cell group of Brodal and Pompeiano; IN, intercalated nucleus of the medulla; Pr, prepositus nucleus. (From Pompeiano, 2001, with permission; reproduced from *Acta Otolaryngolica*.)

heart, blood vessels, and respiration (Figure 4). Because of these projections, the NTS is involved in the central and reflex control of cardiovascular and respiratory activities (Balaban, 1994). On the other hand, the AP is apparently involved, at least in some animal species, in the development of motion sickness (Miller, 1994). Changes in Fos expression in these areas during different times of the spaceflight are shown in Table 1, B.

The effects of different flight conditions on Fos protein expression in the NTS and AP were similar to those observed in the MVe and SpVe, except for time point FD2. In particular, in FLT-FD2 rats, the number of Fos-positive cells in the NTS and AP decreased with respect to control (AGC) rats.

The reduced Fos protein expression described in the NTS and AP at FD2 persisted also in FLT-FD14 rats, indicating that it depended upon exposure to microgravity. In FLT-R+1 rats, the number of Fos-positive cells was greatly increased compared to controls in the NTS and AP; this increased number also affected the medullary reticular areas related to the control of cardiovascular and respiratory functions. These findings were attributed to

the increase in gravity during reentry. These effects were greatly attenuated in FLT-R+13 rats, as shown by the fact that the number of labeled cells found in the NTS and AP of FLT rats was only slightly higher than the number in the AGC controls. Changes in FRA protein expression, similar to those obtained for Fos protein, were also observed.

Results obtained in LP rats were comparable to those obtained in NLP rats. However, the increase in Fos protein expression obtained in the NTS and AP of FLT-R+1 rats exposed to LP was higher than that observed in NLP and was also spread out to all of the ventrally related reticular structures controlling the cardiovascular and respiratory activities. Moreover, the increase in Fos-positive cells, observed in the NTS and AP of LP rats at R+1, persisted also at R+13 during the flight.

The central nucleus of the amygdala showed an increased number of Fos-labeled cells, which was significantly higher in FLT-R+1 rats (about 15 cells/section/side) than in the controls (about one to two cells/section/side); this effect was absent at the other time points of the spaceflight.

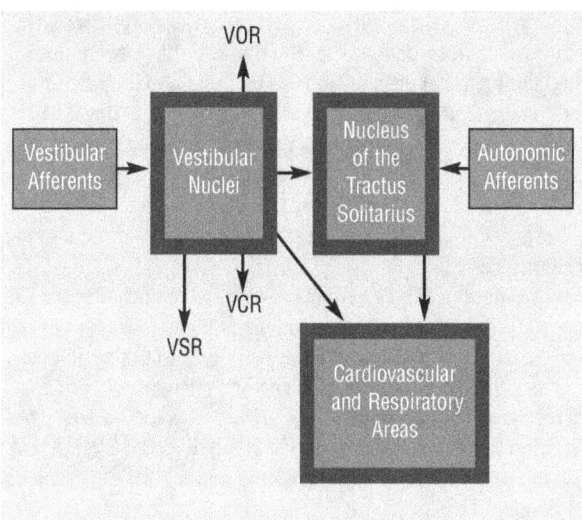

Figure 4. Diagram of central vestibulo-autonomic pathways. Vestibular afferents are signals from the inner ear that contain information on movement, gravity, and position. The vestibular nuclei process this information and use it in reflexes involving the eyes (vestibulo-ocular reflex (VOR)), the neck (vestibulocollic reflex (VCR)), and the spinal cord (vestibulospinal reflex (VSR)). The vestibular nuclei also connect to areas involved with cardiovascular, respiratory, and gastrointestinal functions (the autonomic functions). In particular, the vestibular nuclei connect to the nucleus of the tractus solitarius (NTS), which receives input from the cardiovascular, respiratory, and gastrointestinal systems (autonomic afferents).

Fos protein expression in brain areas concerned with posture and the control of movement: the inferior olive (IO) and lateral reticular nucleus (LRt).

Changes in Fos protein expression occurred during the different spaceflight conditions in the IO and LRt, which send fibers to the cerebellar cortex (Figure 1). The cerebellar cortex, in turn, is critical for the smooth control of movement.

In the IO (Table 1, C), a prominent increase occured in Fos-positive cells in FLT-FD2 rats with respect to the AGC controls. The areas affected all connect to the cerebellum, which coordinates movement. Also, these zones may control posture. A few labeled cells were observed in the subnucleus β (IOBe), but no labeling was seen in either the dorsal cap of Kooy (IOK) or the dorsomedial cell column of the inferior olive (IODM). These areas are believed to integrate inner ear and visual signals and project to the vestibulocerebellum, where eye movements are controlled (Wilson, 1979).

In FLT-R+1 rats that were studied 24 hours after landing, (i.e., when the gravity force increased from zero-G to 1.5–1.8-G before stabilizing to one-G), there was an almost complete absence of Fos-positive cells in the above-mentioned spinal areas of the IO. No labeled cells were found in the remaining regions of the IO, including IODM. Similar results were also observed at R+13.

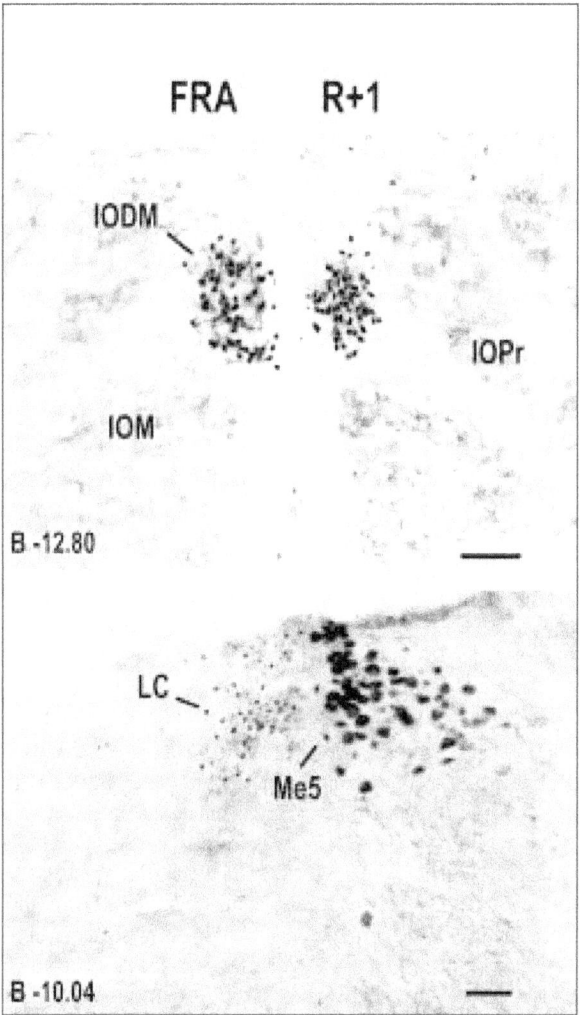

Figure 5. Visualization of FRA-positive cells observed in the dorsomedial cell column of the inferior olive (IODM) (upper figure), and in the locus coeruleus (LC) (lower figure) in a rat of the FLT group (NLP rat) studied at R+1. Upper figure: Prominent labeling in IODM, but not in the neighboring olivary regions (principal inferior olive (IOPr) and medial inferior olive (IOM)). This labeling, which was absent in control rats, was attributed to the gravity force that increased at R+1 from zero-G to 1.5-G to1.8-G, before returning to one-G at reentry. Lower figure: A large number of FRA-positive cells appeared in the LC of the same rat studied at R+1. The effect is attributed to the increase in gravity force from zero-G to 1.5-G to 1.8-G, before stabilizing to one-G after landing. This effect is also associated with a waking state, which probably occurred at reentry. Thus, the gravity force produced a long-lasting activation with persistence of the FRA protein in IODM, where integration of visual and inner ear signals occurs, as well as in the LC, which sends projections throughout the brain. Me5 = mesencephalic trigeminal nucleus. Scale bars = 100 μm.

One surprising finding is that the increase in gravity force, which occurred during the launch (at FD2) and during landing (at R+1), did not increase the number of Fos-positive cells in the IODM; i.e., in the vestibular area of the IO, where we might expect to see activity. This is probably due to the fact that the brain structures were examined 24 hours or more after the short periods of hypergravity produced during the launch or landing. If the FRA protein (which persists for days) is examined instead of the Fos protein (which lasts for hours), a different result is found. While no particular increase in the FRA protein expression was found in the IODM of both LP and NLP rats at FD2, the increase in gravity signals (followed by exposure to one-G) that occurred at R+1 led to a prominent increase in FRA-positive cells in IODM (Figure 5, upper part). These findings contrast with the absence of labeled cells in the olivary regions of the AGC controls.

The LRt is a medullary structure that receives inputs from both the spinal cord and the brain (neocortex). Its neurons respond also to inputs from the gravity sensors in the inner ear—the macula. Table 1, D, illustrates the changes in the number of the Fos-positive cells that occurred during the different space-flight conditions with respect to the AGC controls. In these controls, Fos-positive cells were usually more numerous in the parvicellular (LRtPC) than in the magnocellular part of the nucleus (LRt). This finding is of interest because of the more prominent connections from the spinal cord to the LRtPC.

In FLT-FD2 rats, a bilateral and symmetrical increase in the number of Fos-positive cells with respect to AGC controls occurred, particularly in the LRtPC and to a lesser extent also in the LRt. A similar result was also obtained in FLT-FD14. Most interestingly, FLT-R+1 rats showed a slight decrease in the number of labeled cells in LRtPC with respect to the AGC controls, which persisted even in FLT-R+13 rats.

In contrast with the results obtained in LRt, changes in the Fos protein expression with respect to the AGC controls were not observed in areas that send or process sensory signals from the muscles and joints (the external cuneate nucleus (ECu), the nucleus X, the gracilis nucleus, and the cuneate nucleus).

Similar results were also obtained during the different spaceflight conditions in both the NLP and the LP rats.

DISCUSSION

We have shown the changes in Fos and FRA protein expression —changes that affected various brain structures during different spaceflight conditions. Since Fos persists only for hours after the induction while FRAs persist for days, these methods allowed us to identify the neuronal systems that showed short-lasting (Fos) and long-lasting (FRA) molecular changes related to launch (FD2), followed by adaptation to zero-G (at FD14), as well as the changes that occurred during the symmetric time points at the reentry (R+1) followed by re-adaptation to one-G (R+13).

Changes in the locus coeruleus

Ground-based experiments have shown that the increase in resting discharge of the noradrenergic LC neurons during

waking or stress, as well as its decrease or suppression during slow wave sleep or REM sleep (Barnes, 1991), were associated with parallel changes in the Fos protein expression in the LC. These findings suggest that activation of LC during waking is responsible for the increase in the Fos protein expression that occurred in the corresponding targets, such as the cerebral cortex. The decrease of LC discharge during sleep is responsible for the reduced expression of Fos protein in the cerebral cortex.

Changes in Fos protein expression, which occur in the LC during different spaceflight conditions, do not simply depend upon changes in the sleep-waking activity. Electrophysiological experiments performed in our laboratory have shown that LC neurons respond to stimulation of gravity receptors with changes in firing rate that depend upon static and dynamic (Pompeiano, 1990) changes in head position during rat tilt. This effect can be attributed to the existence of a direct VN projection to the LC (Figure 1; Barnes, 1991). In addition to these findings, we demonstrated that noradrenergic LC neurons are involved in the modulation and adaptation of vestibular reflexes (Pompeiano, 1994), as well as in the changes in Fos expression that lead to the plastic changes responsible for compensation of the vestibular deficits following the loss of the inner ear on one side (Cirelli, 1996). These findings are of great importance if we consider that elderly people, who often have equilibrium disturbances, show a decrease in number and density of noradrenergic LC neurons. These equilibrium disturbances develop progressively with age.

In this study, the expression of FRA proteins slightly increased with respect to the AGC controls at FD2 (i.e., when the gravity force increased from one-G to three-G during the launch before reaching zero-G), while they decreased at FD14 when the rat had reached a condition of microgravity. In these instances, the rats were apparently mostly awake.

The FRA expression, however, was greatly increased at R+1, due to the increase in gravity force from zero-G to 1.5-G to 1.8-G before the animal returned to one-G after landing. This increase, which was likely associated with a state of active waking, was much higher than that developed at FD2 because, after launch, the rats reached a condition of zero-G while, after landing, the rats reached the level of one-G. These findings indicate that an increase or a decrease in the gravity force represents a condition that increases or decreases, respectively, the expression of FRA proteins in the LC neurons. The level of forced waking that occurred early after reentry was more prominent than that observed at FD2, since it was followed by a rebound period of deep sleep, as revealed by the great reduction in Fos-positive cells observed within the neocortex. These findings indicate that a relationship exists between the gravity signals driving the LC neurons and the waking level.

It is known that the noradrenergic LC neurons also project to several brain structures (Barnes, 1991), such as the VN, the NTS, and the central nucleus of the amygdala (Figure 1). These structures undergo an increase in Fos protein expression in FLT-R+1 rats. This finding supports the hypothesis that activation of the LC neurons by the gravity signals, such as occurs during landing, contributes to the increased expression of Fos protein in the above-mentioned structures. In particular, an increase or a decrease in the gravity force may, through activation or inactivation of the LC neurons, modify the Fos protein expression in the corresponding targets.

The persistent induction of FRAs in the LC at FD2 and, more prominently, at R+1 makes FRAs an attractive candidate to mediate some of the long-term changes involved in the adaptation of the rat to zero-G, as well as in the animal's re-adaptation to one-G. Similar to Fos protein, the FRA proteins induced by activation of LC neurons may operate as third messengers to activate other genes. This could finally lead to molecular changes in the LC target regions, contributing to brain plasticity in different spaceflight conditions.

Changes in the VN controlling posture and movements

Fos protein expression slightly increased in the VN at FD2, which may be due to the increase in the gravity force from one-G to three-G occurring during the first 10 minuies after the launch. This increase was small and may have been attenuated by the condition of microgravity (~zero-G) that occurred during the remainder of the 24-hour period after launch. In fact, a reduction of the Fos-positive cells with respect to the AGC controls was found in the VN at FD14; i.e., when the gravity force decreased to zero-G. An additional finding was an increase in Fos protein expression that occurred at R+1; i.e., during reentry. This effect was more prominent than that observed at FD2, probably due to the fact that the effect of the increased gravity force, from zero-G to 1.5-G to 1.8-G, was reinforced by exposure of the rat to one-G after landing. On the other hand, the process of re-adaptation to Earth's gravity prevented the Fos protein expression from decreasing below the control levels at R+13. A similar pattern of expression was also observed for FRA proteins. These results demonstrate that: (1) the gravitational stimulus at one-G following a prolonged period of microgravity contributes to re-adaptation to Earth conditions, thereby maintaining some levels of Fos and FRA expression in VN; and (2) Fos and FRA protein expression increases or decreases when the gravity force increases or decreases, respectively.

The molecular changes described above affected mainly the MVe and SpVe, and to some extent also the SuVe. These nuclei, which apparently respond to gravity changes, are known to project in part to the nuclei-controlling eye muscles and in part to the cervical segments of the spinal cord, where neck motor neurons are located (Wilson, 1979). Our results, therefore, suggest that the plastic changes responsible for adaptation to zero-G following launch and reentry involve the neuronal systems contributing to reflexes involving the eyes and neck.

In contrast to these findings, we observed no changes in Fos or FRA protein expression in LVe at both FD2 and R+1. This result is surprising, given the anatomical and physiological evidence that lateral vestibulospinal neurons, which project to whole segments of the spinal cord, receive inputs from gravity receptors in the inner ear (the macula) (Wilson, 1979). It appears, therefore, that gravity signals may modify nerve discharges in the LVe without affecting Fos and FRA

protein levels. A similar result was also observed in studies performed after the inner ear had been damaged on one side (Cirelli, 1996). This damage produced asymmetric changes in Fos protein expression in MVe and SpVe, but not in LVe.

Changes in medullary structures controlling cardiovascular, gastrointestinal, and respiratory functions

The NTS and the AP not only integrate inputs from the VN (MVe and SpVe) (Balaban, 1994; Figure 2), but they also integrate signals that originate from cardiovascular, gastrointestinal and respiratory systems—the pressure and chemical receptors in blood vessels, and the receptors located in the oral and gastrointestinal system. The NTS and the AP are also involved in the appearance of space motion sickness (Miller, 1994). These structures showed changes in Fos and FRA protein expression during different time points of the flight with respect to the AGC controls, which were similar, to some extent, to those occurring in the MVe and the SpVe.

The only difference was represented by the fact that the slight increase in Fos protein expression, which occurred in the VN of FLT-FD2 rats, was not observed in the NTS and the AP. The expression of Fos protein decreased in the NTS and the AP during exposure of the rat to zero-G at FD14. This expression, however, greatly increased at R+1 (i.e., when the gravity force increased from zero-G to 1.5-G to 1.8-G during reentry and before reaching one-G after landing). In this instance, the increase in both Fos and FRA expression could involve not only the NTS and the AP, but also several other structures controlling cardiovascular and respiratory functions. Thus, changes in gravity force determined not only the short-lived increase in Fos protein expression, but also the long-lived increase of FRAs, which are likely to mediate some of the long-lasting transcriptional changes particularly involved in the re-adaptation to one-G. These findings were greatly attenuated at R+13.

In addition to the structures indicated above, a prominent increase in Fos protein expression occurred in FLT-R+1 rats in the central nucleus of the amygdala. The central nucleus of the amygdala may thus contribute to the adaptive changes affecting vegetative functions (cardiovascular, respiratory, and gastrointestinal functions) both during and after reentry.

Changes in medullary structures involved in posture and movement

The IO and LRt connect to the cerebellum to induce adaptation to microgravity after launch, as well as re-adaptation to the terrestrial environment after landing (Figure 1). Both of these nuclei receive signals from different sources (such as position sense from the muscles and joints and inputs of vestibular and autonomic origin). These signals are then transmitted to different regions of the cerebellar cortex, which are important for the control of movement.

In our experiments, FLT rats showed an increase in Fos-positive cells with respect to the AGC controls both in the IO and the LRt. These regions control posture, as well as the activity of spinal reflexes. In the LRt, both the magnocellular and the parvicellular part of this structure showed an increased number of Fos-positive cells in FLT with respect to the AGC controls. This increase was most prominent in the parvicellular part, which receives particularly spinal inputs and projects to the spinocerebellum.

The effects described above were observed at FD2 and, to a lesser extent, at FD14. The increase in the gravity force that occurred prior to FD2 during launch did not play a role in these changes since the changes persisted until FD14, when the rats were exposed only to zero-G. On the other hand, Fos protein expression did not increase but actually decreased with respect to the AGC controls both in the IO and, to a lesser extent, in the LRt at R+1; i.e., when the gravity force increased from zero-G to 1.5-G to 1.8-G, before reaching one-G. This effect persisted, even at R+13.

The increased expression of Fos protein in the spinal areas of the IO and the LRt at FD2 and FD14 could be attributed to sensory signals related to the active and/or passive movements that occurred in mostly awake rats during and after the launch; i.e., under a condition of microgravity. Thus, the increased expression of Fos protein in the spinal areas of the IO and the LRt could be due to a mechanism of sensory substitution—i.e., where sensory signals from the body (i.e., position sense in the muscles and joints) induce plastic changes in structures responsible for the adaptation to zero-G.

The great reduction in Fos protein expression observed in the spinal areas of the IO and the LRt at R+1 could be attributed to the suppression of posture and movements that occurred during the episode of deep sleep that followed the period of forced waking related to reentry. This attenuation persisted even at R+13. An additional possibility is that the reduced activity of noradrenergic LC neurons, during the episode of deep sleep indicated above, might have contributed to a decrease after reentry.

Interestingly, the dynamic increase in gravity force that occurred during the launch (FD2), as well as during landing (R+1), did not give rise to a detectable increase in Fos protein expression in the olivary regions that are under the direct influence of inner ear signals. This effect was probably missed since, at the time points of the flight indicated above, the rats were studied 24 hours after exposure to gravity changes. By substituting the marker of the short-lived Fos with that of the long-lived FRA, we found that a selective increase of FRA proteins did occur in the IODM and, to a lesser extent, also occurred in the IOBe and IOK of flight rats studied at R+1 but not at FD2. It appears, therefore, that at R+1 the olivary regions, which integrate inner ear and visual signals and project to the vestibulocerebellum, undergo a prominent and long-lasting increase in FRA protein expression. This finding is likely to produce long-term plastic changes not only in the olivary regions indicated above, but also in the corresponding cerebellar areas that may thus contribute to re-adaptation to the terrestrial environment (one-G).

CONCLUSION

We have identified the important brain areas activated during the adaptation to weightlessness and the re-adaptation to Earth's gravity. This knowledge may be important in future for the development of countermeasures and in the planning of further research.

Acknowledgements

We thank the Committee of Cellular and Molecular Biology of the NIH, Bethesda, MD, for selecting our research project for the Neurolab mission, and also the Agencia Spatiale Italiano (ASI), Rome, for financially supporting our study (Research Grants ARS 98-21 and ARS 99-70). We are also particularly grateful to Dr. M.J. Iadarola (NIDR, NIH, Bethesda, MD) for providing us with the antibody for FRAs. We also thank all of the Members of the Neurolab Program, who contributed to the development of our project. They include Arnauld E. Nicogossian, Mary Ann Frey, Jerry Homick, Chris Maese, Louis Ostrach, Tom Howerton, Didier Schmitt European Space Agency (ESA); the Ames Research Center (ARC) Assistant Payload Project Scientists: Laurie Dubrovin, Paula Dumars; the Experiment Support Scientists: Lisa Baer, Erin Genovese, Gail Nakamura, Shari Rodriguez; the Mission Support Scientists: Angelo de la Cruz, Karin Berg, Peggy Delaney, Catherine Katen, Leann Naughton; and the veterinarian, Joe Bielitski. We are also grateful to the astronauts who have greatly contributed to the scientific success of the mission, namely the Orbiter crew: Richard Searfoss, Scott Altman, and Kay Hire; and the payload crew: Richard Linnehan, Dave Williams, Jay Buckey, Jim Pawelczyk, Alexander Dunlap, and Chiaki Mukai.

REFERENCES

Mammalian Vestibular Physiology. V.J. Wilson and G. Melvill Jones. Plenum Press, New York and London, pages XI–365; 1979.

Responses of Locus Coeruleus and Subcoeruleus Neurons to Sinusoidal Stimulation of Vestibular Receptors. O. Pompeiano, D. Manzoni, C.D. Barnes, G. Stampacchia, and P. d'Ascanio. Neurosci., Vol. 35, pages 227–248; 1990.

Neurobiology of the Locus Coeruleus. Edited by C.D. Barnes and O. Pompeiano. Prog. Brain Res. Amsterdam, Elsevier, Vol. 88, pages XIV–642; 1991.

Emetic Reflex Arc Revealed by Expression in the Immediate-Early Gene c-fos in the Cat. A.D. Miller and D.A. Ruggiero. J. Neurosci., Vol. 14, pages 871–888; 1994.

Noradrenergic Control of Cerebello-Vestibular Functions: Modulation, Adaptation, Compenstation. O. Pompeiano in: F. E. Bloom (ed.) Neuroscience: From the Molecular to the Cognitive. *Prog. Brain Res.*, Amsterdam, Elsevier, Vol. 100, pages 105–114; 1994.

Vestibular Nucleus Projections to Nucleus Tractus Solitarius and the Dorsal Motor Nucleus of the Vagus Nerve: Potential Substrates for Vestibulo-Autonomic Interactions. C.D. Balaban and G. Beryozkin. *Exp. Brain Res.*, Vol. 98, pages 200–212; 1994.

c-fos Expression in the rat Brain After Unilateral Labyrinthectomy and Its Relation to the Uncompensated and Compensated Stages. C. Cirelli, M. Pompeiano, P. d'Ascanio, P. Arrighi, and O. Pompeiano in *Neurosci.*, Vol. 70, pages 515–546; 1996.

Inducible and Constitutive Transcription Factors in the Mammalian Nervous System: Control of Gene Expression by Jun, Fos and Krox, and CREB/ATF Proteins. T. Herdegen, J.D. Leah. *Brain Res. Rev.,* Vol. 28, pages 370–490; 1998.

Physiological Evidence That the Vestibular System Participates in Autonomic and Respiratory Control. B.J. Yates and A.D. Miller. *J. Vestibul. Res.,* Vol. 8, pages 17–25; 1998.

The Rat brain in Stereotaxic Coordinates. G. Paxinos and C. Watson. Academic Press, 4th Ed.; 1998.

Immediate Early Gene Expression in the Vestibular Nuclei and Related Vegetative Areas in Rats During Space Flight. O. Pompeiano, P. D'Ascanio, C. Centini, M. Pompeiano, C. Cirelli, and G. Tononi. *Acta Otolaryngol.*, Vol. 545, Suppl., pages 120-126, 2001.

Ribbon Synaptic Plasticity in Gravity Sensors of Rats Flown on Neurolab

Authors

Muriel D. Ross, Joseph Varelas

Experiment Team

Principal Investigator: **Muriel D. Ross**

Technical Assistants: **Joseph Varelas**
Heidi Harbaugh
Nicole Gomez-Varelas

NASA Ames Research Center, Moffett Field, USA

ABSTRACT

Previous spaceflight experiments (Space Life Sciences-1 and -2 (SLS-1 and SLS-2)) first demonstrated the extraordinary ability of gravity sensor hair cells to change the number, kind, and distribution of connections (synapses) they make to other cells while in weightlessness. The number of synapses in hair cells in one part of the inner ear (the utricle) was markedly elevated on flight day 13 (FD13) of SLS-2. Unanswered questions, however, were whether these increases in synapses occur rapidly and whether they remain stable in weightlessness. The answers have implications for long-duration human space travel. If gravity sensors can adapt quickly, crews may be able to move easily between different gravity levels, since the sensors will adapt rapidly to weightlessness on the spacecraft and then back to Earth's gravity when the mission ends. This ability to adapt is also important for recovery from balance disorders. To further our understanding of this adaptive potential (a property called neuronal synaptic plasticity), the present Neurolab research was undertaken. Our experiment examined whether: (a) increases in synapses would remain stable throughout the flight, (b) changes in the number of synapses were uniform across different portions of the gravity sensors (the utricle and saccule), and (c) synaptic changes were similar for the different types of hair cells (Type I and Type II).

Utricular and saccular maculae (the gravity-sensing portions of the inner ear) were collected in flight from rats on FD2 and FD14. Samples were also collected from control rats on the ground. Tissues were prepared for ultrastructural study. Hair cells and their ribbon synapses were examined in a transmission electron microscope. Synapses were counted in all hair cells in 50 consecutive sections that crossed the striolar zone. Results indicate that utricular hair cell synapses initially increased significantly in number in both types of hair cells by FD2. Counts declined by FD14, but the mean number of synapses in utricular Type II cells remained significantly higher than in the ground control rats. For saccular samples, synaptic number in Type I and Type II cells declined on FD2, but returned to near-baseline values by FD14. These findings indicate that: (a) synaptic plasticity occurs rapidly in weightlessness, and (b) synaptic changes are not identical for the two types of hair cells or for the two maculae.

INTRODUCTION

There are two gravity sensors on each side of the head. These are the utricular and saccular maculae located in the vestibular (balance organ) part of the inner ear. The two maculae are situated at roughly right angles to one another, the utricular macula being approximately horizontal and the saccular macula nearly vertical. The paired maculae have sensory hair cells that detect linear accelerations (gravitational and translational) acting on the head. However, the functional orientation of the hair cells relative to a stripe (called a "striola") that bisects each macula differs between the two maculae. In the utricular maculae, hair cell orientation (and greatest sensitivity) is in the direction of the striola; in the saccular maculae, hair cell orientation (and greatest sensitivity) is away from the striola. These anatomical distinctions also have physiological consequences (Baird, 1986; Goldberg, 1990a,b).

The hair cells communicate via ribbon synapses with primary vestibular neurons (afferents). These afferent neurons carry information to the brain about the direction and force of linear accelerations. This information is used centrally to coordinate eye movements and antigravity muscle activity, to maintain balance whether one is at rest or in motion, and to sustain eye focus on a target (tracking) during head movement. Disturbances in the reflex pathways lead to motion sickness and to the balance disorders frequently encountered in the aged. Space adaptation syndrome (space motion sickness) is a further manifestation of a disturbance in the reflex pathways that occurs when the peripheral gravity sensors are challenged by weightlessness.

Previous spaceflight experiments, Space Life Sciences-1 and -2 (SLS-1 and SLS-2), were the first to demonstrate the extraordinary ability of the ribbon synapses of gravity sensor hair cells to change in number, kind, and distribution when the gravitational environment was perturbed by weightlessness (Ross, 1993, 1994, 2000). The mean number of synapses doubled in Type II hair cells of utricular maculae collected on flight day 13 (FD13) when all cells in a 100 two-dimensional section series were considered, and tripled in completely three-dimensionally reconstructed Type II cells taken from the same series (Ross, 2000). Synaptic increments were lower in Type I hair cells in the same series, and were insignificantly different from controls in complete cells.

A rise in the number of synaptic ribbons, whether by clustering at a synaptic site or by establishing new sites, would augment the possible number of vesicles available for releasing neurotransmitters. This could then increase primary vestibular afferent nerve activity to the brain both under resting conditions and when the cell is stimulated. Synaptic increments were, therefore, interpreted to indicate an adaptive response to improve hair cell output under a condition of a reduced stimulus of gravity (10^{-3} to 10^{-5} G).

The present Neurolab research was undertaken to further our understanding of the potential of the hair cells to alter their ribbon synapses (a property called neuronal synaptic plasticity). A question to be answered by the Neurolab experiment was whether the numerical increments in ribbon synapses noted late in the SLS-2 flight in utricular maculae occurred rapidly upon insertion into weightlessness as an early adaptive response. The answer to this question has implications for human space travel to the Moon and distant planets. That is, if gravity sensors adapt rapidly to a new gravitational environment such as exists on the Moon (1/6-G) or Mars (1/3-G), work could begin quickly upon landing. When such a mission ends, the sensors would re-adapt rapidly again to weightlessness on the spacecraft and, eventually, to Earth's one-G. Thus, crews could move easily among all these different gravity levels.

The present research, therefore, focused on FD2 and FD14 of the Neurolab mission. We wanted to determine whether increments on the order of those seen previously in Type II hair cells on FD13 (Ross, 2000) would be duplicated early in the mission and would be sustained. To expand the previous work, a portion of the macula that included the striolar zone, where sensitivity to phasic linear accelerations increases, was to be included in the study; and saccular maculae were analyzed for the first time. The purpose was to learn whether all portions of a macula, and the two maculae, would respond similarly to weightlessness.

MATERIAL AND METHODS

Rats used in this study were specific pathogen free (SPF) Fischer 344 rats obtained from Taconic Farms, Germantown, New York. Ten rats served as ground controls. Inner ear tissues were obtained on FD2 from four rats and on FD14 inflight from nine rats. Other tissues that are still under study were obtained from six rats on postflight day two (PF2) and from six rats on PF14. The manner of dissection of the labyrinths, of microdissection and tissue preparation for transmission electron microscopy, and of statistical analysis SuperANOVA™ software have been described in detail previously (Ross, 2000). The only differences for the present study were that, with the exception of the basal controls that were fixed by technicians of the Ross laboratory for their own specific use, tissue samples were collected by technical staff on the ground and by crewmembers inflight for shared use by several scientists. All maculae we received were from the right side of the head.

Only one of the utricular maculae fixed in space on FD2 and another on FD14 were suitable for transmission electron microscopy. This necessitated a truncated study of the utricular maculae, which is reported here. The investigation then focused on the saccular maculae.

Inflight data were obtained from one utricular macula and two saccular maculae fixed on FD2 and FD14. Basal samples were obtained on the ground on FD2. The utricular maculae were sectioned from the lateral border; saccular maculae were sectioned from the superior border. One hundred sixty-five sections 1-μm thick cut inward from the utricular lateral border were discarded and then 150 thin sections (~160-nm thick) were collected. In the case of the saccular macula, distance from the superior border to the striolar zone was shorter. Seventy-seven sections were discarded before collecting 250 thin sections. More sections were collected from the saccular maculae to permit later three-dimensional reconstructions. For this study, 50 consecutive (serial) sections were used routinely to obtain counts. However, 100 sections of one FD2 utricular and one FD2 saccular sample were used in two sets of 50 sections each to compare findings between striolar (set 1) and juxtastriolar zones (set 2), with the second set on the internal side of the striola (pars interna). The striolar zone was determined by the presence of M-type terminals in which the myelin reached the base of the calyx and calyceal processes were not present.

Ribbons were classified as rodlike or as spherules. Teardrop-shaped ribbons were considered to be rodlike for counting purposes (Ross, 2000).

RESULTS

Synapses – Ultrastructurally, ribbon synapses are characterized by a central electron-opaque body (ribbon) that is generally rodlike (Figure 1) or spherical in shape (Figure 2) and is surrounded by a halo of vesicles. The vesicles are tethered to the central body by slender filaments. The ribbon is continuous with one or more foot-like processes that proceed into an arc-shaped density that attaches the synapse to the cell membrane. The synaptic vesicles are docked at the synaptic site alongside the arc-shaped density. Synapses are considered to be multiple when the vesicles are shared by more than one ribbon (Figure 3). Figure 4 illustrates two synapses, one sphere-like and one rodlike (arrows), ending separately on a collateral process terminating on a Type II hair cell.

Utricular hair cell synapses; Type I – The total number of ribbon synapses in the 151 Type I cells was 448. The ranges in synaptic number per cell for the samples were: basal, 1-5; FD2, set one, 1-11; FD2, set two, 1-9; FD14, 1-11.

By FD2, the mean number of ribbon synapses in Type I cells had risen from the basal value (Table 1) and was still slightly higher than the basal value on FD14. The FD2 mean values differed significantly from the control (Table 1). The only other significant differences in kind or distribution of the ribbon synapses were the increases in spherical synapses in both FD2 samples ($p < 0.0082$ for set 1; and $p < 0.0309$ for set 2) and on FD14 ($p < 0.0093$) (not illustrated).

Utricular hair cell synapses; Type II – There were 1035 ribbon synapses in the 149 Type II hair cells counted. The ranges in synaptic number per cell for the data sets were: basal, 1-14; FD2, set one, 1-23; FD2, set two, 1-17; FD14, 1-25.

The mean numbers of synapses in the FD2 and FD14 samples were significantly higher than the basal values (see Table 2). Although not illustrated, spherical synapses had increased significantly in all of the inflight samples compared to the basal values: FD2, set one, $p < 0.0034$; FD2, set two, $p < 0.0008$; FD14, $p < 0.0093$ (not illustrated).

Saccular hair cells; Type I – There were 602 synapses in the 167 Type I hair cells counted. The ranges in number of synapses per data set were: basal, 1-13; FD2, set one, 1-5; FD2, set two, 1-8; FD2 total, areas with striola, 1-11; FD14, 1-6.

The mean number of synapses in Type I cells had declined significantly on FD2 ($FD2_t$, total data, areas with striola), but had returned to a near-normal mean value on FD14

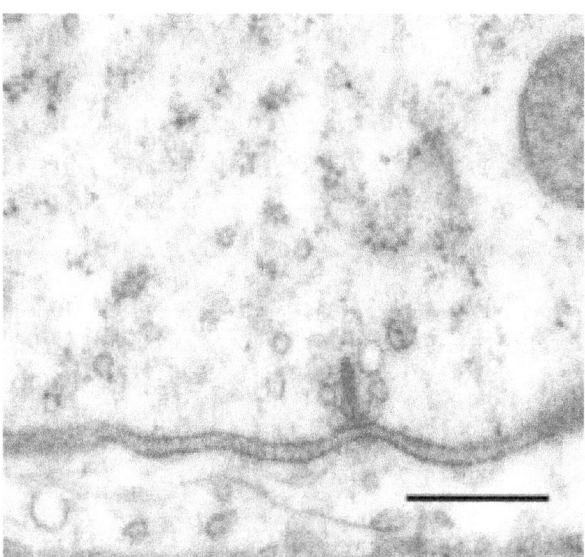

Figure 1. This figure shows a transmission electron micrograph of a synapse. The double black lines at the bottom of the figure are where the Type I cell and its calyx intersect. The black vertical structure along the lines illustrates a rodlike ribbon synapse between the Type I cell and its calyx. This picture was taken from an FD2 utricular sample, striolar zone. The bar equals 0.5 μm.

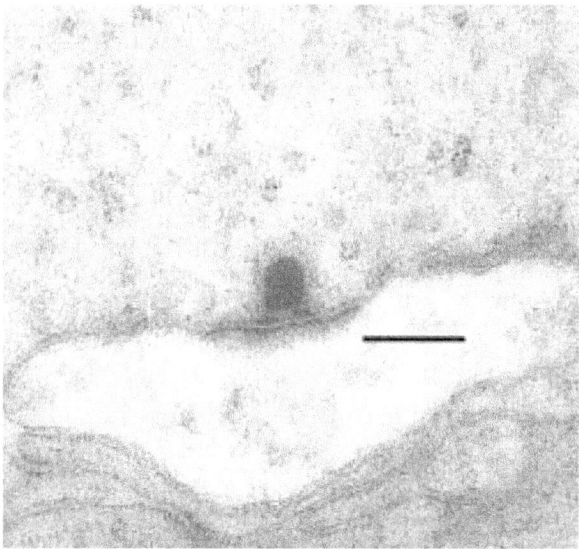

Figure 2. This figure illustrates a sphere-like ribbon synapse in a Type II hair cell of the utricle. Data from FD14. The bar equals 0.5 μm.

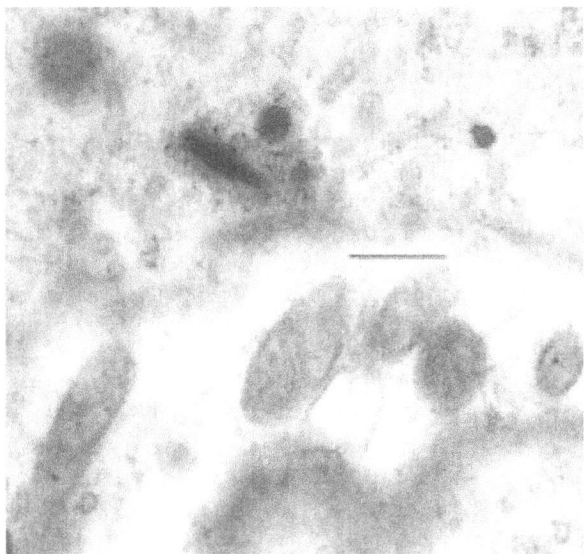

Figure 3. A multiple synapse is illustrated here. The rodlike ribbon synapse on the left is flanked by two sphere-like ribbons on the right. The vesicles containing neurotransmitter are shared. Picture from a ground control rat saccular macula. The bar equals 0.5 μm.

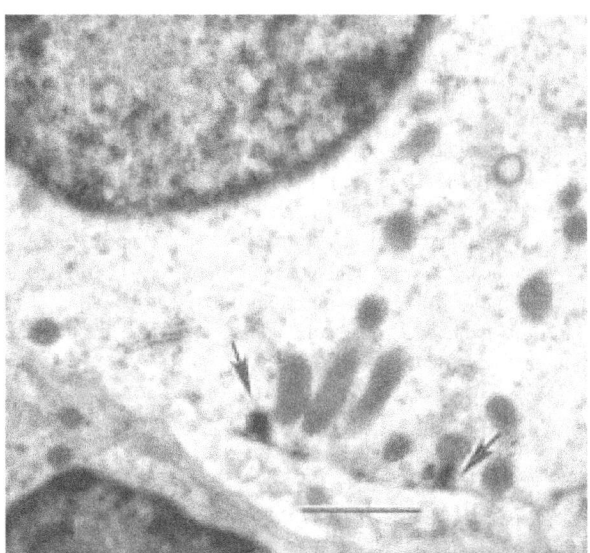

Figure 4. Some processes ending on Type II hair cells receive more than one synaptic input. The arrow on the left indicates a sphere-like ribbon synapse, and the arrow on the right indicates a rod-like synapse with a vesiculated process. Picture from a ground control rat saccular macula. The bar equals 1.0 μm.

(Table 1). There was a statistically significant decline in mean values of sphere-like ribbons (p>0.0170) in the FD2 total data (FD2$_t$, not illustrated). No other significant differences were noted.

Saccular hair cells; Type II – There were 208 Type II hair cells with 1434 synapses. The ranges in number of synapses per cell were: basal, 1-14; FD2 set one, 1-18; FD2 set two, 1-18; FD2 total, areas with striola, 1-23; and FD14, 1-16.

The synaptic mean value in the Type II inflight FD2 areas with striola declined as compared to ground controls (FD2$_t$, Table 2). There was also a significant decline in the mean value of rods (p<0.0376) in the FD2$_t$ sample (not illustrated).

Table 1. Type I Hair Cells

		Utricle				Saccule		
Day	N	MVS	SD	S	N	MVS	SD	S
Basal	29	2.034±1.239		*	45	4.133±3.334		*
FD2₁	41	3.512±2.491		*0.0044	28	3.036±2.349		ns
FD2₂	46	3.543±2.208		*0.0013	22	3.545±2.483		ns
FD2ₜ	41	3.512±2.491		*0.0044	48	2.646±2.037		*0.0105
FD14	35	2.400±2.103		ns	52	3.788±2.953		ns

This table compares data for Type I hair cells from utricle and saccule. Experimental day is given in the left-hand column. Basal = data from ground control rat, FD2₁ = data from one sample, striolar zone, on FD2, FD2₂ = data from same sample, juxtastriolar zone (internal side of the striola) on FD2, FD2ₜ = FD2 total data, striolar zones only, N = hair cell number, MVS = mean value, synapses, SD = standard deviation, S = significance compared to the basal, (*) = difference between basal and listed value is significant, ns = not significant.

Table 2. Type II Hair Cells

		Utricle				Saccule		
Day	N	MVS	SD	S	N	MVS	SD	S
Basal	43	4.744±3.793		*	61	6.939±3.968		*
FD2₁	35	9.000±5.122		*0.0001	37	5.865±3.441		ns
FD2₂	28	7.929±4.626		*0.0023	29	7.207±5.081		ns
FD2ₜ	35	9.000±5.122		*0.0001	63	6.317±3.750		*0.0435
FD14	43	6.884±5.128		*0.0306	55	6.382±5.237		ns

This table compares data for Type II hair cells from utricle and saccule. Experimental day is given in the left-hand column. Basal = data from ground control rats, FD2₁ = data from one striolar zone on FD2, FD2₂ = data from same sample, juxtastriolar zones (internal side of the striola) on FD2, FD2ₜ = FD2 total data, N = hair cell number, MVS = mean value, synapses, SD = standard deviation, S = significance compared to the basal, (*) = difference between basal and listed value is significant, ns = not significant.

DISCUSSION

The most important findings of the Neurolab research were the early elevations in the mean number of synaptic ribbons in utricular hair cells, particularly in Type II cells; and the relative stability of mean values in hair cells of the saccular macula. The results in the utricular macula support previous findings that, in flight, synapses are elevated numerically in hair cells of this macula, particularly in Type II hair cells (Ross, 1993, 1994, 2000). However, the extent of synaptic change is affected by macular location. Increments in Type II hair cell synapses in the posterior part of the utricular macula were greatest there on FD13 of the SLS-2 14-day flight (11.4±7.2, p<0.0001) for all Type II cells analyzed (Ross, 2000). On FD14 of the present Neurolab experiment, the mean value of Type II hair cell synapses in utricular striolar zone was 9.000±5.122 (Table 2). Mean values also differed in the FD2 samples, depending on striolar or juxtastriolar (pars interna) zones (Table 2). Hair cell type (I or II), macular location, and macula all provide internal controls for the findings since important differences in synaptic plasticity exist in each case.

The generation and loss of synapses (synaptogenesis and synaptic deletion) in weightlessness may be reactive responses to altered influences from the nerves connecting to the hair cells (Ross, 2000). This effect can be seen during normal development. Initially during development, more synapses than are needed are generated. These are reduced in number (pruned) once innervation to the brain is established (Sobkowicz, 1982; Sobkowicz, 1992). Results from a similar system (the developing organ of Corti) indicate that an interaction between the afferent innervation and the hair cells determines final synaptic number. This is likely in the case of macular synapses in an adult animal as well.

In an adult, the level of normal synaptic activity and turnover has been established for a particular environment through development and maturation. When a decline in feedback occurs (due to a lack of stimulation from gravity, for example), this should stimulate synaptogenesis. An opposite effect would occur when discharge rates are high, such as in a high-gravity environment (hypergravity). That is, synaptogenesis would be shut down until an efficacious number of synapses is reached and balance is restored. Pilot research on hypergravity effects on utricular hair cells supports this thesis, since exposure to two-G for 14 days lowered the synaptic number in Type II hair cells of the rat utricular macula but left Type I cells unaffected (Ross, 1994).

A question raised by macular differences in responses to weightlessness is whether the synaptic changes reported here have relevancy to possible physiologic differences between the utricular and saccular macular afferents. In a series of papers, Fernandez et al. (Fernandez, 1972) and Fernandez and Goldberg (Fernandez, 1976a,b) dealt with differences in response characteristics between superior vestibular nerve (utricular) and inferior nerve (saccular) afferents. Fernandez et al. (Fernandez, 1972) originally described saccular afferents as having unexplained much lower resting discharge rates and lower sensitivity to static tilt. In a later paper (Fernandez and Goldberg, 1976b), use of a larger sample indicated that resting discharge rates did not differ much between the two maculae. Both end organs responded to linear accelerations. Fernandez and Goldberg reported that utricular afferents were most responsive to static tilt in the X-direction (side-to-side) while saccular afferents were primarily sensitive to Z-direction (vertical) tilt. Both kinds of afferents were less sensitive to tilt in the Y-direction (nose forward, backward). The results on tilt in the sacculus are supported by recent findings of Uchino et al. (Uchino, 1997), who reported that saccular but not utricular influences on neck muscles stabilize relative head and body positions against the vertical linear acceleration of gravity.

Overall, our findings indicate that utricular macular synaptic number, kind, and distribution were altered early in weightlessness. Additionally, there were differences in degree and kind of synaptic change based on macula, intramacular location, and day during flight. We conclude that synaptic plasticity occurs rapidly in weightlessness; and that the synaptic changes are not identical for the two types of hair cells or for the two maculae. The changes probably result from a profound change in vestibular input to the macula that occurs in weightlessness.

Impact on space travel – The results continue to be good news for space travelers. Trips to the Moon (1/6-G) or to Mars (1/3-G) should not have a major or a long-lasting effect on the vestibular system, as prior Moon missions have already indicated. The periphery should readily adapt to a novel partial G-force, and space motion sickness should not be a major problem.

Acknowledgements

We are grateful to the Neurolab crew for obtaining inner ear tissues in flight for our use. Without their willingness to carry out tedious dissections under less-than-ideal circumstances, we would not have maculae to study. We also thank the staff of Payload Operations at NASA Ames Research Center and, in particular, Lisa Baer for her cooperation during mission planning and execution. We thank Heidi Harbaugh and Nicole Gomez-Varelas for their assistance with transmission electron microscopy and synapse counting. This research was supported by NASA and by NIH Grant # 5U01NS33448.

REFERENCES

RESPONSE TO STATIC TILTS OF PERIPHERAL NEURONS INNERVATING OTOLITH ORGANS OF THE SQUIRREL MONKEY. C. Fernandez, J.M. Goldberg and W.K. Abend. *J. Neurophysiol.,* Vol. 35, pages 978–997; 1972.

PHYSIOLOGY OF PERIPHERAL NEURONS INNERVATING OTOLITH ORGANS OF THE SQUIRREL MONKEY I. RESPONSE TO STATIC TILTS AND TO LONG-DURATION CENTRIFUGAL FORCE. C. Fernandez and J.M. Goldberg. *J. Neurophysiol.,* Vol. 39, pages 970–984; 1976a.

PHYSIOLOGY OF PERIPHERAL NEURONS INNERVATING OTOLITH ORGANS OF THE SQUIRREL MONKEY II. Directional Selectivity and Force-Response Relations. C. Fernandez and J.M. Goldberg. *J. Neurophysiol.,* Vol. 39, pages 985–995; 1976b.

RIBBON SYNAPSES IN THE DEVELOPING INTACT AND CULTURED ORGAN OF CORTI IN THE MOUSE. H. Sobkowicz, J.E. Rose, G.L. Scott, S.M. Slapnick, *J. Neurosci.,* Vol. 2, pages 942–957; 1982.

CORRESPONDENCES BETWEEN AFFERENT INNERVATION PATTERNS AND RESPONSE DYNAMICS IN THE BULLFROG UTRICLE AND LEGENA. R.A. Baird and E.R. Lewis. *Brain Res.,* Vol. 369, pages 48–64; 1986.

THE VESTIBULAR NERVE OF THE CHINCHILLA. V. RELATION BETWEEN AFFERENT DISCHARGE PROPERTIES AND PERIPHERAL INNERVATION PATTERNS IN THE UTRICULAR MACULA. J.M. Goldberg, G. Desmadryl, R.A. Baird and C. Fernandez. *J. Neurophysiol.* Vol. 62, pages 791–804; 1990a.

THE VESTIBULAR NERVE OF THE CHINCHILLA. V. RELATION BETWEEN AFFERENT DISCHARGE PROPERTIES AND PERIPHERAL INNERVATION PATTERNS IN THE UTRICULAR MACULA. J.M. Goldberg, G. Desmadryl, R.A. Baird and C. Fernandez. *J. Neurophysiol.,* Vol. 67, pages 791–804; 1990b.

THE DEVELOPMENT OF INNERVATION IN THE ORGAN OF CORTI. H. Sobkowicz In: Development of auditory and vestibular systems, R. Romand (ed), Elsevier, Amsterdam, pages 59–100; 1992.

MORPHOLOGICAL CHANGES IN RAT VESTIBULAR SYSTEM FOLLOWING WEIGHTLESSNESS. M.D. Ross, *J. Vestib. Res.,* Vol. 3, pages 241–251; 1993.

A SPACEFLIGHT STUDY OF SYNAPTIC PLASTICITY IN ADULT RAT VESTIBULAR MACULAS. M.D. Ross, *Acta Otolaryngol.* (Stockholm), Suppl., Vol. 516, pages 1–14; 1994.

SACCULOCOLIC REFLEX ARCS IN CATS. Y. Uchino, H. Sato, M. Sasaki, M. Imagawa, H. Ikegami, N. Isu, W. Graf. *J. Neurophysiol.,* Vol. 77, pages 3003-3012; 1997.

CHANGES IN RIBBON SYNAPSES AND IN ROUGH ENDOPLASMIC RETICULUM OF RAT UTRICULAR HAIR CELLS IN WEIGHTLESSNESS. M.D. Ross, *Acta Otolaryngol.* (Stockholm), Vol. 120, pages 490–499; 2000.

Neural Re-adaptation to Earth's Gravity Following Return from Space

Authors

Richard Boyle, Allen F. Mensinger, Kaoru Yoshida,
Shiro Usui, Anthony Intravaia, Timothy Tricas,
Stephen M. Highstein

Experiment Team

Principal Investigator: **Stephen M. Highstein[1]**

Co-Investigators: **Richard Boyle,[2,5] Allen F. Mensinger,[1]
Kaoru Yoshida,[3] Shiro Usui,[4]
Timothy Tricas[6]**

Engineer: **Anthony Intravaia[5]**

[1]Washington University School of Medicine, St. Louis, USA
[2]Oregon Health Sciences University, Portland, USA
[3]University of Tskuba, Tskuba, Japan
[4]Toyohashi University of Technology, Toyohashi, Japan
[5]NASA, Ames Research Center, Moffett Field, USA
[6]Florida Institute of Technology, Melbourne, USA

ABSTRACT

The gravity sensors in the inner ear (the utricle and saccule) no longer receive strong gravitational signals in weightlessness. In an effort to compensate for the reduced input, these sensors may become more sensitive using a process called up-regulation. If this happens, the output of the vestibular nerve from the inner ear would be greatly increased after a space mission, when the sensitized gravity sensors are reexposed to gravity. We believe that information gathered by studying the inner ear of a lower animal, such as a fish, will provide the information needed to prove this. Despite evolution, the balance and equilibrium functions of the inner ear have not appreciably changed since their appearance in the earliest vertebrates. As a result, the fish vestibular system compares favorably in both structure and function with that of mammals. By chronically recording the output of the vestibular system in the fish (specifically the output from the utricle), the question of how microgravity affects the output of the inner ear could be answered precisely. For five days following two NASA Shuttle flights, we recorded from the vestibular nerves supplying the utricle the responses to inertial accelerations (head movements) in four oyster toadfish (*Opsanus tau)*. Within the first day postflight, the magnitude of response to an applied translation (side-to-side) movement was on average three times greater than for controls. The reduced gravitational acceleration in orbit apparently resulted in an up-regulation of the sensitivity of the utricle. By 30 hours postflight, responses were statistically similar to control. The time course of return to normal sensitivity parallels the reported decrease in vestibular disorientation and improvement in balance in astronauts following their return from space.

INTRODUCTION

It is fundamentally important that organisms remain orientated within their terrestrial environment. Vertebrates possess a gravity-sensing system, the utricular and saccular organs, that senses the sum of forces due to head movements and transforms the sum of these accelerations into a neural code. This code is combined with acceleration signals from the semicircular canals and with information derived from other sensory modalities (such as vision and position sense) to compute a central nervous system representation of the body in space that is called the gravitoinertial vector. In this way, the central nervous system resolves the ambiguity of signals due to gravity and to self-motion, thereby maintaining balance and equilibrium under varying conditions.

Exposure to microgravity imposes an extreme condition to which the traveler must adapt. Many, if not most, human travelers experience some disorientation and motion sickness during the first few days in microgravity. This effect is called the space adaptation syndrome. This syndrome is akin to terrestrial motion sickness (Reason, 1975). From studies on the earliest space crews, it was evident that adjustments to the microgravity environment occur in flight and then reverse upon return to Earth's gravity (Black, 1999). The specific adaptation mechanisms are conjectural and could range from neural to structural changes or both. We studied the neural re-adaptation to Earth's gravity using electrophysiological techniques to measure the nerve signals from the inner ear (utricular nerve afferents) in fish upon return from exposure to microgravity.

METHODS

Six oyster toadfish (*Opsanus tau*), weighing 150–700 gm, were individually housed in seawater tanks aboard two NASA Shuttle missions (Neurolab and STS-95). The fish were returned to the laboratory where all experiments were performed within ~10 hours of the Shuttle landing. Surgical procedures, similar to those described by Boyle and Highstein (Boyle, 1990), were performed in accordance with the American Physiological Society Animal Care Guidelines and approved by the international Animal Care and Use Committee. Fish were anesthetized with MS-222 (Sigma) and secured in a Plexiglas tank placed atop an experimental table. The utricle and its afferent nerve were exposed, and extracellular potentials were recorded from individual nerves. The experimental apparatus allowed for a variety of accelerations and movements to be applied to the fish (manual yaw rotation about Earth vertical, translational acceleration parallel to Earth horizon, and/or static tilt with respect to gravity). The fish could be repositioned in a 360-degree circle such that: (a) the translational acceleration was delivered along any direction in the horizontal head plane, and (b) the acceleration was specifically directed; e.g., nose-down (pitch) or side-down (roll). Static sensitivity of otolith afferents to gravity was observed both in control and in postflight fish.

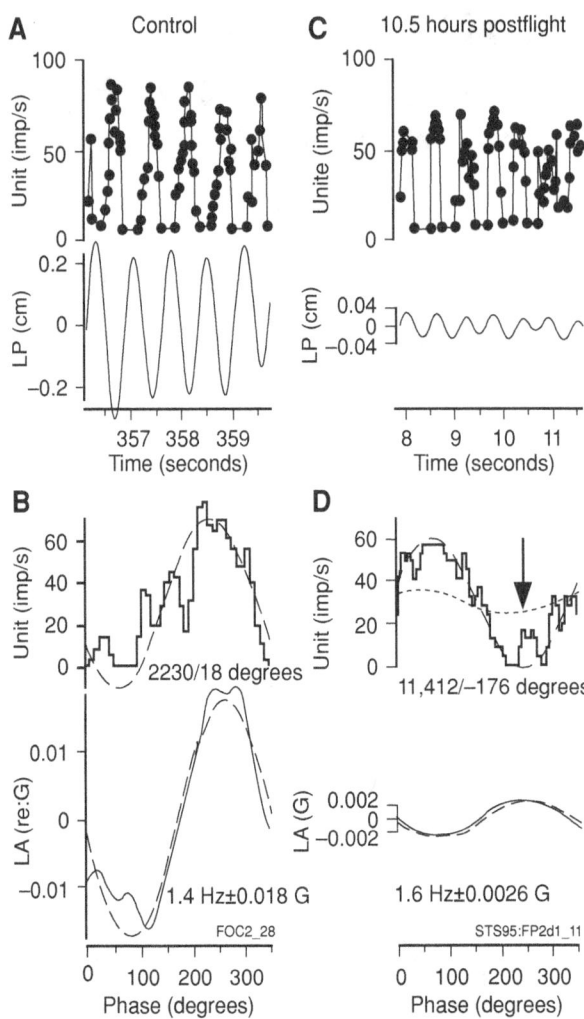

Figure 1. Control (A, B) and 10.5-hour postflight of STS-95 Shuttle (C, D) responses of oyster toadfish utricular afferents to translational (side-to-side) accelerations. The nerves recorded shortly after return to Earth exhibit a profound hypersensitivity to translational accelerations. The amplitude of the applied 1.6-Hz stimulus in C and D was almost negligible (±0.0026 G) but induced a firing rate modulation of ~±30 impulses/second (or 11,412 impulses/second/G). An average control afferent would exhibit only ~±4 impulses/second modulation for this weak stimulus. The control response of the preflight afferent (shown in A and B) is modeled as the dashed curve marked by the arrowhead in the histogram of the postflight afferent in D.

A, C: Afferent firing rate (in impulses/second; upper trace) is sinusoidally modulated to an applied sinusoidal change in linear position ((LP), in cm; lower trace). In both records, the fish were rotated counterclockwise by 90 degrees about the vertical axis, resulting in a maximum rate increase for an acceleration directed rightward along the inter-labyrinth axis, on the same side as the recorded control afferent in A and in the opposite sense for the postflight afferent in C.

B, D: Averaged response (upper histogram) of records shown in A and C. Lower trace shows the averaged linear acceleration ((LA) re:G) of each stimulus. Ordinates in each panel are scaled equally to illustrate the primary finding. (From Boyle, 2001, with permission, reproduced from the *Journal of Neurophysiology*.)

RESULTS

Four toadfish were flown on the STS-90 Neurolab mission (16 days), and two toadfish were flown aboard the STS-95 mission (nine days). Two fish survived Neurolab, and both fish survived STS-95. Responses of utricular nerves to gravitational (tilt) and inertial (translation) accelerations were recorded from four flight fish.

Control responses were obtained from 32 utricular nerves in three fish. Figure 1 shows the firing rate response (upper trace, in impulses/second) to a sinusoidal change in linear position ((LP), lower trace). In Figure 1A, the fish was first rotated about the vertical axis in a counterclockwise step to a 90-degree head (and body) angle. The resulting linear acceleration ((LA), Figure 1B) was directed to the right on the same side as the recorded afferent along the inter-labyrinth axis. This maximally excited the afferent. The averaged response to five stimulus cycles had a maximal sensitivity of 2230 impulses/second/G (Figure 1B; see Table 1).

Hair cell bundles are morphologically polarized, and hair cell receptor potentials are directionally sensitive to bundle displacement. Directional selectivity of utricular nerves is distributed in a fanlike shape (Fernández, 1976), as expected from hair cell orientations in the utricular macula (Spoendlin, 1966). Figure 2 shows the test used to determine the directional selectivity of individual afferents in control (A) and postflight (B) fish. A sinusoidal translational acceleration along an Earth-parallel plane was delivered at successive 15 degree positions after the fish was stepped around a 360-degree circle. Head angle (degree) was defined using a right-hand rule relative to the laboratory: a positive acceleration at zero degree represents a forward movement directed out the fish's mouth, and one at 90 degrees represents a movement directed out the fish's right ear. Directional selectivity was determined by plotting the response sensitivity and phase relative to head angle. The data were fit by a rectified cosine function (dashed lines in A and B); and correspondence between the tested and predicted responses reflects the sharpness of directional tuning. The control nerve in A was sharply tuned and directionally selective to acceleration directed along the inter-labyrinth axis. All control nerves were directionally selective, and the maximal response vectors spanned 360 degrees of head angle.

In early postflight fish, utricular nerves were hypersensitive to translational acceleration (Figures 1-3) and directionally selective (Figure 2B). One of the first nerve fibers recorded at 10.5 hours postflight illustrates the striking increase in response sensitivity when stimulated at ±0.0026-G acceleration or ±0.025-cm displacement (Figures 1C, 1D). The discharge modulation was ~±30 impulses/second, yielding a maximal sensitivity of 11,412 impulses/second/G, nearly seven-fold greater than the control mean and ~three-fold greater than the maximum response obtained in any individual control (4136 impulses/second/g; Table 1).

Data for both STS-90 and STS-95 Shuttle missions are presented in the form of a probability plot (Figure 3) to show the initial increase and recovery of response sensitivity. The data in this figure and in Table 1 are divided into groups based on time postflight. Maximum sensitivity of each nerve is plotted as a

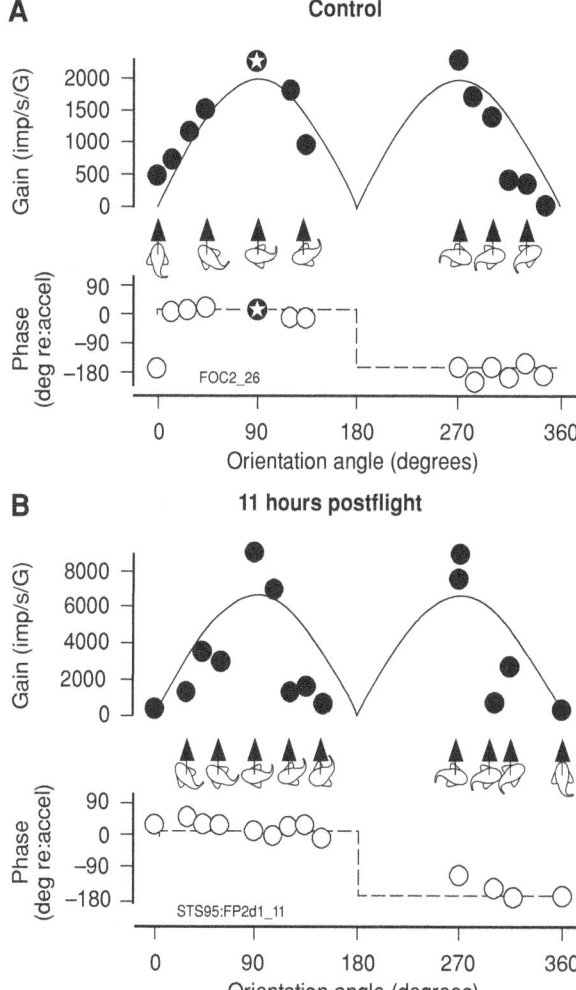

Figure 2. Directional selectivity of control (A) and 11 hours postflight of STS-95 Shuttle (B) toadfish utricular afferents to translational (side-to-side) accelerations. A, B: Sensitivity (solid symbols) and phase (open symbols) are plotted as a function of orientation angle (in degrees) of the animal. The small fish drawn between the sensitivity and phase plots graphically depict the orientation as viewed from above for selected tests; by convention, forward motion of the test sled is denoted as an arrowhead. The afferents recorded exhibited an enhanced sensitivity to applied linear accelerations and directional selectivity. (Note the difference in scale for gain between the control and the 11-hour-postflight graphs.)

The individual response indicated by the star in panel A represents the response shown in Figure 1A, B. Forward acceleration out the fish's mouth is given as zero degree, and the fish is rotated counterclockwise about the Earth's vertical axis and the designated angles follow a right-hand rule. For example, 90 degrees is an acceleration directed rightward along the inter-labyrinth axis, 180 degrees is a backward acceleration out the fish's tail, and 270 degrees is an acceleration directed leftward along the inter-labyrinth axis. The response of utricular afferents follows a rectified cosine function (dashed curves fit to the empirical data) with respect to orientation angle, indicating directional selectivity. A: Control afferent. B: afferent recorded 11 hours postflight of STS-95 Shuttle. (From Boyle, 2001, with permission, reproduced from the *Journal of Neurophysiology*.)

Table 1. Summary of control and postflight utricular afferent data.

Fish	Hours Postlanding	Mean Max. Sensitivity (impulses/second/G)
3 Controls		168 ± 1195 (61–4136; n=32)
STS-95: 1+2	10–16	3772 ± 3607* (142–13290; n=24)
STS-95: 2	29.5–32	1582 ± 1750 (101–4992; n=16)
STS-95: 2	52–55	1195 ± 1478 (126–5861; n=18)
STS-90: 2	53–59	1399 ± 1599 (116–7819; n=29)
STS-95: 1+2	70–76	1337 ± 1076 (100–4738; n=30)
STS-90: 3	112–117	1476 ± 951 (154–3685; n=28)

*p<0.01

Comparison of maximum response sensitivity (impulses/second/G) of utricular afferents recorded under control conditions to those at different times postflight of STS-90 (two fish labeled 2 and 3) and STS-95 (two fish labeled 1 and 2). At 10-16 and 70-76 hours postflight, the results obtained from the two STS-95 fish were comparable and are combined. The number of afferents at 10-16 hours is eight (STS-95: 1) and 16 (STS-95: 2). At 70-76 hours, the number of afferents is 10 (STS-95: 1) and 20 (STS-95: 2). Mean and one standard deviation, with range of smallest to largest and number (n), are given in parentheses. Results show that the sensitivity was significantly greater than control (p<0.01) for the earliest recording session (10–16 hours) from the two fish flown on STS-95. Results obtained from flight fish at later periods were not significantly different from control. The first records taken from fish flown on STS-90 began 53 hours postflight after re-adaptation to Earth's gravity.

percentage of population sensitivity with a value less than the individual sensitivity. For roughly 60% of the nerves (14/24) in both fish on STS-95 (filled red circles), labeled STS-95: 1+2, the sensitivity recorded 10–16 hours postflight was dramatically enhanced relative to control (crosses). Within this time group, the sensitivity of the entire sample (n=24) was roughly triple that of the control (p<0.01). The sensitivity returns to near normal at the recorded time of 29.5–32 hours postflight and remains within normal range after five days postflight.

To examine for possible recording bias, all measured parameters were compared between control and postflight nerves. No statistical difference was found in the range and mean of discharge rate (impulses/second) and regularity of discharge (standard deviation of the interval divided by the mean interspike interval) between postflight and control fish. An equal distribution of head angles was also found, evoking maximum and minimum response modulations postflight, similar to those observed in control fish, and response phase (degree).

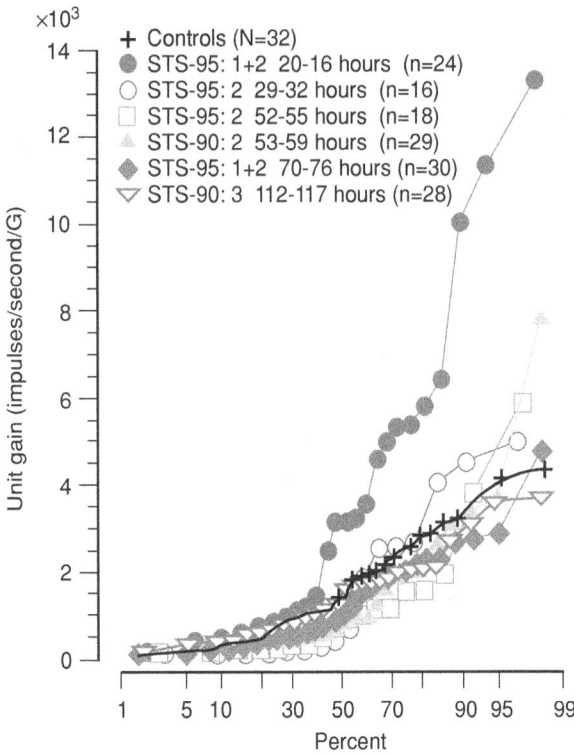

Figure 3. Hypersensitivity of toadfish utricular afferents to translational (side-to-side) accelerations within the first day postflight of STS-90 and STS-95. Groups are formed with respect to the time postflight, from the earliest time of 10–16 hours for both fish (1+2) of STS-95 (filled circles) to the latest time of 112–117 hours for fish labeled 3 of STS-90 (inverted open triangle). Maximum sensitivity of each nerve is plotted as a percentage of the population, with sensitivity less than that of the measured nerve. For example, for roughly 60% of the nerves (14/24) in both fish on STS-95 (filled red circles), labeled STS-95: 1+2, the sensitivity recorded 10–16 hours postflight was dramatically enhanced relative to control (crosses). (From Boyle, 2001, with permission, reproduced from the *Journal of Neurophysiology.*)

DISCUSSION

The increase in sensitivity of certain inner ear nerves following space travel is most likely due to the animal being exposed to microgravity. The results, however, should be regarded as preliminary because single nerve fibers were not studied sequentially. The findings could be explained by changes at a number of locations within the inner ear (see illustration at the beginning of this section for a description of the anatomy). Possible explanations include: (1) an increase in the sensitivity of the hair cells, (2) a temporary structural alteration affecting the ability to convert otolith movement of the stones into nerve impulses, (3) a change in coupling between the otolith and hair cells, causing enhanced deflection of hair cell bundles for a given movement, or (4) an alteration in the strength of synaptic transmission. The number of synaptic ribbons in certain Type II hair cells in rodents is labile, increasing after exposure to microgravity (Ross, 2000) (see science report by Ross et al. in this publication). Toadfish possess only Type II hair cells. The number of synaptic boutons on these cells and the response sensitivity to vestibular stimulation are correlated (Boyle, 1990). Thus, an increase in number of synaptic ribbons in toadfish otolith hair cells following exposure to microgravity could potentially explain the present results.

If the inner ear structure is arranged for optimal responses in one-G, mechanical alterations may also occur in microgravity. For example, loss of gravitational force might displace the otolithic membrane relative to the macula, thus affecting neural responses. Altered gravity conditions might also trigger an adaptive response of the weight-lending structures (Weiderhold, 1997) (see science report by Weiderhold et al. in this publication). It is clear that more experiments under controlled states of altered gravity are required to determine the structural and developmental response of the inner ear and the consequence of spaceflight on inner ear function.

Adaptation of hair cell receptor potentials occurs to prolonged deviation of their sensory hair bundle (Eatock, 1987). Unweighting of the otolith mass in microgravity might cause an adaptation of receptor potentials. That the enhanced sensitivity remained for at least 24 hours, substantially longer than the suggested time course of adaptation, is inconsistent with this view.

Otolith sensors provide a major input to the internal representation of the gravitoinertial vector. Thus, an abnormal utricular or saccular component should have profound effects upon the orientation of the organism, and has been hypothesized to be causal in vestibular disorientation or space adaptation syndrome. The demonstrated time course in the altered responses parallels the time course of disorientation experienced by space travelers, and gives support to this hypothesis.

The earliest recordings began 10 hours after STS-95 landing. To what extent this delay affects the interpretation of the data is indeterminate. Because of enhanced afferent sensitivity, the initial postflight results were limited to fewer stimulus frequencies (1–2 Hz) and to lower amplitudes than were delivered in control tests. These restrictions in stimulus parameters were required to minimize discharge nonlinearity (Boyle, 1990). Sensitivity on average declined from day two to day four, and larger stimulus amplitudes could be progressively applied. The first single-unit recordings after the STS-90 flight began 53 hours after landing, well after the postflight recovery time observed in the STS-95 fish. Significantly, the directional tuning of nerves remained unchanged after exposure to microgravity. We therefore would not expect to find that significant remodeling of the spatial extent of dendritic arbors within the sensory epithelium has occurred.

To date, the toadfish utricular nerve has only been crudely evaluated at one frequency with a hand-powered linear sled.

If the fish utricle bears any resemblance to similar epithelia studied in other species, more complete functional evaluations of afferents will no doubt demonstrate considerable diversity. It therefore remains to be tested whether a specific population demonstrated increased sensitivity or whether this finding is a general feature of all cells.

Acknowledgements

This work was supported by grants from NASA Ames # 2-945 and NIH/NIDCD PO-1 DC 01837 (USA) and NASDA (Japan).

REFERENCES

THE VESTIBULAR SYSTEM AND ITS DISEASES. H.H. Spoendlin. edited by R.J. Wolfson. Philadelphia: University of Pennsylvania Press, pages 39–68; 1966.

MOTION SICKNESS. J.T. Reason and J.J. Brand. Academic Press, London; 1975.

PHYSIOLOGY OF PERIPHERAL NEURONS INNERVATING OTOLITH ORGANS OF THE SQUIRREL MONKEY. II. Directional selectivity and force-response relations. C. Fernández and J.M. Goldberg. *J. Neurophysiol.,* Vol. 39, pages 985–995; 1976.

ADAPTATION OF MECHANOELECTRICAL TRANSDUCTION IN HAIR CELLS OF THE BULLFROG'S SACCULUS. R.A. Eatock, D.P. Corey, and A.J. Hudspeth. *J. Neurosci.,* Vol. 7, pages 282–236; 1987.

RESTING DISCHARGE AND RESPONSE DYNAMICS OF HORIZONTAL SEMICIRCULAR AFFERENTS OF THE TOADFISH, OPSANUS TAU. R. Boyle and S.M. Highstein. *J. Neurosci.,* Vol. 10, pages 1557-1569; 1990.

DEVELOPMENT OF GRAVITY-SENSING ORGANS IN ALTERED GRAVITY CONDITIONS: OPPOSITE CONCLUSIONS FROM AN AMPHIBIAN AND A MOLLUSCAN PREPARATION. M.L. Wiederhold, H.A. Pedrozo, HJ.l. Harrison, R. Hejl, and W. Gao. *J. Grav. Physiol.,* Vol. 4, pages P51–P54; 1997.

DISRUPTION OF POSTURAL READAPTATION BY INERTIAL STIMULI FOLLOWING SPACE FLIGHT. F.O. Black, W.H. Paloski, M.F. Reschke, M. Igarashi, F. Guedry, and D.J. Anderson. *J. Vestib. Res.,* Vol. 9, pages 369–78; 1999.

CHANGES IN RIBBON SYNAPSES AND ROUGH ENDOPLASMIC RETICULUM OF RAT UTRICULAR MACULAR HAIR CELLS IN WEIGHTLESSNESS. M.D. Ross. *Acta Otolaryngol.,* Vol. 120, pages 490–499; 2000.

NEURAL READAPTATION TO EARTH'S GRAVITY FOLLOWING RETURN FROM SPACE. R. Boyle, A.F. Mensinger, K. Yoshida, S. Usui, A. Intravaia, T. Tricas, and S.M. Highstein. *J. Neurophysiol.,* Vol. 86, pages 2118-2122; 2001.

Section 2 Sensory Integration and Navigation

Background *How the nervous system combines different senses to control location, posture, and movement*

INTRODUCTION

Amusement park visitors in the 19th century didn't have roller coasters with upside-down loops, but they enjoyed an attraction that offered the same kinds of thrills. In this device, riders sat in a gondola that swung back and forth. Surrounding the gondola was a large room that could rotate completely around the riders. The room was furnished with nailed-down tables, chairs, pictures, etc., so it appeared to have a definite floor, ceiling, and walls. While the people swung back and forth, the entire room rotated around them. The sensation the riders had was not that they were swinging, but that they were rolling completely around—similar to making a full loop in a roller coaster. The unique visual scene (the rotating room), combined with the input to the inner ear (swinging) was enough to override the seat-of-the-pants sensations (position sense) that the body was upright the whole time.

This example demonstrates how the brain can be fooled. The brain creates a sense of body orientation from information it receives from the eyes, inner ear, and position sense. In the case above, the strong sensations of rotation the eye received were enough to overcome information from position sense. The brain must combine information from a variety of senses to maintain balance, control posture, and navigate successfully. When doing this, however, the brain can alter the weight it gives to different senses and can use assumptions about how the world works to simplify its task.

INTERNAL MODELS AND NAVIGATION STRATEGY

To help with executing movements, the brain can use "internal models" in addition to the senses. These are predictive models in the brain that allow people to predict future events (such as when a ball will reach the hand) and react to them accordingly. With an internal model, the brain takes into account the laws of motion and uses them to improve performance.

For navigation, the brain can create maps by using visual information about landmarks (landmark navigation strategy). If, for example, tourists in a strange city park their car near a landmark, such as a clock tower, they can later use the clock tower to navigate their way back to their car. Another strategy would be to use vestibular information and position sense to sum up the turns taken and distances traveled to determine location (path integration strategy). The parked car above could also be found by tracing back the turns and distances traveled from the parking lot.

The hippocampus plays an important role in the ability to use these different navigation strategies. Within the hippocampus there are cells that can encode the direction the head is moving (head direction cells) and cells that fire at particular locations (place cells). These cells are likely part of the mental map and compass the brain uses to navigate.

SENSORY INTEGRATION IN WEIGHTLESSNESS

Weightlessness provides a novel environment for studying sensory integration. The inner ear, which ordinarily provides a strong sense of up and down, does not provide this information in space. As a result, information from other senses (like vision) may become more important for determining body orientation. In other words, the balance between senses may need to shift. Similarly, in space the limbs are weightless, which means some movements may need to be recalibrated. This might affect the speed or accuracy of movements.

The low-gravity environment in orbit also can be used as a tool to study basic questions. For example, when baseball players catch a ball on Earth, their nervous systems have to anticipate where the ball will be in the future so their arms can be there for the catch. Whether the ball is caught—or falls to the ground—may depend on the player's internal model of gravity's effects. Weightlessness provides an environment to test whether this internal model is in place and to determine how it works.

Weightlessness also provides a unique way to put the two main navigation strategies (landmark vs. path integration) into conflict and determine how the hippocampus responds. On the Neurolab mission, a series of experiments studied sensory integration and examined changes in the hippocampus, evidence for internal models, changes in the hand-eye coordination, and shifts in the balance between senses.

SENSORY INTEGRATION AND NAVIGATION

The upper left panel in Plate 3 shows the equipment needed to study the evidence for internal models. The crewmember observes the falling ball. Electrodes on the arm show when muscle activity is generated in anticipation of the ball's arrival. These signals provide the information needed to study whether an internal model of gravity's effects exists.

The panel on the upper right shows the location of the hippocampus. The hippocampus is critically involved in spatial memory. A tragic example of this is in patients who have Alzheimer's disease. These patients, who can sustain significant damage to the hippocampus, lose the ability to navigate. They can leave home and walk into a neighborhood they have known for years, but cannot find their way back since they have lost their spatial memory. On Neurolab, four rats were flown that had miniature electrodes in place for continuously measuring nerve activity in the hippocampus while they performed a navigation task. This task put the two main navigation strategies outlined above into conflict.

The bottom panel shows how visual stimulation can produce a sensation of movement. For example, when people see a pattern of lines move past their visual field, they may feel that they themselves are moving, not the lines. The sensation is called vection and can be experienced in everyday life. When, for instance, passengers on a stationary train see the train on the next track pull away, they may feel that their train is moving. If visual cues become more important for determining orientation in space, sensations of vection may become more powerful. More powerful sensations of vection in space would be evidence that the balance between senses is shifting while in orbit.

THE NEUROLAB SENSORY INTEGRATION AND NAVIGATION EXPERIMENT

Dr. Berthoz and his colleagues showed that an internal model of gravity was used when catching a ball, and this model was still evident in space. The experiment from Dr. McNaughton's laboratory showed that when path integration and landmark navigation strategies come into conflict, the cells in the hippocampus have difficulty forming strong links to particular locations. This difficulty disappears later in the flight, suggesting an adaptive process. The evidence from Dr. Bock's study showed that visual-motor (movement) coordination was good in space, but to keep accuracy and speed high during simple movements the crewmember had to devote more processing (brain) power to the task. The experiment from Dr. Oman and his colleagues showed that crewmembers became more dependent on visual cues in space.

The results of these experiments offered insights into the brain's adaptation during weightlessness. Also, the results showed how weightlessness could be used as a tool to study basic questions in physiology and neuroscience.

Plate 3: Sensory Integration and Navigation

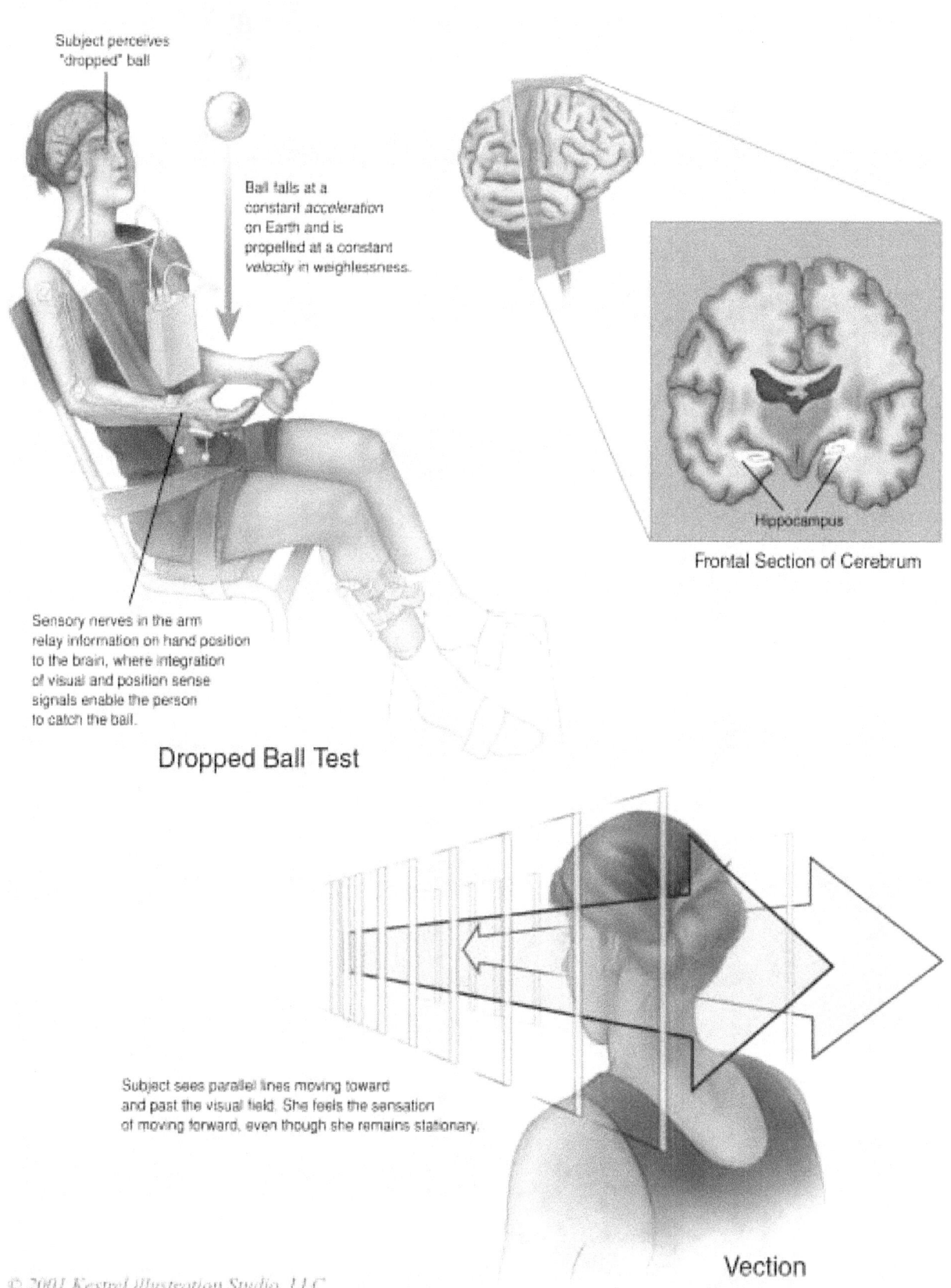

Subject perceives "dropped" ball

Ball falls at a constant *acceleration* on Earth and is propelled at a constant *velocity* in weightlessness.

Sensory nerves in the arm relay information on hand position to the brain, where integration of visual and position sense signals enable the person to catch the ball.

Dropped Ball Test

Hippocampus

Frontal Section of Cerebrum

Subject sees parallel lines moving toward and past the visual field. She feels the sensation of moving forward, even though she remains stationary.

Vection

The Brain as a Predictor: On Catching Flying Balls in Zero-G

Authors
Joseph McIntyre, Myrka Zago,
Alain Berthoz, Francesco Lacquaniti

Experiment Team

Principal Investigators: **Alain Berthoz**[1,2]
Francesco Lacquaniti[1,3,4]

Co-Investigators: **Joseph McIntyre,**[1,2,3] **Myrka Zago**[1,3]

Engineers: **Michel Venet,**[4] **Brian Marchiel,**[5]
Angelene Lee,[7] **Felix Silvagnoli**[6]

Technical Assistants: **Diane McMahon,**[6] **Daniela Angelini**[3]

[1]The European Laboratory for the Physiology of Action, Paris, France, and Rome, Italy
[2]Collège de France, Paris, France
[3]IRCCS Fondazione Santa Lucia, Rome, Italy
[4]Università di Roma Tor Vergata, Rome, Italy
[5]Centre National d'Etudes Spatiales (CNES), Toulouse, France
[6]Lockheed Martin Space Operations Company, Houston, USA
[7]NASA Johnson Space Center, Houston, USA

ABSTRACT

The art of intercepting a flying ball has preoccupied athletes and sports fans for centuries. Moreover, the ability to intercept or avoid a moving object—whether it be to catch a ball, snatch prey, or avoid the path of a predator—is a skill that has been acquired by almost every evolved species. This behavior, remarkable in that it requires *anticipation* of future events, is thus a fundamental property of biological sensorimotor systems. Yet the mechanisms have yet to be clearly identified that allow the brain to predict the future positions of a moving object, rather than simply perceiving where that moving object is now. In this study, we addressed an unresolved question in the fields of psychology and neuroscience: Does the brain rely strictly on sensory information to estimate the movements of an object, or can we improve our performance on interceptive tasks by using knowledge about how the world works? We used the microgravity conditions of the Neurolab Space Shuttle mission to examine this question as it applies to catching a "falling" ball. Astronauts caught a ball projected downward from the ceiling with one of three different speeds. The astronauts triggered anticipatory motor responses slightly earlier (with respect to impact) in flight than on the ground, with a greater timing shift for lower ball velocities. Data derived from this show that the central nervous system uses an *internal model* of gravitational acceleration, in addition to sensory information, to predict the time-to-contact between the ball and the hand and applies this model by default—whether on Earth or in microgravity.

INTRODUCTION

On Earth, we expect to see certain features in the movements of objects in the environment. For instance, an apple falling from a tree will accelerate downward (leading, as the story goes, to Sir Isaac Newton's discovery of the laws of gravity). Similarly, a tennis ball will bounce off the ground in a direction that depends on its angle of incidence, the speed of the ball, etc. In building these expectations, the brain is implicitly creating a model of the laws of motion that determines how the world works, even if we cannot necessarily describe these laws in words. Such "internal models" (jargon used to describe predictive processes in the brain) allow us to anticipate future events and react to them accordingly. Thanks to such predictive mechanisms, a tennis player need not wait for the ball to bounce before moving towards a good place to hit it. Similarly, if Newton had seen his proverbial apple, he would presumably have moved sideways to avoid being hit on the head. But the models used by the brain to predict future events are not 100% accurate, and sometimes mistakes are made. This may be because the true laws of motion are too complex to be computed quickly, or because they are not sufficiently general to cover all conditions and circumstances. Thus, an unexpected gust of wind might push the falling apple sidewise, causing poor Newton to be hit on the head anyway. And even world-class tennis players need to practice between the French Open and Wimbledon to adjust for differences in the way a ball bounces on clay vs. grass. With these ideas in mind, we undertook a study of ball catching in the absence of gravity to see how the human brain adapts to this highly unusual situation. By studying not only how humans successfully perform sensorimotor tasks in familiar situations, but also how, when, and why they adjust their responses in novel conditions, we can better understand how the brain uses internal models to anticipate.

We chose the simple task of catching a ball because catching a falling object is a common, everyday occurrence. Many a glass object has been saved by quickly anticipating the direction and timing of a fall. Internal models of how gravity affects falling objects may be firmly established in the brain because we see these effects so often. And, unlike the highly developed skills of a tennis player that must be learned and practiced, models of what it means to fall may, in fact, be innate to the human perceptual system. Infants as young as one year old already anticipate gravity's effects. When gravity is absent, however, "up" and "down" no longer exist and free-falling objects will no longer accelerate toward the "ground." When faced with such an unusual situation for the first time, a human subject might be surprised at the unexpected movement of objects. This is illustrated by an anecdote reported to us by Dr. Mark Lipshits of the Russian Academy of Science:

> "A novice cosmonaut, arriving for the first time on the Mir space station, raised his arm to greet his colleagues who were already on board. As he did so, the camera he was carrying slipped from his grasp. The cosmonaut quickly reached 'down' to prevent the camera from falling, but of course, in the absence of gravity, the camera instead moved in a straight line forward."

Note that this response occurred despite the fact that the cosmonaut had already been living and working without feeling the effects of gravity for at least 48 hours.

How, then, do we control movements to catch a falling ball? To perform this task, the brain must resolve a number of questions, based on information about the ball and about the current position and movement of the body. First, it is necessary to know where the ball is, how fast it is moving, and in what direction it is moving. Once the brain has this information, it can then predict the future motion of the ball and intercept it by placing the hand in the right place at the right time. In our Neurolab experiment, we concentrated on the timing aspects of this task. Specifically, we studied how the brain estimates the time it will take a ball to reach the hand, even if the hand lies already on the ball's flight path. Thus, the problem to be resolved by the brain is not "Where will the ball go?" but rather "When will the ball get there?"

Theories about Time-To-Contact

Although scientists and sports fans alike have studied these questions for years, the answers are far from definitive and a spirited scientific debate continues. According to one theory, the brain should be able to estimate the time-to-contact (TTC) with an approaching object by using only sensory signals that can be perceived directly, without resorting to higher levels of neural processing. A prime example of such a purely sensor-based strategy was proposed by David Lee and colleagues (Lee, 1983). They showed that for an object approaching along the sight-line, the ratio of the size of the object's retinal image (r) to its rate of change (dr/dt) is equivalent to a first-order estimate of the TTC based on its current distance (d) and velocity (v).[1] Lee called this ratio τ:

$$\tau \equiv \frac{r}{\frac{dr}{dt}} = -\frac{d}{v} \qquad (1)$$

Under this hypothesis, the central nervous system (CNS) could trigger anticipatory actions such as catching when the optic variable τ (derived from the readily available sensory signals r and dr/dt) reaches a certain threshold without explicitly measuring the object's true size, distance, or speed of movement.

Since its introduction, the τ hypothesis has attracted considerable attention for its simplicity and biological plausibility; it has been a standard-bearer of the Gibsonian school of ecological perception. But, the purely retinal-based strategy provided by the τ hypothesis has been questioned in its original form for not being sufficiently general. For instance, even for a head-on approach of the ball (as originally considered by Lee et al. in the formulation of Eq. 1), the optic variable τ is not the only information available to the CNS—a combination of τ and binocular retinal disparity (the slight difference in position of the ball's image in each of the two eyes that gives us

---

[1]This equivalence was in fact identified somewhat earlier in a science-fiction story by Fred Hoyle.

stereo vision) provides more robust information about TTC (Rushton, 1999). Furthermore, additional retinal or sensory and motor signals about eye movements could also contribute to estimates of TTC when the ball moves along an oblique path (Gray, 1998). From another perspective, the original τ hypothesis proposed that motor responses are elicited when the estimated TTC reaches a threshold τ-margin. This implies that subjects must trigger a response at a fixed delay before impact, without the ability to adjust the movement as new information becomes available. Such an "all-or-nothing" response is neither theoretically desirable nor consistent with available data for many interceptive tasks (Peper, 1994). Nevertheless, the more fundamental question of whether motor responses are geared to available sensory signals alone, without resorting to higher-order neural processing, remains open. Many current hypotheses used to explain interceptive behaviors nevertheless invoke first-order (constant velocity) approximations of TTC to trigger and adjust motor responses (Savelsbergh, 1993; Peper, 1994; Rushton, 1999).

First- or second-order models?

A more enduring criticism of first-order hypotheses concerns the ability of such strategies to handle accelerating objects (Lacquaniti, 1989). A first-order estimate of TTC, such as τ, provides an exact prediction of the true TTC only in the case of constant velocity motion. When the approach is accelerated, as is the case for a ball falling in Earth's gravity, τ overestimates TTC. Although it has been argued that catching could still be geared to τ because the resulting timing error would be tolerable (Lee, 1983), a strategy that takes both target velocity and acceleration into account (a second-order estimate) could tackle a wider range of interceptive tasks with a higher degree of precision. However, the CNS seems to discriminate poorly among the magnitude of accelerations based only on visual information. While one can obviously infer that an object is accelerating if it is seen to be moving slowly at one moment and more quickly at another, human subjects are not able to detect differences of acceleration, particularly over short viewing periods (Werkhoven, 1992). While it would seem difficult, therefore, to account for arbitrary acceleration of the target when estimating TTC, some specific accelerations can be foreseen based on *a priori* knowledge about laws of motion and on sensory information about the environment. To take the most pertinent example, gravity is a terrestrial invariant that can be monitored by the brain through the vestibular apparatus of the inner ear and through somatosensory receptors throughout the body. Furthermore, the consequences of gravity's effects on the ball could be learned and stored from experience (Hubbard, 1995). Thus, Lacquaniti and colleagues proposed that when timing anticipatory motor responses used for catching, "*a priori* knowledge on the most likely path and law of motion" may be employed (Lacquaniti, 1993). By combining an internal model of gravity's actions with a perceptual estimate of the target's instantaneous height and velocity, the CNS could account for the acceleration of a dropping object and thus more accurately predict TTC.

TTC estimates in zero-G

Consider the simple task of catching a ball that is "falling" downward from the ceiling toward the outstretched hand of the catcher (Figure 1). To better understand how the brain might estimate TTC according to the theories we described above, imagine what would happen if the effects of gravity were suddenly removed from this situation. The different hypotheses about how the brain synchronizes movements to the arrival of the ball lead to widely different catching movements in zero-G (Figure 2). If subjects could measure the ball's acceleration (Hypothesis 1), they would correctly estimate TTC in both zero-G and one-G. Thus, if the brain wanted to trigger a response at a fixed time prior to contact, it should

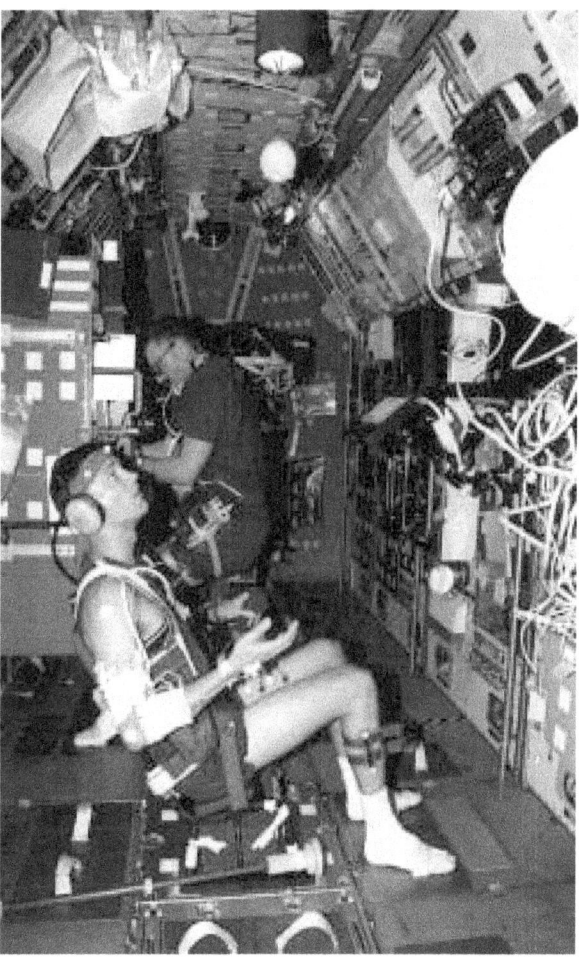

Figure 1. Payload Specialist Jim Pawelczyk performing the ball catch experiment during the Neurolab mission, while Mission Specialist Dave Williams works on another experiment in the background. Subjects caught a 400-g ball projected downward from the ceiling of the Spacelab compartment with one of three randomly assorted initial velocities. Movements of the arm were measured via reflective markers placed at the wrist, elbow, and shoulder. Electrodes on the surface of the skin measured the muscle activity in the biceps and triceps, and in the flexors and extensors of the wrist.

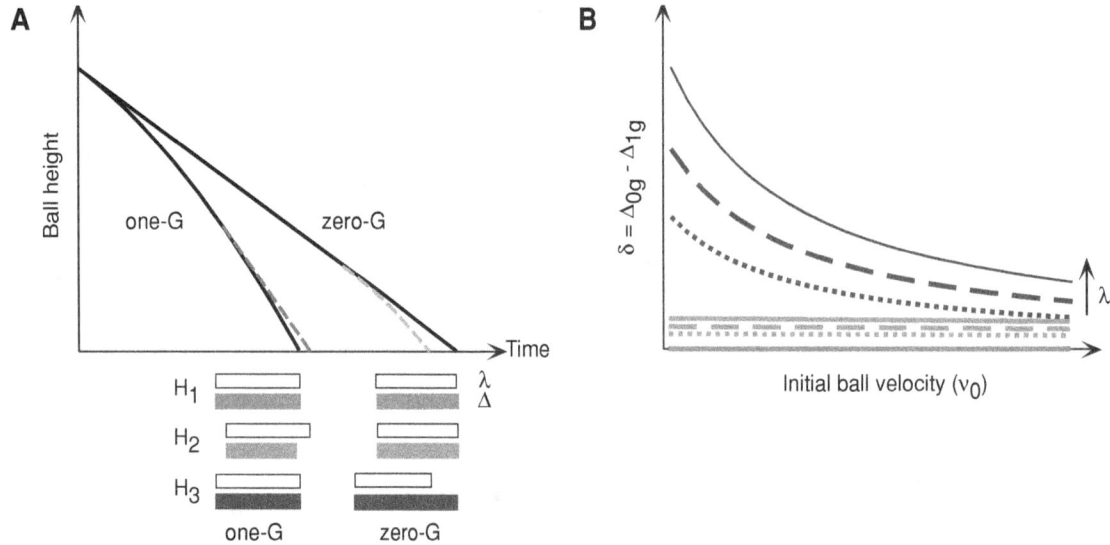

Figure 2. Experiment design. (A) Solid lines indicate the height of a ball projected downward with an initial velocity in zero-G and one-G. Dashed lines indicate the expected impact time for a first-order TTC estimate applied in one-G (red) and a for a second-order one-G internal model applied in zero-G (blue). White bars indicate a fixed TTC threshold (λ). The right edge of each white bar is aligned with the *expected* impact, while the left edge indicates when a motor response will be triggered based on that threshold. Colored bars show the real difference in time (Δ) between the response onset (left edge) and the *actual* impact (right edge) for three different hypotheses. H_1: exact estimate of TTC based on real-time measurements of acceleration—responses are always triggered at the same time prior to impact ($\Delta_{1g}=\Delta_{0g}=\lambda$). H_2: first-order TTC estimate—the ball arrives earlier than expected in one-G ($\Delta_{1g}<\lambda$). H_3: second-order internal model of gravity—the ball arrives later than expected in zero-G ($\Delta_{0g} > \lambda$). We can measure Δ, but we do not know the value of λ, so we cannot simply compare λ and Δ on Earth. We can, however, measure the difference between Δ's in zero-G and one-G. (B) Predicted time-shifts (δ) for zero-G vs. one-G for different values of λ. H_1 predicts no difference ($\delta=0$). Both H_2 and H_3 predict earlier responses with respect to impact in zero-G ($\delta > 0$); increasing λ increases δ. Shifts predicted by H_3 are greater for a given λ and rise more quickly as $v_0 \rightarrow 0$, compared to H_2. This simulation reveals a strong, counterintuitive prediction for the second-order internal model strategy: timing shifts observed in zero-G can exceed the TTC thresholds that generate them. Thus, if H_3 is true, a motor response that is triggered when TTC is less than or equal to 200 ms on Earth will be triggered 600 ms prior to impact in microgravity; i.e., a timing shift of 400 ms. (From McIntyre, 2001, with permission, reproduced from *Nature Neuroscience*.)

be able to do so regardless of the actual gravity level. Both on the ground and in orbit, subjects should initiate catching movements at a fixed time prior to the impact of the ball in the hand. By contrast, estimates of TTC that do not directly measure the true gravitational acceleration lead to systematic differences in timing between zero-G and one-G. A first-order estimate (Hypothesis 2), which assumes no acceleration, predicts TTC correctly in zero-G but overestimates the drop duration at one-G. If such an estimate is used to trigger anticipatory activity at a fixed TTC threshold (λ), anticipatory onset time will occur slightly later, relative to ball contact, in one-G because the ball arrives earlier than expected. Consequently, anticipatory activities will occur earlier in zero-G vs. one-G for a first-order TTC estimate. If instead the brain uses a second-order *internal model* (Hypothesis 3) to account for gravity's accelerating effects on the ball, motor activity will be correctly timed to impact at one-G but the same second-order internal model applied in zero-G will underestimate the true TTC. This will cause responses, in the absence of gravity, to start earlier than necessary. Thus, while Hypothesis 1 predicts no difference in the timing of

responses for catches performed without gravity, the other two hypotheses (2 and 3) predict that anticipatory activities will occur earlier in zero-G than in one-G. Despite this ambiguity, the magnitude of timing shifts observed in zero-G can tell us which of the three hypotheses is more likely to be correct. We simulated mathematically how motor responses would be synchronized with impact for each of the hypotheses if we varied the initial speed of the ball and the gravitational conditions (Figure 2B). For any given threshold (λ), the shift in timing (δ) between zero-G and one-G would be much smaller for the first-order TTC estimate than for the second-order internal model estimate over a wide range of initial ball velocities (v_0). Moreover, for the second-order internal model strategy, δ rises much more rapidly as λ increases. Finally, whereas δ changes very little with v_0 for the first-order estimate, a stronger, inverse-proportional relationship is seen for the second-order model. Thus, we were able to test the three different hypotheses by measuring shifts in timing between zero-G and one-G and by comparing those results with the shape and magnitude of the curves in Figure 2B.

METHODS

In light of these predictions, the 16-day Neurolab mission on board the Space Shuttle *Columbia* (STS-90) provided a unique opportunity to test the three hypotheses described above. Six astronauts caught a ball projected "downward" from the ceiling with one of three different, randomly assorted initial speeds, both in one-G (pre- and postflight) and zero-G (inflight). Subjects were seated in a chair fixed to the floor of the Spacelab module, with the forearm of the dominant hand held horizontally at approximately waist level, palm up. Waist, chest, thigh, and foot straps held the subject in the chair so that the ground and inflight postures were the same. A custom-made electromechanical ball launcher, attached to the ceiling, projected a 400-g ball downward from a point situated approximately 1.6 m above the outstretched hand, with one of three different, randomly assorted initial ball speeds (0.7, 1.7, 2.7 m/second). Muscle activity was monitored through surface electromyelography (EMG) electrodes placed over the biceps, triceps, wrist flexors, and wrist extensors.[2] EMG was band-pass filtered, rectified, and smoothed to compute the intensity of the muscle activation as a function of time. Averages of 10 trials for the same subject, session, and initial ball speed were computed, after first being aligned with impact. We then measured the timing of the anticipatory peak (Figure 3A) with respect to the moment of impact. The orientations of limb segments were calculated from the positions of reflective markers placed at the shoulder, elbow, wrist, and hand. The three-dimensional positions and velocities of these markers were measured by the KINELITE infrared tracking system, which was developed specifically for this mission by the French Space Agency CNES (Centre National d'Etudes Spatiales). Movement onset was detected when the forearm rotation around the elbow surpassed a certain position or velocity threshold. The timing of movement onset was also measured relative to the moment of impact (Figure 3B).

RESULTS

When these experimental tests were performed on the ground prior to flight, the astronaut subjects generated the peak of anticipatory biceps activity that occurs about 40 ms prior to impact, independent of the initial ball speed, as has been seen previously in other Earth-based studies (Lacquaniti, 1989; Lacquaniti, 1993). Subjects also rotated their forearm just prior to contact so as to move the hand upward to meet the ball. This upward movement is *not* generated by the peak of muscle activity described above, because the movement is initiated somewhat earlier than the measured EMG activity. Rather, this measurement represents two separate responses—upward movement to meet the ball and limb stiffening to stiffen the hand—both of which can tell us when the brain expects the ball to arrive. To see whether the brain uses an internal model of gravity when estimating the ball's arrival time, we compared the time course of these two responses between one-G (pre- and postflight) and zero-G (on-orbit) conditions.

Even though in this experiment the ball took much longer to cover the same distance in zero-G vs. one-G (for the same initial velocity and height), we did not expect the astronauts to make very large errors in the timing of their responses. They

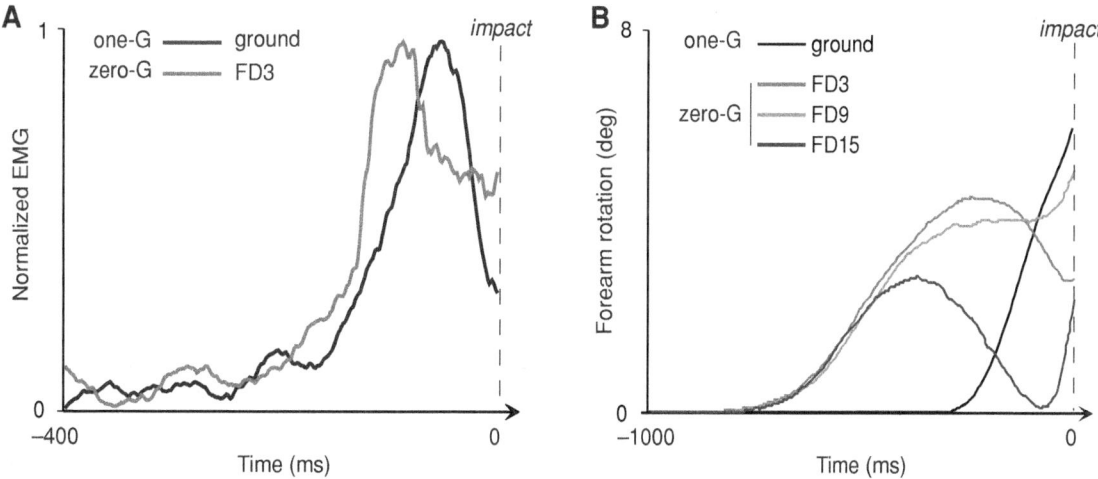

Figure 3. Anticipatory motor responses during catching in zero-G and one-G. Traces represent the average of 10 trials from a single subject and a single initial ball speed for (A) biceps muscle activity (v_0 = 0.7 m/s) and (B) forearm movements (v_0 = 1.7 m/s). Both of these responses occurred earlier (with respect to ball impact) for tests performed in flight vs. on the ground. (From McIntyre, 2001, with permission, reproduced from *Nature Neuroscience*.)

---

[2]EMG measures small differences in electrical potential that are generated on the skin above an activated muscle. The measured signals can be described as bursts of electrical "noise" from 20 to 500 Hz, the amplitude of which is correlated with the force and stiffness of the muscle. To measure the muscle activity, EMG signals are rectified and smoothed to calculate the envelope of the underlying alternating current electrical signal, in much the same way that an AM (amplitude modulation) radio converts high-frequency radio waves into audible sounds.

could easily see that for the lowest initial ball velocity, the ball would arrive after approximately two seconds in zero-G, compared to only about 0.5 second in one-G. Thus, we expected and found that catching responses were more or less synchronized with impact in flight even from the very first trials. In fact, the astronauts never missed the ball. This does not, however, mean that subjects were immediately able to account for the presence or absence of gravity (Hypothesis 1). Even approximate estimates of TTC that ignore gravity (Hypothesis 2) improve as the ball approaches impact. By continually updating the TTC estimate and checking it against the TTC threshold, the difference between prediction and reality is reduced. In this way, a first-order TTC estimate could *implicitly* adjust for changing accelerations. Similarly, a one-G internal model strategy (Hypothesis 3) could partially correct for the lack of acceleration in zero-G. Initial estimates of TTC will grossly underestimate the drop time in zero-G , but later estimates will improve. Following this reasoning, we expected to see only small, but measurable, variations in response timing depending on which hypothesis is correct.

Anticipatory muscle activity occurred earlier relative to impact in zero-G compared to one-G, as shown in Figure 3A. This difference cannot be explained by the repeated practice on this task that the astronauts had leading up to the flight because the timing returned to preflight values upon return to Earth. Nor can the shift simply be explained by the fact that subjects had more time to react in zero-G than in one-G; dropping the ball from higher heights to increase the drop duration on Earth led to no differences in timing. One might also guess that muscle activity must start sooner in flight to overcome the fact that the muscles tend to be more relaxed in zero-G (Lackner, 2000); but we found that supporting the forearm with an externally applied force to counteract gravity on Earth reduced the tonic activation of the biceps (as would be expected in zero-G) while causing no significant changes in timing of the anticipatory EMG peak. Thus, the timing shifts observed in flight are best explained by a failure to fully compensate for the lack of ball acceleration in zero-G.

The timing of forearm movements also changed in flight. On the ground, such movements were monotonically upward and accurately synchronized to impact, but these movements started much too early in zero-G (Figure 3B). Unlike the simple shift observed for muscle activity onset, however, limb movements stopped or even reversed direction after the initial false start. The change from monotonic movement on Earth to the non-monotonic movements observed in flight shows that movements were not simply slower in microgravity, as has often been observed, or simply due to reduced muscle tone in zero-G

(Lackner, 2000). Instead, this evidence suggests that, once triggered, the CNS can correct the ongoing movement by updating estimates of TTC based on visual feedback (Peper, 1994).

While one might predict that catching responses would occur too early when subjects are first faced with microgravity conditions, it would be surprising if the brain was unable to adapt to this novel environment. Some evidence for adaptation exists. Although forearm movements started prematurely for all inflight tests, on flight days (FDs) 9 and 15 the amplitude of the early, erroneous movement diminished and a later upward rotation developed just prior to impact. On the other hand, re-adaptation to Earth conditions occurred almost immediately. Average data from early postflight sessions (R+0,1,2) did not differ significantly from preflight or later postflight sessions. This indicates that subjects could rapidly retune internal models to the one-G environment.

The key experimental test in this experiment (Figure 2B) lies in the magnitude of time shifts for different initial ball velocities (v_0). As shown in Figure 4, the second-order, internal-model hypothesis predicts the nonlinear increases in time shifts (δ) for decreasing values of v_0 using reasonable, fixed values of the TTC threshold (λ). These λ values are biologically plausible. Allowing for the time it takes a neural signal to travel from the brain to the periphery, threshold values of TTC used to initiate a given behavior (λ) should be somewhat greater than the lead time of the behavior itself. That is, if the brain wants to activate a muscle x ms prior to ball impact, it has to use a TTC threshold of $x+y$ ms, where y represents the time it takes for the message to get from the brain to the muscles.

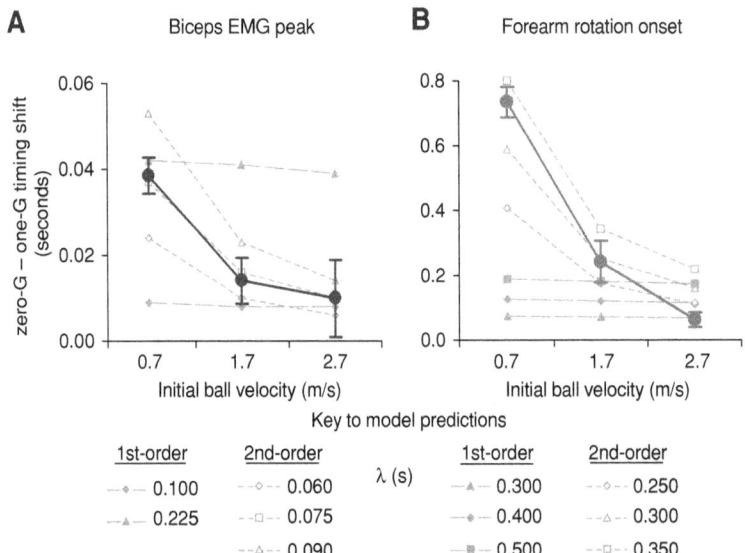

Figure 4. Timing of forearm rotations and biceps EMG compared to first- and second-order model predictions for different TTC thresholds (see Figure 2B). Each data point represents the timing advance (δ) in zero-G for (A) biceps EMG peaks and (B) anticipatory forearm movements, averaged across subjects and sessions (±1 s.e.). The patterns of timing shifts in flight indicate that the brain uses a second-order internal model to anticipate gravity's effects on the falling ball. (From McIntyre, 2001, with permission, reproduced from *Nature Neuroscience*.)

A response initiated at λ=75 ms within the brain is consistent with the arrival of the EMG peak in the muscle 40 ms prior to impact, due to neural transmission delays. Similarly, λ values of 300–400 ms are compatible with movement onset that normally occurs 200 ms prior to impact, given the time necessary to overcome limb inertia. In contrast, a first-order TTC estimate would require unreasonably long λ values ($\lambda \geq 1.2$ seconds; i.e., longer than the total drop time of the ball in one-G) to reproduce the time shifts observed in zero-G. Furthermore, a first-order TTC estimate cannot, with a fixed λ, predict the substantial changes of initiation time as a function of v_0. Finally, had the subjects been able to directly measure the acceleration of the ball in real time, one would expect to see no time shift at all (with respect to impact) between zero-G and one-G, irrespective of λ.

DISCUSSION

The data from these experiments are compatible with our third hypothesis; i.e., that to catch a falling ball the CNS: (1) employs a second-order estimate of TTC, (2) uses an *internal model* to predict gravity's effect on the ball, and (3) applies this internal model initially even in microgravity (McIntyre, 2001). Furthermore, it appears that the brain gradually adapts the internal model with prolonged exposure to zero-G. But, why did the adaptation take so long? The brain could have adjusted more rapidly to zero-G, given that the subjects could see that the ball moves more slowly in zero-G. Furthermore, information given by vestibular organs of the inner ear, pressure cues on the skin, and visual cues from objects floating within the cabin all clearly attest to the microgravity conditions on orbit. On the other hand, the identifiable walls, floor, and ceiling in the Spacelab compartment, directional overhead lighting, and the "upright" posture adopted by astronauts while performing most tasks all provide a strong up-down sense to the working environment. Under these conditions, the brain apparently gives credence to an *a priori* model of the physical world in which a downward moving object will accelerate. The timing errors observed in zero-G indicate a strategy that is "good enough for survival in the world, but which clearly breaks down under highly artificial laboratory conditions" (Hubbard, 1995).

Of course, internal models are not cure-alls for the brain. When accelerations are not measurable or predictable, the CNS may indeed employ clever tricks or approximations (such as the τ hypothesis and other first-order models) as the best means available to program anticipatory responses. Furthermore, the internal models may nevertheless be approximate, applying only to common situations. The inappropriate application of a one-G model in zero-G may be a particular example of "naïve physics," in which the internal model does not fully reflect the true underlying physical principles (Hubbard, 1995). Nevertheless, for the terrestrial conditions to which most people are bound, the CNS improves its chances of success by including a second-order *internal model* of gravity in TTC estimates.

The ability to anticipate is an essential requirement for the normal control of posture and movement. Thus, if the parts of the brain suspected of implementing internal models (such as the cerebellum or the basal ganglia) are injured or diseased, movement-related disorders will follow. Although much work still needs to be done to establish the relationship between internal models and pathology of the motor system, ongoing research about the cues that contribute to these models (sensory and/or cognitive), how they arise (acquired or innate), how they are implemented (neural circuits), and how they adapt (learning and neural plasticity) will lead to further insights into the workings of the human brain.

REFERENCE

VISUAL TIMING IN HITTING AN ACCELERATING BALL. D.N. Lee, D.S. Young, P.E. Reddish, S. Lough, and T.M. Clayton. *Q. J. Exp. Psychol.- A,* Vol. 35, pages 333–346; 1983.

THE ROLE OF PREPARATION IN TUNING ANTICIPATORY AND REFLEX RESPONSES DURING CATCHING. F. Lacquaniti C. and Maioli. *J. Neurosci.,* Vol. 9, pages 134–148; 1989.

VISUAL PROCESSING OF OPTIC ACCELERATION. P. Werkhoven, H.P. Snippe, and A. Toet. *Vision Res.,* Vol. 32, pages 2313–2329; 1992.

THE ROLE OF VISION IN TUNING ANTICIPATORY MOTOR RESPONSES OF THE LIMBS. F. Lacquaniti, M. Carrozzo, and N.A. Borghese. In: *Multisensory Control of Movement,* edited by A. Berthoz. Oxford: Oxford University Press, pages 379–393; 1993.

THE VISUAL GUIDANCE OF CATCHING. G.J. Savelsbergh, H.T. Whiting, J.R. Pijpers, and A.A. van Santvoord. *Exp. Brain Res.,* Vol. 93, pages 148–156; 1993.

CATCHING BALLS: HOW TO GET THE HAND TO THE RIGHT PLACE AT THE RIGHT TIME. L. Peper, R.J. Bootsma, D.R. Mestre, and F.C. Bakker. *J. Exp. Psychol. Human,* Vol. 20, pages 591–612; 1994.

ENVIRONMENTAL INVARIANTS IN THE REPRESENTATION OF MOTION: IMPLIED DYNAMICS AND REPRESENTATIONAL MOMENTUM, GRAVITY, FRICTION AND CENTRIPETAL FORCE. T.L. Hubbard. *Psychon B Rev,* Vol. 2, pages 322–338; 1995.

Accuracy of Estimating Time to Collision Using Binocular and Monocular Information. R. Gray and D. Regan. *Vision Res.,* Vol. 38, pages 499–512; 1998.

WEIGHTED COMBINATION OF SIZE AND DISPARITY: A COMPUTATIONAL MODEL FOR TIMING A BALL CATCH. S.K. Rushton and J.P. Wann. *Nat. Neurosci.,* Vol. 2, pages 186–190; 1999.

Human Orientation and Movement Control in Weightless and Artificial Gravity Environments. J.R. Lackner and P. DiZio. *Exp. Brain Res.,* Vol. 130, pages 2–26; 2000.

DOES THE BRAIN MODEL NEWTON'S LAWS? J. McIntyre, M. Zago, A. Berthoz, and F. Lacquaniti. *Nat. Neurosci.,* Vol. 4, pages 693–694; 2001.

Ensemble Neural Coding of Place in Zero-G

Authors

James J. Knierim, Gina R. Poe, Bruce L. McNaughton

Experiment Team

Principal Investigator: **Bruce L. McNaughton**[1]

Co-Investigators: **James J. Knierim,**[1,2]
Gina R. Poe[1]

Technical Assistants: **Kathy Dillon,**[1] **Shanda Roberts,**[1]
Veronica Fedor-Duys[1]

Engineers: **Casey Stengel,**[1] **Krzysztof Jagiello,**[1]
Vince Pawlowski[1]

[1]University of Arizona, Tucson, USA
[2]University of Texas-Houston Medical School, Houston, USA

ABSTRACT

One of the most disconcerting aspects of moving to a new city is the initial feeling of always being lost—of being unable to find your way in unfamiliar surroundings. Over time, as you explore the new environment, you begin to form a "mental map" that allows you to determine where you are at any given time, and to find where you want to go. Behavioral and physiological evidence implicates a particular structure of the brain, the hippocampus, and its related structures in the formation and maintenance of this cognitive map. The exact mechanisms by which the neurons in the hippocampus generate the map are still unknown. In the rat, hippocampal neurons fire selectively when the rat occupies particular locations in the environment. These "place cells" combine information from self-motion cues and from external landmarks to generate a neural code for spatial location. The nature of this interaction was investigated in the unique environment of microgravity on NASA's Neurolab Space Shuttle mission. Ensembles of 20-40 place cells were recorded as rats freely navigated a three-dimensional track that, in the absence of a gravitational reference framework, was expected to cause the rats' self-motion cues to be in conflict with the external landmark cues. Under these conditions, place cell firing was abnormal in two of three rats tested on the fourth day of flight. The firing patterns of both of these rats returned to normal when tested on the ninth day of flight. These results reveal important aspects of the network interactions underlying place cell firing under normal conditions on Earth and provide a possible central neurophysiological correlate for the disorientation experienced by many astronauts during the first days of spaceflight.

INTRODUCTION

Imagine parking your car in a lot in the center of town. You walk a few blocks to run some errands at different stores, and now it is time to head back to the car. Where is it? If, in your mind, you had continuously kept track of your position relative to the car as you turned and walked along the different streets on your route (the path integration strategy), you could simply turn in the direction that the car is located (even if you could not see it yet) and walk the proper distance toward it. If you lost track along the way, however, another method would be to look around until you recognized a familiar landmark—for example, the large bank across the street from the parking lot (the landmark navigation strategy). People (and animals) use a combination of these and other types of navigational strategies to keep themselves oriented as they move around their environments (Gallistel, 1990). A part of the brain called the hippocampus plays a crucial role in this ability. Neurons in the rat hippocampus are active when the rat occupies specific places in an environment (Figure 1A). These "place cells" encode a mental map of the environment, which is used to remember important locations and to calculate routes between them (O'Keefe, 1978). It has been hypothesized that the spatial tuning of these cells (i.e., the linking of the cells to a particular place) may also act as an index that ties together the neural activity of the different brain regions that are active during a particular event. This index then allows the reactivation of these regions at a later time when the event is recalled (episodic memory). The objective of the E100 "Escher Staircase" experiment was to use the unique environment of microgravity as a research tool to understand how the place cells in the hippocampus combine the path integration and landmark

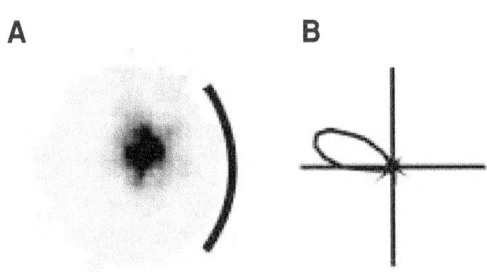

Figure 1. (A) Typical hippocampus "place field" recorded as a rat foraged in a cylindrical enclosure with a cue card covering 90 degrees of the wall (black arc). This firing rate map was constructed by dividing the number of spikes fired by the cell at each location by the amount of time the rat spent in that location. This cell fired strongly when the rat occupied a location near the center of the cylinder (dark area) and was inactive when the rat occupied other locations (light areas). (B) Typical head direction cell tuning curve. This plot was constructed by dividing the number of spikes by the amount of time the rat's head faced each direction, in 10-degree increments. This cell fired strongly when the rat's head was facing the 10 o'clock direction, regardless of the rat's position in the cylinder.

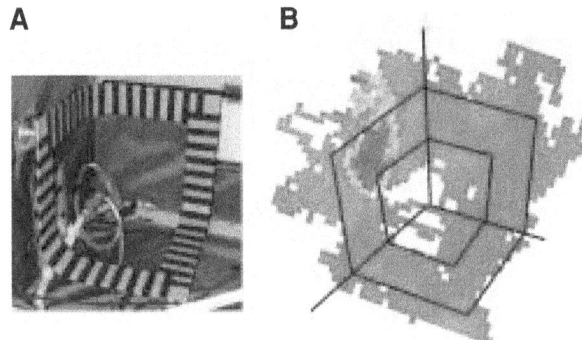

Figure 2. (A) Escher Staircase track. The rat makes only three 90-degree turns in yaw (side-to-side horizontal movement) instead of the usual four to return to the starting place. (B) Example of a normal place field as a rat ran around the Escher Staircase in space. In this average firing rate map, red denotes high firing rate, whereas blue denotes locations where the cell was silent. Thus, this particular cell fired strongly only when the rat was making the 90-degree yaw turn on the left-hand wall. (From Knierim, 2000, with permission; reproduced from *Nature Neuroscience*.)

navigation strategies to create a mental map under normal conditions here on Earth—and to understand how these mechanisms may be altered in space.

Another class of neurons that is associated with spatial orientation is the "head direction" (HD) cells of the thalamus and limbic cortex (Figure 1B) (Taube, 1998). The firing properties of place cells and HD cells are tightly coupled, and one function of the HD system may be to orient the cognitive map in the hippocampus. Under normal gravity, HD cells are sensitive to the horizontal component of head direction (yaw) only; changes in pitch and roll attitude do not affect their firing. As an animal moves on a horizontal surface, input from self-motion cues (e.g., vestibular signals, position sense, vision, etc.) are thought to keep the HD system aligned with the external landmarks long enough for stable associations to form between these landmarks and the place and HD cells. As a result of the formation of these associations, the external landmarks eventually gain control over these cells and become a major determinant of their firing properties (McNaughton, 1996). In three-dimensional space, however, body movements are not commutative (i.e., a 90-degree pitch followed by a 90-degree yaw does not result in the same three-dimensional orientation as a 90-degree yaw followed by a 90-degree pitch). Assuming that HD cells continue to respond only to yaw during spaceflight, this predicts inconsistent associations between HD and landmark information and a consequent inconsistency in the hippocampal place code.

During the Neurolab Space Shuttle mission of April 1998, ensembles of place cells were recorded from three rats implanted with a multi-electrode recording array. The rats negotiated a three-dimensional track (the "Escher Staircase") in which three 90-degree turns in yaw were interleaved with three 90-degree turns in pitch (Figure 2). The result of these movements was that the rat completed a full circuit of the track and returned to its starting location/direction after having made only three right turns (270-degree total yaw). The spatial information provided by external landmarks was thus presumably in conflict with the spatial information from self-motion cues, which under normal conditions require a fourth 90-degree turn (360-degree total yaw) to signal a return to the starting location. If the self-motion cues drive place cells exclusively in this novel environment, one would predict that the hippocampal map would indicate a return to the starting location only after this fourth 90-degree turn. As a result, the map would appear to lag the rat's behavior by 1/3 of the track on each lap, thus exhibiting a precession effect. Alternatively, if the external landmarks provide the primary basis for the assignment of cells to locations, one would predict that the hippocampal map would appear stable on all laps. If both types of cues exert a degree of control over the cells, then this might result in an overall loss of spatial selectivity as the coherency of the map breaks down due to the constant conflict between the different sources of spatial information.

METHODS

Prior to launch, 15 adult male rats were trained to run clockwise along a rectangular track, the longer sides of which were tilted 35 degrees to 45 degrees such that the rat was running up one side of the rectangle and down the other side. After three weeks' training, rats were implanted with a multiple-electrode recording drive over the right hippocampus. Rats also were implanted bilaterally with bipolar stimulating electrodes aimed at the medial forebrain bundle (MFB). Stimulation of the MFB is a powerful reinforcing stimulus for the rat, and after surgery the rats were retrained to run the tilted rectangle for MFB stimulation as a reward. Individual tetrodes were slowly advanced to about two mm below the cortical surface to record place cells of the CA1 layer of the hippocampus. Three days before launch, the four rats with the best combination of stable, high-quality neuronal recordings and robust behavioral performance were selected for flight. The rats were loaded on board the Space Shuttle *Columbia* Orbiter in individual cages approximately two days prior to launch. On the fourth and ninth days of flight (approximately 70 and 189 hours after launch), the rats were secured in an 8×8×18-cm canvas pouch and were placed into a gloved workstation. The recording cable and headstage were attached and 15–20 minutes of baseline data were recorded. Each rat was then placed individually on the Escher Staircase track (see Figure 2) and induced to move clockwise along the track by stimulation of the MFB while the neuronal signals from the place cells were recorded on a computer. After 7–20 laps around the track, the rat was removed and replaced into the canvas pouch.

After the landing of the Shuttle, videotapes of each rat's behavior were run through a tracking program that calculated the rat's position based on the location of a ring of bright light-emitting diodes (LEDs) on the headstage centered over the rat's head. The rat's moment-by-moment position was correlated with the signals from individual neurons. The spatial tuning of each cell was measured by constructing firing rate maps for each cell (see Figure 1) and by calculating quantitative measures of the spatial selectivity of each cell. All animal care and experimental procedures conformed to National Institutes of Health (NIH) guidelines and were approved by institutional Animal Care and Use Committees at NASA Ames Research Center and at the University of Arizona.

RESULTS

Figure 3B shows the spatial firing patterns of 12 of the 37 cells recorded simultaneously from Rat 2 on its first exposure to the track, which occurred on the fourth day of flight (FD4); the remaining 25 cells were silent or fired only a few spikes during the behavioral test session. While this proportion of activity in a hippocampal ensemble is normal for a track of this size, none of the cells that fired on the track showed strong spatial tuning. This lack of tuning was evident from the very first lap around the track, and it did not improve over the nine

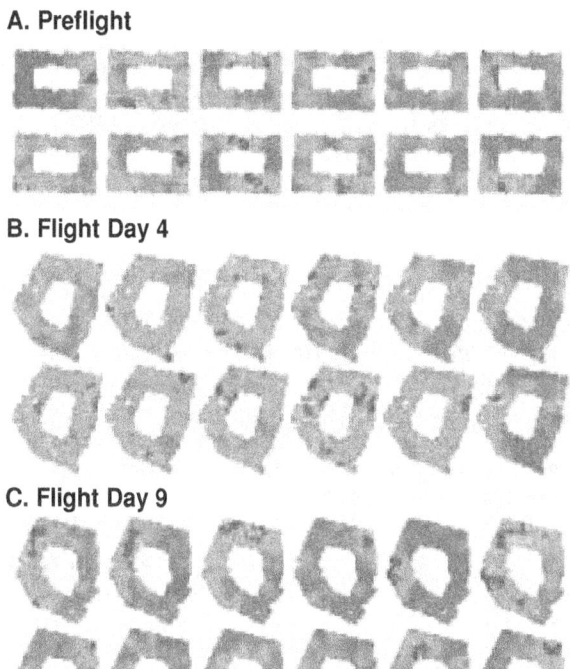

A. Preflight

B. Flight Day 4

C. Flight Day 9

Figure 3. (A) Preflight place fields from Rat 2 as the rat ran clockwise on a rectangular eight-cm-wide track (62×38 cm). Firing rate maps for 12 of the 19 cells that fired at >0.05 Hz during the session (out of 35 cells recorded during baseline) are shown here. (B) Lack of spatial tuning on FD4. Firing rate maps for all 12 cells that fired >0.05 Hz during the session are shown here. Only two of the cells had a statistically significant (p<0.01) information score. (C) Normal place fields on FD9. Firing rate maps for 12 of the 16 cells that fired >0.05 Hz during the session are shown here. (From Knierim, 2000, with permission; A, C reproduced from *Nature Neuroscience*.)

A. Flight Day 4

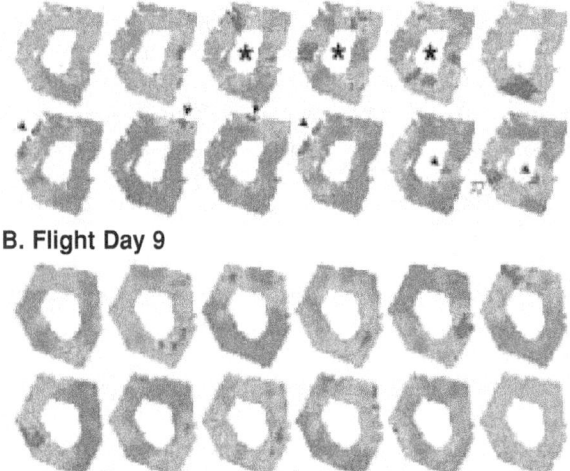

B. Flight Day 9

Figure 4. (A) Abnormal spatial tuning for Rat 1 on FD4. Firing rate maps for the 12 cells that fired >0.05 Hz during the session are shown here. Most of the highly specific place fields occurred when the rat's head was off the apparatus, investigating the off-track environment (black arrows). The cells that fired when the rat was moving along the track tended to have radially symmetric place fields (asterisks), firing either at each of the yaw turns or each of the pitch turns. (B) Normal place fields on FD9. Firing rate maps for 12 of the 21 cells that fired >0.05 Hz during the session. (From Knierim, 2000, with permission; B reproduced from *Nature Neuroscience*.)

laps run by the rat in this session. For comparison, Figure 3A shows the spatial firing patterns of 12 of the 19 cells that fired as this same rat ran clockwise on a flat, rectangular track four days before launch. Most cells that fired on the track had normal, highly specific place fields. A spatial tuning index, which quantifies the amount of information about the rat's location conveyed by the firing of the cells, was significantly different between the preflight recording and the FD4 recording. Moreover, only two of the 12 cells on FD4 had a statistically significant spatial information score, compared to 16 of the 19 cells on the preflight day.

Rat 1 displayed a different pattern of abnormal spatial selectivity. Figure 4A shows the firing patterns of the 12 cells (out of 31) that fired strongly on the track on FD4. Almost all of the spatially selective firing in this rat occurred when the rat's head was off the track (black arrows), while the rat was either turning around or investigating the off-track environment. Only one sharply tuned place field occurred when the rat was moving forward on the track performing the behavioral task (white arrow). Three other cells that fired when the rat was moving forward along the track displayed a radial symmetry in their place fields (asterisks), firing at three symmetric locations on the track. This was not the place field precession effect that was predicted, however, as the cells fired reliably at all three locations on each lap around the track. Note that two of the cells fired at each of the pitch locations

(where the rat was usually rewarded), whereas the third cell fired at the yaw locations. Thus the symmetric firing cannot be explained solely by the cell responding to the reward. This type of symmetric firing was not seen during the preflight data collection on the flat rectangle, in which MFB stimulation rewards also occurred at symmetric locations on the track, nor is it typically seen when rats traverse a flat triangular track under normal laboratory conditions.

In contrast to Rats 2 and 1, Rat 3 displayed normal spatial tuning on FD4, as six of seven cells that fired on the track had robust place fields. Unfortunately, these data came from this animal's second run on the Escher Staircase, as the data from the first run were lost due to technical problems. (A third run on FD4 was also performed on this rat, and it was shown that the cells maintained the same, stable place fields in Sessions 2 and 3.) It is thus not known whether on its first run this rat had abnormal spatial tuning, which improved as the rat gained more experience on the track in the second run 25 minutes after the first. This possibility is consistent with the second set of recordings from Rats 1 and 2 on the ninth day of flight (FD9). In both cases, these rats displayed normal spatial selectivity on FD9 (Figure 3C and Figure 4B). Thus some type of adaptation might have occurred in these rats that allowed the hippocampus to form a normal, coherent map of the track on FD9 after displaying abnormal firing patterns on FD4. It was not possible to control all variables between FD4 and FD9 in the Spacelab environment. For example, the rats performed the behavioral task in microgravity better on FD9 than on FD4; the ongoing activities of other Neurolab experiments outside the experimental chamber were different on the two days; and a miniature flashlight generated a bright spot on one segment of the track on FD4 for Rats 1 and 2. There were a number of salient visual cues common to all recording sessions on both days (including two large windows that provided the ambient illumination into the chamber), however; and it is improbable that the difference in spatial selectivity between FD4 and FD9 was the result of the uncontrolled differences.

Hippocampal electroencephalogram (EEG) activity was recorded during baseline and behavioral sessions in all rats on FD4 and FD9. Although the small number of subjects precluded a statistical analysis, normal theta rhythm was apparent during active locomotion, and normal sharp waves and ripples were apparent during quiet inactivity in the canvas pouch on both flight days. The rats were also run on a +-shaped track in which they were subjected to various pitch and roll rotations, in addition to the Escher Staircase. Although the poor behavioral performance of the rats on this task made a thorough analysis unwarranted, visual inspection of the data suggests that the firing of the hippocampal cells on this track resembled that on the Escher Staircase. That is, the spatial selectivity of the cells of Rats 1 and 2 was poor on FD4 but improved on FD9, and the spatial selectivity of Rat 3 was good on FD4.

Place cells normally demonstrate a stereotypical temporal signature when rats run on a closed track (Figure 5A). Periods of low firing rate are interrupted by brief bursts of activity as the rat runs through the place field of the cell. The cells of Rat 2 displayed a very different pattern of activity on FD4 (Figure 5B).

A. Preflight

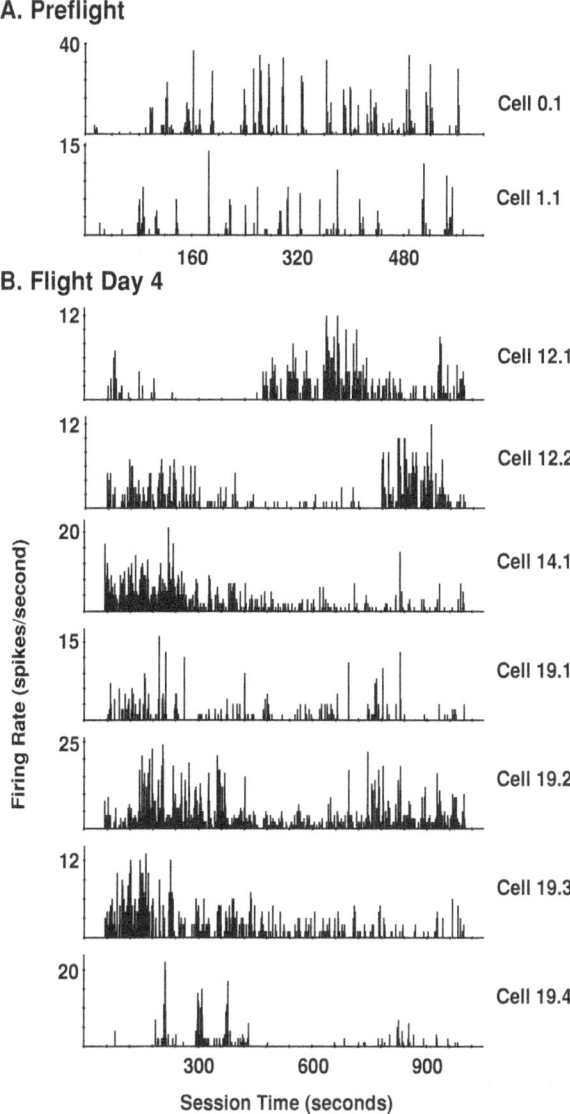

B. Flight Day 4

Firing Rate (spikes/second)

Session Time (seconds)

Figure 5. (A) Rate vs. time plots for two typical cells from Rat 2 during preflight data collection on the rectangular track. Periods of relative inactivity are interrupted by bursts of high activity as the rat runs through the place field of the cell (shown as the first and sixth graphs of Figure 3A). (B) Abnormal rate vs. time plots for the seven most active cells of Rat 2 on FD4. These cells, which were not spatially selective, did not fire at a uniform rate throughout the session, nor did they fire with the characteristic temporal pattern shown in (A). Rather, different subsets of cells fired at high rates for extended periods of time, with apparently abrupt transitions between active sets of neurons.

These cells tended to fire at high rates for a few minutes and then to become less active for a comparable time. For example, Cell 12.1 was mostly quiet at first, then became highly active midway through the session. Other cells (12.2 and 19.2) were active for approximately the first five minutes, then became quiet at approximately the same time that Cell 12.1 became active; these cells became active again at approximately the

same time that Cell 12.1 became less active (~800 seconds). Other cells (14.1 and 19.3) were active at first, became quiet, but then never became highly active again. Thus the abnormal spatial tuning of these cells (Figure 3B) was accompanied by abnormal temporal firing properties. In contrast to Rat 2, the temporal signatures of Rats 1 and 3 were apparently normal.

DISCUSSION

The prediction that the place cells would demonstrate a precession effect on the Escher Staircase was based on the notion that place cells in a novel environment are updated on the basis of relative position, determined by summing the linear and angular motion, and that external landmarks gain control only after a period of associative learning. It was hypothesized that the place cells would be subject to a 90-degree mismatch between the signals from the HD system and the external landmarks during each lap of the track, thus preventing unique associations between vestibular-based head direction signals and external landmarks. This would result in a precession of the place fields as the rat moved along the track. Recent results demonstrate, however, that only a few minutes of exposure to a new environment are required for the salient visual landmarks to begin to exert some control over HD cells (Goodridge, 1998). Thus the lack of the precession effect may have been the result of the visual landmarks having gained some degree of control over the HD system in the few minutes during which the rats were exposed to the environment before the recording began. The subsequent conflicts between the external landmark input and the self-motion input to the HD cells may have resulted in different patterns of abnormal behavior of the HD cell system, which were reflected in the different patterns of activity of the hippocampal cells of Rats 1 and 2. In the case of Rat 2, the HD system might have totally broken down and the cells might have lost all selectivity for head direction. This loss of selectivity has been reported in normal rats under certain conditions in the terrestrial laboratory and under certain conditions during weightlessness in parabolic flight (Knierim, 1998; Knierim, 1995; Taube, 1999). Because lesions to the HD system can alter the spatial tuning of hippocampal cells, it may be that the loss of normal HD cell input caused the hippocampal network to enter a completely anomalous state on FD4, in which no cells demonstrated robust spatial tuning.

The abnormal, three-fold symmetric firing fields exhibited by the hippocampal cells of Rat 1 (Figure 4A) might be explained if the HD system had reset to a single default direction each time the rat moved from one orthogonal surface to another. Such an outcome would be consistent with the "subjective vertical" phenomenon experienced by astronauts early in spaceflight, in which they perceive that the horizontal reference frame is based on body orientation (i.e., the floor is where the feet are (Oman, 1988)). HD cells have been proposed to be an important "orthogonalizing" input for place cells; that is, when the external sensory cues available at two given locations are similar, the differential input from HD cells may cause the hippocampus to represent the two locations with an uncorrelated ("orthogonal") distribution of place fields. When a rat is on a symmetric track under normal conditions, it may be the differential firing of the HD system that allows the hippocampus to form a unique representation of each of the symmetric locations. If the HD system becomes reset to a default value as the rat moves from one surface to another on the Escher Staircase in space, however, then the firing of the HD system would be identical at each of the symmetric locations on the track. Without the orthogonalizing input of the HD cells, the hippocampal place cells would respond similarly at each of the locations, ignoring the off-track cues that could disambiguate the locations. When the rat's head was off the track, investigating the asymmetric off-track environment, the cues on the different surfaces of the recording enclosure may have been distinct enough to generate unique firing patterns on each segment of the track without the need for the symmetry-breaking input from the HD system.

The hippocampus was able to adapt eventually to the demands of three-dimensional navigation in space, as the place cells displayed highly specific spatial tuning on FD9. Thus hippocampal cells can form unique, reliable representations of three orthogonal surfaces in microgravity; however, they appear to require either a period of adaptation to microgravity or more experience with the environment than is typically required in normal gravity. The adaptation may entail both bottom-up mechanisms, such as changes to the vestibular system that are known to occur in spaceflight, and top-down mechanisms, such as shifts in cognitive orientation strategies or attentional mechanisms where the rat (and astronaut) learns to attend to external landmarks to keep itself oriented. It remains to be determined whether the hippocampal code in microgravity can fully represent three dimensions, or whether the system adapts by developing independent two-dimensional representations for each orthogonal surface.

Astronauts frequently report a feeling of disorientation during spaceflight. When they are "upside down" relative to the orientation of the environment learned during preflight training, they often report that the environment suddenly feels somewhat unfamiliar. This perception may fade over time as they adapt to motion in weightlessness and learn to perceive and recognize the spatial arrangement of the environment from many different viewing angles. The data presented here may reflect a neurophysiological correlate of this adaptation phenomenon. As astronauts maneuver in three-dimensional space, the conflict between their internal reckoning system of place and direction and the spatial information provided by visual landmarks may result in abnormal patterns of firing in the hippocampus. In addition to memory for locations and spatial relationships, the hippocampus is thought to be important in humans for "episodic" memory—the memory of specific events in one's past. It is an important and open question whether other cognitive abilities that are dependent on the normal operation of the hippocampus are also affected during the first days of spaceflight, and whether such changes recover as the brain adapts to weightlessness. The present results are an important first step towards addressing such issues at the level of nerve cells working as an ensemble. These results demonstrate the feasibility of performing such complex neurophysiological and behavioral experiments in the microgravity environment, and the ability of such experiments to elucidate not only the effects of prolonged spaceflight on the nervous system but also the normal functioning of brain systems on Earth.

REFERENCES

THE HIPPOCAMPUS AS A COGNITIVE MAP. J. O'Keefe and L. Nadel. Clarendon Press; 1978.

THE ROLE OF STATIC VISUAL ORIENTATION CUES IN THE ETIOLOGY OF SPACE MOTION SICKNESS. C.M. Oman in *Proceedings of the Symposium on Vestibular Organs and Altered Force Environment*, edited by M. Igarashi and K. Nute. NASA Space Biomedical Research Institute, pages 25–37; 1988.

THE ORGANIZATION OF LEARNING. C. R. Gallistel. MIT Press; 1990.

PLACE CELLS, HEAD DIRECTION CELLS, AND THE LEARNING OF LANDMARK STABILITY. J.J. Knierim, H.S. Kudrimoti, and B.L. McNaughton. *J. Neurosci.,* Vol. 15, pages 1648–1659; 1995.

DECIPHERING THE HIPPOCAMPAL POLYGLOT: THE HIppocampus AS A PATH INTEGRATION SYSTEM. B.L. McNaughton, C.A. Barnes, J.L. Gerrard, K. Gothard, M.W. Jung, J.J. Knierim, H. Kudrimoti, Y. Qin, W.E. Skaggs, M. Suster, and K.L. Weaver. *J. Exp. Biol.*, Vol. 199, pages 173–185; 1996.

CUE CONTROL AND HEAD DIRECTION CELLS. J.P. Goodridge, P.A. Dudchenko, K.A. Worboys, E.J. Golob, and J.S. Taube. *Behav. Neurosci.*, Vol. 112, pages 749–61; 1998.

HEAD DIRECTION CELLS AND THE NEUROPHYSIOLOGICAL BASIS FOR A SENSE OF DIRECTION. J.S. Taube. *Prog. Neurobiol.,* Vol. 55, pages 225–256; 1998.

INTERACTIONS BETWEEN IDIOTHETIC CUES AND EXTERNAL LANDMARKS IN THE CONTROL OF PLACE CELLS AND HEAD DIRECTION CELLS. J.J. Knierim, H.S. Kudrimoti, and B.L. McNaughton. *J. Neurophysiol*, Vol. 80, pages 425–446; 1998.

RAT HEAD DIRECTION CELL RESPONSES IN 0-G. J.S. Taube, R.W. Stackman and C.M. Oman. *Soc. Neurosci. Abstracts,* Vol. 25, pages 1383; 1999.

THREE-DIMENSIONAL SPATIAL SELECTIVITY OF HIPPOCAMPAL NEURONS DURING SPACE FLIGHT. J.J. Knierim, B.L. McNaughton, G.R. Poe. *Nat. Neurosci.*, Vol. 3(3), pages 209-210; 2000.

The Role of Visual Cues in Microgravity Spatial Orientation

Authors

Charles M. Oman, Ian P. Howard, Theodore Smith,
Andrew C. Beall, Alan Natapoff, James E. Zacher,
Heather L. Jenkin

Experiment Team

Principal Investigator: **Charles M. Oman[1]**

Co-Investigators: **Ian P. Howard,[2] Theodore Smith,[1] Andrew C. Beall[1]**

Scientific Assistants: **Alan Natapoff,[1] James E. Zacher,[2] Heather L. Jenkin[2]**

[1]Massachusetts Institute of Technology, Cambridge, USA
[2]York University, Toronto, Canada

ABSTRACT

In weightlessness, astronauts must rely on vision to remain spatially oriented. Although gravitational "down" cues are missing, most astronauts maintain a "subjective vertical"—a subjective sense of which way is up. This is evidenced by anecdotal reports of crewmembers feeling upside down (inversion illusions) or feeling that a floor has become a ceiling and vice versa (visual reorientation illusions). Instability in the subjective vertical direction can trigger disorientation and space motion sickness. On Neurolab, a virtual environment display system was used to conduct five interrelated experiments, which quantified: (a) how the direction of each person's subjective vertical depends on the orientation of the surrounding visual environment, (b) whether rolling the virtual visual environment produces stronger illusions of circular self-motion (circular vection) and more visual reorientation illusions than on Earth, (c) whether a virtual scene moving past the subject produces a stronger linear self-motion illusion (linear vection), and (d) whether deliberate manipulation of the subjective vertical changes a crewmember's interpretation of shading or the ability to recognize objects.

None of the crew's subjective vertical indications became more independent of environmental cues in weightlessness. Three who were either strongly dependent on or independent of stationary visual cues in preflight tests remained so inflight. One other became more visually dependent inflight, but recovered postflight. Susceptibility to illusions of circular self-motion increased in flight. The time to the onset of linear self-motion illusions decreased and the illusion magnitude significantly increased for most subjects while free floating in weightlessness. These decreased toward one-G levels when the subject "stood up" in weightlessness by wearing constant force springs. For several subjects, changing the relative direction of the subjective vertical in weightlessness—either by body rotation or by simply cognitively initiating a visual reorientation—altered the illusion of convexity produced when viewing a flat, shaded disc. It changed at least one person's ability to recognize previously presented two-dimensional shapes. Overall, results show that most astronauts become more dependent on dynamic visual motion cues and some become responsive to stationary orientation cues. The direction of the subjective vertical is labile in the absence of gravity. This can interfere with the ability to properly interpret shading, or to recognize complex objects in different orientations.

INTRODUCTION

When an astronaut ventures into space, the response of the body's gravity-sensing organs is profoundly altered. As soon as orbit is achieved, the spacecraft is literally falling around the Earth. The tiny stones in the inner ear balance organs—the otoliths—float into unusual positions, and tilting the head produces no sustained otolith displacement the way it does on Earth. The unusual signals from the otolith organs in zero-G are apparently not sufficient to produce major changes in the inner ear to eye reflexes that allow the eyes to stay fixed on an object while the body is moving. Also, although astronauts typically don't report sensations of "falling," they are susceptible to illusions about body orientation.

Some describe a paradoxical sensation of feeling continuously upside down ("inversion illusion"), often while seated upright in the cabin immediately after reaching orbit. The fact that fluids normally shift from the legs to the upper body and the viscera elevate upon entering weightlessness may also contribute to this sensation of being upside down. The inversion illusion can also occur with the eyes closed. Fortunately, susceptibility to this illusion usually only lasts a day or so before it subsides. Another much more common type of illusion—the "visual reorientation illusion" (VRI)—can occur when a crewmember is working upside down inside the spacecraft. It can also happen when a person is working upright, if the person sees another person floating upside down. In both situations, the ceiling of the spacecraft suddenly changes its subjective identity and seems somehow like a floor. The perceived port/starboard and forward/aft directions of the spacecraft may also reverse. VRIs usually happen spontaneously, but they can also be initiated or reversed by cognitive effort—i.e., by imagining a change in position. The sudden change in perceived orientation associated with VRIs is known to trigger attacks of space motion sickness during the first week in orbit. It is reportedly more difficult to keep your sense of direction when moving between spacecraft modules with differently oriented visual verticals. Mir station crewmembers say VRI susceptibility and sense-of-direction difficulties persist for months. Since dropped objects in space don't fall, it may seem surprising that astronauts have a sense of the vertical—as revealed by inversion illusions and VRIs—but most apparently do. Inversion illusions were first reported by Cosmonaut Titov in 1961, and VRIs were first described by Skylab and Spacelab crews (Oman, 1986) in the 1970s and 1980s. Some crews refer to both phenomena as "the downs."

What causes VRIs? On Earth, all of us occasionally visually reorient our sense of direction; for example, when we emerge from a subway station, catch sight of a familiar building, and realize we are facing in a different direction than we thought. Familiar objects provide important directional cues, but our sense of direction usually shifts only in a horizontal plane because gravity anchors our sense of which way is down. Laboratory experiments conducted on Earth in specially built tumbling rooms (Howard, 1994) have shown that tilting the room away from the normally upright position shifts the direction a subject will set a "down" pointer. The subjects tend to

point not toward the gravitational down but instead toward the principal visual axes of symmetry of the environment. If the room is furnished with familiar objects that have a clearly recognizable "top" and "bottom," such as a chair or a table, and both the room and the subject are tilted 90 degrees, many people report they still feel upright even though they are gravitationally supine. If the number of polarized objects is reduced and the room is slowly tumbled around the subject, most people initially feel tilted opposite to the direction of room rotation. Eventually, as a wall or a ceiling rotates into a position beneath their feet, that surface suddenly seems like a floor. The subject instantly feels tilted in the opposite direction.

This illusion corresponds to the VRIs described by astronauts. Rotating a strongly polarized room typically produces a sensation of full head-over-heels tumbling, with no VRIs. The hypothesis that emerged from these and other experiments (reviewed by Oman (Oman, 2002)) is that the subjective vertical (SV) direction—and the identity of surrounding surfaces—is determined by the interaction between signals from the body's gravity receptors and visual cues. Gravity direction cues come not only from the otolith organs, but also from receptors in the kidneys and the cardiovascular system. Individual subjects show a small but consistent headward or footward bias (Mittelstaedt, 1992). Visual cues include the principal directions defined by the major surfaces and symmetries of the surrounding environment (such as walls and ceilings), with the up/down axis of the visual environment identified based on two factors: (1) the gravitational polarity of familiar objects, and (2) a tendency to perceive the visual vertical as oriented along the body axis in a footward direction (known as an idiotropic orientation (Mittelstaedt, 1983)). If there are minor directional differences between the gravity receptor and visual cues to the vertical, the brain apparently compromises and the SV points in an intermediate direction. The remaining component of gravity is then perceived as a mysterious force, pulling the body to one side. (This illusion can be readily experienced in houses tilted by an earthquake.) If the disparity in direction of the gravity receptor and visual verticals is large, one sensory modality or the other typically captures the SV. Tilting the head away from the gravitationally erect position enhances the effect of visual cues. There seem to be consistent differences between individuals in the relative weighting assigned to visual vs. gravity receptor cues. Older individuals appear more susceptible to visually induced tilt. Scene motion enhances visually induced tilt for most subjects.

What determines the direction of the SV in weightlessness? The body's gravity receptors are unweighted in space, but the individual subject's headward or footward bias presumably remains. The bias may increase in a headward direction because of the zero-G fluid shift, though this effect may only last a few days. We hypothesized that with eyes open, the SV should align with the body axis if the crewmember has a strong idiotropic tendency. In a more visually dependent individual, the SV should align with one of the principal environmental axes of symmetry, depending on which way the person's feet are pointing (idiotropic effect) and on the orientation of polarized objects in the visual scene. Other crewmember's bodies are strongly gravitationally polarized, since they have a readily recognizable top and bottom, and are consistently encountered upright in normal life. Hence VRIs should not occur in a visually familiar environment if everyone on board remains upright with respect to the deck. However, if the viewer floats sideways or upside down or another crewmember does so, the viewer may experience a sudden change in the direction of the SV. If unanticipated changes in the relative direction of the SV contribute to space motion sickness, one could speculate that idiotropic crewmembers should be less prone to space motion sickness, since they "carry down around with them." These people may have more difficulty keeping track of objects and surfaces as they move about. If this is really so, it has potential implications for astronaut selection, training, spacecraft architecture, and space sickness prevention. Do crewmembers show consistent interindividual differences in idiotropic vs. visual dependency? How can this be assessed? Harm et al. (Harm, 1999) retrospectively classified crewmembers based on postflight debriefings concerning illusions experienced in flight. Young and coworkers (Young, 1986; Young, 1996) had crewmembers insert their heads into a polka-dotted drum that rolled about the visual axis and report the amount of illusory angular self-motion (circular vection) they experienced. On Earth, upright subjects reported a paradoxical rolling/tilting sensation. In orbit, most astronauts felt continuous rotation. Wearing a bungee cord harness that pulled the subject to the deck inhibited the strength of circular vection in some subjects. Young concluded that astronauts become more visually dependent in weightlessness since they generally experience stronger sensations of angular speed in response to visual scene rotation. However, display limitations did not permit assessment of responses to linear vection stimuli, or to rotating and statically tilted, gravitationally polarized scenes. Scientific study of these illusions and sensations requires experiments on human subjects using controlled visual stimuli, which has been impractical on previous Shuttle flights. The virtual environment generator (VEG) display flown on Neurolab provided the first opportunity to study VRIs and related phenomena.

On Earth, the process of object recognition and shading interpretation depends on the gravitational orientation of the objects. For example, Rock (Rock, 1957) found that people more easily recognize nonsensical doodles if the doodles are shown in the same gravitational orientation as when previously seen. Howard and colleagues (Howard, 1990) showed that the illusory concavity or convexity people normally perceive when interpreting shading on a truly flat surface depends on a "light comes from above" assumption, where "above" depends on the relative orientation of the dark-to-light shading gradient to head orientation and to gravity. We predicted that if a subject experienced a VRI in weightlessness, it would not only change the subjective identity of surrounding surfaces but should also influence the recognition of complex figures, and the perceived convexity of gradient-shaded circles. Since many crewmembers say that they can cognitively initiate a VRI in weightlessness (i.e., "whichever way I decide is down becomes down"), on Neurolab we tested the hypothesis that figure and shading gradient recognition could even be changed just by cognitively altering the SV, without any physical movement or change in the visual scene content.

METHODS

Five related experiments were performed by four male Neurolab crewmembers, coded alphabetically A–D, who were aged 37–43 years old and had no visual or vestibular abnormalities. One of the subjects had participated in a 16-day flight two years earlier; the other three were making their first orbital flight. General experimental design and methods are described here. Specific procedures and stimuli for each experiment are detailed in the next section.

During all experiments, the subject wore a color stereo head-mounted display (Kaiser ProView-80; 640×480 resolution and 65 degrees×48 degrees field-of-view in each eye, 100% binocular overlap) equipped with a visor that completely excluded exterior views. Subjects viewed computer-generated visual scenes rendered by the NASA VEG, a custom Pentium/Windows NT real-time graphics workstation, described elsewhere in this book (see technical report by Oman et al. in this publication). A head tracker was not used to stabilize the visual environment, so subjects remained motionless during all trials. Experiment instructions were presented on virtual cue cards. Subjects initiated the trials and made subjective reports using a joystick strapped to their thigh. To eliminate directional sound cues, an area microphone mounted on the VEG front panel fed back ambient sounds monophonically to the subject's binaural headphones.

Subjects practiced with the apparatus and performed portions of the experiments in mission simulations several times during the six months prior to flight. They completed each experiment as a subject three times preflight (approximately 90, 60, and 30 days before launch); once on the third or fourth day of orbital flight; again on landing day or the first day after landing, the second day after landing, and on the fourth and fifth days after landing. (The 90-day preflight session served as a training session for one subject, whose data were discarded. Limited testing was also performed on some subjects on the 16th flight day and on the day of return, as noted.) For purposes of statistical analysis, results from different sessions were usually pooled into epochs: preflight ("PRE"), inflight ("IN"), postflight days 0–2 ("EARLY"), and postflight days 4 and 5 ("LATE"). In each test session, the subject was tested under multiple conditions. During ground testing, the subject either was seated erect in a padded chair or lay supine or left shoulder down in a padded gurney bed. Inflight, the subject was tested free floating upright in the virtual visual environment, and also (depending on the experiment) either floating left shoulder down relative to the visual environment or while "standing" in a restraint harness. The harness was connected to a pair of deck-mounted constant force springs, which provided a "downward" position sense cue to the shoulders and hips. Conditions were necessarily nested by epoch. For operational reasons, the upright floating condition was always tested first. Results were analyzed statistically (Systat v.10) by subject, epoch, condition, and scene type or speed.

RESULTS

Tilted Room Experiment

The four subjects viewed the interior of a virtual spacecraft module, 7.1 m long, 2.1 m high, and 2.1 m wide. In successive trials, the scene was presented in different tilt orientations with respect to their head/body axis by an angle that varied over ±180 degrees in four-degree increments, in randomized order. The presentation alternated between a scene (Figure 1A) that had identical left and right walls, and a ceiling and floor with similar (though not identical) details, and a second more visually polarized scene with readily distinguishable ceiling, wall, and deck surfaces, and an astronaut figure floating upright (Figure 1B) 2.5 m away from the viewpoint. Scene lighting was completely even, to eliminate directional effects. Subjects were instructed to look quickly around the scene at the beginning of each trial, and decide which surface seemed most like a "floor," and which way objects would fall if gravity were present. Subjects then clicked a joystick button to make two green balls appear (as shown in Figure 1B, for example). The subjects indicated the SV by moving the outer ball into a position where the center ball would hit the subjective floor if it fell. If no falling direction was discernable, they were to point the outer ball at the subjective floor. In each of the pre- and postflight sessions, subjects performed 96 trials in both the upright and the supine conditions. In flight, subjects performed the same series of 96 trials in the floating condition. In the inflight restrained condition using the spring-loaded harness, 39 trials using the more polarized scene (Figure 1B) were performed.

To assess whether subjects had consistent differences in static (i.e., motionless scene) visual dependence, SV indications from individual subjects were plotted vs. the absolute value of scene presentation angle (Figure 2). Indications generally clustered along the diagonal (i.e., SV aligned with the head-to-foot axis) and/or pointed toward one of the four surfaces in the scene. Subjects sometimes gave responses in an intermediate direction, particularly in preflight testing, and for scene angles near 0, 90, or 180 degrees. For a particular range of scene angle, a subject's indications could fall in several different clusters. We think this is because when the subject glanced at the scene, several competing interpretations of the sensory cue set were possible, and which one that the subject chose in any given trial was probabilistic in nature. Indications within ±5 degrees of the scene floor (angle= 0), walls (angle=90), or ceiling (angle=180) surfaces were classified as visually captured responses. Those within ±5 degrees of body axis (indication=angle) were classified as body dominated, presumably due to the effects of idiotropic and gravity receptor bias. In upright one-G testing, this occurs because of the gravitational down cue. Indications that were intermediate or ambiguous, in that they met both visual and body criteria, were classified as "other." A static visual dependence (VD) coefficient was defined as the number of visual responses, minus the number of body axis dominated responses, divided by the total number of all responses. VD=1 indicated strong visually dependent response, while VD=−1 indicated a strong visual independence. Figures 3A–D show the mean VD coefficient for each of the four subjects, by epoch, for all trials.

Subjects B and C were strongly visually independent and remained so throughout the trials. They were not reliably affected by scene or posture manipulations. Subject A was moderately visually independent preflight but did show preflight sensitivity to postural and scene manipulations. This subject's responses shifted dramatically toward strong visual dependence when floating in zero-G. When the constant force spring restraint harness was worn, responses became visually independent. After return to Earth, responses eventually returned to near-preflight levels. However, we noted some postflight carryover of visual dependence during tests on the first two postflight days, suggesting that the earlier inflight visual dependence was not the instantaneous result of the absence of gravity receptor and fluid shift effects. Subject D was moderately visually dependent preflight, and gradually became more so. None of the four subjects indicated or reported inversion illusions during our FD3 testing, though Subject A did experience an inversion illusion on FD1, and also on FD2 while performing another experiment in darkness.

Figure 1A. Tilted room experiment, weakly polarized scene.

Figure 1B. Tilted room experiment, strongly polarized scene.

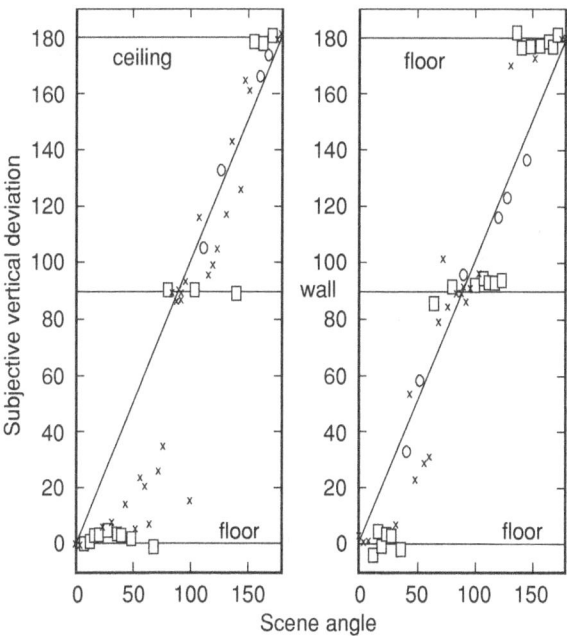

Figure 2. Scene tilt angle vs. absolute value of subjective vertical deviation from body axis: squares: visual responses; circles: body axis responses; crosses: other.

Overall, our tests confirmed that crewmembers show consistent differences in dependence on static visual cues to the subjective vertical. Clearly the effects of spaceflight on visual dependence varied among crewmembers. At least one of our subjects showed the hypothesized increase in dependence to static visual cues with some postflight carryover, and none of the four subjects exhibited increased visual independence inflight.

Tumbling Room Experiment

The four subjects viewed scenes, which rolled about their visual axis at 18 degrees/second. Scene rotation direction was alternated in successive trials. Three scene types were used: two spacecraft interior scenes (Figures 1A and 1B) and a polka-dotted cylinder interior (Figure 4). The latter was comparable to that used in earlier zero-G vection studies by Young et al. (Young, 1986; Young, 1996), and was expected to produce circular vection but not VRIs. The three scenes were presented in a fixed order that was repeated for a total of 12 trials. Subjects were instructed to indicate the onset of visual reorientation illusions by pushing the joystick trigger at the onset of each VRI; i.e., each time a scene surface changed subjective identity. Trial durations were 80 seconds for the polarized rooms, and 20 seconds for the dotted cylinder. At the end of each trial, subjects used a virtual indicator to report the magnitude of illusory rolling circular vection relative to perceived stimulus motion using a five-level ordinal scale (1 = no self-motion/full scene motion to 5 = full self-motion and no scene motion). Data from all three scenes were analyzed for the relative magnitude of circular vection experienced.

Data trials using the two spacecraft scenes were analyzed for the frequency of VRIs and the phase of their onset.

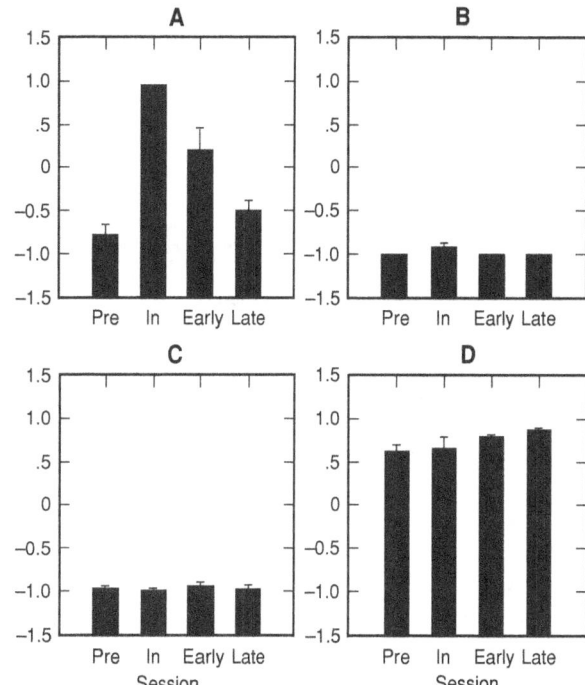

Figures 3A-3D. Static visual dependence (VD) coefficient vs. epoch for Subjects A–D. VD=1 indicates strong visual dependence, VD=−1 indicates strong visual independence. Error bars show ±1 standard error.

Average roll circular vection magnitude of the four subjects is shown in Figures 5A–5D. Preflight roll vection was only moderate for three of the subjects, perhaps due to the relatively small visual angle subtended by the display. Subject D (who was the most visually dependent subject in the tilting room tests) had the strongest circular vection in preflight tests. Roll vection increased dramatically when free floating inflight as compared to preflight values for most of the crew. The apparent lack of effect for Subject D was probably because his preflight reports

Figure 4. Tumbling room experiment. Dotted cylinder scene.

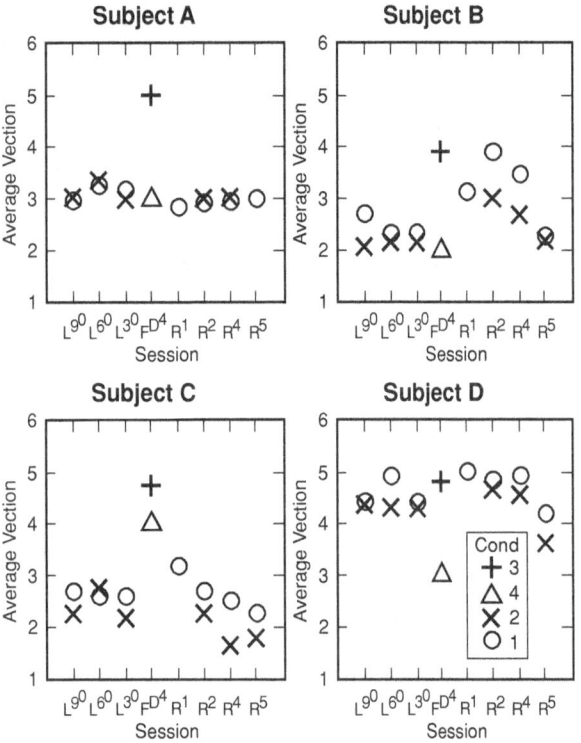

Figures 5A-5D. Tumbling room experiment average circular vection magnitude.

were also at a high level. Postflight vection returned preflight levels. Analysis of variance (ANOVA) demonstrated that vection reports were significantly ($p<0.025$) higher with the dotted cylinder stimulus ($F(1,3)=30.8$) than with the room scenes. Wearing the spring-loaded restraint significantly ($p<0.025$) decreased inflight vection ($F(1,3)=18.3$) as compared to the floating condition. Changing from the upright to the supine posture in one-G testing did not produce a reliable effect. Overall, results indicate an increased reliance on visual cues and a strong effect of position sense cues in flight in three of the four subjects.

Subject A almost always experienced complete tumbling under all conditions both in one-G and zero-G, so his data could not be analyzed for VRI effects. Changes in the clustering of VRI phase angles from Subjects B, C, and D were examined using a two-step procedure. Data from each trial were first screened for significant 90-degree and 180-degree tendencies by multiplying the phase angle data by two and four, and testing the resulting circular distributions for significant ($p<0.05$) nonuniformity using a Rayleigh test. Changes in the percentage of 90-degree clustered data vs. 180-degree clustered data were then examined using a Kruskal-Wallis test. Significant 90- and 180-degree tendencies were consistently found. The effect of spaceflight (epoch) on VRI modal angle approached significance ($U=14$; $p<0.058$). Individual differences may have masked effects. Ninety-degree tendency was greater inflight and early postflight, particularly for B and D. Tilted room VD for the individual subjects did not predict which tumbling room subjects would experience full tumbling or VRIs using identical scenes, which emphasizes the importance of scene motion cues in VRIs.

We were unable to demonstrate reliable effects of epoch, condition (upright/supine or floating/restrained), or scene type on VRI frequency for the remaining subjects (B, C, and D) as a group using Kruskal-Wallace, nonparametric rank ANOVA tests. However, effects may have been masked by the small group's heterogeneous response. Looking at the subjects individually, Subject C frequently reported VRIs in one-G but had very few in zero-G. Subjects B and D reported VRIs, but we could find no clear effects of epoch on VRI frequency.

Looming Linear Vection Experiment

Looming linear vection is the illusion of self-motion induced when the surrounding visual scene translates towards the viewer. The goal of this experiment was to determine whether subjects become more susceptible to looming linear vection in orbit. Each subject viewed a virtual corridor (Figure 6) that was 1.5 m wide and 3.0 m high from an eye height of 1.5 m. Background scene motion is more effective for eliciting motion illusions, so a black frame that did not move relative to the subject was provided in the foreground. In each trial, the corridor moved towards the subject at a constant speed for 10 seconds. Five different scene speeds (0.4, 0.6, 0.8, 1.1, and 1.6 m/second) were tested in randomized order, 12 repetitions each for 60 trials. At the start of each trial when the visual scene began to move, visual motion cues momentarily conflicted with gravity receptor cues, which indicated no physical acceleration. This cue conflict is believed to delay vection onset. We hypothesized that if subjects learned to respond more to visual than to gravity receptor cues in adapting to weightlessness, the latency of vection onset should be reduced. During the remainder of the constant velocity scene motion, no further cue conflict would be expected. Subjects may also become accustomed to constant velocity motion without physical effort in zero-G. If increased weight was given to visual cues in weightlessness, the vection sensation should seem more compelling and therefore be reported as a greater percentage of scene speed, thereby showing fewer spontaneous interruptions of vection (referred to as "dropouts"). We expected that using the constant force spring to provide a strong body axis force cue that firmly anchors the subject to the deck would inhibit vection, because such cues are entirely absent in weightlessness. Subjects were tested in both the upright and supine positions during pre- and postflight sessions, since cues from position sense and the relative orientation of gravity differed in the two positions.

Subjects were instructed to deflect the joystick in proportion to their perceived speed of self-motion. Immediately before starting the trials in each session, subjects practiced deflecting the joystick to specific numeric values, initially with feedback and then without feedback. Analysis revealed no evidence of consistent nonlinearities in joystick setting performance. Subjects were tested pre- and postflight in both upright and supine conditions. Inflight, subjects were tested on the fourth day of flight in both floating and spring-restrained conditions. Subject B got relatively little vection in preflight testing or when tested on FD4, but when retested on FD16 the subject reported strong vection. Subject D reported scene motion rather than self-motion in most preflight trials, so his data for this test were set aside.

Figure 6. Looming linear vection scene. Background scene moved toward subject. Black frame in foreground was stationary relative to the subject.

Typically (in coarse approximation), each subject would begin to deflect the joystick after a latency of about one second. The deflection would then increase to a maximum level that was often maintained for the rest of the trial, but was sometimes punctuated by vection dropouts. Dependent measures analyzed included latency to onset of joystick deflection (seconds), peak joystick deflection (percent), and the time integral of joystick deflection during the trial (seconds).

As we had hypothesized, inflight free-floating vection latency was shorter than in preflight testing on the ground. Latencies at the lowest speeds tended to be the most variable. Pairwise comparisons were significant (Mann Whitney test, $p<0.05$) at all speeds for Subject A, at the two highest speeds for Subject B, and at the highest speed for Subject C's preflight upright data. Latency to vection onset decreased monotonically with scene speed (across all subjects, and for all epochs and conditions; Page test, $p<0.05$). In our preflight tests, latencies did not differ consistently between the upright and the supine positions. Previous one-G studies on linear vection have collectively not shown consistent effect of upright vs. supine posture (Kano, 1991; Tovee, 1999).

Integrated joystick deflection and maximum joystick deflection both increased monotonically with scene speed under all conditions, but not in a linear way. Both responses were consistently reduced at the higher scene speeds. Though this could be a perceptual phenomenon, it could also be due to a speed-dependent change in joystick deflection strategy. Subjects told us that when indicating self-motion during a trial, they often tended to deflect the joystick in proportion to their vection as a percentage of the stimulus speed rather than using a consistent modulus across all speeds. We normalized peak joystick deflection values (Y) across scene speeds by dividing the values by the function $(1-\exp(-v/V))$, where v was scene speed, and V was a constant parameter for each subject. V was taken as the median of all values of $-v/[\log(1-Y)]$ calculated for each preflight trial. V ranged from 1.9 m/second to 3.4 m/second. Since all scene speeds were less than V and all trials were 10 seconds in duration, integrated joystick deflection divided by $10*(1-\exp(-v/V))$ is a measure of average linear vection speed during each trial, normalized to that subject's preflight, low scene speed response. We refer to this metric as "normalized velocity."

Normalized velocity for Subjects A, B, and C is shown in Figure 7. As we had anticipated, inflight floating normalized velocity was generally greater in flight free floating than in preflight erect or supine. Pairwise comparisons of normalized velocity (or the integrated area measure) were significant (Mann Whitney test, $p<0.05$) at all but the highest speed for Subject A, at the highest speed for Subject B, and at all speeds for Subject C. Inflight floating was significantly greater than inflight restrained at four out of five speeds for Subject A, and all speeds for Subject C. Both sets of results confirmed our hypotheses. The positive $p<0.05$ finding in even two out of three (independent) subjects is significant at the $p<0.00725$ level for a family of three subjects. Pairwise comparisons of preflight vs. early postflight showed significant effects at the highest speed for Subjects A and B, and the lowest speed for Subject C. Preflight erect vs. preflight supine, and preflight vs. late postflight showed no reliable effects. Subjects B and C tended to show larger responses at lower speeds in flight and postflight.

The quantitative analysis results are consistent with the subject's inflight and postflight oral debriefing reports.

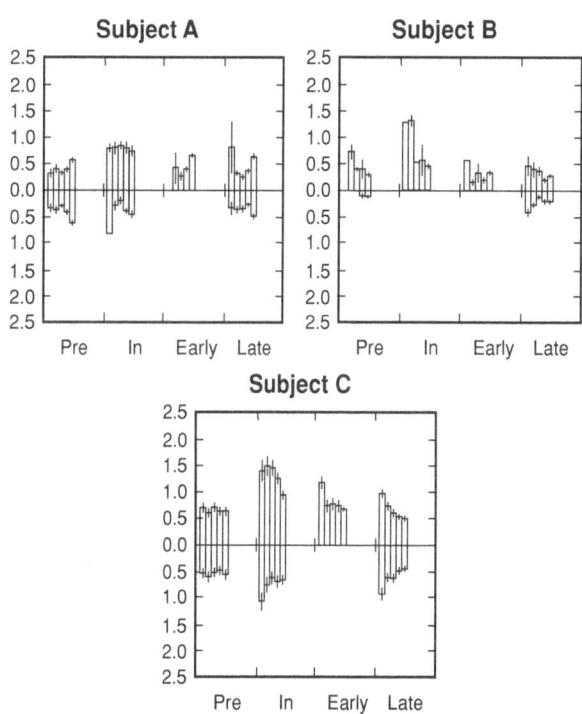

Figure 7. Normalized velocity response for Subjects A, B, and C. Data bars show mean values, error bars represent standard error of the mean. Data grouped by scene velocity for each of the four epochs (preflight, inflight, early postflight, late postflight). Erect and floating data are plotted upwards; supine and restrained data are plotted downwards. Subject A and B data from FD4 session, Subject C data from FD16.

On FD4, Subject A said: " In the floating scenario, I got vection virtually instantaneously ... In the restrained position ... I felt that it was pretty similar to doing it on Earth. In fact my ability to get vection was probably less than it was in the one-G environment. It's kind of like standing with your legs in a cement boot and you get the feeling that you're supposed to be moving but you're not really feeling any vection...." In a landing day debriefing, he added: "To me there was a striking difference doing the (floating) vection experiment in orbit as compared to the one-G controls prior to flight. Inflight, on FD4, I got vection virtually instantaneously at all speeds. It was strikingly impressive. It was a very cool sensation—felt like a ghost flying down a corridor—like in *Ghostbusters*—you even vect with low velocities.... Back here on Earth the tactile cues feel like being ... in a two-G pullup with a heavy weight on my head. This is a very strong cue saying you are in a stationary mode. Today (back on Earth) I didn't get much in the way of compelling vection unless things are flying by. I felt 40% vection saturation today, 60–80% preflight, and 100% saturated vection in flight."

On FD16, Subject B reported "linear vection much stronger than on the ground.... When floating I felt as if I would move forward into the rack." On the day of return, he added: "Normally I don't vectate much, but I had considerably more vection by the end of the flight."

In a landing day debriefing, Subject C added: "Vection sitting upright today very different than inflight—it is like the preflight trials—not nearly as pure vection as inflight. The slow-speed vection was really dramatic inflight. It was fully saturated vection inflight. The slow-speed vection stimulus was basically the speed you move in the Spacelab—very much in place with what we were doing at the time. The tactile cue and the weight of hand may be inhibiting my vection today. The fast speeds today, I notice that I have an initial lurch forward and then settle down into a speed. Today I felt the slow speeds are in the 80–100% range and for the faster speeds in the 20–40% range.... Some of the slow ones were a little stronger than preflight. Slow speeds today felt more saturated."

Shape-from-Shading Experiment

The goal of this experiment was to see whether the illusion of three-dimensional shape produced by two-dimensional shading depends on the direction of the perceived vertical, even in the absence of gravity. For hundreds of years, artists have used shading to create an illusion of concavity or convexity on a flat surface. The illusion presumably occurs because a real protuberance illuminated from one side produces the same retinal image as an indentation that is illuminated from the other side. Light normally comes from above, making the upper part of truly convex surfaces and the lower part of concave surfaces bright. If you position a truly flat gradient-shaded disc (Figure 8A) so that it is relative to your head, the light part of the disc is "above" the dark part and the disc seems convex. Rotating the disc 180 degrees makes the disc appear concave. The dominant factor in the shape illusion is clearly the orientation of the shading gradient with respect to your head. If the light and dark portions of the disk are oriented to your left and right, the disc appears flat (Figure 8B).

Figure 8A. Shape-from-shading experiment. Upright scene orientation. Shading gradient aligned with viewer body axis. Upper disc appears concave, and lower disc appears convex. Turn the page upside down. Does the illusion reverse?

However in this "neutral" orientation of the disc relative to the head, gravity has been shown to play a role. If the head and disc are then both tilted together, so that the light part becomes gravitationally above, the disc will again appear convex. Does this effect remain in weightlessness when the head is tilted, or if the subject does not physically move but simply cognitively initiates a VRI?

The experiment was conducted using the same spacecraft module scene as in the tilted room experiment (Figure 1A). Stimuli were pairs of gradient-shaded discs shown in various orientations, rendered on a 1.5-m-diameter gray circular easel located in the middle of the module, 3.0 m away from the subject (Figures 8A, 8B). The gradient-shaded discs, each subtending 20 degrees of visual angle, appeared alternately to the left and right or above and below each other. A forced-choice procedure was used: the subject decided which disc of the pair appeared more convex and moved an indicator square over the corresponding disc using the joystick. Each trial required about three seconds.

The experiment was conducted under three successive conditions: "upright," "left-side-down," and "VRI." The first condition served as the control for manipulations of the SV in the latter two conditions. In the upright condition, the subject's body axis was parallel to the walls of the virtual environment. In one-G tests, the subject sat upright in the laboratory. In zero-G tests, the subject floated upright in the Neurolab module. In the left-side-down condition, the virtual environment was rotated 90 degrees clockwise, so that the subject's body axis was parallel to the floor and ceiling of the virtual environment (Figure 8B). In one-G tests, the subject lay left shoulder down on a gurney bed. In zero-G tests, the subject floated parallel to the deck of the Neurolab module. In the VRI condition, the visual background environment was again upright with respect to the subject (Figure 8C), but the subject was instructed to cognitively initiate a VRI so that the wall to his left seemed like a floor and he felt

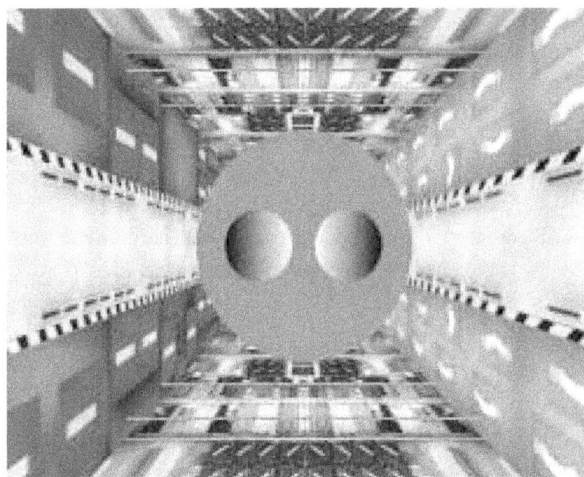

Figure 8B. Shape-from-shading experiment. Left-side-down background scene orientation. Shading gradient in neutral orientation as in Figure 8B. The convexity illusion is usually absent until you rotate your head and the page together, left shoulder down. Does the illusion of convexity reappear?

Figure 8C. Shape-from-shading experiment. Upright scene orientation. Shading gradient in neutral orientation, perpendicular to subject. In the VRI condition, when the subject simply decided the left wall was down, the illusion of convexity often reappeared—without any actual head tilt.

in a left-side-down condition. In one-G tests, the subject lay supine on a gurney bed; and in zero-G, the subject floated upright in the Neurolab module. At the beginning, middle, and end of the trials in each condition, the subject confirmed the direction of the subjective vertical using a green ball pointer, as in the tilted room experiment.

Forty-eight trials were conducted under each condition. Different sequences were used; but in each condition, 12 disc pairs had their shading gradients in the neutral direction perpendicular to the body axis. Our hypothesis was simply that if the subject changed the direction of his SV by changing from the upright to either the left-side-down or VRI condition, the percentage of convex discs oriented with the light side on the right should increase markedly, since in the latter two conditions up was to the subject's right. Twelve other disc pairs had shading parallel to the body axis. Responses to these allowed us to verify that the illusion was present in the upright condition, even in zero-G. Sixteen other pairs had one disc aligned with the body and one perpendicular to the body. These served as distracters and allowed us also to examine responses to conflicting stimuli.

With the shading axes perpendicular to the body, the percentage of light left responses for the group of four subjects increased in both the left-side-down and VRI (imaginary left side down while supine) both preflight, inflight, and postflight confirming that the cognitive reference frame, rather than gravity, contributes to the illusion of convexity when the shading gradient is perpendicular to the body (head) axis. Looking at subject responses individually, Subject B showed a strong light right bias over light left in the upright position, which biased group statistics. When the SV was manipulated to the left-side-down condition, light right responses increased relative to upright for Subjects A, B, and C preflight and inflight as hypothesized. In the VRI condition, light right responses increased for Subject A for all epochs, and for Subjects B and

C preflight but not inflight. One possible explanation is that subjects sometimes were not always able to initiate the required VRI. Green ball subjective vertical indications in flight were not always in the expected direction inflight. Subject D reported he imagined the floor to his right rather than to his left. His light right responses to neutral stimuli consistently decreased for both manipulations, as one might expect. Subject A commented that VRIs were harder to initiate early postflight than preflight, because the direction of true gravitational down felt unusually strong; but Subject C found the reverse.

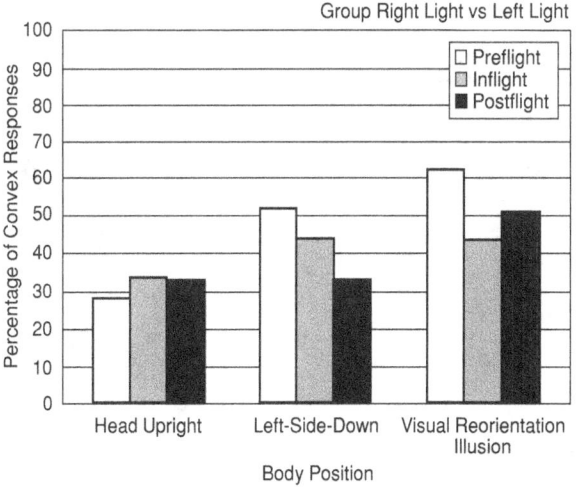

Figure 9. Shape-from-shading results. Percent of light right stimuli perceived as convex when paired with light left for Subjects A-D. In preflight testing (white bars), group responses showed an increase in light right responses, compared to light left, for both left-side-down and VRI conditions, as anticipated.

With the disc-shading axes parallel to the body, all four subjects reliably (95–100%) chose the light-on-top shading pattern in the upright condition regardless of gravity level, demonstrating that the shape-from-shading illusion persists even in zero-G. They also reliably chose the neutral (light-left) stimulus over convex (light-bottom) in all conditions.

Complex Figure Recognition Experiment

The goal of this experiment was to show that a subject's ability to recognize a previously memorized complex figure in weightlessness depends on the orientation of the test figure to the perceived vertical, even in the absence of gravity. The experimental paradigm was analogous to the shape-from-shading experiment previously described in that subjects learned sets of figures in an "upright" condition, and were then asked to recognize the sets after the direction of the SV was manipulated to left-shoulder-down and VRI conditions. As in the previous experiment, the subject indicated the direction of the subjective vertical using green ball pointers at the beginning, in the middle, and at the end of the trial sets under each condition.

Figures used in the experiment were black line drawings, subtending approximately 20 degrees of visual angle. Examples are shown in Figures 10A–D. Figures were drawn with either straight or curved lines that were either open or closed. The number of sides, branches, or loops varied. Different figure sets were used for each subject and session, with equal figure-type representation.

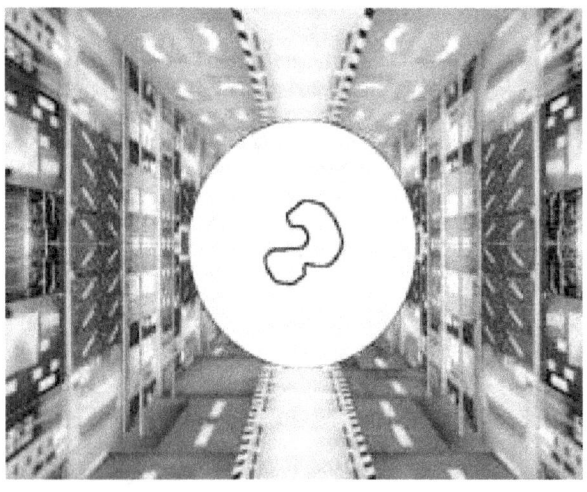

Figure 10A. Complex figure recognition experiment. Training figure example (closed curved type). Upright orientation. Subject was sequentially shown four such figures for five seconds each.

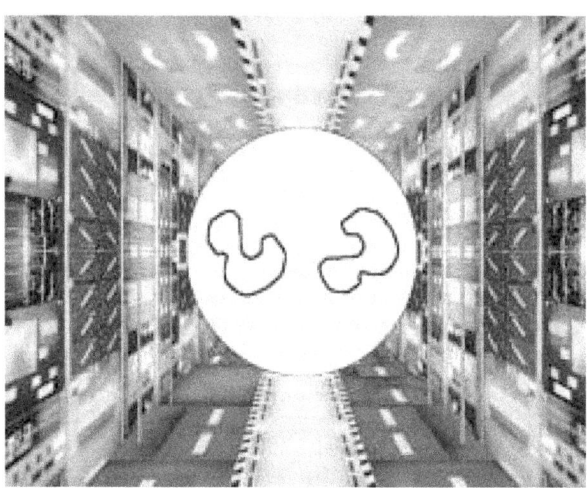

Figure 10B. Complex figure recognition experiment. Trial figure example. Upright orientation. One figure has been rotated 90 degrees clockwise. Shown for 0.5 second, then each figure was replaced with a black dot.

Figure 10C. Complex figure recognition experiment. Upright or VRI condition. After viewing each trial pair (e.g., Figure 10B), subject indicated which figure seemed most familiar by moving the indicator box with the joystick.

Figure 10D. Complex figure recognition experiment. Left-side-down condition. Open curved trial figure example.

The experiment was conducted using the same spacecraft module scene as in the shape-from-shading experiment, except that the circular easel provided a white rather than a gray background. In the "upright" condition, a series of four training figures (Figure 10A) was first shown for five seconds each. The subject was instructed to memorize these four shapes, without giving them names. Next, the subject fixated on a central cross, and a pair of test shapes (Figure 10B) appeared to the left and right or above and below the cross. Each test pair consisted of two versions of the training figure, one in the original orientation and the other rotated 90 degrees clockwise. After 0.5 second, the figures were replaced with marker dots. As in the shape-from-shading experiment, a forced-choice procedure was used. The subject had to indicate which figure looked "most like" one of the training figures using the joystick indicator (Figure 10C). Each trial required about three seconds. The subject completed 48 trials.

Next, the subject was again given a new set of four training figures to memorize for five seconds each. Then the subject physically turned into the same left-side-down position as was used in the shading experiment, and was repeatedly tested using a second set of 48 pairs of test shapes (Figure 10D). The visual environment in this orientation was thus rotated 90 degrees from the upright condition. The figure pairs were oriented so that one was in the same orientation with respect to the visual environment as when memorized (but therefore rotated 90 degrees clockwise with respect to the subject), and the other is in the same orientation with respect to the subject as when memorized (but therefore rotated 90 degrees counterclockwise with respect to the subject).

Finally, the subject and the visual scene were rotated back to the upright position, and the subject was given a third set of training figures to memorize. Then—without physical movement or a change in the visual scene orientation—the subject was instructed to initiate a VRI so that the left wall seemed like a floor, and he felt in a left-shoulder-down condition. The subject was again tested using a third set of 48 pairs of test shapes, one in the same orientation as during the memorization step and the other rotated 90 degrees counterclockwise with respect to the subject. Our working hypothesis was that in both one-G and zero-G, subjects recognize complex figures by recalling their features relative to the perceived "top" and "bottom" of the scene. We expected that the percentage of test figures that were presented upright with respect to the head in each condition would decrease in the left-shoulder-down and VRI conditions.

When tested on FD3, the percentage of body-upright test figures recognized decreased as expected from 78% (the control condition) to 55% in the left-side-down condition to 68% in the VRI condition. However, looking at the subjects individually, only Subject C's data (Figure 11) truly followed the hypothesized pattern both on the ground and inflight, with a general trend to prefer body axis presentations. Unlike most other subjects in similar studies, Subject A consistently chose the body axis figure under all conditions in one-G, and there was no clear effect of the SV manipulation in any condition. Subjects B and D showed no effect of the SV manipulations in

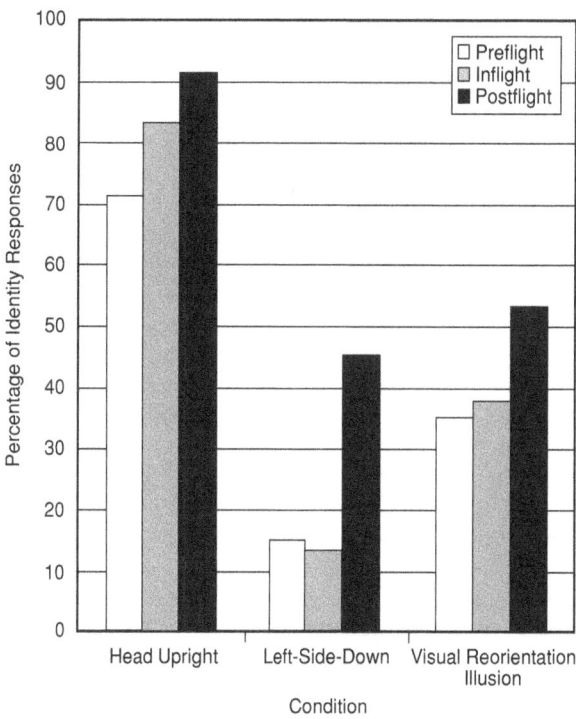

Figure 11. Complex figure recognition. Percent of previously memorized figures recognized by Subject C when presented upright with respect to head, by condition, and by epoch. A decrease was expected in left-side-down and VRI conditions relative to Head Upright, both in one-G and zero-G.

ground testing. We cannot generalize from these data, except to say that the expected effect of SV manipulation was clearly present in one subject preflight, inflight, and postflight.

DISCUSSION

With a very limited number of subjects, of nearly the same age and gender, usually tested only once in orbital flight, and always under conditions of time pressure and fatigue, we must be cautious in generalizing. However, our principal conclusions are:

Crewmembers differed in the extent to which their judgment of the subjective vertical was influenced by the orientation of motionless (static) surrounding surfaces and objects. Some subjects consistently reported subjective down as beneath their feet, while others reported down in a direction consistent with the surrounding visual environment. Subjects who were strongly statically visually dependent or independent on the ground remained so inflight. One moderately visually independent subject became more visually dependent during flight, and then returned to his preflight characteristic. None of our subjects became more statically visually independent as a result of spaceflight. The variety of responses among crewmembers suggests that there may be several ways to perceptually adapt to weightlessness.

In tests with tilted and tumbling scenes, most subjects experienced visual reorientation illusions so that the floor was the surface beneath the crewmember's feet. With tumbling scenes, the angles at which VRI onset occurred tended to be 90 degrees or 180 degrees apart. Individual responses to motionless tilted scenes did not predict tumbling responses, which suggests that a scene's angular motion cues play an important role.

The magnitude of angular self-motion illusions produced by a rolling visual scene increased in zero-G for most subjects. One crewmember recalled occasionally making reaching direction errors during the first few days of flight because of VRIs.

The preponderance of our data on linear self-motion illusion (linear vection) produced by a scene translating toward the viewer shows small but statistically significant decreases in latency, and an increase in average vection magnitude consistent with the hypothesis that visual cues to linear motion become more compelling in weightlessness. This may account for anecdotal astronaut reports that spacecraft orbital motion relative to the Earth seems surprisingly vivid. Crews should be aware that moving visual images would produce motion sensations less effectively in ground simulators than in actual spaceflight, due to the vection-inhibiting effect of gravity. Because we had no opportunity to test most of our subjects more than once in flight, we can offer no firm conclusions as to how rapidly vection susceptibility increases upon reaching orbit. That Subject B had little vection on FD4 but was susceptible when retested on FD16 suggests that the effect may not necessarily be immediate, however.

Downward restraint reduces susceptibility to visually induced motion illusion, probably because it provides an unambiguous cue that the subject is not moving. Downward force cues are also reportedly effective in suppressing symptoms of space motion sickness (Oman, 1984).

Changing the relative direction of the SV in weightlessness either by body rotation or by simply cognitively initiating a visual reorientation illusion altered the illusion of convexity produced when viewing a flat, shaded disc for several subjects. It also changed at least one subject's ability to recognize previously memorized two-dimensional figures. This supports our hypothesis that there is a correlation between the surface the subject cognitively perceives as down and precognitive figure recognition and shading interpretation.

To measure these changes, we used the VEG head-mounted display to provide controlled, repeatable visual stimuli in both ground and flight testing since there was no other practical alternative—particularly for inflight testing. To what extent might our results have been affected by using this system? The VEG lacked a head tracker, which may have compromised the subject's sense of immersion. Although we used photorealistic textures, the spacecraft interior scenes had a cartoonish quality. The head-mounted display introduced some pixelization and visual distortion, and it constrained the subjects' field-of-view in a way that may have limited the effectiveness of the visual flow cues in our vection experiments. On the other hand, our subjects were physically motionless during all trials, so the lack of head tracking during the trials

themselves was arguably not important. Scenes were rendered in color stereo, which made pixelization less apparent and provided foreground/background motion cueing for the vection experiments. Most of the perceptual effects we were studying are mediated by peripheral vision, or other systems where visual acuity probably plays only a minor role.

Several of our subjects experienced symptoms of space motion sickness during their first days in orbit. Because of the limited number of subjects available and the uncontrolled nature of their daily activities, we made no attempt to correlate individual crewmember results with space motion sickness susceptibility under operational conditions. Although our experiments were deliberately designed to change the direction of the subjective vertical or the sense of self-motion sensation, our subjects described themselves as less susceptible than average to motion sickness, and none of our subjects reported symptoms during preflight or inflight testing. On the day of return to Earth, one subject had re-adaptation sickness, which was slightly exacerbated during the linear vection experiment.

Despite these limitations, results show that most astronauts become more dependent on dynamic visual motion cues and some also become more responsive to static orientation cues. The direction of the SV is labile, and shifts can cause disorientation and influence recognition of objects as well as interpretation of shading and shadows, and hence impact astronaut performance.

The scientific results and methods developed for this Neurolab experiment are currently being employed in the National Space Biomedical Research Institute to better understand spatial disorientation in orbit, and to develop countermeasures. A follow-on experiment to study the effects of long-duration flight aboard the International Space Station is in development. Our results broaden the understanding of how elderly people and vestibular patients who have altered inner ear balance function rely on visual and position sense cues to determine the direction of the subjective vertical, and why they find certain situations in daily life disorienting.

Acknowledgements

We thank our astronaut subjects, the VEG engineering team, the Neurolab mission and E136 experiment support cadre, our colleagues at MIT and York, and of course our families, without whose support this experiment would not have been successful—or even possible. Special thanks to: A. Al-Hajas, C. Amberboy, R. Allison, H. Anderson, F. Booker, L. Braithwaite, J. Buckey, M. Buderer, J. Cheng, G. Dalrymple, J. DeSouza, D. Grounds, H. Guiterrez, W. Hutchison, J. Homick, S. Johnston, J. Krug, K. Lawrence, A. Lee, A. Leyman, G. Lei, L. Li, G. Lutz, D. Merrick, T. Mills, A. Mortimer, K. Nguyen, M. Pickering, H. Rahman, J. Rummel, S. Sawyer, N. Skinner, A. Skwersky, C. Tovee, and J. van Twest. Supported by NASA Contract NAS9-19536 and Canadian Space Agency Contract 9F007-5-8515.

REFERENCES

THE EFFECT OF RETINAL AND PHENOMENAL ORIENTATION ON THE PERCEPTION OF FORM. I. Rock and W. Heimer. *Am. J. Psychol.. Vol.* 70(4), page 493; 1957.

A NEW SOLUTION TO THE PROBLEM OF SUBJECTIVE VERTICAL. H. Mittelstaedt. *Naturwissenschaften,* Vol. 70, pages 272–281; 1983.

SYMPTOMS AND SIGNS OF SPACE MOTION SICKNESS ON SPACELAB-1. C. M. Oman, B. K. Lichtenberg and K. E. Money. NATO-AGARD Aerospace Medical Panel Symposium on Motion Sickness: Mechanisms, Prediction, Prevention and Treatment, Williamsburg, Va, NATO AGARD, pages 35/1–35/21; 1984.

MIT/CANADIAN VESTIBULAR EXPERIMENTS ON THE SPACELAB-1 MISSION: 4. SPACE MOTION SICKNESS: SYMPTOMS, STIMULI, AND PREDICTABILITY. C. M. Oman, B. K. Lichtenberg, K. E. Money and R. K. McCoy. *Exp. Brain Res.*, Vol. 64, pages 316–334; 1986.

MIT/CANADIAN VESTIBULAR EXPERIMENTS ON THE SPACELAB-1 MISSION: 2. VISUAL VESTIBULAR TILT INTERACTION IN WEIGHTLESSNESS. L. R. Young, M. Shelhamer and S. Modestino. *Exp. Brain Res.*, Vol. 64, pages 299–307; 1986.

SHAPE FROM SHADING IN A DIFFERENT FRAMES OF REFERENCE. I. P. Howard, S. S. Bergstrom and M. Ohmi. *Perception,* Vol. 19, pages 523–530; 1990.

THE PERCEPTION OF SELF-MOTION INDUCED BY PERIPHERAL VISUAL INFORMATION IN SITTING AND SUPINE POSTURES. C. Kano. *Ecol. Psychol.,* Vol. (3), pages 241–252; 1991.

SOMATIC VERSUS VESTIBULAR GRAVITY RECEPTION IN MAN. H. Mittelstaedt. Sensing and Controlling Motion: Vestibular and Sensorimotor Function. B. Cohen, D. L. Tomko and F. E. Guedry, Eds. New York, *Ann. N.Y. Acad. Sci.*, pages 124–139; 1992.

THE CONTRIBUTION OF MOTION, THE VISUAL FRAME, AND VISUAL POLARITY TO SENSATIONS OF BODY TILT. I. P. Howard and L. Childerson. *Perception*, Vol. 23, pages 753–762; 1994.

TACTILE INFLUENCES ON ASTRONAUT VISUAL SPATIAL ORIENTATION: HUMAN NEUROVESTIBULAR EXPERIMENTS ON SPACELAB LIFE SCIENCES–2. L. R. Young and Mendoza. *J. Appl. Physiol.,* 1996.

ADAPTATION TO A LINEAR VECTION STIMULUS IN A VIRTUAL REALITY ENVIRONMENT. C. A. Tovee. Cambridge, MA, Mass. Inst. of Tech. *Aeronaut. Astronaut.,* page 113; 1999.

VISUAL-VESTIBULAR INTEGRATION MOTION PERCEPTION REPORTING. D. L. Harm, M. R. Reschke and D. E. Parker. In: C.F. Sawin, G.R. Tayler, W. L. Smith Eds. *Extended Duration Orbiter Medical Project,* NASA SP-534. NASA Lyndon B. Johnson Space Center, Houston, TX, pages 5.2-1–5.2-12, 1999.

HUMAN VISUAL ORIENTATION IN WEIGHTLESSNESS. C. M. Oman. In: *Levels of Perception.* M. Jenkin and L. R. Harris, Eds, Springer Verlag., pages 375–398, 2003.

Visual-motor Coordination During Spaceflight

Authors
Otmar Bock, Barry Fowler,
Deanna Comfort, Susanne Jüngling

Experiment Team

Principal Investigator: **Otmar Bock**[1]

Co-Investigator: **Barry Fowler**[2]

Co-Authors: **Susanne Jüngling,**[1]
Deanne Comfort[2]

[1]German Sport University, Köln, Germany
[2]York University, Toronto, Canada

ABSTRACT

We applied a battery of tests to study the motor skills of astronauts. An intriguing pattern of findings was revealed. In weightlessness, subjects slowed down when performing a pointing and a grasping task, but not when executing a tracking and a reaction-time task. The accuracy of their responses was little affected by weightlessness in any of our tasks. To explain this task-specific pattern of changes, we propose a three-component hypothesis, according to which manual dexterity depends on the required speed and accuracy, and also on the amount of processing power allocated by the brain to each given task. Depending on actual or presumed task constraints, subjects can keep accuracy high by reducing speed, with no change in processing cost, or they can keep both accuracy and speed high by allocating more processing power.

INTRODUCTION

The role of humans on space missions

Why do we send humans into space? Why don't we replace them with robots? A main reason is the superior ability of people to observe their environment intelligently, and to turn their observations into purposeful, dexterous actions. These skills will play an ever-increasing role as space missions of longer and longer durations are launched, such as a trip to Mars. For long-term missions, less can be planned and automated ahead of time, and less can be resolved by ground control (due to excessive signal delay times), so more and more will depend on the skills of astronauts aboard the spaceship.

In this context, it could become potentially dangerous if human skills are degraded during spaceflight. A number of anecdotal observations by astronauts and cosmonauts, as well as semi-quantitative crew surveys, indicate that spatial disorientation, visual illusions, clumsiness, and slowing down may be encountered during space travel. It is important for the safety and success of future missions to quantify and interpret this possible decay of visual-motor performance.

What aspects of manual skills are affected at which intervals during a mission—and why? Answering these questions should allow us to develop appropriate countermeasures, such as preflight training, inflight skill exercises, or improved equipment and procedure design that are geared towards the changed skill levels.

Previous findings on visual-motor deficits during spaceflight

Over the last few years, several experiments have provided quantitative evidence for a degradation of visual-motor performance during space missions. Two groups of studies measured the speed and accuracy of pointing movements: When subjects pointed with their arm at stationary visual targets, their responses slowed down in weightlessness while accuracy remained unchanged (Berger, 1997; Sangals, 1999). In other work, subjects had to use a joystick to keep a cursor in the center of a visual display; this task required considerable skill, since the cursor position was inherently unstable (like balancing a pen at the tip of the finger). The authors found that during the mission, centering performance was distinctly lower than preflight (Manzey, 1998; Manzey, 1993; Schifflett, 1995). Other investigators addressed the ability of astronauts to carry out two tasks at the same time by combining the above centering task with a concurrent memory-search task, where subjects had to decide whether a displayed letter was or was not contained in a previously learned list of letters. When the centering and the memory-search tasks were administered together, performance was poorer than during single-task testing. Most importantly, this task interference was substantially more pronounced in weightlessness (Manzey, 1998; Manzey, 1993; Schifflett, 1995). To interpret the latter finding, it was argued that a space mission is a stressful environment, and that stress is known to degrade our ability to spread attention across several tasks.

Possible reasons for performance deficits

Previous studies have proposed several reasons why visual-motor performance could be degraded in space. Firstly, the absence of gravity could result in sensory and motor changes (Bock, 1992). For example, the sensory organs for balance located in the inner ears change their function in weightlessness (see balance section, this volume), which could result in visual illusions and mislocalizations. Furthermore, the limb position sense could be degraded due to a mismatch between muscle length and tension. Finally, automated movements acquired on Earth and played back in space could include a "built-in" compensation for gravitational force, which is no longer adequate.

Besides sensory and motor dysfunctions, an alternative explanation holds that the observed deficits could be symptoms of the adaptation to the space environment (Bock, 1998). According to this view, subjects exposed to weightlessness must learn anew how to interpret sensory signals and how to produce adequate motor commands; this restructuring of the brain's processing patterns could well slow down manual skills or make them more erratic, as it does during motor learning on Earth.

A third explanation, outlined above, attributes the observed deficits not to weightlessness per se, but rather to the presence of various stressors during space missions.

Objectives of present investigation

One problem in the interpretation of previous studies is the diversity of experimental procedures used. Methodological differences could influence the subjects' stress level, their postural stability, and their understanding of the task, thus making comparisons across experiments difficult. One objective of the present work was, therefore, to design and evaluate the utility of a standardized skill-testing device. To this end, we have developed the Visual-motor Coordination Facility (VCF), a compact, versatile, and easy-to-use piece of hardware, which we describe elsewhere in this publication (see technical report by Bock et al. in this publication).

A second objective of the present study was to apply a battery of different visual-motor tests to the same subjects under the same experimental conditions before, during, and after the Neurolab Space Shuttle mission, and compare the outcome across tasks. The rationale for selecting these particular tasks (see below) was that they are well established in the literature and so a wealth of normative data is available. Also, they presumably engage different processing mechanisms in the brain. The pattern of findings yielded with these different tasks might provide a particularly revealing insight into the gravity-dependence of motor control.

Our hypothesis was that different motor tasks are based on different control strategies, which are affected differently by weightlessness. We expected to find a task-specific pattern of inflight deficits, which will tell us more about the underlying mechanisms than the outcome of a single paradigm.

METHODS

Apparatus

We used the VCF to implement a battery of visual-motor coordination tasks. In short, the VCF presents visual targets to which subjects respond with their preferred hand. Viewing the hand while performing the test can be allowed or prevented, depending on the experimental paradigm. Finger position is registered by a video-based technique at a frame rate of 50 Hz, with an accuracy of three mm. The subjects' body is stabilized with respect to the VCF by a head- and chin-rest, a waist belt, footloops, and a handle grasped by the non-preferred hand. The handle has a built-in microswitch, which allows the registration of reaction-time responses by that hand.

The VCF is self-calibrating and uses a series of automatic diagnostic routines, which releases the subjects from operator duties and allows them to focus on the experimental task. Also, the VCF occludes vision of other activities in the spacecraft, which also helps subjects focus on their task.

Testing battery

A number of visual-motor coordination tests were applied in a mixed order within a session time of 30 minutes. Seven sessions were administered preflight, three in flight (flight days 1, 7, 14) and three postflight (postflight days 0, 2, 9). Six crewmembers of the Neurolab mission volunteered to participate. Unfortunately, two of them were not available for data collection during the first inflight session due to critical crew duties. A complete analysis could therefore be only done with the remaining four subjects. However, a look at the recorded data confirmed that all six subjects exhibited similar inflight trends.

In the pointing task, subjects pointed with their index finger from a common starting dot at targets presented six cm to the left, right, above, or below. The target remained on for one second, and the starting dot then reappeared. Each session included 96 pointing trials.

In the grasping task, subjects grasped luminous discs between thumb and index finger. A starting disc of four cm diameter appeared near the left display edge, and was replaced by a target disc of one- or seven-cm diameter located 13.4 cm away from the start in a randomly varying direction; the starting disc reappeared 1.2 seconds later. Fifty-six grasping trials were presented in each session.

In the tracking task, subjects followed with their index finger a target moving at constant speed along a circular path. Nine target frequency (0.50, 0.75, or 1.25 cycles/second) × movement diameter (3, 7.5, or 12 cm) combinations were presented in each session in a mixed order, for 12 to 16 seconds each. Three conditions were employed: tracking hand was visible (VF), not visible (NVF), and not visible with an additional reaction-time task for the other hand (DUAL).

The reaction-time (RT) task was embedded in the tracking task. Subjects grasped a contoured handle with their non-dominant hand and pressed a button on top of the handle with their thumb when the stimulus disc changed into an annulus, consisting of an outer luminous circle and a darker inner core. The annulus remained on the screen for 400 ms, and its onset was controlled by a rectangular distribution of 1300, 1600, and 1900 ms interstimulus intervals. Sixty-seven to 70 RT responses were collected in each session; responses occurring more than 1000 ms after stimulus onset (but before appearance of the next stimulus) were defined as errors. The RT task had three levels of difficulty. The easiest task required RT responses to a stationary stimulus in the display center (RT Alone). In the more demanding task, the stimulus moved in the same way as in the tracking task; subjects had to watch it without tracking, and to produce RT responses to stimulus changes (RT Watching). In the most difficult task, subjects had to track the target with their dominant hand, and produce RT responses with their other hand (RT Tracking).

RESULTS AND DISCUSSION

Pointing task

Figure 1 illustrates the paths and velocity profiles of pointing movements executed before and during the mission. The pre- and inflight paths look similar, but the velocity profiles are distinctly different in weightlessness, with lower peaks and extended durations. The latter observation is confirmed by the data in Figure 2, which show mean values across subjects for all sessions. Movements executed during inflight sessions had a lower peak finger velocity, and a longer duration than pre- and postflight responses. It is of particular interest to note that response speed didn't normalize during the mission; on the contrary, responses slowed down even more as the mission progressed.

For a more detailed analysis, we subdivided each movement into four episodes: movement onset to one-half peak velocity, one-half to full peak velocity, full to one-half peak velocity, and one-half peak velocity to movement end. Only the third and fourth episodes were significantly prolonged in weightlessness. Thus, response slowing was not uniform throughout the movement; rather, it evolved as the finger approached the target.

No effects of gravity on movement amplitude were found, which indicates that weightlessness affects primarily the timing and not the accuracy of pointing movements.

Grasping task

From the recorded data, we calculated the grip aperture as three dimensional distance between thumb and index fingertips. Figure 3 illustrates typical pre- and inflight grip aperture pro-

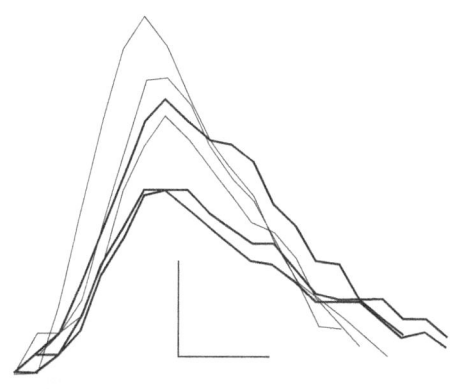

Figure 1. Original pointing data. Left: Fingertip paths to targets at the left, right, top, and bottom; calibration bars represent five cm. Right: Fingertip velocity profiles; calibration bars represent 50 cm/second and 0.1 second. Thin curves were recorded preflight, and bold curves were recorded inflight.

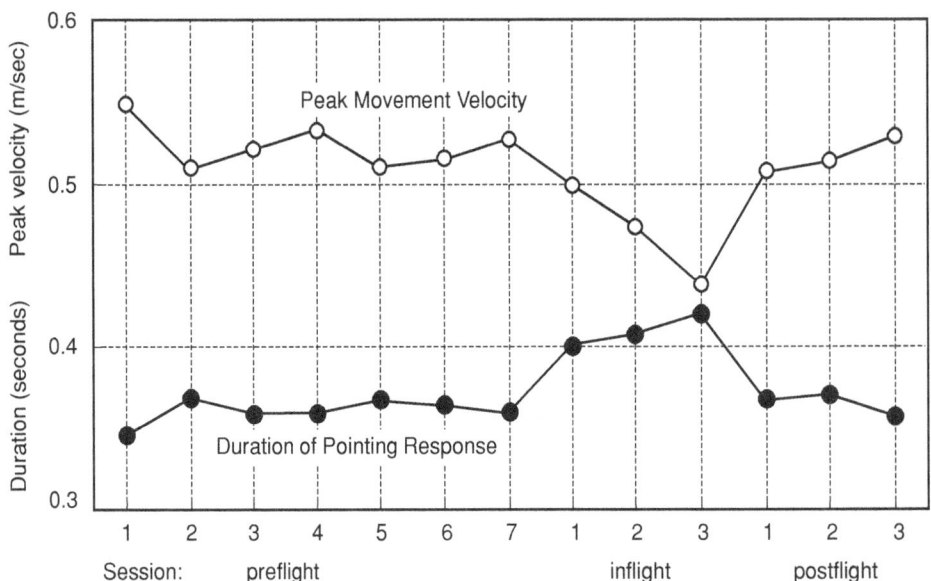

Figure 2. Peak movement velocity and duration of pointing responses in different experimental sessions. Each symbol represents the mean value across four subjects and four target positions. (From Bock, 2001, with permission; reproduced from *Aviation Space and Environmental Medicine*.)

files in response to a large object. We found that the mean duration of those profiles increased from 0.40 second preflight to 0.44 second inflight, and dropped again to 0.41 second postflight. The mean amplitude (i.e., difference between final and initial grip aperture) was not significantly different pre-, in-, and postflight. Thus, our grasping data replicate the main outcome of the pointing task—inflight slowing with no overt change of accuracy.

Tracking task

Figure 4 illustrates original tracking data from a preflight and an inflight session. Even though the target moved along a circle, subjects produced ellipsoid response paths, a phenomenon

that has been described before. We fitted an ellipse to the recorded data, thus yielding the response parameters' size, shape, and inclination with respect to the vertical. Size and shape remained similar in all experimental sessions, but inclination changed. Responses were less inclined during the mission (mean: 6.5 degrees) than they were preflight (mean: 20.4 degrees).

We also evaluated the speed of tracking movements by comparing response frequency (in cycles/second) to target frequency. The two values were remarkably similar in all sessions. As target frequency increased from 0.5 to 0.75 to 1.25 Hz, mean response frequency increased from 0.50 to 0.75 to 1.24 Hz. Even when the inflight sessions were considered alone, response and target frequency differed by less than 1%.

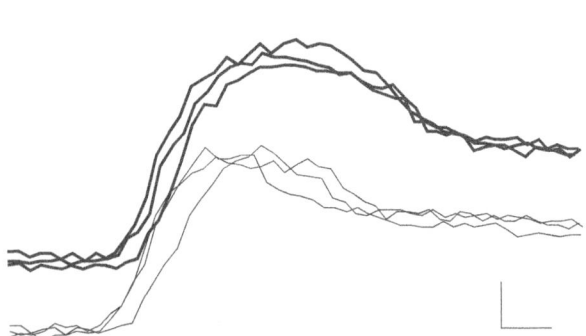

Figure 3. Original grasp aperture profiles of preflight (thin) and inflight (bold) responses. Calibration bars represent one cm and 0.1 second.

Figure 4. Original tracking data with fitted ellipse. Preflight recordings are at the left, and inflight data are at the right. Calibration bars represent five cm. (From Bock, 2001, with permission; reproduced from *Aviation Space and Environmental Medicine*.)

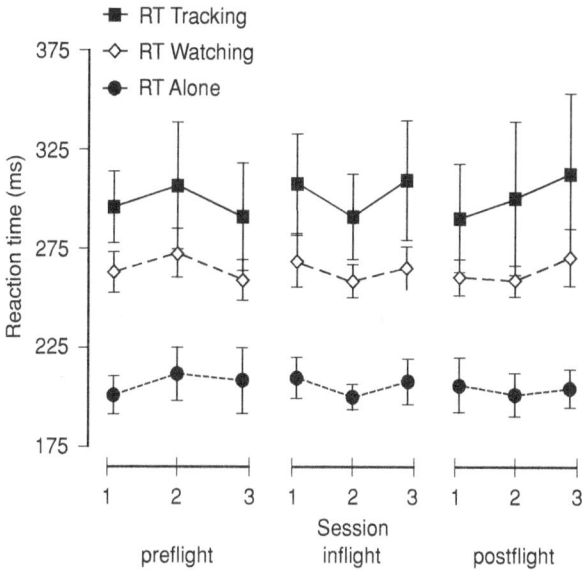

Figure 5. Mean reaction times in the three RT task conditions. No change in reaction time was noted in space. (From Fowler, 2000a, with permission; reproduced from *Human Factors*.)

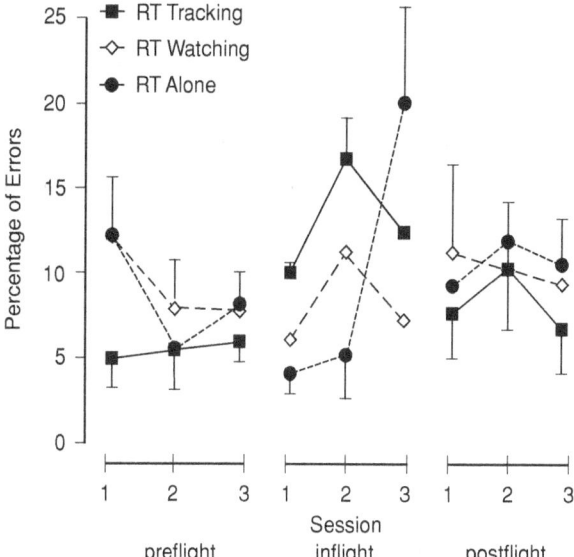

Figure 6. Mean errors in the three RT task conditions. Errors increased in the RT tracking task in space. (From Fowler, 2000a, with permission; reproduced from *Human Factors*.)

Dual task

Our dual-task results have been presented in detail elsewhere (Fowler, 2000a). In summary, we found that the response latency on the secondary RT task didn't increase during the flight (Figure 5), but the percentage of errors became more variable (Figure 6). Two points can be noted about the latter figure. First, an extreme increase in errors for RT alone, which we cannot account for, and second, a general loosening of the error limits, especially for RT tracking. We attribute this loosening to a decrease in strategic control on the part of the subjects.

GENERAL DISCUSSION

Summary of Main Findings

Our pointing findings confirm previous reports on the slowing of pointing responses during spaceflight (Berger, 1997; Sangals, 1999). Also, our grasping data replicate the main outcome of the pointing task—inflight slowing with no overt change of accuracy. As one possible interpretation, it has been suggested that in weightlessness, the motor system depends more heavily on time-consuming visual feedback. We can now discard this view, since we found slowing to occur when the pointing hand is not visible. We can also exclude other explanations discussed in literature, such as fatigue, stress, and the effects of antiemetic drugs: When the same subjects performed other tasks in the same session, only minutes apart from the pointing task, no comparable slowing was observed (see findings on RT and tracking below). The reduction of movement speed in weightlessness therefore appears to be specific for certain tasks. A possible interpretation of our data emerges when considering the progressive nature of response slowing: progressive effects are likely to reflect adaptive rather than disruptive processes in the visual-motor system. We therefore posit that response slowing is a symptom of adaptive restructuring in the brain.

We interpret the inflight change of response inclination on the tracking task as a stronger dependence on an egocentric frame of reference. Principally, astronauts can base their orientation both on an extrinsic (visual) or an egocentric reference frame (Reschke, 1998), but the latter is likely to produce less discomfort in weightlessness (Kornilova, 1997) and may therefore well become the preferred frame during space missions.

The most important finding from the tracking paradigm is the excellent match of response and target frequency across all sessions. Thus, tracking speed was not reduced during spaceflight, a finding that is in sharp contrast to the marked slowing of pointing movements. Since both tasks employed the same subjects in the same apparatus, only minutes apart, the observed discrepancy can be explained only by differences in the nature of the two tasks.

Two hypotheses have been proposed to explain the dual-task deficits that have been observed in previous space experiments: the direct effects of weightlessness on the central nervous system, or multiple stressors in the space environment (Manzey, 1998; Manzey, 1993). Our dual-task results tend to support the latter interpretation. It is important to note, however, that the marginal effects found with our dual-task test suggest that it should be possible to avoid or ameliorate performance deficits in space by paying careful attention to task design and by providing inoculation training against stressors (Fowler, 2000b).

Pattern of main findings

Our battery of visual-motor tests revealed an intriguing pattern of findings. Subjects' responses slowed down in weightlessness in the pointing and grasping task, but not in the tracking and RT task (single or dual). The slowing was not due to an increase of corrections based on visual feedback, since no such feedback was available in our study. Response accuracy was little affected by weightlessness in any of the three tasks where accuracy could be measured; i.e., in the pointing, grasping, and tracking tasks. To explain this task-specific pattern of findings, we posit that some tasks put a stronger emphasis on response speed than others did.

In our pointing and grasping tasks, targets were presented at intervals that were long enough for subjects to complete their responses within a fraction of the allotted time. Subjects therefore didn't start lagging behind the target sequence when moderately slowing down, and thus experienced no negative consequences of reduced speed. In contrast, any response slowing in our tracking task would bring the finger gradually out of phase with the target, which should be noticed as a performance deficit by the subjects. Similarly in the RT task, where response speed is the most relevant measure, any slowing should be considered by the subjects as a performance deficit. In consequence, it is quite conceivable that our tracking and RT tasks contained an implicit instruction to the subjects to keep up their normal working pace, while the pointing and grasping tasks didn't.

A hypothesis on visual-motor performance in space

If we accept that astronauts may keep up their normal working pace, or may slow down, depending on actual or perceived task constraints, a three-component hypothesis on human visual-motor control in space emerges. Manual performance will depend on the trade-off between speed and accuracy (Fitts, 1954), but also on the allocation of neural processing resources to the task (Navon, 1979). When adaptive restructuring during spaceflight makes motor control more difficult, we can keep response accuracy relatively high by reducing speed, with no change in resource investment. Such a strategy could explain our pointing and grasping data. Alternatively, task constraints may be in conflict with speed reduction, in which case we can keep both accuracy and speed high by allocating more resources to the task. This strategy would fit with our tracking and RT data. Further experiments will be needed to substantiate this hypothesis.

The proposed interpretation calls for a stricter control of experimental variables in future studies of visual-motor performance in space. It should not be left up to the subject to decide how much emphasis to place on response speed, accuracy, and the investment of resources. When these three factors are well under the experimenter's control, more consistent and more dramatic space-related deficits should be expected.

It should be noted that our conclusions are not in conflict with the fact that astronauts participating in this (see other papers in the present volume) and previous missions performed delicate and challenging motor acts with a high level of skill. According to our hypothesis, subjects can keep both speed and accuracy high if they allocate more resources to visual-motor processes, as they presumably did in our tracking task.

Acknowledgements

This work was supported by the Canadian and German Space Agencies, CSA and DLR. We would like to thank I. Benick, L. Geisen, J. Kaiser, J. Lipitkas, M. Schulirsch, and N. Wenderoth for their help in software development and data analysis; NASA support staff for their continuous patience and help with even the most unexpected and untimely requests; and, last but not least, the Neurolab crew who did an excellent job in operating the VCF and volunteered their time to serve as subjects. A more detailed account of our findings from the pointing and tracking task has been published in Bock et al. (Bock, 2001).

REFERENCES

THE INFORMATION CAPACITY OF THE HUMAN MOTOR SYSTEM IN CONTROLLING THE AMPLITUDE OF MOVEMENT. P.M. Fitts. *J. Exp. Psychol.,* Vol. 47, pages 381–391; 1954.

ON THE ECONOMY OF THE HUMAN-PROCESSING SYSTEM. D. Navon, D. Gopher. *Psychol. Rev.,* Vol. 86, pages 214–255; 1979.

ACCURACY OF AIMED ARM MOVEMENTS IN CHANGED GRAVITY. O. Bock, I.P Howard, K.E. Money, K.E. Arnold. *Aviat. Space Envir. Med.,* Vol. 63, pages 994–998; 1992.

BEHAVIORAL ASPECTS OF HUMAN ADAPTATION TO SPACE: ANALYSES OF COGNITIVE AND PSYCHOMOTOR PERFORMANCE IN SPACE DURING AN 8-DAY SPACE MISSION. D. Manzey, B. Lorenz, A. Schiewe, G. Finell, G. Thiele. *Clin. Investigator,* Vol. 71, pages 725–731; 1993.

PERFORMANCE ASSESSMENT WORKSTATION (PAWS). S. Schifflett, D. Eddy, R. Schlegel, J. French, D. Eddy. (1995). *Final Science Report*, NASA Marshall Space Flight Center; 1995.

ORIENTATION ILLUSIONS IN SPACEFLIGHT. L.N. Kornilova. *J. Vestib. Res.,* Vol. 7, pages 429–439; 1997.

POINTING ARM MOVEMENTS IN SHORT- AND LONG-TERM SPACEFLIGHTS. M. Berger, S. Mescheriakov, E. Molokanova, S. Lechner-Steinleitner, N. Seguer, I. Kozlovskaya. *Aviat. Space Envir. Med.,* Vol. 68, pages 781–787; 1997.

MENTAL PERFORMANCE IN EXTREME ENVIRONMENTS: RESULTS FROM A PERFORMANCE MONITORING STUDY DURING A 438-DAY SPACEFLIGHT. D. Manzey, B. Lorenz, V. Poljakov. *Ergonomics,* Vol. 41, pages 537–559; 1998.

POSTURE, LOCOMOTION, SPATIAL ORIENTATION, AND MOTION SICKNESS AS A FUNCTION OF SPACE FLIGHT. M.F. Reschke, J.J. Bloomberg, D.L. Harm, W.H. Paloski, C. Layne, V. McDonald. *Brain Res. Rev.,* Vol. 28, pages 102–117; 1998.

PROBLEMS OF SENSORIMOTOR COORDINATION IN WEIGHTLESSNESS. O. Bock. *Brain Res. Rev.,* Vol. 28, pages 155–160; 1998.

CHANGED VISUOMOTOR TRANSFORMATIONS DURING AND AFTER PROLONGED MICROGRAVITY. J. Sangals, H. Heuer, D. Manzey, B. Lorenz. *Exp. Brain Res.*, pages 378–390; 1999.

IS DUAL-TASK PERFORMANCE NECESSARILY IMPAIRED IN SPACE? B. Fowler, O. Bock, D. Comfort. *Human Factors*, pages 318–326; 2000a.

A REVIEW OF COGNITIVE AND PERCEPTUAL-MOTOR PERFORMANCE IN SPACE. B. Fowler, D. Comfort, O. Bock. *Aviat. Space Envir. Med.,* Vol. 7, pages A66–68; 2000b.

HUMAN SENSORIMOTOR COORDINATION DURING SPACEFLIGHT: AN ANALYSIS OF POINTING AND TRACKING RESPONSES DURING THE "NEUROLAB" SPACE SHUTTLE MISSION. O. Bock, B. Fowler, D. Comfort. *Aviat. Space Envir. Med.*, Vol. 72, No. 10, pages 877-883; 2001.

Section 3 Nervous System Development in Weightlessness

Background *How gravity affects nervous system development*

INTRODUCTION

In a classic study, investigators David Hubel and Thorsten Weisel asked: If one eye were covered during development and uncovered when the animal was an adult, would the adult animal see normally? Their research found that there are critical times during development when vision from the eye must be present for the brain serving that eye to develop normally. Otherwise, the brain will be irreversibly changed. Uncovering the eye after the critical period will not restore normal vision.

If an environmental factor like vision is crucial for normal brain development, could other factors also be important? Is gravity, or the loading that gravity brings, essential for normal development? If animals or people were raised in weightlessness, would they later be able to adapt to gravity, or would their bodies be fundamentally different? Would their muscles be able to tolerate the force of gravity? Could they maintain their balance back on Earth?

The Neurolab mission lasted only 16 days, but for some animals this represents a significant time in their development. Over the course of 16 days, an eight-day-old rat changes from an infant to almost an adolescent. The rat learns to walk and its nervous system matures considerably. Other simple organisms like snails, crickets, and fish progress through several developmental stages over 16 days. As a result, basic questions about nervous system development can be answered over the course of a Shuttle mission. Studies in animals provide insights into what might happen if humans experienced weightlessness during development.

GRAVITY AND DEVELOPMENT

Plate 4 shows the main areas of the body affected by weightlessness. At the top of the panel is the inner ear and the connections that it makes throughout the nervous system. Many animals use small crystals, like the otoliths shown in Plate 1, within their gravity sensors. These provide a mass that can move in the direction of gravity. Snails use crystals for gravity sensing just as humans do, and one question is whether developing snails would produce either more or larger crystals when gravity is not present.

The gravity sensors have nerve cells within them. These nerve cells in turn make connections throughout the nervous system to help with posture, balance, and navigation. It is possible that nerve cells within the gravity sensors and the connections they make might be altered by weightlessness. The inner ear and nervous system of the rat provide a way to answer this question. By studying gravity sensors and their connections within the developing rat, the nervous systems of rats developed in space could be compared to those that developed on the ground.

Since the rat nervous system is very complex, a simpler organism also may help in answering these questions. Crickets have gravity sensors and a simple nervous system. By comparing crickets that developed in weightlessness to those that developed in a one-G centrifuge on the spacecraft, a very well-controlled experiment on the effects of gravity can be performed.

The center panel of Plate 4 shows another part of the body affected by gravity. Whenever the head is above the heart, the cardiovascular system has to work against gravity to keep blood flowing to the brain. Without this blood pressure control system, the animal would faint when upright. The blood pressure control system has pressure receptors within it that sense changes in blood pressure (such as might occur when the head is above the heart) and signal the brain to make the appropriate corrections. It is not known if these pressure sensors and the nerves that serve them are affected by the absence of gravity.

Approximately 60% of the muscles in the body work against gravity. As the bottom panel of Plate 4 shows, the development of muscle is also an important consideration in weightlessness. Without the loading that gravity supplies, will the antigravity muscles develop normally?

THE NEUROLAB
DEVELOPMENT EXPERIMENTS

Dr. Weiderhold and his colleagues showed that snails reared in weightlessness produced significantly more and larger gravity-sensing crystals. The experiment from Dr. Horn and his colleagues showed that the gravity sensors of the cricket developed normally in space, but there were changes in the connections made farther on in the nervous system. Similarly, in the rat, Dr. Raymond and her colleagues demonstrated that the lack of gravity did have a notable effect on the development of nerve connections in the brain, but not in the gravity sensor itself.

These structural changes in the brain may explain part of the findings from Dr. Walton's research team. They showed that rats that developed in space showed definite differences in how they righted themselves when compared to Earth-raised rats. These changes appeared to be permanent. Dr. Kosik and his laboratory, however, showed no change in the development of the hippocampus—the part of the brain concerned with navigation. Dr. Nowaskowski's laboratory showed that there were significant differences in the development of the brains of prenatal mice in space compared to those maintained in identical conditions at the Kennedy Space Center.

Dr. Shimizu and his colleagues demonstrated significant differences between the pressure receptors in the cardiovascular systems of space-flown rats as compared to those that developed on Earth. Muscle development was also affected. The team from Dr. Baldwin's laboratory showed that weightlessness impaired growth in an antigravity muscle and changed the types of proteins used in that muscle. Dr. Riley and his colleagues showed fewer nerve terminals and decreased muscle growth in the antigravity soleus muscle, suggesting that normal nerve and muscle development depends on loading during the formative period.

On Neurolab, the crickets, snails, mice, and 14-day-old (at launch) rats were healthy. The rats that were eight days old at launch, however, had significant problems, and many of these rats died inflight. This had an effect on the experiments involving these rats. Taken together, however, the results from the experiments suggest that there are critical periods when gravity is important to development. These findings could have long-term importance for space exploration, especially concerning the future colonization of animals or people outside of Earth's gravity.

Plate 4: Nervous System Development

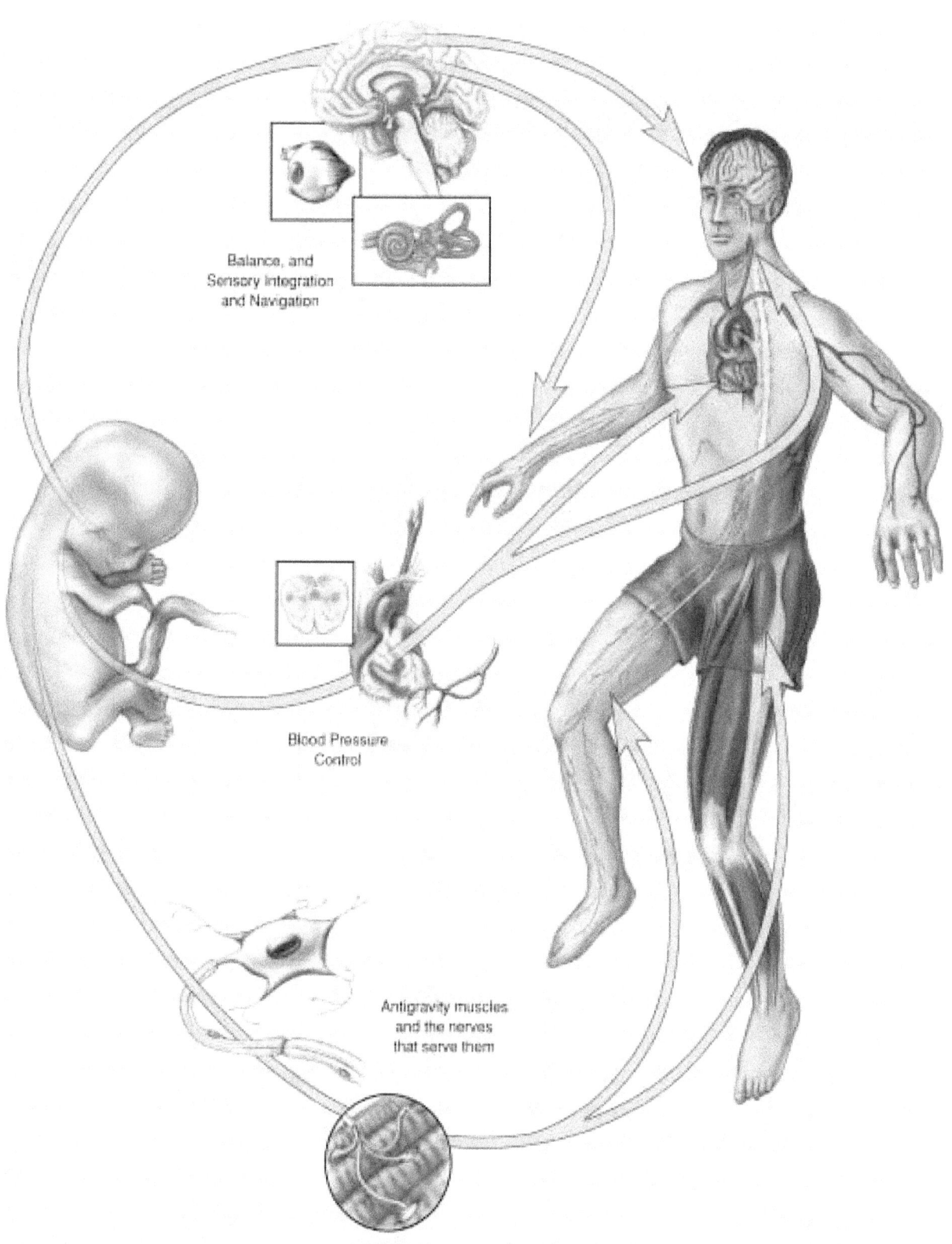

Balance, and
Sensory Integration
and Navigation

Blood Pressure
Control

Antigravity muscles
and the nerves
that serve them

©2001 Kestrel Illustration Studio, LLC

Motor System Development Depends on Experience: A microgravity study of rats

Experiment Team

Principal Investigators: **Kerry D. Walton[1]**
Rodolfo R. Llinás[1]

Co-Investigators: **Robert Kalb[2]**
Dean Hillman[1]
Javier DeFelipe[3]
Luis Miguel Garcia-Segura[3]

Research Assistants: **Shannon Harding[1]**
Dan Sulica[1]
Luis Benavides[1]

[1]New York University School of Medicine, New York City, USA
[2]Children's Hospital of Philadelphia, Philadelphia, USA
[3]Instituto Cajal (CSIC), Madrid, Spain

Authors

Kerry D. Walton, Robert G. Kalb,
Javier DeFelipe, Luis Miguel Garcia-Segura,
Dean Hillman, Rodolfo R. Llinás

ABSTRACT

Animals move about their environment by sensing their surroundings and making adjustments according to need. All animals take the force of gravity into account when the brain and spinal cord undertake the planning and execution of movements. To what extent must animals learn to factor in the force of gravity when making neural calculations about movement? Are animals born knowing how to respond to gravity, or must the young nervous system learn to enter gravity into the equation? To study this issue, young rats were reared in two different gravitational environments (the one-G of Earth and the microgravity of low Earth orbit) that necessitated two different types of motor operations (movements) for optimal behavior. We inquired whether those portions of the young nervous system involved in movement, the motor system, can adapt to different gravitational levels and, if so, the cellular basis for this phenomenon.

We studied two groups of rats that had been raised for 16 days in microgravity (eight or 14 days old at launch) and compared their walking and righting (ability to go from upside down to upright) and brain structure to those of control rats that developed on Earth. Flight rats were easily distinguished from the age-matched ground control rats in terms of both motor function and central nervous system structure. Mature surface righting predominated in control rats on the day of landing (R+0), while immature righting predominated in the flight rats on landing day and 30 days after landing. Some of these changes appear to be permanent.

Several conclusions can be drawn from these studies: (1) Many aspects of motor behavior are preprogrammed into the young nervous system. In addition, several aspects of motor behavior are acquired as a function of the interaction of the developing organism and the rearing environment; (2) Widespread neuroanatomical differences between one-G- and microgravity-reared rats indicate that there is a structural basis for the adaptation to the rearing environment. These observations provide support for the idea that an animal's motor system adapts for optimal function within the environment experienced during a critical period in early postnatal life.

INTRODUCTION

Professional baseball players can hit a small, round object traveling over 100 miles an hour with a stick. If their timing is just right and they put the full force of their bodies into the swing, they can hit this object (a ball) over 500 feet through the air. Hitting a speeding ball requires that batters match the trajectories of the ball and the bat perfectly. That is, they must predict when the ball and bat will occupy the same space. Even the smallest error means that they will miss the ball completely.

To hit a home run, a batter must predict the effect of gravity on both the ball and the bat—and have exquisite motor control. Not only adults have such abilities. A professional baseball player probably first hit a home run at age eight or nine years. This skill was developed during youth as the player learned to control and coordinate movements in Earth's gravitational field. We are interested in understanding the processes underlying the development of such controlled movement to determine whether there is demonstrable "tuning" of the central nervous system (CNS) to the Earth's gravitational field during development.

Movement control begins for baseball players, as for all people, soon after birth. At birth, a baby can kick its arms and legs and lift its head. Within months it can roll over, sit up, and begin to crawl. Between its first and second birthdays, a toddler will take its first steps. Running and jumping are not far behind. During all this time, planning and controlling movements must include calculations for the effects of gravity on the body as well as on the world. Thus, it seems that gravity is as essential to the development of motor function as light is to the development of vision.

We expected that if gravity were reduced or removed, motor system development would be profoundly altered. Indeed, the microgravity of spaceflight represents a fundamental alteration in an animal's environment. The constellations of sensory signals stimulated by gravity, such as righting reactions (turning upright when placed upside down), are not elicited. The perceptual realm has been modified, but motor experience is also altered. In the presence of gravity, animals must negotiate a two-dimensional space. In microgravity, animals must develop a strategy to negotiate a three-dimensional space. The sensory-motor transformation in these animals must change. Further, the internal representation of reality built as the animal explores and interacts with its environment must differ from the internal representations of animals that are reared on the ground. This view is supported by the results of our ground-based studies as well as by studies of the postnatal development of sensory systems. Yet, our view can be tested directly only if young animals are studied after they spend a period of development in microgravity.

It is well known that vision and hearing, for example, do not develop normally if animals are deprived of normal input for those sensory systems. The reduction or absence of stimuli must occur, however, during a particular period of development—one during which the nervous system is particularly sensitive to environmental stimuli. Hubel and Weisel, who identified such a period in their studies of the cat visual system, named this time a "critical period of development" (Hubel,

1970). Over the last 30 years, such periods of development have been identified for many sensory systems (Katz, 1996). There is even a critical period for the acquisition of primary songs in songbirds and of language in humans (Doupe, 1999).

Anatomical correlates have been identified for behaviorally defined critical periods of development. Indeed, as a young animal interacts with its environment, the patterns of activity in the animal's CNS profoundly influence CNS structure. This has been shown most extensively for the visual system (Katz, 1996). Work from our lab also supports the view that the development of motor skills is fundamentally linked to experience. If the hindlimbs do not support an animal's weight during the first weeks of its life, coordinated movements such as swimming and walking are profoundly altered (Walton, 1992). Modifications in electrical or chemical signals reaching the spinal cord during this same period of development lead to changes in the structure of the CNS (i.e., to alterations in the motor neuron dendritic architecture and the expression of surface molecules) (Kalb, 1994).

The Neurolab mission provided an opportunity to test our hypothesis that postnatal development incorporates an experience-dependent adaptation of the CNS to Earth's gravity. Thus, the nervous system of animals that spend a critical period of development in space would be adapted to the microgravity environment. When these animals returned to Earth, this adaptation would be evident in the way the animals moved and in the structure of those regions of their nervous system that support motor function.

METHODS

Our investigations aimed to study systematically the effect of microgravity exposure in early life on the motor system using behavioral, anatomical, and molecular studies.

Animals

Three groups of rats were studied; flight, ground control, and basal control. These were divided again into two groups according to age. The younger rats were launched on postnatal day 8 (P8) and landed on P24. The older rats were P14 at launch and P30 at landing. The flight rats spent the intervening 16 days in low Earth orbit. The ground control rats remained at the Kennedy Space Center (KSC). Half of the control rats were housed in standard shoebox-sized (vivarium) cages. The other half of the control rats were housed in smaller cages that were matched to those of the flight rats (asynchronous ground control (AGC)). The basal groups were studied when they were P8 or P14; that is, they were age-matched to the day of launch.

Behavioral studies

For the behavioral studies, the older flight and AGC rats were videotaped during the mission on flight day 6 (FD6). The younger rats were videotaped on FD15. The rats were studied in a general-purpose workstation (GPWS) in the Spacelab module (spaceflight) or at KSC (AGC). They were videotaped at

60 frames per second (fps) while they moved about an "animal walking apparatus." This comprised a platform that was covered with foam on one side and a grid on the other.

After landing, three behaviors were used to evaluate the effect of microgravity on motor system function: swimming, free walking, and surface righting. The flight and control rats were videotaped at 60 fps during swimming, and at 200 fps during walking and surface righting. In all cases, the hindlimb joints were marked with a felt pen. For swimming, each rat was carefully placed near one end of a tank of warm water. When the rat reached the other end of the tank, it could leave the water. The rat was then picked up and placed in the water for another trial. Each session lasted two minutes and comprised up to 15 trials. Walking was tested following a similar paradigm using a walking stage. Walking sessions could last up to 10 minutes per rat. Each surface righting trial began by holding the rat on its back with its head straight. Once the animal had relaxed, it was released from this position. Direct frame-by-frame analysis or computer-assisted motion analysis (Motus 5, Peak Performance, Inc.) was used. Statistical analysis used Statview software (SAS Institute, Inc.).

Anatomical studies

Neurons in the cervical spinal cord, neocortex (hindlimb area), and supraoptic nucleus of the hypothalamus were studied.

The spinal cord studies focused on the younger group of rats. The spinal cords were collected on FD7 and FD14 and a few hours after landing on recovery day zero (R+0). They were preserved by perfusion fixation. Motor neurons were labeled retrogradely with a dye (DiI) applied to the ventral roots of the cervical spinal cord. Tracings of neurons and measurements of neuronal characteristics were made using computer-assisted camera lucida software (Neurolucida) (Inglis, 2000).

Our studies of the brain regions focused on the older rats. Brain tissue collected on landing day and 18 weeks after landing was fixed by immersion. Sections were made through the cerebral cortex or hypothalamus. In the case of the cortical tissue, once the cortical layers were identified, serial ultra-thin sections were cut, stained with uranyl acetate and lead citrate, and examined in a Jeol-1200 EX electron microscope. Synaptic density per unit area was estimated from 10 electron microscope samples of neuropil from each layer (layers I, II/III, IV, Va, Vb, and VI) from each rat. The cross-sectional lengths of synaptic junctions (synaptic apposition length) of all synaptic profiles were measured in the prints using a magnetic tablet (SummaSketch III) and the Scion Image image analysis program (Scion Corporation, Frederick, MD) (DeFelipe, 2002).

The right and left half halves of the hypothalamus were processed separately. A small block containing the supraoptic nucleus was dissected from the one block and processed for electronmicroscopy. Serial sections were cut from the other block using a Vibratome. These were processed for oxytocin and vasopressin immunohistochemistry (García-Ovejero, 2001). In all our studies, differences with $p<0.05$ were considered to be significant.

RESULTS

Behavioral and anatomical studies found clear differences between both ages of flight and control rats on the day of landing. Some of these differences were transient, while other differences persisted for as long as the rats were studied. There were no consistent differences, however, between the two groups of control rats.

Growth

We first examined the effect of microgravity and cage design on the growth of the rats. Figure 1 is a plot of body weight as a function of days after landing for flight (●), AGC (○), and vivarium (◆) rats. The younger flight rats (Figure 1A) were significantly smaller than the cage-matched AGC rats from the day of landing (R+0) through R+7. Flight animals were significantly smaller than the vivarium control rats through R+12. The cage itself influenced the growth of these rats. The control rats housed in flight-like cages (AGC rats) were significantly smaller than vivarium rats R+0 through R+5. Spaceflight did not influence the growth of the older rats (Figure 1B). In fact, there was no significant difference between flight and AGC rats from R+0 through R+30. Again, the cage itself did influence their growth. The AGC control rats were significantly smaller than vivarium rats between R+1 and R+9. The vivarium rats were significantly larger than flight rats between R+5 and R+8.

Behavior

Clear differences between flight and control rats were observed in all of our behavioral tests. Similar changes were seen in the younger and older groups of rats. This report will focus on our findings in the older rats. Further, although the older AGC rats were lighter than their age-matched vivarium control rats, this did not influence the behavioral parameters that we studied.

The most marked difference between flight and control rats was the combination of movements they used to right themselves. When a control rat is placed on its back and released, it rights itself in a controlled sequence as shown in Figure 2. First, it rotates its head, then the forelimbs, and then the hindlimbs. This tactic, called "axial righting," is typical of adult animals. Young animals do not always use this tactic (Pellis, 1991). For example, when young animals are released, they may rotate their head and forelimbs in one direction and their hindlimbs in the opposite direction. This is called "corkscrew righting." Another major righting tactic that we saw we called "ventroflexion." In this case, the animal lifts the front and back legs toward each other so that the body forms a U-shape. The front of the animal then falls in one direction and the hindlimbs follow.

Before launch, the rats used all three tactics to right themselves. This is shown in the first column of Figure 3 where the three tactics are color-coded. Trials using axial righting are blue, corkscrew righting trials are green, and trials incorporating

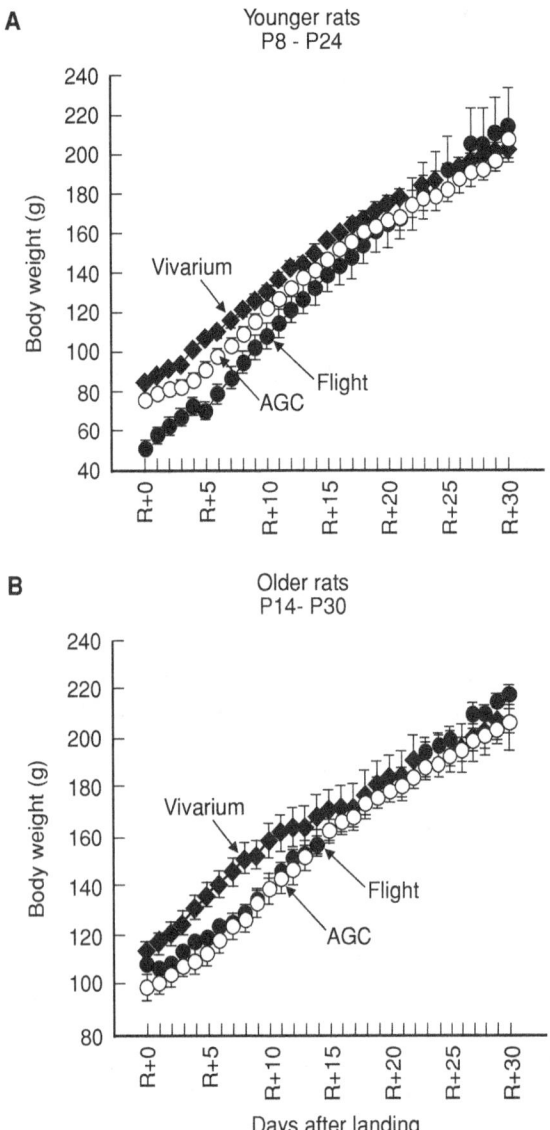

Figure 1. Plot of body weight as a function of days after landing in flight (●), AGC (○), and vivarium (◆) rats. (A) Younger animals. Flight rats were lighter than cage-matched control rats until R+8 and lighter than vivarium controls until R+13. Cage effects were also seen. AGC animals were lighter than vivarium control rats until R+6. (B) Older animals. Spaceflight did not affect the weight of this group, but there was a cage effect.

ventroflexion are red. Axial or corkscrew tactics were used in over 80% of the trials on P14. Sixteen days later, when the rats were P30, axial righting predominated in the control rats. This is shown in the second column. The same acquisition of mature righting tactics did not occur in the flight rats. In fact, as shown in the third column in the figure, axial and corkscrew righting tactics were used in fewer cases than at launch. Ventroflexion, the most immature tactic, predominated. Furthermore, mature righting tactics did not predominate in the flight rats even 30 days after landing.

Flight rats were able to walk and swim when they were tested a few hours after landing. Only walking will be discussed here. When the flight rats were placed on the stage on R+0, they walked freely and explored their immediate surroundings. At first glance, they seemed to be quite similar to the control animals. On careful inspection, however, it appeared that the flight rats stayed close to the ground and seemed to tilt their hindquarters from side to side as they walked. To examine this more closely, the position of the hindlimb was entered into a computer using markings on the skin of the animal.

Seven points represented the hindlimb: tip of the toes, metatarsophalangeal joint, ankle, knee, head of the femur, ischial tuberosity, and iliac crest. Line segments connected these points to generate a stick figure of the leg as shown in Figure 4A. The stick figures show five positions of the right leg as a flight rat takes one step from left to right. First the rat puts its foot on the ground (foot contact (FC)), This marks the beginning of what is called the stance phase of the step cycle. Next, the knee bends as the rat places weight on the leg. Third, the rat extends the right leg and shifts its weight to the left leg (not shown). This is in preparation for foot lift (FL). Foot lift marks the end of the stance phase and the beginning of the swing phase. Once the foot is off the ground, the rat swings the leg forward. Finally, the foot is placed on the ground to begin another step.

We measured the distance between the head of the femur and the walking surface in the flight and control rats on R+0. Results are shown in Figure 4B for one flight (red) and one control (blue) rat. The distance is plotted as a function of time for two steps. The phases of the step cycle are marked by the broken lines. During the stance phase, the flight rat brings its hip closer to the ground than the control rat does. This leads to a large hip excursion in the flight rat (double arrow, Figure 4B) compared to the control rat. This finding is consistent with our initial observation that the flight rats seemed to stay closer to the ground and to tilt from side to side more than the control animals. We calculated the hip excursion in all the flight and AGC control rats on R+0. We found that the mean excursion was significantly larger in flight rats than in control animals on (p<0.05, unpaired t-test). This is shown in Figure 4C.

Anatomy

Our morphological studies demonstrate that structures in the CNS of flight rats differed from those of rats reared on Earth.

In the first study, we examined the dendritic architecture of cervical motor neurons using the labeling dye DiI (Figure 5). There were no differences between the dendritic trees of flight and control motor neurons when the tissue was obtained before landing (on FD7 or FD14). In contrast, the dendritic trees of some motor neurons from flight rats sacrificed on R+0 were significantly smaller and less branched than the dendritic trees of age- and cage-matched control rats (Inglis, 2000).

We sought to determine why some motor neurons in flight animals were affected by microgravity while other motor neurons were not. Toward this end, we divided the motor neurons

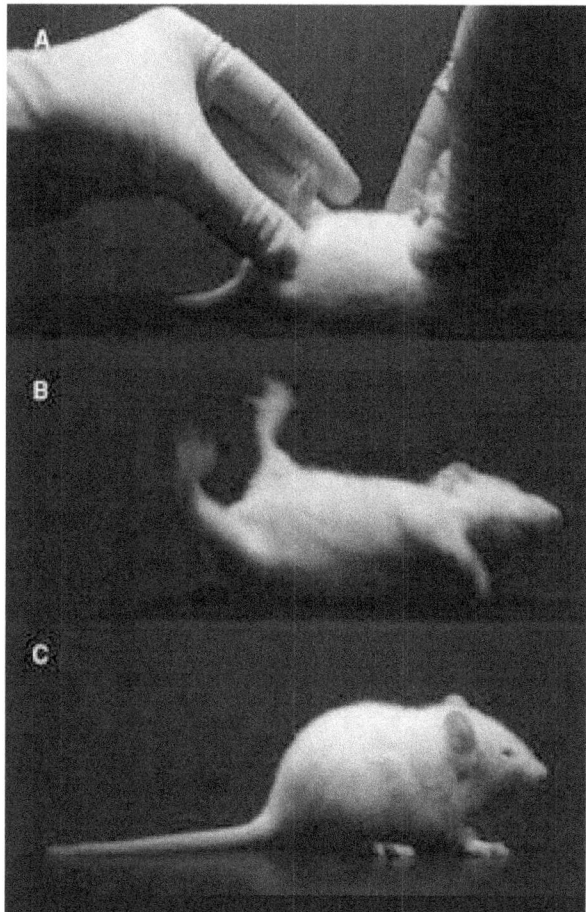

Figure 2. Axial righting in a control rat. (A) The rat is held upside down. (B) When it is released, first the head, then the forelimbs, and finally the hindlimbs rotate in the same direction. (C) This achieves the right-side-down posture shown here.

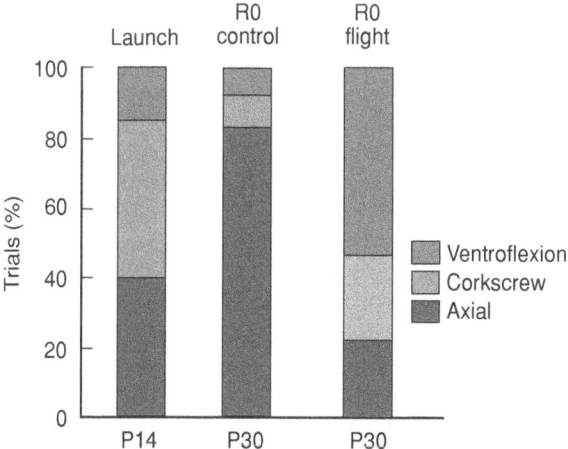

Figure 3. Tactics used in surface righting before launch and on the day of landing. Rats right themselves using: ventroflexion (red), corkscrew (green), or axial (blue) tactics. Mature (axial) righting tactics predominate on R+0 in control rats but not in flight rats.

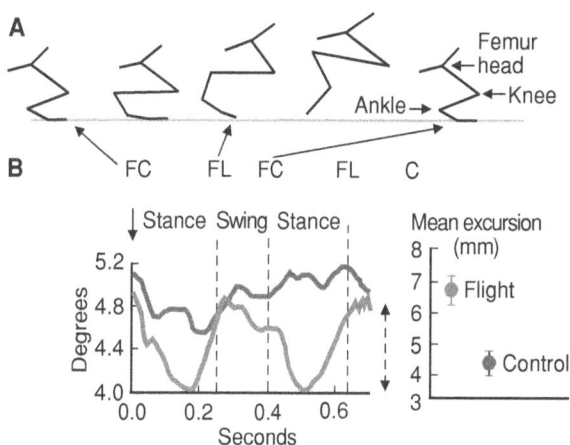

Figure 4. Walking on R+0. (A) Stick figure showing the position of the hindlimb of a rat moving from left to right. (B) Plot of distance of the hip from the ground during two steps for a flight rat (red) and a control (blue) rat. (C) Excursion of the hip (broken vertical line) in flight rats and control rats on R+0 (mean±s.e.m.) showing significantly greater movement in the flight rats. (FC, foot contact; FL, foot lift)

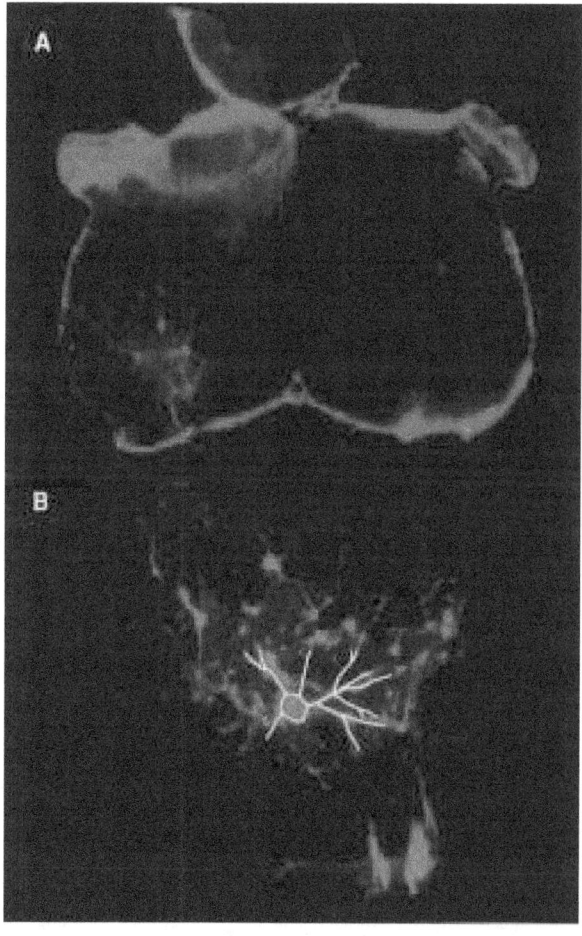

Figure 5. Motor neurons in the cervical spinal cord labeled with the dye DiI. (A) Low-power micrograph showing location of motor neurons in the ventral horn. (B) High-power micrograph of one DiI-filled motor neuron. Superimposed white lines indicate location of the soma and some dendritic branches.

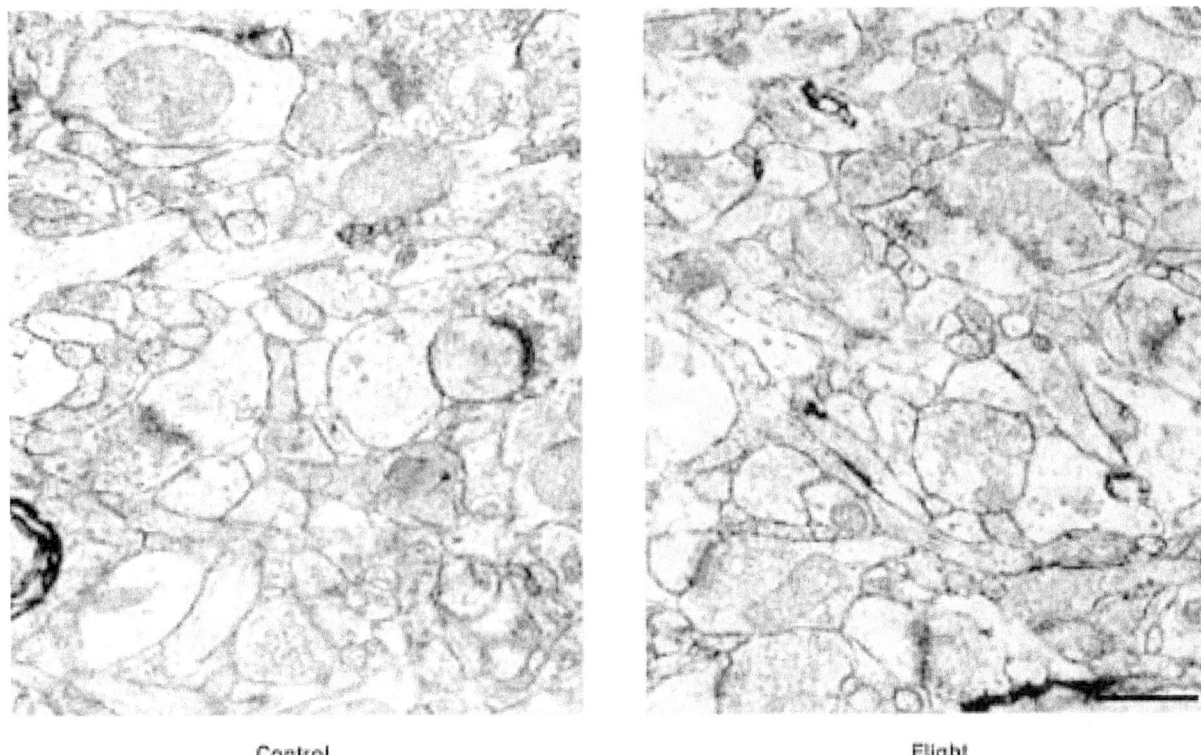

Control   Flight

Figure 6. High-power electron micrographs illustrating the neuropil (interwoven nerve cell processes) of layer II/III from a control and a flight rat. Scale bar: 0.5 microns.

into subgroups based on their location within the ventral horn of the spinal cord. We performed this analysis because it is known that the topography of motor neuron cell bodies reflects the pattern of muscle innervation. When subjected to this analysis, we found that motor neurons located in the medial portions of the ventral horn were very profoundly influenced by rearing in microgravity, while laterally located motor neurons were identical in flight rats and control animals. Medially located motor neurons innervate the muscles involved in postural control and surface righting, so this result is consistent with our behavioral findings.

Differences between flight rats and control rats were also seen in the region of cerebral cortex that is related to hindlimb function. This investigation focused on the older rats and examined the characteristics of individual synapses. The quality of the images, as seen in the electron microscope, was excellent (Figure 6). It was thus possible to measure accurately the cross-sectional length of synaptic contacts between neurons. We were also able to determine whether synapses were symmetrical (densities on both sides of the synaptic cleft) or asymmetrical (density on one side only). Development in microgravity leads to changes in the number and morphology of cortical synapses in a laminar-specific manner. In tissue obtained on R+0, the synaptic cross-sectional lengths were significantly longer in flight rats than they were in ground control rats in layers II/III and layer Va. Differences were only seen in asymmetrical synapses, which are known to be excitatory.

During the 18-week period after landing, synapse cross-sectional length increased in both control rats and flight rats. This was statistically significant in all layers in both groups of rats with one exception. There was no change in layer IV synapses in flight rats.

The increase was greater in control rats than in flight rats. This unequal growth had two consequences: (1) Synapse cross-sectional length in control rats increased to reach that seen in flight rats on R+0. Thus, the significant differences in layers II/III and Va seen on R+0 were abolished. (2) Synaptic cross-section length in layer I was significantly greater in control rats than in flight rats by R+18 weeks.

Finally, we examined neurosecretory neurons in the supraoptic nucleus of the hypothalamus at R+0 and R+18 weeks. This study focused on the large cells that secrete the hormones oxytocin or vasopressin. Several signs of enhanced transcriptional and biosynthetic activity were observed in magnocellular supraoptic neurons of flight rats on R+0 compared to control rats. These include increased *c-fos* expression, larger nucleoli and cytoplasm, and higher volume occupied in the neuronal perikaryon by mitochondria, endoplasmic reticulum, Golgi apparatus, lysosomes, and cytoplasmic inclusions known as nematosomes. In contrast, the volume occupied by neurosecretory vesicles in the supraoptic neuronal perikarya was significantly decreased in flight rats. This decrease was associated with a significant decrease in oxytocin and vasopressin immunoreactive levels, suggestive of an increased hormonal release.

Vasopressin levels, cytoplasmic volume, and *c-fos* expression returned to control levels 18 weeks after landing. However, oxytocin levels were still reduced in flight rats compared to control rats.

DISCUSSION

Our results demonstrate that: (a) motor system development clearly depends on the interaction of the animal with its environment, (b) the CNS is sensitive to changes in the gravitational field, and (c) there is a critical period for motor system development.

Development can be considered as incorporating four processes: growth, maturity, adaptation, and learning. Our results concern all four processes.

Growth

Microgravity itself did not affect body growth. This is seen in the older rats where there were no significant differences in the body weight of AGC rats and flight rats (Figure 1B). The low weight of the younger flight animals on R+0 was a secondary result of the microgravity environment. It seemed to be due to a reduced interaction between the mother and younger rats in microgravity. This was critical for the younger rats since their eyes were still closed at launch and for the first days of the mission. Nevertheless, this did not have long-lasting effects since the flight rats had reached the weights of the AGC rats by a week after landing (Figure 1A). The nature of the cages did influence body growth, however. This is evidenced by the greater weight of the vivarium rats compared to the AGC control rats in both age groups (Figure 1). Nevertheless, in the behavioral tests, no differences were seen between vivarium rats and AGC control rats. One anomaly is worthy of comment. The older vivarium rats were significantly larger than the flight rats between R+5 and R+8. As indicted by the shape of the graph (Figure 1B), this was probably due to weight gain by the vivarium rats rather than to a five-day delayed influence of spaceflight on the flight rats.

Surface righting

Maturation is the acquisition of adult characteristics. We have shown that the acquisition of adult movement tactics, such as how to turn over, is not hardwired into the CNS. Rather, it is an example of an adaptation to the environment and requires specific goal-directed activity. When placed on its back in a gravitational field (Figure 2A), a rat will do whatever is needed to turn over. This can be accomplished in a number of ways as seen in young rats (first column, Figure 3). As rats grow older, the most efficient tactic, axial righting, comes to predominate (second column, Figure 3). This maturation does not occur in flight rats (third column, Figure 3), however, because righting is not an appropriate movement in microgravity. In fact, it does not even occur in microgravity. When flight rats are released from a supine position, they float up without rotating their bodies. This is not surprising since there is no vestibular input to signal "up" and "down." These results indicate that

mature surface righting tactics arise as an adaptation to gravity and require performance to acquire.

It is worth noting that the flight rats were able to right themselves the first time they were tested on R+0. That is, the signals from the peripheral vestibular apparatus were sufficient to signal that the rat was upside down. This indicates that the critical period for the establishment of reflexes between the inner ear and the neck (vestibulo-collic reflexes) ends before P14.

Furthermore, even 30 days after their return to Earth, the flight rats had not acquired mature righting characteristics. That is, there is a critical period for the development of mature surface righting tactics. This period ends on or before P30.

Walking

When they launched, the older rats had their eyes open and could walk and run on a flat surface. However, mature locomotion was not yet achieved. Thus, any changes in locomotion during the mission would reflect continued maturation of the motor system and adaptation to the environment. The flight rats remained close to the ground and tilted from side to side when they walked on R+0. That downward tilt was due to a flexion of the leg during the stance phase (Figure 4). This suggests that the supporting leg flexed when weight was shifted from the other leg. It seemed that the rat was not prepared to support its weight on its hindlimbs. This is consistent with locomotion in a microgravity environment where the hindlimbs do not support weight. In fact, as in astronauts, locomotion in animals is accomplished largely by the forelimbs (or arms).

On Earth, limb extension is used when the animal pushes against the ground at the end of the stance phase. In fact, extensors are sometimes called "antigravity" muscles because they exert force against gravity. This does not occur in microgravity. If there is reduced joint extension as an animal walks, it will not move away from the surface. Rather, it will appear to be walk close to the ground as seen in the flight rats. This type of walking reflects an adaptation of the flight rats to the microgravity environment.

Anatomy

The structural changes in the brain and spinal cord demonstrate that neurons subserving motor functions undergo activity-dependent maturation in early postnatal life in a manner that is analogous to sensory systems such as vision or hearing. These changes are consistent with, and provide a structural basis for, the behavioral results. This part of our study was carried out on the younger rats that did not grow well in microgravity. It is reasonable to ask if the weight differences between flight rats and control rats could account for the difference seen in dendritic structure. We do not think this is the case since differences between flight rats and control animals were only present in specific areas—i.e., in the motor neurons in the medial and not the lateral aspect of the ventral horn.

There are functional consequences of the anatomical changes found in the dendritic tree. A reduction in motor neuron dendritic tree size and complexity will influence the compartmentalization of electrical and chemical signals received by dendrites. In addition, this reduction is likely to alter the number and types of synaptic inputs these cells receive. The alterations in neuronal information processing will be reflected in the output of the cell. As a consequence of changing the firing pattern of the neuromuscular unit—particularly of postural musculature—animal behavior is affected (Inglis, 2000). For example, a lack of vestibular input in microgravity could alter the development of dendritic architecture. This could, in turn, contribute to changes in motor neuron activity when the input is restored, as when the flight rats return to Earth. These experiments were not designed to provide a direct link between our anatomical and behavioral data. Nonetheless, they do suggest that the rearing environment will influence the operation of the neurons involved in motor function.

Our findings in the cerebral cortex also support this view (DeFelipe, 2002). On R+0, the cross-sectional lengths of excitatory synapses were longer in flight rats than in control rats. This increase may be interpreted in terms of the characteristics of the microgravity environment. Microgravity provides a three-dimensional space for locomotion. Thus, it is "enriched" with respect to the two-dimensionality experienced by rats on the ground. Indeed, flight rats are free to "walk" along all six surfaces of their cages, not just on one surface (the bottom) as on Earth. This may stimulate an increased contact area between neurons as has been found when rats are raised in a cage that contains toys and additional walking surfaces.

After landing, synapses in the flight rats continue to grow, but not as fast as in the control rats. Thus, after 18 weeks, the control rats were able to "catch up." It is possible that the two-dimensional environment of the Earth was not sufficient to stimulate growth of the cortical neurons. The critical period for cortical development had not ended. But, when the flight rats landed, they were deprived of the necessary stimuli (locomotion in three dimensions) needed for development. Only neurons in layer IV did not change significantly after landing. This suggests that critical periods of development may differ among the cortical layers.

The spinal cord and hindlimb area of the cortex were chosen for study because they have a direct role in the organization and control of movement. In contrast, alterations in hypothalamic neurosecretory neurons reflect effects of microgravity on the volume of body fluids outside of cells (which has both a direct and an indirect effect on CNS function) as well as the stress associated with spaceflight. We thought it was important to carry out such a study in the same rats examined in our other studies.

Our results, showing increased activity on R+0, are in agreement with studies in adult rats, suggesting an increased cellular activity in the hypothalamic-pituitary neurosecretory system after spaceflight. These effects are probably associated with osmotic stimuli resulting from modifications in the volume and distribution of body fluids and plasma during spaceflight and landing, and they may reflect normal physiological adaptations (García-Ovejero, 2001). These changes, and the resulting modifications in vasopressin regulation, may be involved in the transient alterations in renal function observed in experimental animals and astronauts after spaceflight. In addition, spaceflight in young animals resulted in permanent modifications in oxytocinergic neurons, probably related to stressful conditions associated with spaceflight and landing. These irreversible modifications may have permanent effects in different functional, behavioral, and neuroendocrine parameters.

The present observations provide strong support for the idea that the motor function of an animal adapts for optimal operation within the environment experienced during a critical period in early postnatal life. Further studies should be designed to explore the possibility of permanent behavioral, neuroanatomical, and endocrine changes associated with exposure to microgravity during critical postnatal developmental periods.

Acknowledgements

The authors would like to thank the Neurolab crew, particularly Drs. David Williams and Jay Buckey who carried out the inflight portion of the study. Special thanks to Mrs. Carol Elland for her patience, hard work, and effective advocacy. Supported by NINDS, NICHD NS/HD33467, NASA, NAF2-662, NAG2-951, and the Department of the Navy NAG2-978 (to KW & RL).

REFERENCES

THE PERIOD OF SUSCEPTIBILITY TO THE PHYSIOLOGICAL EFFECTS OF UNILATERAL EYE CLOSURE IN KITTENS. D.H. Hubel, T.N. Wiesel. *J. Physiol.* (Lond), Vol. 206, pages 419–436; 1970.

A DESCRIPTIVE ANALYSIS OF THE POSTNATAL DEVELOPMENT OF CONTACT-RIGHTING IN RATS (*rattus norvegicus*). V.C. Pellis, S.M. Pellis, P. Teitelbaum. *Dev. Psychobiol.*, Vol. 24, pages 237–263; 1991.

IDENTIFICATION OF A CRITICAL PERIOD FOR MOTOR DEVELOPMENT IN NEONATAL RATS. K.D. Walton, D. Lieberman, A. Llinas, M. Begin, R.R. Llinas. *Neurosci.*, Vol. 51, pages 763–767; 1992.

ELECTRICAL ACTIVITY IN THE NEUROMUSCULAR UNIT CAN INFLUENCE THE MOLECULAR DEVELOPMENT OF MOTOR NEURONS. R.G. Kalb, S. Hockfield. *Dev. Biol.*, Vol. 162, pages 539-548; 1994.

SYNAPTIC ACTIVITY AND THE CONSTRUCTION OF CORTICAL CIRCUITS. L.C. Katz, C.J. Chatz. *Science,* Vol. 274, pages 1133–1138; 1996.

BIRDSONG AND HUMAN SPEECH: COMMON THEMES AND MECHANISMS. A.J. Doupe, P.K. Kuhl. *Ann. R. Neurosci.,* Vol. 22, pages 567–631; 1999.

EXPERIENCE-DEPENDENT DEVELOPMENT OF SPINAL MOTOR NEURONS. F.M. Inglis, K.E. Zuckerman, R.G. Kalb. *Neuron,* Vol. 26, pages 299–305; 2000.

SPACE FLIGHT AFFECTS MAGNOCELLULAR SUPRAOPTIC NEURONS OF YOUNG PREPUBERAL RATS: TRANSIENT AND PERMANENT EFFECTS. D. Garcia-Ovejero, J.L. Trejo, I. Ciriza, K.D. Walton, L. Garcia-Segura. *Dev. Brain Res.,* Vol. 130, pages 11–205; 2001.

SPACEFLIGHT INDUCES CHANGES IN THE SYNAPTIC CIRCUITRY OF POSTNATAL DEVELOPING NEOCORTEX. J. DeFelipe, J.I. Arellano, A. Merchan-Perez, M.C. Gonzákez-Albo, K. Walton, R. Llinás. *Cereb. Cortex,* Vol. 12, pages: 883–891; 2002.

Neuromuscular Development Is Altered by Spaceflight

Authors
Danny A. Riley, Margaret T.T. Wong-Riley

Experiment Team

Principal Investigator: **Danny A. Riley**

Co-Investigator: **Margaret T.T. Wong-Riley**

Technical Assistants: **James Bain, Sandy Holtzman, Wendy Leibl, Angel Lekschas, Thomas Miller, Paul Reiser, Glenn Slocum, Kalpana Vijayan**

Medical College of Wisconsin, Milwaukee, USA

ABSTRACT

Weightbearing skeletal muscles, also known as antigravity muscles, lift the body's weight against gravity. Studies show that periodic weightbearing activity is important for muscle maintenance in adults, but it is not known how critical weightbearing is for the early development of muscles and the nerves that serve them. We predicted that raising young rats in microgravity to eliminate weightbearing by the hindlimbs would compromise the nerve and muscle development of the soleus—a weightbearing leg muscle. The nonweightbearing extensor digitorum longus (EDL) leg muscle, however, should develop normally. To test this, we examined the spinal cords, soleus muscles, and EDL muscles collected from six young rats on inflight day 15 of the Neurolab mission. These were compared with ground samples from age-matched controls in vivarium and flight cages. Interpretation of the findings was complicated by the fact that the space-flown rats grew significantly less than normal, in part related to decreased interaction with the nursing mother rats. Soleus development was the most disrupted as evidenced by decreased muscle fiber growth, increased sensitivity to muscle reloading injury, reduced growth of motor neuron terminals as well as a lowered ability of the nerves to use oxygen. Surprisingly, the normal process of eliminating multiple connections to muscle fibers proceeded without weightbearing activity. The importance of loaded contractile activity for maintenance of adult muscle mass has been recognized for many years. The present findings suggest that normal neuromuscular development depends upon loaded contractions during the formative period.

INTRODUCTION

Muscle weakness and increased susceptibility to muscle damage are observed in humans and animals following brief exposures (circa two weeks) to microgravity unloading of weightbearing leg muscles. Weightbearing, or "antigravity", skeletal muscles lift the body's weight against gravity. If periodic weightbearing activity is important for muscle maintenance in the adult, how critical is weightbearing for the development of normal nerves and muscles? The neuromuscular system consists of motor units, which are the motor neurons in the spinal cord, their axons in the peripheral nerves, and the nerve muscle synapses (motor endplates) in the muscles (Figure 1). A significant component of human neuromuscular development is accomplished in utero and continues throughout the first year of life. The developmental events covering this period are condensed into the first month of life for rats.

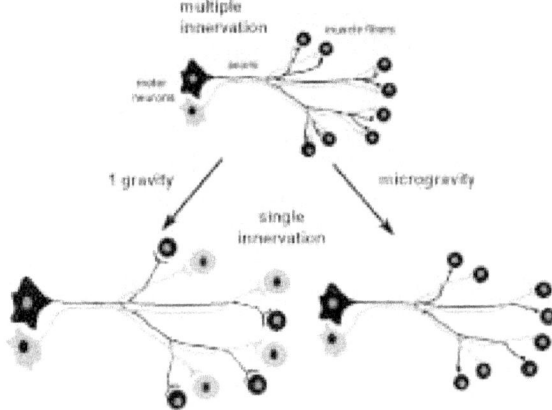

Figure 1. The diagram depicts the immature, multiply-innervated state of two motor units in an eight-day-old rat at launch. The elimination of multiple innervation was achieved during 15 days in microgravity. However, the growth of muscle fibers and motor neuron endings was retarded during spaceflight compared to rats reared in Earth-normal one-G.

Young rats are born with a very immature neuromuscular system. For the four weeks after birth, rapid neuromuscular development occurs, and the effects of removing weightbearing can be well studied. Ground-based experiments indicate that the early removal of muscle loading markedly disrupts the neuromuscular development of weightbearing muscles (Huckstorf, 2000; Nakano, 2000).

Subjecting eight-day-old rats to unloading during the 16-day Neurolab spaceflight mission provided an opportunity to investigate the requirement for weightbearing on neuromuscular development. The present study suggests that loading is necessary for normal development of motor neurons, motor endplates, and muscle fibers in the weightbearing soleus muscle, but not in the nonweightbearing extensor digitorum longus (EDL) muscle.

METHODS

The spinal cords, soleus, and EDL were obtained from six young rats on inflight day 15 and compared with those removed from identical numbers of age-matched ground controls in vivarium or flight cages on comparable mission days. The inflight acquisition of tissues was an extremely important accomplishment because even a brief exposure of space-flown animals to gravity loading postflight can result in marked changes in the neuromuscular system. The resulting postflight changes could have obscured the effects of microgravity. The inflight collection, however, required challenging microsurgery. The delicate microsurgical dissection of the tiny tissues in the young rats demonstrated the feasibility of conducting microsurgery in space (see technical report by Buckey et al. in this publication). Tissues were chemically preserved for histochemistry, immunohistochemistry, and *in situ* hybridization. Electron microscopy was used for analyzing motor neuron oxidative metabolism, motor endplate maturation, and muscle fiber structure (Hevner, 1993; Riley, 1990).

RESULTS

At launch, the mean body weights of the rats in the flight, flight-cage ground control, and vivarium-cage ground control litters were tightly matched (21.7±0.9-G). Space-flown rats returned, on the average, about one half the size of vivarium ground controls (44.2±8.8-G and 90.5±3.6-G, respectively). The flight-cage ground controls were 93% of vivarium control, indicating a slight negative cage effect on growth. Soleus muscle fibers of flight rats were noticeably smaller (31%) than those in the vivarium controls (Figure 2). Reloading damage of the contractile apparatus was present in the soleus fibers of the flight rats (Figure 3). At launch, greater than 75% of the motor endplates were innervated by multiple motor nerve terminals (Figures 1, 4). After 15 days in space, all but one terminal per endplate (synapse) was eliminated as occurred in the ground controls (Figures. 1, 5). However, only 16±1% of soleus motor nerve endings achieved complex branching. This was significantly lower than those with complex branching in the vivarium (56±4%) and cage (44±3%) ground controls (Figure 5). The EDL had lower percentages of complex endings in the cage (30±2%) control compared to vivarium (40±3%), but flight (26±3%) was not different from the cage control. Compared to vivarium, the somas of motor neurons of the flight rats were similar to control but possessed lower levels of mitochondrial cytochrome oxidase (CO) activity and messenger RNAs for subunits III and IV (Figure 6A-I).

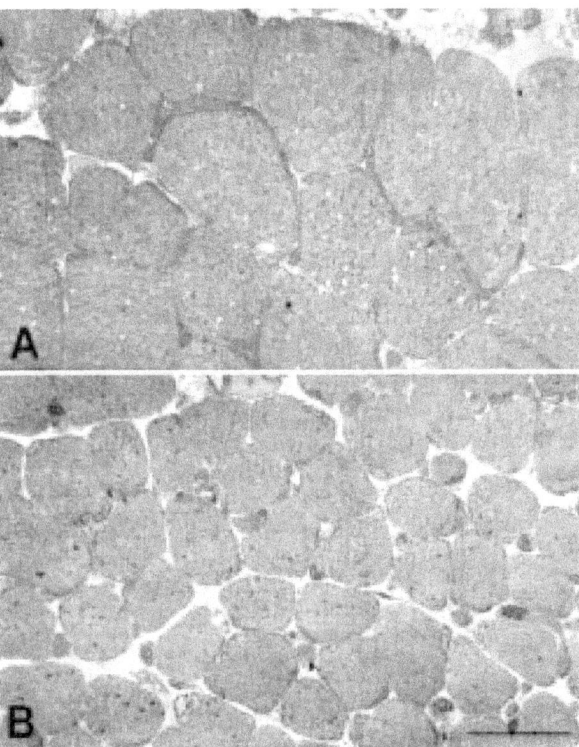

Figure 2. On landing day, the soleus muscle fibers of rats maintained during the Neurolab mission in ground control cages were much larger in cross-sectional area (A) than those of the flight animals, which exhibited significantly reduced growth (B). Bar equals 25 μm.

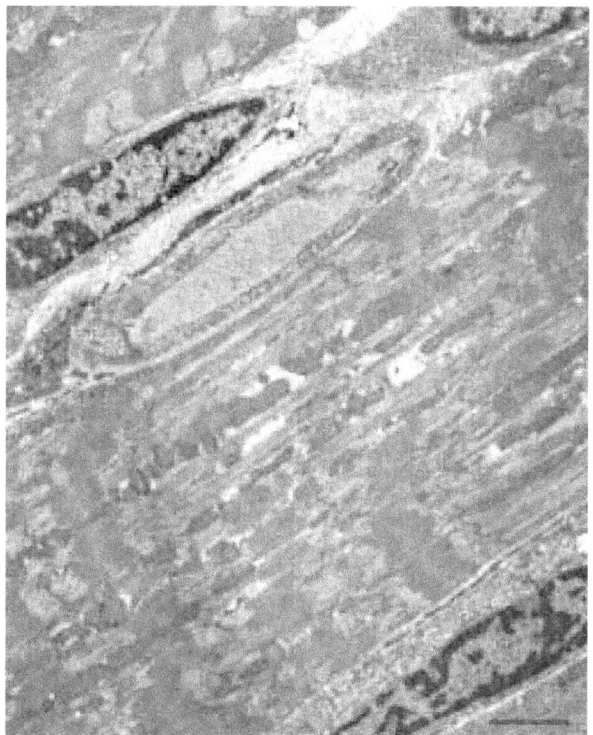

Figure 3. Electron microscopy shows that soleus muscles of flight rats exhibited damage of the contractile proteins following return to gravity weightbearing contractions. Bar equals 2 μm.

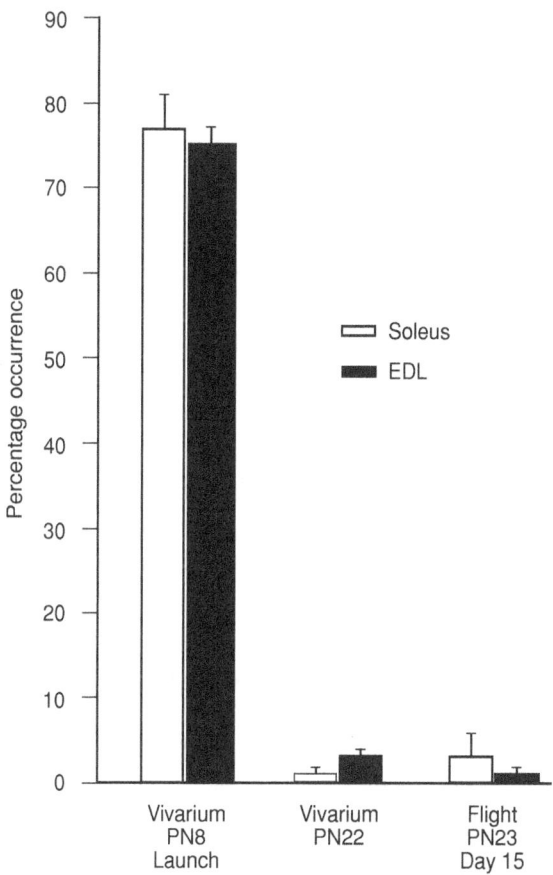

Figure 4. At launch, >75% of the endplates sampled in the normal vivarium control rats were innervated by more than one motor axon in the soleus and EDL muscles. Muscles obtained inflight on flight day 15 achieved reduced levels of multiple innervation comparable to that of the ground controls.

DISCUSSION

Young rats did not grow in the spacecraft environment as large as in the vivarium and flight cages on Earth. Whether the reduced growth represented undernutrition, elevated stress when the nursing rats floated away from the mother rat, an influence of microgravity on the hormonal growth axis, or combinations of these and other factors requires additional flight studies. The full impact on neuromuscular development of removing the environmental stimulus of gravity is not known. The eight-day-old rats experienced gravity preflight, and this may have been sufficient to trigger developmental events. In the future, by having rats born in microgravity in a space station laboratory, it would be possible to eliminate all gravity-induced weightbearing.

Astronauts returning to gravity-induced weightbearing experience a delayed onset of muscle soreness one to two days postflight (Riley, 1999). Space-flown adult rats exhibit extensive muscle tearing in the weightbearing slow muscles after return to Earth (Riley, 1995). The lack of gravity loading apparently deconditions the muscle of the adult rat. Microgravity promotes development of a structurally weaker muscle in the immature rat, leading to a similar end result. Electron microscopic studies of leg muscle tissue of astronauts indicate that reduced amounts of the contractile protein actin predispose fibers to structural failure during muscle contractile activity (Riley, 2002). Failure to

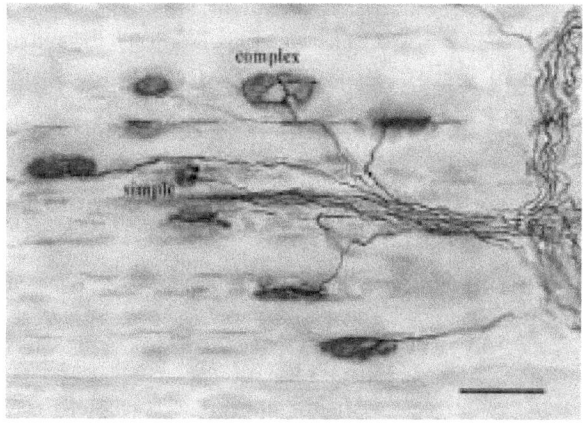

Figure 5. This silver-stained motor axon and cholinesterase-stained endplate section of an EDL muscle from a 23-day-old rat illustrates minimally branched simple and more extensively branched complex motor endplates. The simple endings represent the immature condition. Spaceflight resulted in fewer motor endings achieving the complex state. Bar = 25 μm.

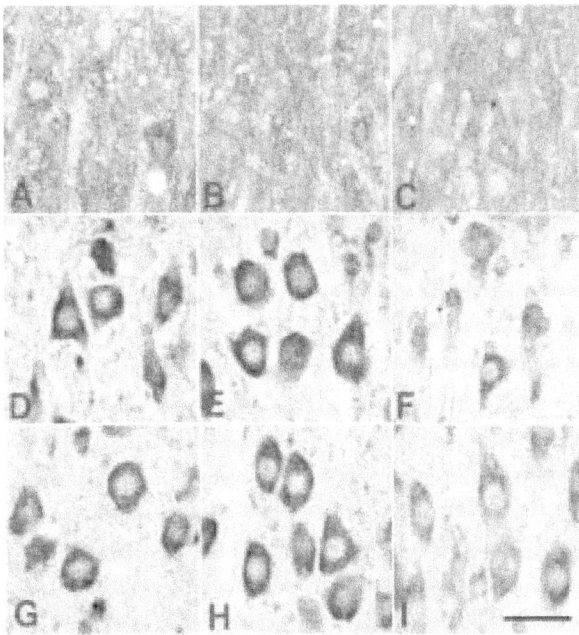

Figure 6. Sections through the ventral horn region in the lumbar portion of the spinal cords of vivarium (A,D,G), ground flight-cage control (B,E,H), and flight rats (C,F,I) were stained histochemically for mitochondrial cytochrome oxidase activity (A,B,C) and probed by *in-situ* hybridization for cytochrome oxidase subunits III (D,E,F) and IV (G,H,I). The motor neurons in the flight rats exhibit reduced cytochrome oxidase activity and subunit message levels (C,F,I). Bar equals 70 µm for all panels.

produce sufficient amounts of contractile proteins during spaceflight may render the developing muscle unable to withstand the workloads imposed by gravity.

The motor endplates on muscle fibers have frequently been investigated as model synapses for developmental studies (Keller-Peck, 2001). As in the immature central nervous system, each muscle fiber synapse transiently receives multiple functional neuronal inputs, which are then subsequently reduced in number to achieve the mature state. For muscle, this is one input per fiber. The process of reducing motor endplate multiple innervation to single innervation was achieved during spaceflight. At launch, the majority (>75%) of endplates received two or more axons. This was reduced to one axon 15 days later, just as occurred in the muscles of the ground controls. It was surprising that the elimination of multiple innervation was not delayed. Ground-based experiments by others reported that reducing motor activity and muscle growth caused persistence of the immature state (Greensmith, 1991). Video recording during the NIH.R3 Shuttle mission showed that the weightless neonatal rats moved their limbs often and vigorously when floating in microgravity (Riley, unpublished). Thus, it appears that, as in the central nervous system, nerve impulses in the motor axons, generated during limb movements, are sufficient for establishing the adult patterns of innervation; and for muscle, weightbearing activity is not essential to eliminate multiple innervation. Another unexpected

finding was that the single innervation pattern was established even though the motor neurons in the flight animals: (a) had lower levels of cytochrome oxidase activity, (b) had lower amounts of CO protein, and (c) had synaptic nerve endings that did not elaborate the normal complex branching present in the ground controls (Figures 1, 5). What inhibited the motor nerve endings from achieving complexity? The most likely explanation is that the postsynaptic area of the endplates failed to enlarge because the muscle fibers of flight rats grew less than normal. Motor endplate area and muscle fiber size are directly related (Wærhaug, 1992). These results imply that the muscle fiber regulates the size of the motor endplate that, in turn, constrains the degree of branching of the motor nerve terminals.

Development of the neuromuscular system was delayed for the weightbearing slow muscle (the soleus) and proceeded normally for the nonweightbearing fast muscle (the EDL). It was not determined whether the developmental delays were permanent by allowing flight rats to return to terrestrial gravity and grow to adulthood with weightbearing. It remains unproven whether a critical period of development exists in the first month of a rat's life during which the neuromuscular system of weightbearing muscles requires gravity-loaded stimulation to complete development. This important issue can be examined by returning one-month-old rats, born on the Space Station free of gravity loading, to Earth and observing the quality of maturation achieved in adulthood.

Acknowledgements

The preflight development and inflight conduct of this experiment succeeded in large measure because of the expert intellectual and hands-on participation of Neurolab crewmembers, Drs. Jay Buckey, Richard Linnehan, James Pawelczyk, and Dafydd Williams, and the numerous contributions of NASA personnel.

REFERENCES

SKELETAL MUSCLE FIBER, NERVE, AND BLOOD VESSEL BREAKDOWN IN SPACE-FLOWN RATS. D.A. Riley, E.I. Ilyina-Kakueva, S. Ellis, J.L. Bain, G.R. Slocum, and F.R. Sedlak. *FASEB J.* Vol. 4, pages 84–91; 1990.

NEUROMUSCULAR CONTACTS IN THE DEVELOPING RAT SOLEUS DEPEND ON MUSCLE ACTIVITY. L. Greensmith and G. Vrbová. *Dev. Brain Res.,* Vol. 62, pages pages 121–129; 1991.

POSTNATAL DEVELOPMENT OF RAT MOTOR NERVE TERMINALS. O. Wærhaug. *Anat. Embryol.,* Vol. 185, pages 115–123; 1992.

MITOCHONDRIAL AND NUCLEAR GENE EXPRESSION FOR CYTOCHROME OXIDASE SUBUNITS ARE DISPROPORTIONATELY REGULATED BY FUNCTIONAL ACTIVITY IN NEURONS. R. F. Hevner and M. T. Wong-Riley. *J. Neurosci,* Vol. 13, pages 1805–1819; 1993.

REVIEW OF SPACEFLIGHT AND HINDLIMB SUSPENSION UNLOADING INDUCED SARCOMERE DAMAGE AND REPAIR. D.A. Riley, J.L.

Thompson, B.B. Krippendorf, and G.R. Slocum. *BAM,* Vol. 5, pages 139–145; 1995.

IS SKELETAL MUSCLE READY FOR LONGTERM SPACEFLIGHT AND RETURN TO GRAVITY? D.A. Riley. *Adv. Space Biol. Med.,* Vol. 7, pages 31–48; 1999.

EFFECTS OF HINDLIMB UNLOADING ON NEUROMUSCULAR DEVELOPMENT OF NEONATAL RATS. B.L. Huckstorf, G.R. Slocum, J.L.W. Bain, P.M. Reiser, F.R. Sedlak, M.T.T. Wong-Riley, and D.A. Riley. *Dev. Brain Res.,* Vol. 119, pages 169–178; 2000.

NON-WEIGHTBEARING CONDITION ARRESTS THE MORPHOLOGICAL AND METABOLIC CHANGES OF RAT SOLEUS MOTONEURONS DURING POSTNATAL GROWTH. H. Nakano and S. Katsuta. *Neurosci. Lett.,* Vol. 290, pages 145–148; 2000.

ASYNCHRONOUS SYNAPSE ELIMINATION IN NEONATAL MOTOR UNITS: STUDIES USING GFP TRANSGENIC MICE. C.R. Keller-Peck, M.K. Walsh, W.-B. Gan, G. Feng, J.R. Sanes, and J.W. Lichtman. *Neuron,* Vol. 31, pages 381–394; 2001.

THIN FILAMENT DIVERSITY AND PHYSIOLOGICAL PROPERTIES OF FAST AND SLOW FIBER TYPES IN ASTRONAUT LEG MUSCLES. D.A. Riley, J.L.W. Bain, J.L. Thompson, R.H. Fitts, J.J. Widrick, S.W. Trappe, T.A. Trappe, and D.L. Costill. *J. Appl. Physiol.,* Vol. 92, pages 817–825; 2002.

Gravity Plays an Important Role in Muscle Development and the Differentiation of Contractile Protein Phenotype

Experiment Team

Principal Investigator: **Kenneth M. Baldwin**

Co-Investigators: **Gregory R. Adams, Fadia Haddad**

University of California, Irvine, USA

Authors

Gregory R. Adams, Fadia Haddad, Kenneth M. Baldwin

ABSTRACT

Several muscles in the body exist mainly to work against gravity. Whether gravity is important in the development of these muscles is not known. By examining the basic proteins that compose muscle, questions about the role of gravity in muscle development can be answered. Myosin heavy chains (MHCs) are a family of proteins critically important for muscle contraction. Several types of MHCs exist (e.g., neonatal, slow, fast), and each type is produced by a particular gene. Neonatal MHCs are produced early in life. Slow MHCs are important in antigravity muscles, and fast MHCs are found in fast-twitch "power" muscles. The gene that is "turned on" or expressed will determine which MHC is produced. Early in development, antigravity skeletal muscles (muscles that work against gravity) normally produce a combination of the neonatal/embryonic MHCs. The expression of these primitive MHCs is repressed early in development; and the adult slow and fast MHC genes become fully expressed. We tested the hypothesis that weightbearing activity is critical for inducing the normal expression of the slow MHC gene typically expressed in adult antigravity muscles. Also, we hypothesized that thyroid hormone, but not opposition to gravity, is necessary for expressing the adult fast IIb MHC gene essential for high-intensity muscle performance. Groups of normal thyroid and thyroid-deficient neonatal rats were studied after their return from the 16-day Neurolab mission and compared to matched controls. The results suggest: (1) Weightlessness impaired body and limb skeletal muscle growth in both normal and thyroid-deficient animals. Antigravity muscles were impaired more than those used primarily for locomotion and/or nonweightbearing activity. (2) Systemic and muscle expression of insulin-like growth factor-I (IGF-I), an important body and tissue growth factor, was depressed in flight animals. (3) Normal slow, type I MHC gene expression was markedly repressed in the normal thyroid flight group. (4) Fast IIb MHC gene expression was enhanced in fast-twitch muscles of normal thyroid animals exposed to spaceflight; however, thyroid deficiency markedly repressed expression of this gene independently of spaceflight. In summary, the absence of gravity, when imposed at critical stages of development, impaired body and skeletal muscle growth, as well as expression of the MHC gene family of motor proteins. This suggests that normal weightbearing activity is essential for establishing body and muscle growth in neonatal animals, and for expressing the motor gene essential for supporting antigravity functions.

INTRODUCTION

Skeletal muscles comprise the largest organ system in the body, accounting for ~50% of the body's mass. They can change their structure and function in response to (a) chronic physical activity and/or (b) various hormonal/growth factors. Muscles used extensively in either opposing the force of gravity or in performing locomotion, i.e., leg muscles, are highly sensitive to the lack of gravity, as occurs when individuals are exposed to spaceflight. In space, affected muscle fibers undergo marked atrophy and lose strength and endurance (Baldwin, 1996; Caiozzo, 1994; Caiozzo, 1996; Widrick, 1999). Collectively, this can impair performance capability. The full extent of the atrophy in space has not been delineated in humans due, in part, to the fact that astronauts perform physical exercise to counteract muscle atrophy. These activities curtail, but do not prevent, muscle atrophy. Studies on adult rodents in which no countermeasures are performed clearly show that certain muscles normally used in opposing gravity can atrophy by ~50% over several days (Baldwin, 1996; Caiozzo, 1994).

The contractile machinery of a muscle fiber, which regulates muscle contraction, is particularly sensitive to gravity. One of the major proteins comprising the contractile machinery is myosin (referred to as myosin heavy chain (MHC)). This complex protein serves as an important structural and regulatory molecule that precisely regulates the contraction process, particularly how fast or intensely it occurs. MHC proteins exist in different forms called isoforms, and each MHC isoform is the product of a specific MHC gene. Every muscle is highly organized in terms of how these genes are expressed among the muscle fibers. For example, there are clusters of muscle fibers innervated by a common neuron (called a motor unit) that express a slow type of MHC (designated as type I; see Figure 1). Motor units that express this slow MHC are more effective (e.g., they function with less energy expenditure) in enabling the muscle to oppose gravity and perform sustained movement patterns such as walking and jogging. Other motor units express faster forms of MHC in different combinations, and these fast MHCs are designated as fast IIa, fast IIx, and fast IIb in increasing order of their impact on the speed of contraction (Figure 1). Thus, motor units that express these faster MHCs contract faster and generate greater power and speed of movement compared to the slow motor units—although at the expense of greater energy expenditure. In microgravity, the slow muscle fibers lose their capacity for slow MHC gene expression; and the fibers express greater than normal amounts of the fast isoforms, while at the same time the affected fibers atrophy; e.g., they become smaller in diameter, weaker, and faster contracting (Caiozzo, 1994; Caiozzo, 1996; Ohira, 1992; Widrick, 1999).

Following birth, muscles are in an undifferentiated state in terms of their relative size, functional properties, and pattern of MHC gene expression (Adams, 1999). Instead of expressing the typical slow and fast forms of MHC in the various motor units as presented in Figure 1, in the infant state various motor units express immature forms of MHC, designated as embryonic and neonatal isoforms. These are thought

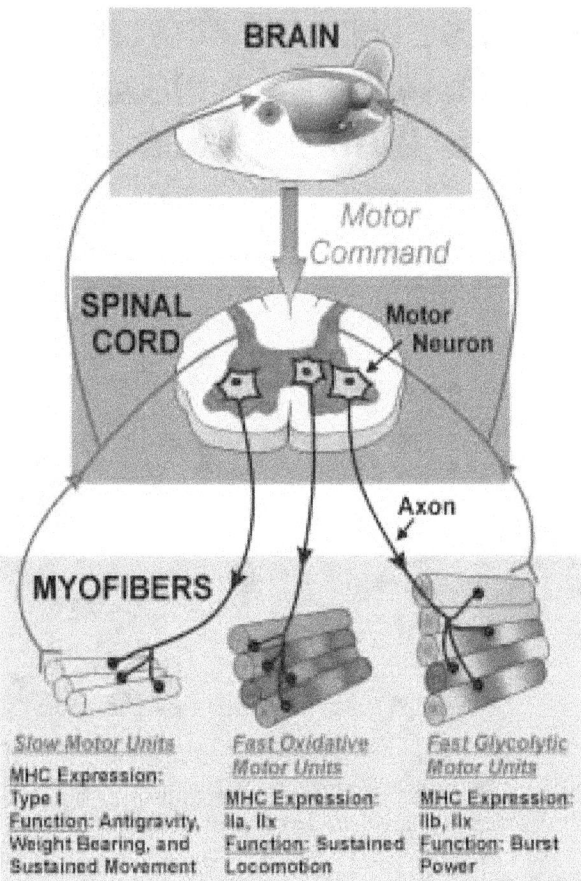

Figure 1. Functional specialization of skeletal muscle. The brain initiates the motor command for a motor neuron to fire and stimulate a group of muscle fibers to contract. A motor unit consists of a single motor neuron together with all the muscle fibers it innervates. Motor units vary in the number of myofibers included; this number can vary several hundred fold in fast vs. slow motor units. Different types of motor units express different MHC phenotypes, having a specialized function. Note that each myofiber can express either a single MHC isoform, or a hybrid mix of two or more isoforms. A skeletal muscle contains thousands of myofibers that belong to different groups of motor units.

to be ineffective in opposing gravity and producing locomotion (Adams, 1999). Therefore, we postulate that with weightbearing activity, and in combination with developmentally induced surges in factors such as growth hormone, insulin-like growth factor-I (IGF-I), and thyroid hormone (T3), muscles become stimulated to undergo marked enlargement and differentiation. This transforms the muscle system to the adult state. Furthermore, accumulating evidence strongly suggests that IGF-I treatment may be an important therapeutic strategy in the treatment of individuals with debilitating muscle-wasting diseases such as muscular dystrophy and amyotrophic lateral sclerosis (ALS). Thus, fundamental studies on the role that IGF-1 might play in regulating skeletal muscle growth in infants and adults are of clinical importance. Figure 2 depicts the proposed cascade of events impacting body growth, muscle growth, and fiber differentiation.

Thus, while gravity has been suspected to play an important role in the control of muscle structure and function, relatively little is known about its role in neonatal development. Since it is neither logistically nor ethically feasible to expose human infants to the environment of spaceflight, animal research is necessary. In animals, however, the effects of gravity (and especially its lack) on muscle growth and development have not been studied extensively. This is because ground-based models cannot continuously remove gravity or weightbearing stimuli during critical stages of development to impact the musculoskeletal system. In rats, the first month of postnatal life

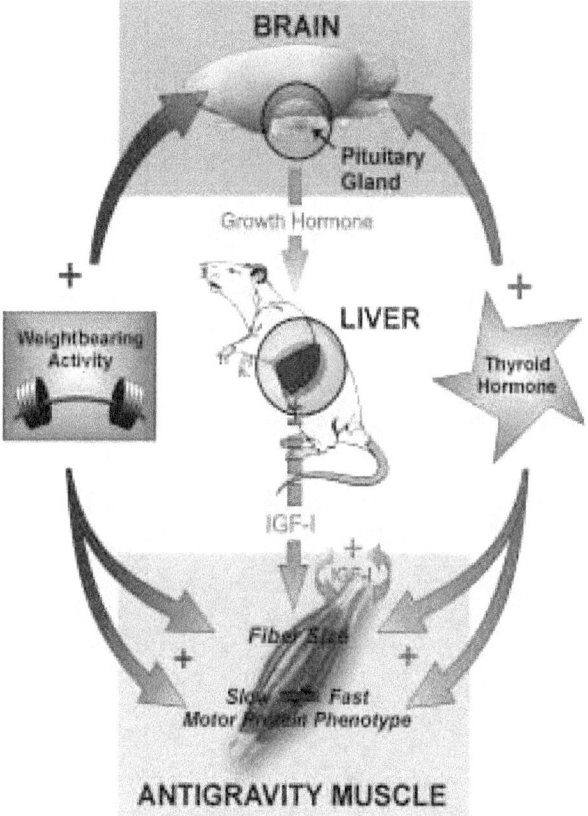

Figure 2. Proposed cascade of events impacting muscle growth and fiber differentiation. Growth hormone (GH) secreted by the pituitary gland in the brain stimulates the liver to produce systemic IGF-I, which circulates in the bloodstream. This GH-IGF-I production is stimulated by circulating thyroid hormone. Also shown is that the muscle has the capacity to express IGF-I independently of the liver. This response is influenced by the level of activity and functional demand imposed on the muscle. Thyroid hormone affects muscle differentiation and MHC phenotype. Muscle fiber size and biochemical properties (MHC phenotype) are influenced by the interaction between thyroid hormone, weightbearing activity, and growth factors such as IGF-I.

encompasses a period of rapid growth and development when these animals increase in size almost exponentially (Adams, 1999). Compared to humans, the rate of rodent development is compressed such that the first 25 days of growth roughly approximate the first 5–7 years for children. For example, starting at just 6–7 days of age, young rats begin pelvic weightbearing and by day 10 they walk on all fours (Clarac, 1998). During this time the growth of some muscles, particularly those limb muscles that oppose gravity and bear the animal's weight, will outpace the growth of the body as a whole. Since these muscles can be expected to experience the same milieu of circulating hormones and growth factors as the other tissues in the body, the question arises as to how this differential muscle growth is controlled. The 16-day Neurolab spaceflight mission enabled us to critically examine the continuous absence of weightbearing activity during key stages in the development of rodent muscle. During this time, primitive motor genes are being turned off while adult motor protein (MHCs) genes are being turned on. Thus, the role that gravity plays in this muscle growth and differentiation process, in the context of other factors such as thyroid hormone and IGF-I, could be examined for the first time in a systematic way.

We tested the following hypotheses: (1) In the absence of either weightbearing activity or an intact thyroid state, both body and muscle growth become impaired due to reduced IGF-I expression, both systemically and intramuscularly. (2) Younger neonates are more sensitive than older neoanates to unweighting interventions, because their normal regulatory mechanisms are interrupted earlier in development. (3) Weightbearing activity is essential for the normal expression of slow, type I MHC gene expression. (4) Thyroid hormone is necessary for transforming the neonatal/embryonic MHC into the fast adult MHC type of a typical fast skeletal muscle—and this process occurs independently of gravity.

METHODS

Experimental design and litter formation

To study the interactive effects of spaceflight and thyroid status on muscle development, we needed to generate groups of rats having either an intact functioning thyroid gland (designated as euthyroid) or a defective thyroid gland (designated as thyroid deficient (TD)). To accomplish this, timed-pregnant female rats were obtained from Taconic Farms (Germantown, NY), and were housed initially in standard rodent cages in the vivarium at Kennedy Space Center in Florida. The litters produced were adjusted to contain eight rats with equal gender distribution. Study litters were selected based on: (a) normal body growth during the first five days of age, and (b) normal water/food consumption and normal rat retrieval behavior among the mothers (Adams, 2000a). Litters were then randomly assigned to the following experimental groups: (1) euthyroid-ground-control (N=16); (2) euthyroid-flight (N=8); (3) TD-ground-control (N=16); and (4) TD-flight (N=8). In actuality, there were two separate groups of ground controls used for comparisons for both the euthyroid- and TD-flight animals (Adams, 2000a).

Table 1. Litter and treatment characteristics.

Age at Launch ↓	Group Assignment ↓
	Euthyroid
	Basal
Eight Days	NC8-Ground*
	NC8-Flight
14 Days	NC14-Ground*
	NC14-Flight
	Thyroid Deficient
Eight Days	TD8-Ground*
	TD8-Flight

NC, normal (euthyroid) control; TD, thyroid deficient.
*: Each ground group was a pool of two separate groups that differed in that they were housed in either standard vivarium rat cages or in smaller cages that matched the configuration of cages used on the Space Shuttle.

These two groups were housed in either standard rat cages or in smaller cages matching the configuration of the cages used on Neurolab. Since the results obtained from these two groups were similar, they were combined under a single "ground" heading to simplify data presentation. The flight groups were launched into space at eight days of age.

An additional three litters from timed-pregnant rats were randomly assigned to experimental groups for launch at approximately 14 days of age. They were designated as: (1) euthyroid-ground-controls-14 (N=12) and (2) euthyroid-flight-14 (N=six). This older flight group was launched at approximately 14 days of age. As in the case for the two younger ground control groups, a single large ground control group was formed from the merging of the two separate older ground control groups as described above (Adams, 2000a). The older group did not include a TD group since there was insufficient housing on board the Shuttle. In addition to the above experimental groups, three litters also were selected randomly and studied on the day of launch (i.e., ~eight days of age) for baseline analyses. All experimental procedures were approved by both the NASA Institutional Animal Care and Use Committee (IACUC) and the University of California, Irvine IACUC. Table 1 summarizes the various assigned litters and their subsequent grouping for tissue analyses and data presentation.

Manipulations and tissue processing

Thyroid deficiency was induced in those designated litters by surgically implanting an osmotic pump into the abdominal cavity of the nursing mother rat. The pump delivered a continuous dose of the antithyroid drug, propylthiouracil (PTU), which made the young rats hypothyroid via the delivery of PTU via the mother's milk. The details of this procedure are provided elsewhere (Adams, 2000a).

Sixteen days after launch, the rats were returned to Kennedy Space Center, which was the Orbiter landing site. Five hours after landing, the rats were euthanized and dissected for tissue procurement as previously described (Adams, 2000a).

Analytical procedures

Cardiac native myosin analyses utilizing non-denaturing polyacrylamide gel electrophoresis were performed as previously described (Haddad, 1993; Swoap, 1994). Skeletal MHC isoforms were separated using an SDS-PAGE technique (Adams, 1999; Talmadge, 1993). MHC mRNA analyses utilized a semiquantitative reverse transcriptase polymerase chain reaction procedure as described previously (Adams, 2000b). Muscle and plasma IGF-I, muscle protein, and DNA content were determined as previously described (Adams, 2000a).

Statistics

All values are reported as mean and standard error of the mean. For each time point, treatment effects were determined by analysis of variance (ANOVA) with post-hoc testing using the Prism software package (GraphPad Software, Inc., San Diego, CA). Pearson's correlation analyses of relationships were performed using the Prism package. The 0.05 level of confidence was accepted for statistical significance.

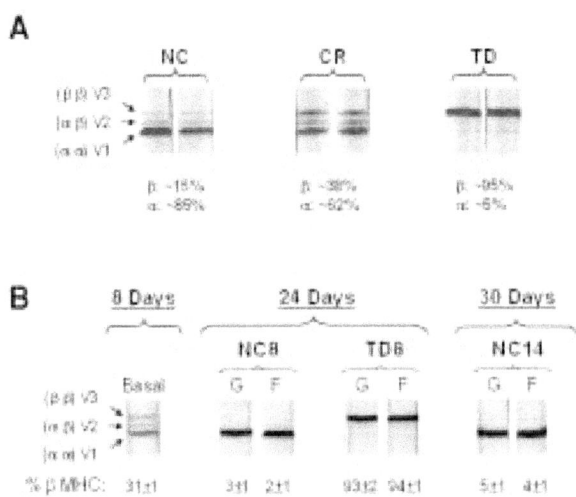

Figure 3. Heart myosin isoforms as separated by native gel electrophoresis and stained with coomassie blue. (A) Cardiac isomyosin profile of normal control (NC), 50% caloric restricted (CR), and TD adult rats. (B) Cardiac isomyosin composition in the different Neurolab groups. Eight days (basal group), 24 days (NC8, TD8), and 30 days (NC14) refer to the age of the rats at the time of tissue procurement. G: Ground, F: Flight. Also shown is the % MHC in each group as determined by laser scanning densitometry of the actual gels (Molecular Dynamics, Inc., Sunnyvale, CA).

Table 2. Body weight and muscle wet weight from basal, control thyroid-deficient, and space-flown young rats.

	N	Body	Ventricles	Soleus	Plantaris	Medial Gastroc.	Lateral Gastroc.	Vastus Inter.	Vastus Lateralis	Vastus Medialis	Tibialis Anterior
		(g)					(mg)				
Basal	24	17±1#	77±4#	4±0.3#	7±1#	12±1#	—	—	—	—	15±1#
NC8-Ground	16	80±2	340±13	31±1	64±2	135±4	155±6.	19±1	121±4	61±3	128±4.
NC8-Flight	6	40±5*	230±36*	7±1*	19±4*	48±8*	57±1*	7±1*	53±8*	25±5*	53±6*
% change (F vs. G)		−50	−32	−77	−70	−64	−63	−63	−56	−59	−58
TD8-Ground	16	39±1#	137±3#	10±1#	20±1#	38±3#	46±3#	6±0.4#	41±2#	21±1#	42±2#
TD8-Flight	4	31±3#	104±3#	5±1*#	13±2#	27±3#	29±3#	6±1#	50±9#	14±2#	28±3*#
% change (F vs. G)		−21	−24	−50	−35	−29	−37	−6	22	−33	−33
% change (TD vs. NC)		−51	−60	−68	−69	−72	−70	−66	−66	−66	−67
NC14-Ground	12	114±4#	425±19#	46±2#	99±3#	216±8#	234±12#	27±1#	181±7.#	100±5#	192±5#
NC14-Flight	6	98±3*#	365±8*	27±2*#	78±3*#	180±8*#	203±12*#	19±1*	163±6*#	90±4#	151±6*#
% change (F vs. G)		−14	−14	−-41	−21	−17	−13	−30	−10	−10	−21

Basal rats studied at eight days of age; NC8, euthyroid neonates eight days old at liftoff; TD8, thyroid-deficient neonates eight days old at liftoff; NC14, euthyroid neonates 14 days old at liftoff. Ground (G), ground-based rats; Flight (F), space-flown rats. Gastroc, gastrocnemius; Inter, intermedius. *, P<0.05 Flight vs. matched ground-based control; #, P<0.05 vs. NC8-Ground group.

RESULTS

Body weights

Body weights of the euthyroid and TD ground control groups in this study closely matched the growth curve established for these two groups in previous ground-based experiments (Table 2, Adams, 1999). As is typical for TD animals, body weights of the TD rats were significantly lower than those of the euthyroid group. Body weight of the euthyroid and TD flight groups were significantly lower than their ground-based counterparts. The effect of flight on body weight in the older rats (NC14 group) was much less dramatic than in the younger group (NC8) (Table 2).

On the Neurolab mission, many of the young (eight days old at launch) rats died on the flight. One concern was that these rats received inadequate nutrition. The reduced weight gain in the flight groups, however, is less likely the result of lack of nutrition. This is based on myosin isoform profiles in the heart. The MHC genes in the heart are very sensitive to both energy intake restriction and thyroid state. Cardiac β-MHC expression becomes predominant in response to either hypothyroidism or reductions in energy intake (Figure 3A). Cardiac myosin analysis shows, however, that cardiac β-MHC is not expressed in young euthyroid rats (NC8) exposed to spaceflight (Figure 3B). If these animals were energy deprived, we would have expected to see significant β-MHC gene expression in the hearts of the euthyroid-flight group (NC8-flight). Further, since the hearts of the both the ground- and flight-based TD rats responded with almost exclusive (>93%) expression of the β-MHC isoform (which is a key marker of thyroid dependency), the flight-exposed TD rats must have received relatively the same amount of nourishment (milk containing the antithyroid drug) as compared to their ground-based counterparts, so as to make both groups TD to the same degree.

Skeletal muscle growth

Figures 4 and 5 show a schematic representation of some key hindlimb muscles, soleus and plantaris, as they grew during the Neurolab experiment. Note the dramatic difference between the size of the muscles in the baseline controls studied at launch vs. the ground-control animals at the completion of the mission. Consistent with the body weight data, absolute muscle weights were markedly reduced in all flight groups relative to age-matched ground controls (Table 2). Also, muscle weights of the TD groups were lower than the euthyroid groups (both flight and ground control) (Table 2). In the antigravity slow-twitch soleus muscle, spaceflight exerted a greater relative growth retardation response compared to its fast-twitch counterpart, the plantaris (compare Figure 4 v. 5). When the data were normalized to body weight (mg muscle/gram body weight), it was apparent that the relative soleus muscle weight in the younger spaceflight groups did not increase beyond basal values (Figure 6). In other words, the relative muscle mass of the soleus muscle of the young spaceflight rats at 23 days of age remained essentially the same as that seen in the baseline group studied at eight days of age; i.e., just prior to the 16-day flight (Figure 6). Nonweight-bearing leg muscles such as the tibialis anterior appeared to be the least affected by spaceflight (see Table 2 and Figure 6). Also, it is important to note that the spaceflight inhibitory effects on muscle growth were less dramatic in the older rodents (NC14) than in the younger groups (NC8) (Table 2, Figure 6).

As an index of muscle growth, muscle total protein and DNA were studied in selected muscles, such as the medial gastrocnemius (MG). The data show significant decreases in total muscle protein and DNA content due to either flight or

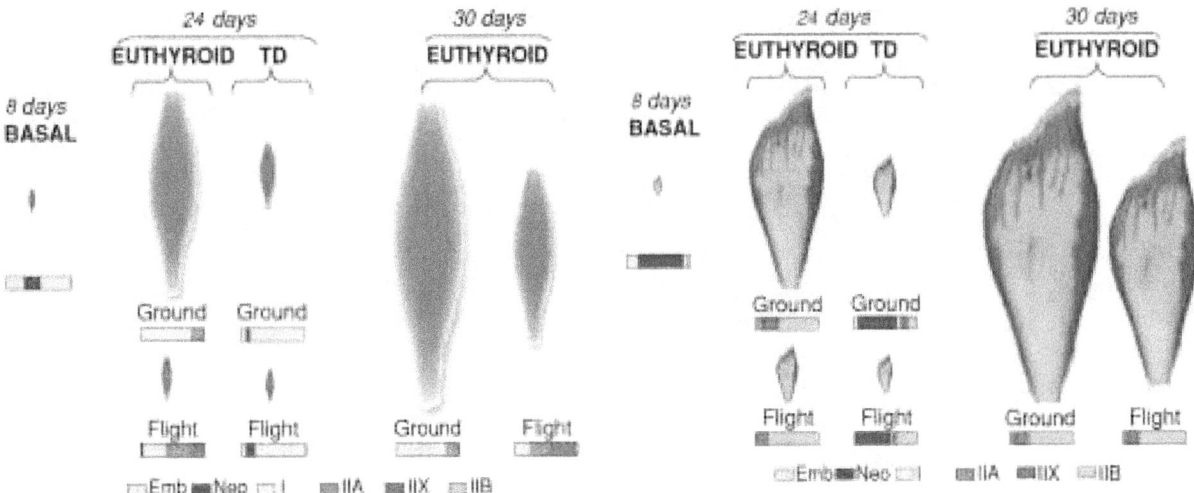

Figure 4. Soleus muscle growth and MHC protein profile across the Neurolab experimental groups. Schematic representation of the soleus muscle as it grows during the time-course of the Neurolab experiment in each of the experimental groups. Note the dramatic difference between the size of the muscle in the baseline controls studied at launch (basal) vs. the ground control animals at the completion of the mission (24 days and 30 days euthyroid). Also note the lack of growth in the soleus muscle of TD flight animals. Below each muscle is a bar representing the distribution of MHC isoform proteins in the corresponding group.

Figure 5. Plantaris muscle growth and MHC protein profile across the Neurolab experimental groups. Schematic representation of the rat plantaris muscle as it grows during the time-course of the Neurolab experiment in each of the experimental groups. Note the dramatic difference between the size of the muscle in the baseline controls studied at launch (basal) vs. the ground control animals at the completion of the mission (23 days and 30 days euthyroid). Below each muscle is a bar representing the distribution of MHC isoform proteins in the corresponding group.

thyroid deficiency, especially in the younger neonates (NC8). These data are consistent with a flight-induced reduction in muscle growth based on muscle weight data (Table 2). Despite these differences in muscle DNA and protein content, a strong relationship was maintained between the DNA and protein content across all the experimental groups. These findings suggest a tightly coordinated relationship between muscle size and both DNA and protein content. This relationship was not disrupted by either spaceflight or thyroid deficiency.

Hormonal data: insulin-like growth factor-I

The reductions in growth, evidenced by the body weight measurements, were paralleled by the decreased levels of plasma IGF-I (Figure 7) such that there was a significant correlation between plasma IGF-I levels and body weight among the experimental groups (Figure 7C). In analyzing IGF-I expression at the muscle level, it is important to note that IGF-I was not determined on soleus muscle since there was insufficient tissue available for such analyses. However, determinations were made on the tibialis anterior and medial gastrocnemius muscles, which are relatively larger muscles than the soleus (Table 2). Results show that IGF-I peptide levels for these muscles were significantly reduced in response to both spaceflight and thyroid deficiency (Figure 8 A and B). Data analyses reveal a significant positive correlation between intramuscular IGF-I and muscle weight (Figure 8 C and D). In the ground-based neonates, an apparent developmental surge in both circulating and muscle

IGF-I can be seen when compared to the basal values (Figures 7 and 8), especially between eight days (basal) and 24 days of age (NC8). This response appears to have been blunted, particularly in the young euthyroid flight-based animals and in both the ground-based and flight-based TD groups such that their levels corresponded more closely to those of the less mature basal group. Also, there was a significant positive correlation between muscle IGF-I peptide concentration and the protein or DNA content of the MG muscle.

Interaction between spaceflight and thyroid deficiency on growth processes

Exposure to spaceflight resulted in a significant reduction in the general growth of young rats (NC8 and NC14) (Figure 7A). The imposition of a TD state resulted in a reduction in body mass growth that was similar in effect to spaceflight exposure and was not significantly altered further by the combination of the two interventions (Figure 7A).

The complex presentation of growth data in Figure 7 makes it difficult to discern the relative impact the different treatments imposed on the rats. To partition the impact of spaceflight vs. thyroid deficiency and to discern potential interactions between these variables, we calculated the relative body growth deficit imposed by each treatment (Flight/TD) either alone or in combination relative to the appropriate group. The data from Figure 7A were used to generate Figure 9. From this analysis, it is evident that both spaceflight (NC8-Flight vs. NC8-Ground) and hypothyroidism (TD8-Ground vs. NC8-Ground) resulted in an ≈50% decrease in somatic growth in the younger rats when each was imposed separately. The flight effect on somatic growth was much reduced when imposed on TD animals (TD-Flight vs. TD-Ground). Likewise, TD effects were also reduced when taken in the context of spaceflight (TD-Flight vs. NC-Flight). The combined effects of thyroid deficiency and of spaceflight on somatic growth (TD8-Flight vs. NC8-Ground) were slightly larger (60% vs. 50%; $p > 0.05$) but not statistically different than the effects of each of the manipulations imposed

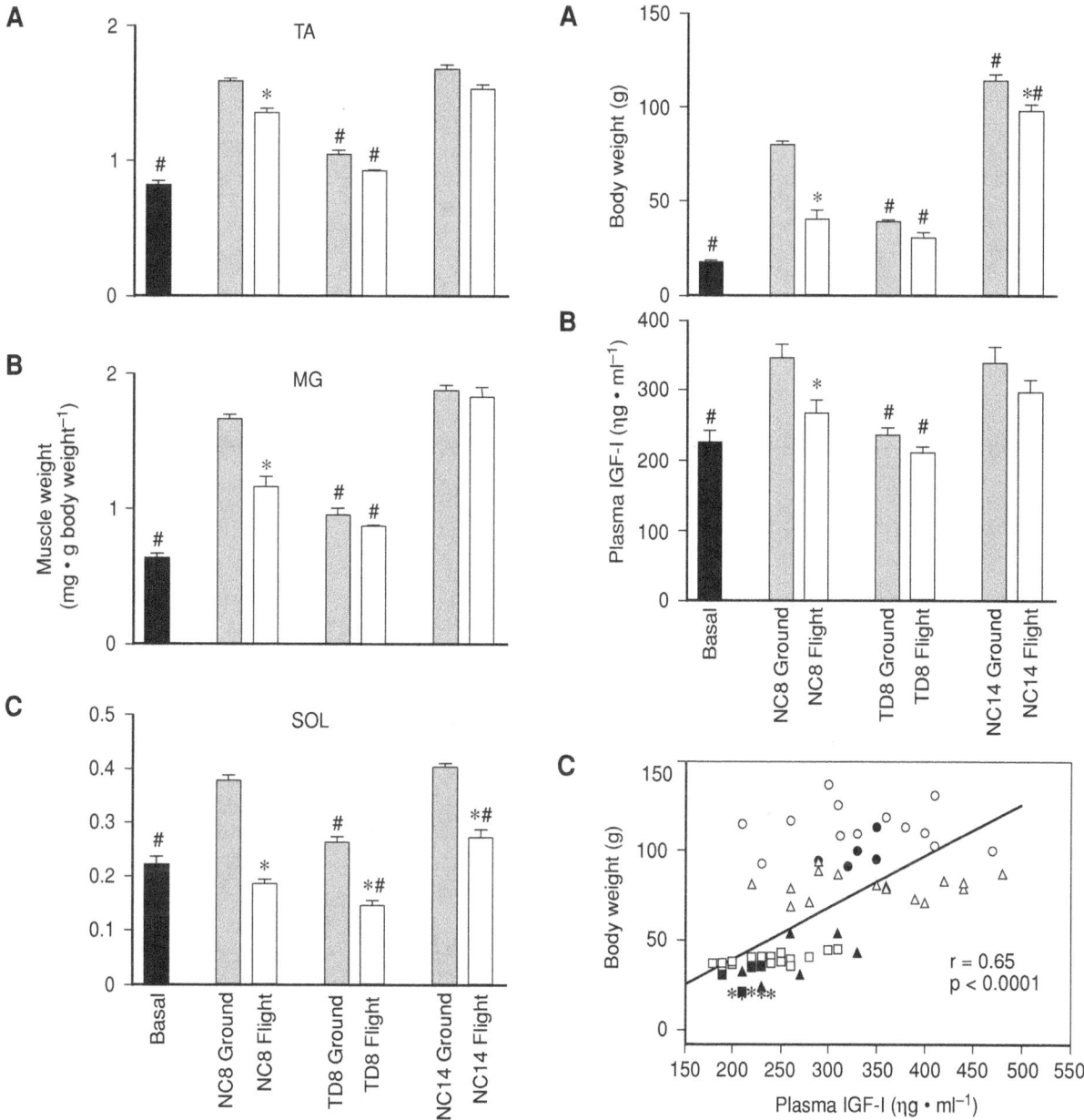

Figure 6. Muscle weight normalized to body weight: the effects of TD and/or spaceflight. (A) TA, tibialis anterior, a non-weightbearing locomotor muscle expressing primarily fast MHC. (B) MG, medial gastrocnemius, a weightbearing locomotor muscle expressing primarily fast MHC. (C) SOL, soleus, a weightbearing postural/locomotor muscle expressing primarily slow MHC. Refer to Table 2 for group designations. *, $p < 0.05$ Flight vs. Ground; #, $p < 0.05$ vs. NC8 Ground. n values as in Table 2. (From Adams, 2000a, with permission; reproduced from the *Journal of Applied Physiology*.)

Figure 7. Body weight and plasma IGF-I concentration: the effects of TD and/or spaceflight. (A) Body weight. (B) Plasma IGF-I concentration. (C) Correlation between body weight and plasma IGF-I concentration with best-fit line. Refer to Table 2 for group designations. *, $p < 0.05$ Flight vs. Ground; #, $p < 0.05$ vs. NC8 Ground; N values as in Table 2. Symbols: *, Basal; △, NC8 Ground; ▲, NC8 Flight; □, TD8 Ground; ■, TD8 Flight; ○, NC14-Ground; ●, NC14 Flight. (From Adams, 2000a, with permission; reproduced from the *Journal of Applied Physiology*.)

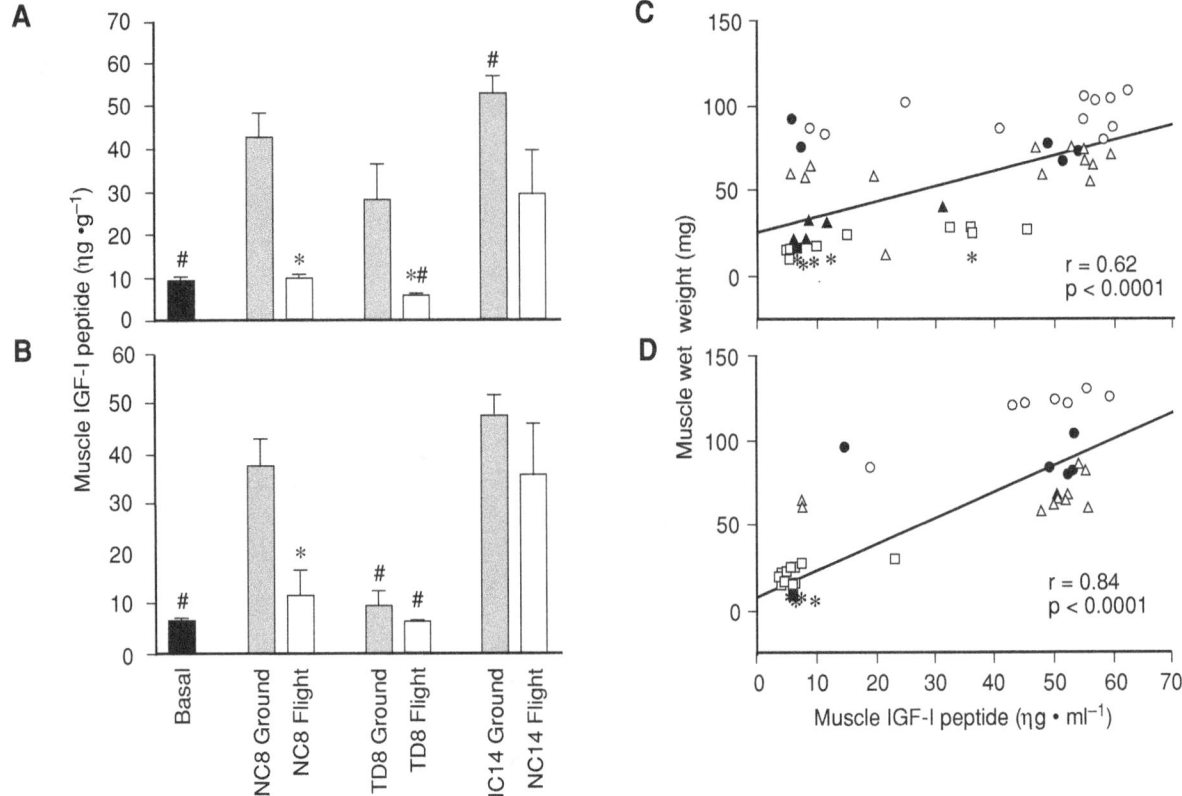

Figure 8. Muscle IGF-I peptide concentrations: effects of TD and/or spaceflight. (A) Tibialis anterior (TA) muscles IGF-I concentration. (B) Medial gastrocnemius (MG) muscles IGF-I concentration. (C) The correlation between muscle wet weight and muscle IGF-I peptide from all TA muscles depicted in A. (D) The correlation between muscle wet weight and muscle IGF-I peptide from all MG muscles depicted in B. Refer to Table 2 for group designations. *, p<0.05 Flight vs. Ground; #, p<0.05 vs. NC8 Ground. Symbols are as in Figure 7, n values are as in Table 2. (From Adams, 2000a, with permission; reproduced from the *Journal of Applied Physiology*.)

separately (Figure 9). This presentation also highlights the lesser sensitivity of the older rats (NC14) to the effects of loss of weightbearing activity (Figure 9, NC14-Flight vs. NC14-Ground). This age-related finding corresponds with data that show a decline in receptors for IGF-I in skeletal muscle IGF-I after about 12 days of age, indicating a potential age-related decrease in the sensitivity of muscles to this growth factor (Shoba, 1999).

The general decrease seen in body growth was also reflected in lower limb muscle weights from the flight vs. ground-based rats (Table 2). As with body weight, we have used the data from Figure 6 to apportion the relative growth deficit imposed by the separate and combined treatments of hypothyroidism and spaceflight (Figure 10). Skeletal muscles expressing primarily fast MHC isoforms (such as the medial gastrocnemius) showed decreased muscle growth in the younger (NC8-Flight vs. NC8-Ground) but not in the older group (NC14-Flight vs. NC14-Ground) (Figure 10). In an antigravity muscle such as the soleus, spaceflight resulted in a significant relative muscle growth deficit in both younger (50%) and older neonates (32%). Thyroid deficiency appeared to have a greater impact on fast-twitch muscle (TA and MG) growth than spaceflight; while

the opposite is true in antigravity slow-twitch muscle (Soleus-Sol) (Figure 10). As with body growth, there did not appear to be a notable additive effect of these treatments in any of the studied muscle types.

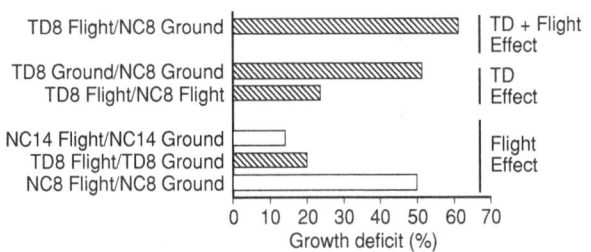

Figure 9. The relative growth deficit imposed by the separate and combined interventions of TD and exposure to spaceflight (Flight). Data are calculated from the body weight data presented in Figure 9A as follows: Flight effect: NC8-Flight/NC8-Ground, TD8-Flight/TD8-Ground, NC14-Flight/ NC14-Ground; TD effect: TD7-Flight/NC7-Flight, TD7-Ground/ NC7-Ground; TD+Flight Effect: TD8-Flight/NC8-Ground. (From Adams, 2000a, with permission; reproduced from the *Journal of Applied Physiology*.)

Spaceflight and thyroid deficiency effects on myosin heavy chain gene expression

Soleus Muscle – At seven days of age, the MHC profile of the soleus muscle consists primarily of embryonic (~27%), neonatal (25%), slow type I (45%), and traces of fast IIa and IIb (~2% each); i.e., the embryonic/neonatal isoforms account for ~50% of total MHC pool (Figure 4). By 24–30 days of age the soleus muscle is transformed into an adult phenotype consisting of ~80% type I MHC and 20% fast IIa MHC such that the embryonic/neonatal isoforms become repressed and replaced by increases in adult slow MHC gene expression (Figure 4). In rodents initially exposed to spaceflight at eight days of age and their muscles subsequently examined at 23 days of age (i.e., after 16 days in space), the soleus MHC profile is 3% neonatal, 36% slow, 42% fast type IIa, 16% fast type IIx, and 3% fast type IIb. Thus, spaceflight not only blunted slow, type I MHC gene expression in the developing soleus muscle, but it also created a profile typically seen in most fast muscles in which the fast MHC isoforms dominate the MHC protein pool. In contrast, thyroid deficiency caused retention of significant relative levels of both the embryonic and the neonatal MHC in the flight-based and ground-based groups, while blunting expression of all fast MHC isoforms as compared to what is typically seen in their euthyroid counterparts. Thus, regardless of gravity status, the soleus muscles of TD neonates expressed predominantly the type I isoform (76–84%) with the rest consisting of the developmental types (Figure 4).

When an MHC gene is expressed, it creates a messenger RNA (mRNA) that carries instructions on how to make the protein. These instructions are translated by the cell to make the MHC protein. The profiles seen at the protein level were essentially mimicked at the mRNA level across the various experimental groups (Figure 11). This suggests that the regulation of MHC gene expression in these developing rats is at the level of how the soleus muscle transcribes and maintains the level of mRNA (a process called pretranslational regulation) prior to translating this message into its encoded protein. Evidence for this type of regulation is the strong positive relationship that exists between MHC mRNA vs. protein for the various isoforms. One exception was that in the euthyroid-flight and ground-based groups, the expression of embryonic mRNA was still manifest throughout the developmental period (Figure 11) even though its protein product appeared to be fully repressed (Adams, 2000b). This is a unique observation, which we don't have a good explanation for at the present time; but the results suggest an uncoupling between the control of transcriptional vs. translational processing of the embryonic mRNA.

Plantaris Muscle – At eight days of age, the MHC profile of the plantaris muscle is even more undifferentiated relative to the adult state than is a slow muscle, since its MHC profile is 20% embryonic, 70% neonatal, 6% slow, and 4% fast IIb (Figure 5). This profile is essentially mirrored at the mRNA level (Figure 11), suggesting that the adult MHC genes have not been "turned on" at this early stage of development. By

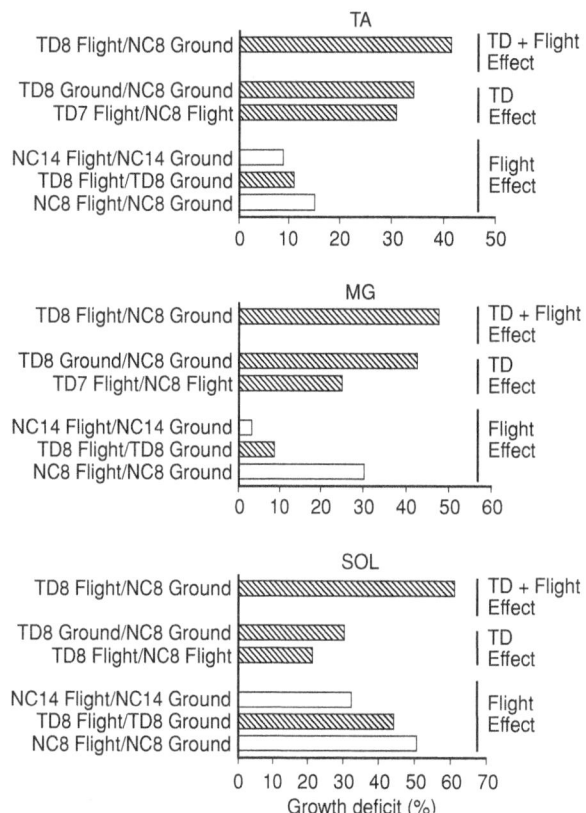

Figure 10. The relative growth deficit imposed on limb skeletal muscles by the separate and combined interventions of thyroid deficiency and exposure to spaceflight. Data are calculated from the normalized muscle weight data presented in Figure 7 (see Figure 9 legend for groupings). (From Adams, 2000a, with permission; reproduced from the *Journal of Applied Physiology*.)

23–30 days of age, the muscle becomes markedly transformed such that the profile consists of 60% fast IIb, 30% fast IIx, 5% fast IIa, and 5% slow MHC (Figure 5). Thus, this muscle, in contrast to the soleus, is characterized by a predominance of the fast type IIb and IIx MHCs. Exposure of young euthyroid rats to spaceflight at eight days of age exerts a subtle effect on this muscle by repressing expression of both the fast type IIa and slow, type I MHCs while augmenting expression of the IIb MHC. Interestingly, thyroid deficiency exerts a unique effect on the plantaris muscle by markedly blunting the transformation process, noted above, whereby the neonatal MHC isoform gene expression is repressed while that of the fast IIb MHC becomes predominant (Figure 5). In essence, thyroid deficiency maintains the plantaris muscle in an undifferentiated state, and this process occurs independently of exposure to microgravity. This process appears to be regulated by a combination of transcriptional, posttranscriptional, and translational processes based on the mRNA data profiles for the different isoforms (Figure 11).

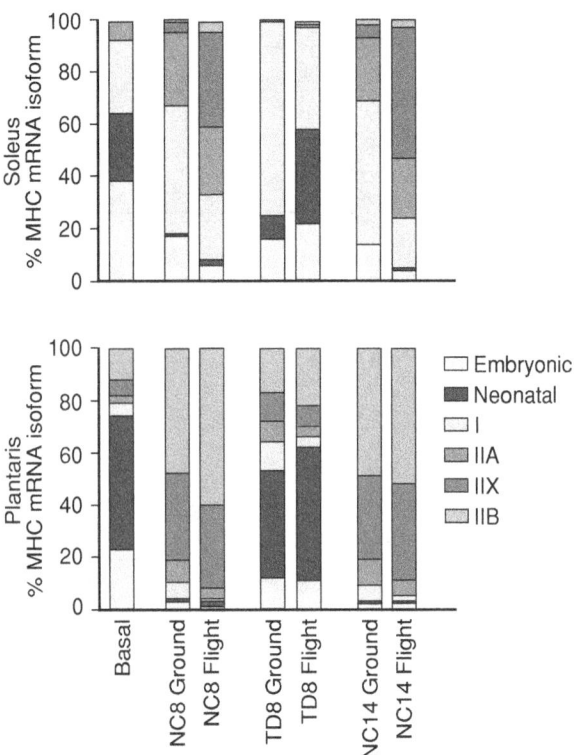

Figure 11. MHC mRNA isoform profiles in the soleus and plantaris muscles. Each MHC isoform is presented by its proportion as percent relative to the total MHC mRNA pool as determined by RT-PCR methods (see Adams, 2000b for details).

DISCUSSION

Factors impacting growth of developing skeletal muscles

The findings of this project clearly show that the separate and combined effects of unloading (as induced by the environment of spaceflight) and thyroid deficiency collectively reduce both body and muscle growth. Also, gene expression of the MHC family of motor proteins responsible for regulating muscle contractile processes is altered.

The reduction in growth appears to be more critical for younger (eight-day-old) vs. older (14-day-old) rodents. Both the body weights and normalized muscle weights of the younger rats were reduced to a markedly greater extent relative to controls compared to the older euthyroid rats exposed to spaceflight (Table 2; Figures 6, 10). This suggests that the latter group, which is further along in development when chronic weight-bearing activity was stopped, may have some protection when developing in the spaceflight environment. Furthermore, hormonal status can play a pivotal role in the growth process, since thyroid deficiency caused approximately equivalent reductions in body and muscle growth of both the ground-based and flight-based TD groups (Table 2).

Moreover, given the equivalence of the growth retardation in both the ground- and flight-based TD groups (and given the similarity of the euthyroid flight-based neonates to the TD groups

irrespective of spaceflight), some other factor(s) also may be playing a role in affecting growth across these groups. One possibility is that the young neonatal euthyroid flight-group was actually TD. We think this is unlikely since both the cardiac and soleus skeletal muscles of the euthyroid flight-group demonstrated either normal or exaggerated fast MHC profiles. This would not have been expected if these flight animals were experiencing thyroid deficiency. That is, these flight animals would have demonstrated a very high relative proportion of the β-MHC (i.e., slow, type I MHC) in both their soleus and heart muscles if they were actually TD. Such was not the case (see Figures 3, 4, 11).

Another possibility is that the euthyroid flight-group was nutritionally (calorically) compromised, which reduced their growth. While spaceflight may have had some impact on the energy intake of these animals, we do not feel that this was extensive enough to account for the growth reductions noted. Otherwise, we would have observed a relatively high level of expression of the β-MHC gene in the hearts of these animals. Instead, we observed very low to nonexistent levels of this MHC in their hearts; i.e., approximately equivalent to that seen in the young and older ground-based control groups as well as in the older euthyroid flight-based group (Figure 5). Further, if the TD flight neonates were energy deprived—i.e., getting insufficient milk and, hence, insufficient antithyroid drug—these rats would not have been hypothyroid to the same degree as their ground-based, nutritionally provided counterparts. The cardiac and skeletal MHC data suggest that they were.

Instead of the above possible scenarios, we feel that in both the younger euthyroid flight-based group and in both TD groups (flight- and ground-based), their body and muscle growth were retarded by an impairment in the normal operation of the thyroid GH-IGF-I axis (Figure 2). This is illustrated by (a) the reduction in both systemic and muscle-specific (weightbearing and nonweightbearing) IGF-I levels in these animals relative to their respective control groups; and (b) the strong positive correlation between body mass and systemic IGF-I (Figure 7). The correlations between muscle mass and muscle-derived IGF-I (Figure 8), as well as between muscle protein accumulation and muscle IGF-I levels, also support this. Taken together, these results point to a complex interaction of weightbearing activity, thyroid hormone levels, and the corresponding expression of both the systemic and muscle-specific levels of IGF-I. Together these are essential for normal body and muscle growth during critical stages of neonatal development; i.e., the first three weeks postpartum. The uncoupling of either weightbearing activity or thyroid hormone from this axis during this key stage of development clearly limits the growth potential of both the animal and the musculoskeletal system in particular.

Factors impacting the myosin heavy chain gene family during development

The MHC isoform gene family of contractile proteins represents the most abundant type of protein expressed in muscle, which accounts for ~20-25% of the protein pool in a typical adult muscle cell. Following birth, all striated muscles (including skeletal muscle) are undifferentiated with regard to the MHC phenotype that is expressed. Those muscles used extensively for weightbearing and locomotion express primarily the embryonic/neonatal isoforms, which are ineffective in supporting the varied contractile intensities that are needed for performing high-intensity activities in the adult state.

The findings reported in this project clearly show that weightbearing activity early in the rodent's first three—four weeks following birth is required for the normal expression of slow, type I MHC, the isoform that predominates in motor units designed for opposing gravity (Figure 1). In the absence of gravity or weightbearing stimuli, full transcriptional activity and hence subsequent translation of the type I mRNA are dramatically reduced in spaceflight-exposed rats relative to their ground-based counterparts (Figures 4, 11). This reduction in slow MHC expression, in combination with the retardation in muscle growth, markedly reduces the mass of skeletal muscle (number and size of the slow motor unit pool) available to support routine antigravity activities and movement. Thus, the motor skills of the animal likely become compromised (Walton, 1998). Hence, in the absence of weight-bearing activity, and in the presence of normal circulating thyroid hormone, all the muscles of the lower extremity become biased to expressing primarily the three fast MHC isoforms (Figures 4, 5, 11). The levels of the slow, type I MHC isoform are insufficient to enable the rats to effectively oppose gravity and perform sustained locomotion.

This experiment also shows the importance of thyroid hormone (T3) in regulating both muscle mass and contractile protein differentiation during neonatal development. Thyroid hormone not only interacts with IGF-I to establish normal growth, but also appears essential to enable the rat to transform its faster contracting muscles from an undifferentiated infant state to a differentiated adult state. This transformation provides a sufficient expression of fast MHCs for the rat to perform activities of a higher intensity and power output. In the absence of sufficient thyroid hormone, while the relative abundance of slow MHC is abundantly expressed independently of weightbearing stimuli, the muscle system remains in an immature state in both size and functional capabilities. Both weightbearing activity and a normal thyroid state combine to ensure that the muscle system develops sufficiently.

One of the shortcomings of the Neurolab mission was the loss of a significant number of young rats during the flight. The reasons are not fully known, but are likely due to the housing facilities, which were not optimal for the mothers to nurture the size of the litters chosen for these experiments. That is, some of the mothers may have purposely ignored weaker rats to adequately care for the remainder of the litter. Whatever the reason, this event limited the availability of rats for analyses concerning their growth and muscle properties during recovery from spaceflight. We can only speculate that over time the young flight-exposed rats would have gained normal body and muscle mass along with a normal MHC phenotype in the absence of any therapeutic activity program. Whether these possible recovery processes would enable appropriate motor performance and exercise capacity remains uncertain.

Acknowledgements

This research was supported by National Institutes of Health NS-33483 and NASA NAG2-555. The authors wish to acknowledge the technical contributions of Mike Baker, Paul Bodell, Sam McCue, Anqi Qin, Li Qin, and Ming Zeng. Also we thank the members of the research support teams from NASA Ames Research Center and the Kennedy Space Center, and in particular, experimental support scientist, Ms. Vera Vizir.

REFERENCES.

RAT SOLEUS MUSCLE FIBER RESPONSES TO 14 DAYS OF SPACEFLIGHT AND HINDLIMB SUSPENSION. Y. Ohira, B. Jiang, R. R. Roy, V. Oganov, E. Ilyina-Kakueva, J. F. Marini, and V. R. Edgerton. *J. Appl. Physiol.*, Vol. 73, pages 51S–57S; 1992.

ELECTROPHORETIC SEPARATION OF RAT SKELETAL MUSCLE MYOSIN HEAVY-CHAIN ISOFORMS. R. J. Talmadge and R. R. Roy. *J. Appl. Physiol.*, Vol. 75, pages 2337–2340; 1993.

FOOD RESTRICTION INDUCED TRANSFORMATIONS IN CARDIAC FUNCTIONAL AND BIOCHEMICAL PROPERTIES IN RATS. F. Haddad, P. W. Bodell, S. A. McCue, R. E. Herrick, and K. M. Baldwin. *Am. J. Physiol.*, Vol. 74, pages 606–612; 1993.

EFFECT OF CHRONIC ENERGY DEPRIVATION ON CARDIAC THYROID HORMONE RECEPTOR AND MYOSIN ISOFORM EXPRESSION. S. J. Swoap, F. Haddad, P. Bodell, and K. M. Baldwin. *Am. J. Physiol.*, Vol. 266, pages E254–E260; 1994.

EFFECT OF SPACEFLIGHT ON SKELETAL MUSCLE: MECHANICAL PROPERTIES AND MYOSIN ISOFORM CONTENT OF A SLOW MUSCLE. V. J. Caiozzo, M. J. Baker, R. E. Herrick, M. Tao, and K. M. Baldwin. *J. Appl. Physiol.*, Vol. 76, pages 1764–1773; 1994.

EFFECT OF SPACEFLIGHT ON THE FUNCTIONAL, BIOCHEMICAL, AND METABOLIC PROPERTIES OF SKELETAL MUSCLE. K. M. Baldwin. *Med. Sci. Sport. Exer.*, Vol. 28, pages 983–987; 1996.

MICROGRAVITY-INDUCED TRANSFORMATIONS OF MYOSIN ISOFORMS AND CONTRACTILE PROPERTIES OF SKELETAL MUSCLE. V. J. Caiozzo, F. Haddad, M. J. Baker, R. E. Herrick, N. Prietto, and OK. M. Baldwin. *J. Appl. Physiol.*, Vol. 81, pages 123–132; 1996.

POSTNATAL DEVELOPMENT UNDER CONDITIONS OF SIMULATED WEIGHTLESSNESS AND SPACEFLIGHT. K. Walton, K. *Brain Res. Rev.*, Vol. 28, pages 25–34; 1998.

ROLE OF GRAVITY IN THE DEVELOPMENT OF POSTURE AND LOCOMOTION IN THE NEONATAL RAT. F. Clarac, L. Vinay, J. R. Cazalets, J. C. Fady and M. Jamon. *Brain Res. Rev.*, Vol. 28, pages 35–43; 1998.

DEVELOPMENTAL REGULATION OF INSULIN-LIKE GROWTH FACTOR-I AND GROWTH HORMONE RECEPTOR GENE EXPRESSION. L. Shoba, M. R. An, S. J. Frank, and W. L. Lowe, Jr. *Mol. Cellular Endocrinol.*, Vol. 152(1-2), pages 125–136; 1999.

EFFECT OF A 17 DAY SPACEFLIGHT ON CONTRACTILE PROPERTIES OF HUMAN SOLEUS MUSCLE FIBRES. J. J. Widrick, S. T. Knuth, K. M. Norenberg, J. G. Romatowski, J. L. Bain, D. A. Riley, M. Karhanek, S. W. Trappe, T. A. Trappe, D. L. Costill, and R. H. Fitts. *J. Physiol., Vol.* 516, pages 915–930; 1999.

THE TIME COURSE OF MYOSIN HEAVY CHAIN TRANSITIONS IN NEONATAL RATS: IMPORTANCE OF INNERVATION AND THYROID STATE. G. R. Adams, S. A. McCue, M. Zeng, and K. M. Baldwin. *Am. J. Physiol.,* Vol. 276, pages R954–R961; 1999.

THE EFFECTS OF SPACEFLIGHT ON RAT HINDLIMB DEVELOPMENT I: MUSCLE MASS AND IGF-I EXPRESSION. G. R. Adams, S. A. McCue, P. W. Bodell, M. Zeng, and K. M. Baldwin. *J. Appl. Physiol.,* Vol. 88, pages 894–903; 2000a.

THE EFFECTS OF SPACEFLIGHT ON RAT HINDLIMB DEVELOPMENT II: EXPRESSION OF MYOSIN HEAVY CHAIN ISOFORMS. G. R. Adams, F. Haddad, S. A. McCue, P. W. Bodell, M. Zeng, L. Qin, A. X. Qin, and K. M. Baldwin. *J. Appl. Physiol.,* Vol. 88, pages 904–916; 2000b.

Early Development of Gravity-Sensing Organs in Microgravity

Authors
Michael L. Wiederhold, Wenyuan Gao,
Jeffrey L. Harrison, Kevin A. Parker

Experiment Team

Principal Investigator: **Michael L. Wiederhold**

Co-Investigators: **Wenyuan Gao,
Jeffrey L. Harrison,
Kevin A. Parker**

University of Texas Health Science Center,
San Antonio, USA

ABSTRACT

Most animals have organs that sense gravity. These organs use dense stones (called otoliths or statoconia), which rest on the sensitive hairs of specialized gravity- and motion-sensing cells. The weight of the stones bends the hairs in the direction of gravitational pull. The cells in turn send a coded representation of the gravity or motion stimulus to the central nervous system. Previous experiments, in which the eggs or larvae of a marine mollusk (*Aplysia californica*, the sea hare) were raised on a centrifuge, demonstrated that the size of the stones (or "test mass") was reduced in a graded manner as the gravity field was increased. This suggests that some control mechanism was acting to "normalize" the weight of the stones. The experiments described here were designed to test the hypothesis that, during their initial development, the mass of the stones is regulated to achieve a desired *weight*. If this is the case, we would expect a larger-than-normal otolith would develop in animals reared in the weightlessness of space. To test this, freshwater snails and swordtail fish were studied after spaceflight. The snails mated in space, and the stones (statoconia) in their statocysts developed in microgravity. Pre-mated adult female swordtail fish were flown on the Space Shuttle, and the developing larvae were collected after landing. Juvenile fish, where the larval development had taken place on the ground, were also flown. In snails that developed in space, the total volume of statoconia forming the test mass was 50% greater than in size-matched snails reared in functionally identical equipment on the ground. In the swordtail fish, the size of otoliths was compared between ground- and flight-reared larvae of the same size. For later-stage larvae, the growth of the otolith was significantly greater in the flight-reared fish. However, juvenile fish showed no significant difference in otolith size between flight- and ground-reared fish. Thus, it appears that fish and snails reared in space do produce larger-than-normal otoliths (or their analogs), apparently in an attempt to compensate for the reduced weight of the stones in space. The fish data suggest that there is a critical period during which altered gravity can affect the size of the test mass, since the larval, but not the juvenile, fish showed the changes.

INTRODUCTION

Most animals sense gravity using dense stones (called otoliths, statoliths, or statoconia) that rest on the sensitive hairs of specialized gravity- and motion-sensing cells. The weight of the stones bends the hairs in the direction of gravitational pull. This excites the hair cells and their associated nerve fibers. The brain uses this information to determine position in the gravity field (i.e., upright vs. lying down) and the direction of movements. In many species, from snails to humans, the stones are either one or a collection of dense calcium carbonate ($CaCO_3$) crystals. These crystals can be separate (in many mollusks), held together by a gelatinous structure (in amphibians and mammals), or fused into three large otoliths in each ear (in fish). Little is known about the mechanisms that control the production and growth of the otoliths and their analogs. In species that grow continually, such as fish, the otoliths continue to grow throughout life; whereas in species that reach a final size, such as humans, the otoliths also stop growing. Thus, there must be mechanisms for initiating and terminating the mineralization of the otoliths.

The bulk of existing data suggests that maintaining *adult* animals in altered gravity, either in hypergravity on a centrifuge or in the microgravity in space, does not substantially affect the otoliths or their analogs (Sondag, 1996). In *developing* animals, however, there may be profound effects. To test the hypothesis that gravity could influence the early development of the stones, we first studied the marine mollusk *Aplysia californica*. This mollusk has a gravity-sensing organ (statocyst) very similar to the pond snail's. In *Aplysia*, there is a single stone (statolith) when the mollusk hatches; multiple stones (statoconia) are not developed until the mollusk undergoes metamorphosis 60 to 100 days later. Developing eggs or isolated statocysts were reared on a centrifuge with G-values ranging from one to five (Pedrozo, 1994). The diameter of the statolith and the volume and number of statoconia were all reduced, in a graded manner, as the centrifuge speed (G-level) was increased.

On a previous Space Shuttle mission (IML-2, 1994), we flew developing eggs of the Japanese red-bellied newt, *Cynops pyrrhogaster*. The eggs were staged so they would reach orbit before any stones were formed. Particularly in the stones made of aragonite (another crystal form of $CaCO_3$), the otoliths were as much as five times the volume of those in ground control newts of the same developmental stage (Wiederhold, 1997 a,b, Koike, 1995 a,b). Lychakov et al. found the utricular otolith to be 30% larger in space-reared frog (*Xenopus*) larvae (Lychakov, 1985); and Anken et al. have reported that in cichlid fish reared on a centrifuge, the saccular otolith was smaller than in one-G controls (Anken, 1998).

These previous experiments suggested that gravity may affect the stones in the developing gravity sensor. Our Neurolab experiments were designed to test the hypothesis that during the initial development of the otolith, the mass of the stones is regulated to achieve a desired *weight*. If this is the case, we would expect a larger-than-normal otolith to be developed in animals reared in the microgravity of space. This modification would produce an average stimulus to the otolith organs in space closer to what they would receive at one-G on Earth. On the Neurolab (STS-90) and the preceding STS-89 Shuttle flights, we had the opportunity to study pond snails reared or conceived in space and juvenile and embryonic fish reared in space.

We used animals that develop rapidly so that they would undergo a significant portion of their development during the 10–8 days usually available on Space Shuttle missions. Like many gastropod mollusks, the pond snail, *Biomphalaria glabrata*, has a very simple gravity-sensing organ, the statocyst. Figure 1 is a schematic drawing of the statocyst. The organ is spherical; the wall is made of 13 sensory receptor cells with smaller supporting cells between them. The lumen of the cyst is filled with fluid (statolymph) and dense stones (statoconia), which fall to the bottom of the cyst in the one-G environment of Earth. The statoconia are produced in the supporting cells and are ejected (exocytosed) into the cyst lumen. The cilia on the receptor cells are motile, beating continuously. On Earth, the statoconia sink to the bottom of the cyst. They interact with the beating cilia of the receptor cells at the bottom. It is this interaction between the beating cilia and the dense statoconia that leads to stimulation of the receptor cells and initiation of impulses in their axons (Wiederhold, 1974, 1978). These impulses give the central nervous system information on the direction of gravity and the direction and magnitude of accelerations due to movement. Larval snails can have as few as two or three statoconia, but adults have 300–400 statoconia in each statocyst (Gao, 1997b). The snails are hermaphrodites, so any pair can mate. They produce

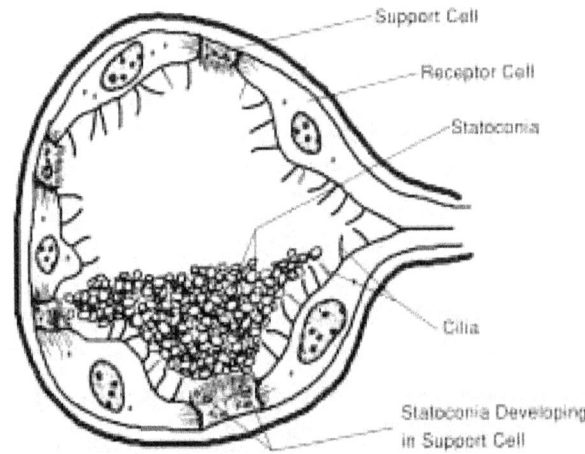

Figure 1. Schematic drawing of the gastropod molluscan statocyst. The cyst is approximately 150 µm in diameter in an adult *Biomphalaria* snail. The cyst wall contains 13 ciliated mechanoreceptor cells, separated by numerous supporting cells, which bear microvilli. The statoconia are produced in the supporting cells and ejected (exocytosed) into the cyst lumen. On Earth, gravity holds the statoconia in the bottom third of the cyst lumen, although the beating cilia keep them in constant motion.

approximately 20 to 30 eggs almost every day. About five days after the eggs are laid, the larvae hatch. The hatched larvae have a fully formed statocyst with three to 10 statoconia.

On Earth, snails tend to crawl downward on an aquarium wall. This is an example of behavior controlled by the statocysts. Instead of crawling in any direction, the snails show a clear preference for crawling downward. After crawling to the bottom, the snails detach from the wall, inflate an air bubble under the shell, and float to the surface. When they contact the wall again, the cycle recurs. This preferential downward crawling is exhibited immediately after they hatch from the egg.

The swordtail fish (*Xiphophorus helleri*) has two gravity-sensing organs similar to those in mammals (the saccule and utricle), as well as a third otolith organ, the lagena, which is often seen in fish, birds, and reptiles. These three organs are sometimes referred to as the sagitta (saccule), lapillus (utricle), and astericus (lagena). The three otoliths are each a solid stone, made of $CaCO_3$, that continues to grow throughout the life of the fish. As in mammals, the otolith lies above a sensory epithelium containing hair cells. Due to the stones' inertia, when the head is accelerated or pulled upon by gravity, the hair bundles of the hair cells are bent, leading to excitation of the cell. Synapses on the hair cells depolarize the endings of the vestibular nerve fibers and increase or decrease the firing of impulses in its fibers, which carry the coded acceleration information to the brain. The location and orientation of the otolith organs and the three otoliths are illustrated in Figure 2.

By studying the pond snails and swordtail fish, our studies addressed the effect of the relative absence of gravity on the early development of the stones in gravity-sensing organs. On Earth, problems such as motion sickness and balance disorders can often be caused by gravity sensors in the inner ear. Basic research into how and when gravity sensors develop will help us better understand inner ear disorders. In space, it will be important to know whether animals born and raised in zero-G will have permanently altered gravity sensors. This knowledge will be especially vital in the future, when planners try to determine whether humans could be born and raised in low gravity.

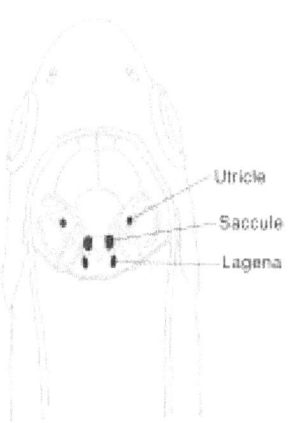

Figure 2. Schematic drawing of fish inner ear, showing the three semicircular canals and the three otolith organs in each ear, with the otoliths represented in solid black.

METHODS

Thirty-five adult pond snails (*Biomphalaria glabrata*), four previously mated adult female swordtail fish (*Xiphophorus helleri*), and 50 to 200 juvenile swordtail fish, approximately

one-centimeter long, were loaded into the closed equilibrated biological aquatic system (CEBAS) (Slenzka, 1999) two days before the Shuttle launch. The CEBAS contains seven liters of spring water and 50 grams of the hornweed plant, *Ceratophylum*. The CEBAS has four chambers. The adult fish and snails were in the front chamber; juvenile fish and adult snails were in the second animal chamber. The third chamber was a plant chamber, containing the hornweed plants and also some adult snails. The fourth chamber was filled with filter material. A system of lights illuminated the plant chamber, and dissolved oxygen (DO) was monitored in the circulating water. When DO dropped below four mg/L, extra lights were turned on until DO reached six mg/L. The system was equipped with a video camera that monitored fish and snail behavior throughout the two Shuttle missions. STS-89 was a 10-day flight, and Neurolab, STS-90, was 16 days. On both flights, many snails mated and produced a large number of spawn packs. Our results came from young snails retrieved on landing day with shell diameters ranging from 0.5 to 2.0 mm. Ground control animals were maintained in a functionally identical CEBAS unit maintained at the Kennedy Space Center (KSC).

The swordtail fish, *Xiphophorus*, reproduce by internal fertilization, and the embryos in the ovary develop to a stage where the otoliths are well formed before the fry are hatched. After a single mating, the female generally produces three broods at one-month intervals. The four mated, adult female fish were selected from previous matings such that the fry would not be hatched until after Shuttle landing. Our results are from fish embryos with spinal cords from three to 10 mm long.

The CEBAS unit was removed from the Shuttle and received in the laboratory at KSC approximately 3.5 hours after Shuttle landing. Half of the small snails were preserved for anatomical study on landing day, and the other half were preserved five days later. Details of fixation of snail statocysts for light and electron microscopy are given in Gao et al. (Gao, 1997a, b). Adult fish ovaries were removed and preserved in alcohol on landing day, as were the juvenile fish. Fish embryos were removed from the ovaries after returning to our home laboratory. Snails were removed from their shell during fixation and sectioned at 1–2 μm. The total volume of statoconia was estimated by outlining the area occupied by statoconia in serial sections through the statocyst, adding the areas, and multiplying by the section thickness.

After removing the otoliths from the fish specimens, the otoliths were rinsed in distilled water, dried, and mounted either on glass slides, in well slides, or on stubs for scanning electron microscopic (SEM) analysis. All otoliths were placed with the flatter side down, and the area of each was measured using either Bioquant or UTHSCSA ImageTool software. Images were obtained with a Spot 2 digital camera. The area of each otolith in these micrographs was measured and plotted against total body length for juveniles or against spine length for embryos. A linear regression analysis was performed for the flight- and ground-reared specimens for each group, relating area to animal or spine length. The flight and ground regression lines were compared by an analysis of covariance for the significance of differences in their slopes. When measurements of both otoliths (of any type) from one animal were available (the usual case), the two were averaged. If the slopes were not significantly different (p>0.05), both groups were fit with a line of the same slope and the difference in mean otolith area at the mean length was tested for significance.

RESULTS

Videotaping of the animal chambers in the CEBAS revealed that, during orbital flight, the snails were easily dislodged from the aquarium walls. On Earth they spend most of their time attached to the walls. Once separated from the wall, the water circulation carried them in a rotating pattern around each chamber. In the front two chambers, it was evident that snails kept their feet extended, apparently searching for a substrate to which to attach. In this situation, they were likely to come into contact with another snail, and mating pairs were often seen floating attached to one another. Thus, it was not surprising that a large number of small snails were retrieved on Shuttle landing. After the STS-89 (10-day) mission, 251 snails from 0.5 to 4.0 mm diameter were retrieved; and on the STS-90 (16-day) mission, 206 snails in the same size range were retrieved. In the ground control run, 308 small snails were retrieved for STS-89 and 155 ground control snails were retrieved during STS-90.

Figure 3 is a photomicrograph of a 1.0-μm section through the statocyst of a two-mm-diameter snail, taken using Nomarsky optics. The statoconia appear brightly colored in the polarized light due to their ability to refract light. This

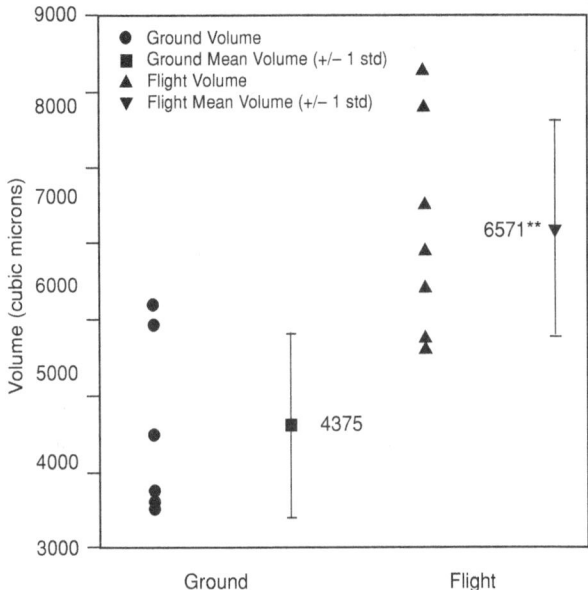

Figure 4. Plot of total statoconia volume for six ground-reared and seven flight-reared snails of two-mm-shell diameter, fixed on landing day after STS-89. The mean and one standard deviation for each group are indicated to the right of the individual data points. The mean total volume is 50% greater for the flight-reared snails in this case. The difference between flight- and ground-reared mean volume is significant at p<0.01.

makes it easy to identify the statoconia and outline the area occupied by statoconia in each section. Figure 4 plots total statoconia volume for flight- and ground-reared two-mm snails fixed on landing day of STS-89. Each measurement for six ground- and seven flight-reared specimens is displayed. It is apparent that considerable variation in total volume exists within both groups, but the difference in mean values for the two groups is statistically significant. In this case, the average volume of statoconia is 50% greater in the flight-reared snails. Sufficient snails were available to make meaningful comparison between flight- and ground-reared animals of one and two or 1.5-mm-diameter snails, both on landing day (R+0) and five days after return to Earth (R+5). The ratios of total statoconia volume for flight and ground-control snails fixed on R+0 and R+5 are shown in Table 1. Note that the mean volume of statoconia is significantly greater in the flight-reared snails in all groups, except for those from STS-90 fixed at R+0, where no significant difference between the groups was seen. Note also that for three of the four groups, excepting only the one- and 1.5-mm snails from STS-90, the ratio of flight- to ground-reared statoconia volume was larger at R+5 than at R+0. That is, the increase in statoconia volume, which occurred in space, was even more pronounced in snails allowed to develop for another five days on Earth before they were fixed.

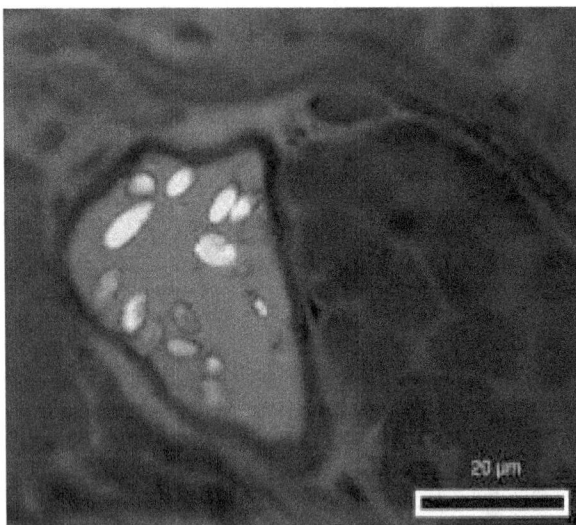

Figure 3. 1.0-μm section through the statocyst of a two-mm shell-diameter pond snail, *Biomphalaria glabrata*. The micrograph was taken with Nomarsky optics. The polarized light makes the statoconia appear bright and multicolored, due to their highly birefringent optical properties. This makes the statoconia easy to identify and outline, from which the cross-sectional area is calculated and multiplied by the section thickness to obtain total statoconia volume.

Table 1. Total Statoconia Volume Ratio: Flight/Ground

STS-89 R+0		R+5	
1 mm:	1.36*	1 mm:	2.16*
2 mm:	1.50*	2 mm:	1.87*
STS-90 R+0		**R+5**	
1 mm:	1.67	1.5 mm:	1.20*
2 mm:	0.95	2 mm:	1.38*
Average:	1.37	Average:	1.73

Overall Average = 1.55
*p≤0.05 to 10^{-6}

Figure 5 shows a plot of the distribution of sizes of statoconia. This is the size distribution for statoconia from a two-mm snail fixed on landing day from STS-90. The general shape of the distribution of areas is similar between flight- and ground-reared specimens. The distribution is slightly shifted to smaller sizes in the flight-reared snails, but by less than two bin widths; i.e., by less than 10 μm^2, with a modal value of 20 to 30 μm^2. In this case, the mean statoconia area is 33 μm^2 for ground-control snails and 29 μm^2 for flight-reared snails. Figure 6 shows the distribution of total number of statoconia in flight- and ground-reared two-mm snails. It is apparent that the flight-reared snails had approximately 50% more statoconia than size-matched, ground-reared snails. These results indicate that rearing in microgravity causes the supporting cells to produce more statoconia, which are of approximately the same size as those in ground-control snails. This implies that the effects are mediated within the supporting cells and probably do not affect the growth of statoconia once they have left the supporting cells and are in the lumen of the statocyst.

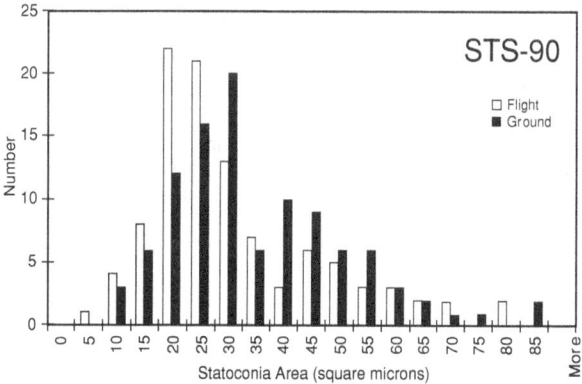

Figure 5. Histograms showing the distribution of areas, as seen in light micrographs, of statoconia lying on their flattest surface, isolated from two-mm-shell-diameter snails, removed on landing day after STS-90. The overall distribution is little different between flight and ground control groups, with the flight-reared statoconia being slightly smaller. The mean area for flight-reared statoconia is 29 μm^2; and for the ground-reared animals, the average is 33 μm^2.

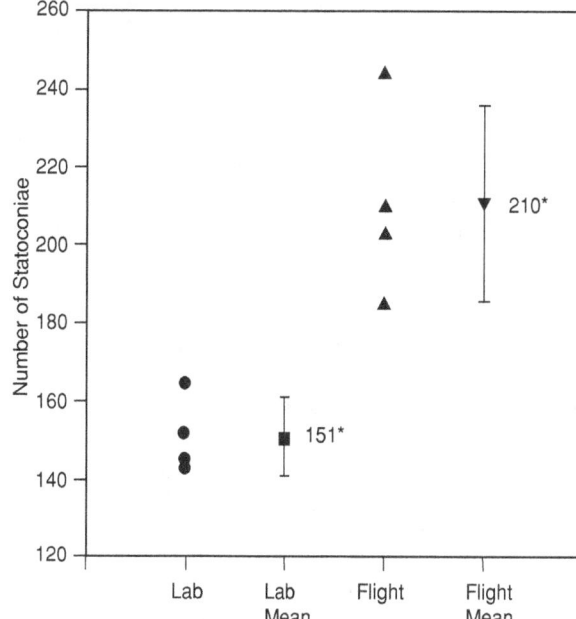

Figure 6. Plot of number of statoconia retrieved from two-mm-shell-diameter snails retrieved from the flight and ground-reared groups from STS-90. In this case, the average number of statoconia was 39% greater in the flight-reared snails.

The data shown above suggest that the process of statoconia development in the supporting cells is modified in snails reared in microgravity. We had only two flight-reared specimens available for transmission electron microscopy (TEM); but the four statocysts from these animals, one fixed on (R+0) and the other on (R+5), demonstrate significant modifications in the flight-reared statocysts. Figure 7a is an electron micrograph from a two-mm flight-reared snail fixed on landing day. It shows three large vacuoles in a thin section through a supporting cell at the ventral side of the statocyst. These vacuoles are from 0.8 to 1.3 μm in diameter. In ground-reared snails, very few vacuoles are seen in the supporting cells, and their maximum size is 0.1 μm in diameter. Thus, the vacuoles in this flight-reared snail have approximately 1000-fold greater volume than those of ground control animals. The flocculent material seen in the upper vacuole suggests that the formation of statoconia is initiated within the vacuoles and that the increased number of statoconia is likely a direct result of the larger volume of vacuoles.

Consistent with the increased number of statoconia in flight-reared snails, more statoconia were seen in the supporting cells of these animals. Figure 7B is an electron micrograph of a thin section through another supporting cell from the same snail as in Figure 7A. This illustration shows three intracellular statoconia. These statoconia appear as "ghosts," probably demineralized during fixation. In an adjacent one-μm section, these statoconia displayed brightly colored birefringence (refraction) as shown in Figure 3, indicating that the statoconia were fully mineralized. In ground control snails, it is generally difficult to find statoconia in the supporting cells; and we have never seen more than one statoconium in a supporting cell.

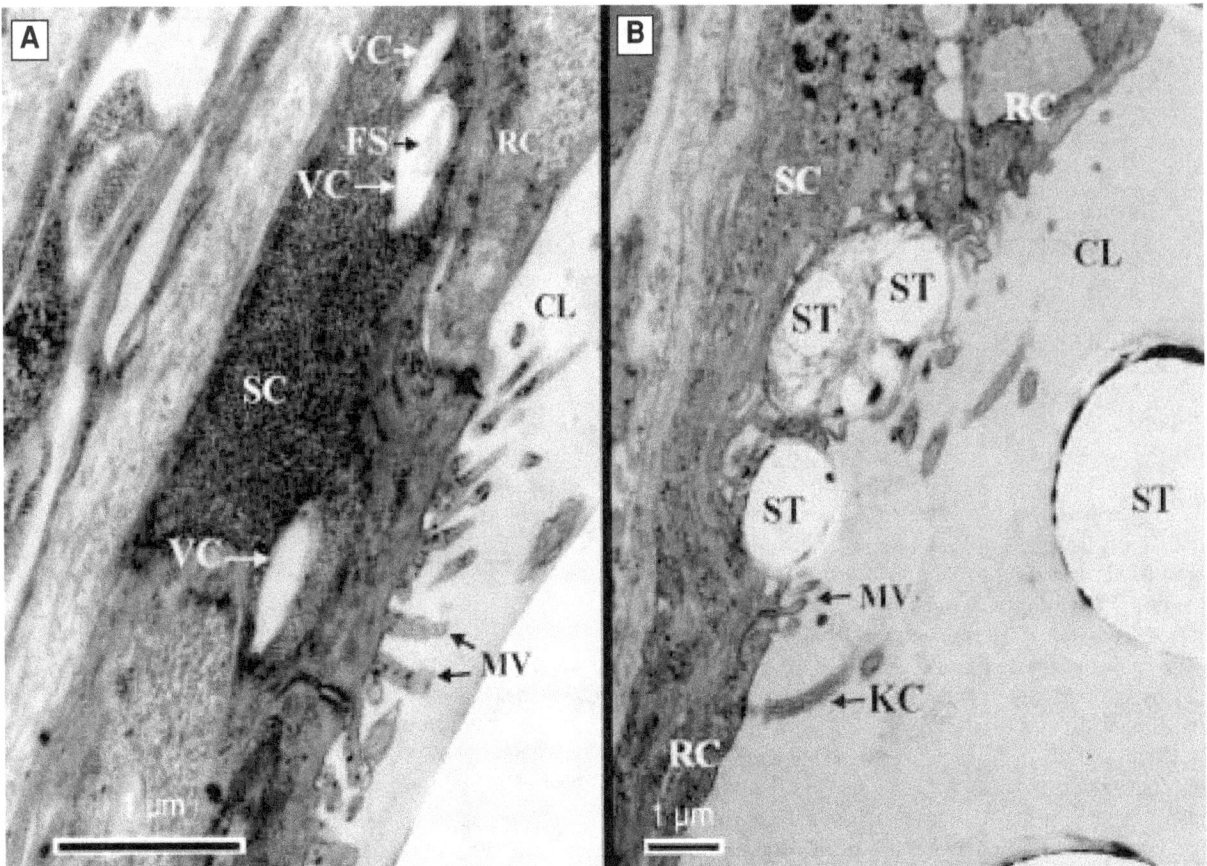

Figure 7. TEMs of supporting and receptor cells from a statocyst of a two-mm flight-reared snail. A. Note three large vacuoles (VCs) in the supporting cell (SC). Flocculent substance (FS) seen in one vacuole. MV: microvilli seen on supporting cells. Receptor cell (RC) is seen adjacent to the middle supporting cell. CL: statocyst lumen. B. Ghosts of three forming statoconia (ST) seen in the supporting cell. Two receptor cells, bearing cilia, are seen on either side of the supporting cell. Ghost of another statoconium is seen is the cyst lumen. Ghosts represent demineralized statoconia.

Thus, increased vacuoles and intracellular statoconia are seen in the supporting cells of flight-reared snails.

Figure 8 shows light micrographs of the three pairs of otoliths from a one-centimeter-long juvenile *Xiphophorus* and from an embryo with a spine length of five mm.

The juvenile fish retrieved from the STS-89 flight ranged from 8.2 to 10.7 mm in length. Figure 9 plots the area of 42 utricular otoliths from 23 ground-reared juvenile fish and 39 otoliths from 20 flight-reared juvenile fish from the STS-89 flight. The area of the otolith (measured from the micrographs of the otoliths resting on their flattest surface) is plotted against the total fish length (measured when the fish were retrieved from the CEBAS.) Neither the slopes nor the mean values for the two regression lines were significantly different between flight and ground control groups (p=0.7904 for slopes, p=0.3068 for means). The same conclusions were drawn for the saccular otolith size for STS-89 juvenile fish (p=0.3974 for slopes, p=0.3553 for means), although the lagenar otoliths were marginally larger in the flight-reared juveniles compared to ground controls, with p=0.0806 for slopes and p=0.0222 for means. The slope for the flight lagenas is

4342±1017 µm²/mm (fish length, mean±SEM). For the ground controls, the slope is 890±1633 µm²/mm.

In contrast to the juvenile fish, there was a significant difference in otolith size between ground- and flight-reared fish embryos from the STS-90 flight. The measurements of utricular otolith area for different size embryos from STS-90 are illustrated in Figure 10. Here the slope of the fit to the flight-reared embryos is significantly larger (p=0.008) than the fit to the ground-reared embryos. Thus, for embryos with spine length >4.5 mm, the flight-reared embryos did have significantly larger utricular otoliths. Similarly, the slope of the best fit to the flight-reared saccular otoliths was significantly larger than that for the ground controls (p=0.035). For the lagenar otoliths, the slope of the ground-reared otoliths was significantly larger than that for the flight specimens (p=0.0028), although there was a very weak relationship between spine length and otolith size in the group of flight animals available. The ground-control group contained three "outliers," with otoliths only a fraction of the area of the smallest otoliths in the rest of the ground and flight specimens. If these three points were deleted from the analysis, there was no significant

The Neurolab Spacelab Mission: Neuroscience Research in Space

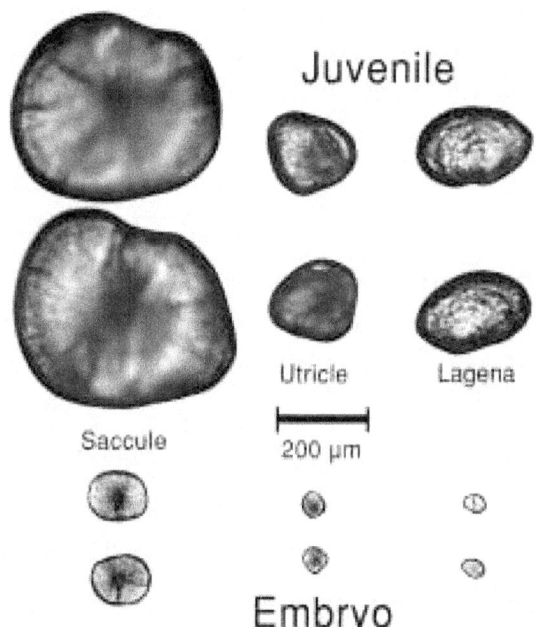

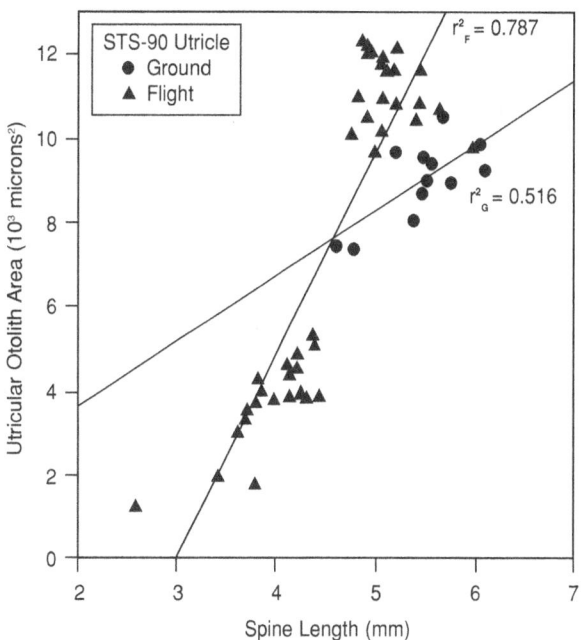

Figure 8. Light micrographs of otoliths of the saccule (sagitta), utricle (lapillus), and lagena (astericus) from a one-cm-long juvenile fish and from an embryo with a spine length of five mm. Both are shown at the same magnification. Scale bar is 200 μm.

Figure 10. Plots of utricular otolith area vs. spine length from flight- and ground-reared fish embryos from STS-90. As in Figure 9, each group is fit with a linear regression line. Here the slope of the fit to the flight-reared group is significantly larger than that for the ground-reared controls (p=0.008). For flight-reared specimens, 82 otoliths from 42 embryos, and for ground-reared specimens, 24 otoliths from 12 embryos, were analyzed. For fish in which two utricular otoliths were retrieved and measured, the area is the mean of the two.

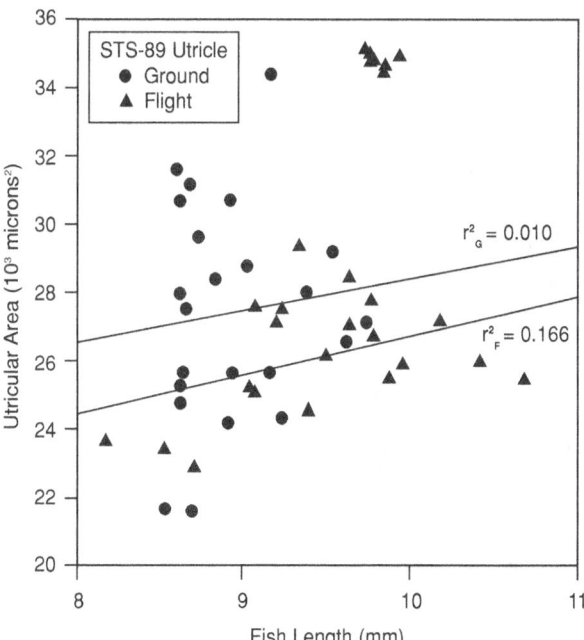

Figure 9. Plot of utricular otolith vs. total fish length for ground- (filled circles) and flight- (filled triangles) reared juvenile fish retrieved from STS-89. Each group was fit with a linear regression line. The square of the correlation coefficient for each regression line is given. An analysis of covariance indicated no significant difference in either slope or mean values between ground and flight groups.

difference in slope between the flight and ground-reared groups.

Embryos were also retrieved from adult female fish on STS-89. For this nine-day flight, the embryos were considerably smaller than those obtained from STS-90 (16 days). Plots of utricular otolith area vs. spine length for the two groups of embryos from STS-89 are presented in Figure 11. In this case, the slope of the regression line fit to the flight-reared embryos is smaller than that for the ground control group. This difference is significant (p=0.0160). For the saccular otoliths in the STS-90 embryos, the growth of otolith area with spine length is also significantly greater for the ground-reared specimens (p=0.0054).

Some insight to the mechanisms that might underlie the production of larger otoliths in the larger embryos reared in space can be gained from an examination of the surface of the otoliths. This was done using an SEM. Figure 12 shows SEM images of saccular otoliths from a flight-reared embryo with a spine length of 5.5 mm and a ground-reared embryo with a spine length of 5.2 mm. The surface of the flight-reared otolith is covered with octahedral crystals with a square base of approximately one μm on a side. Such crystals were found

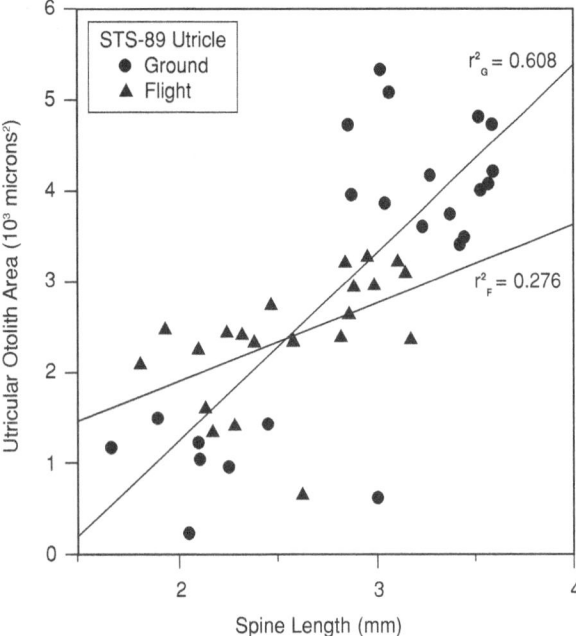

Figure 11. Plots of utricular otolith area vs. spine length from flight- and ground-reared fish embryos from STS-89. Regression lines as in Figures 9 and 10. Here the slope of the fit to the flight-reared group is significantly smaller than that for the ground-reared controls (p=0.0160). Note that embryos in this experiment were smaller than those retrieved from STS-90. For flight-reared specimens, 40 otoliths from 21 embryos, and for ground-reared specimens, 41 otoliths from 23 embryos were analyzed. Where both otoliths were measured, the mean was plotted.

on 110 of 207 flight otoliths (53%) but not on any of 75 ground otoliths examined. Among the flight otoliths, 39/88 (44%) of saccular otoliths, 39/72 (54%) of utricular otoliths, and 32/47 (68%) of lagenar otoliths exhibited these crystals. Thus, there is little difference among the three otoliths in the likelihood of forming surface crystals.

DISCUSSION

The finding of larger otoliths in later-stage embryos from STS-90 is consistent with our own experience with pond snails (Wiederhold, 1997 a,b), *Aplysia* (Pedrozo, 1994), and Anken's report (Anken, 1998) on centrifuged cichlid fish. Here the mass of the otoliths was increased in animals reared in the microgravity of space. The fact that significant differences were not seen in the juvenile fish (Figure 9) is compatible with Anken's conclusion that centrifugation did not affect older (juvenile) fish otoliths. The finding of smaller otoliths in flight-reared, early-stage embryos (Figure 11) is surprising. It is difficult to imagine a mechanism by which gravity (or the relative lack of gravity) could affect mineralization of the smallest otoliths, whose weight would be minimal. It is even more surprising to find an effect opposite to the effect on otoliths in embryos approximately one week older. These findings clearly illustrate that the effects of rearing in microgravity depend critically on the developmental stage at which the animal is exposed to microgravity.

The appearance of crystals on the surface of flight-reared otoliths suggests that the mechanism by which mineral is added to the otolith differs in the low gravity of space.

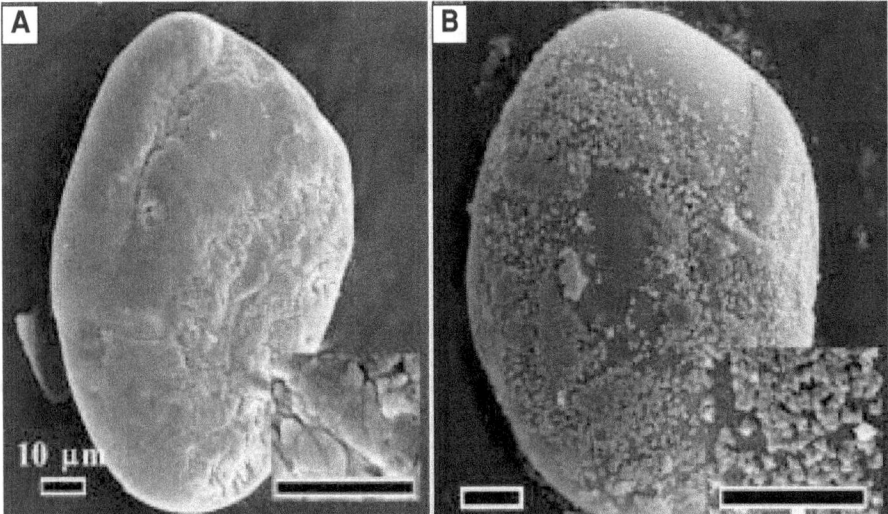

Figure 12. Scanning electron micrographs of saccular otoliths from a ground- (A) and a flight-reared (B) fish embryo with spine length of 5.5 mm. Insets in each panel show surface at 4× higher magnification. Note octahedral crystals seen on the flight-developed otolith. All scale bars are 10 µm.

It has been suggested that the supporting cells in the macula secrete the organic material in the otoliths (Harada, 1998). This secretion could be enhanced in low gravity, due to a reduced otolith input to the macula. The actual crystallization of the mineral component could also be altered in microgravity.

The results presented here and those previously published on newt larvae reared in space all indicate that early growth of the stones (test mass) in the gravity-sensing organs is regulated to produce an otolith, or its analog, of the appropriate *weight*. Thus, in microgravity, where a given mass has almost no weight, a larger mass is produced in an attempt to compensate for the reduced weight. Similarly, animals reared on a centrifuge produce smaller otoliths, since the increased G-force produces a greater weight with the same mass. The intriguing results with the fish embryos (Figures 9–11) suggest that there is a critical period, relatively early in development, in which the regulatory mechanism, which optimizes otolith weight, can act. The very early-stage fish data (Figure 11) suggest that the control mechanisms can even be reversed at the earliest stages. At the earliest stages, when the otoliths do not have sufficient mass to activate the postulated control mechanisms, we had expected to see no effect of altered gravity, so the reverse effect is all the more surprising. The statistical significance of the effects at the early stages suggests that as soon as the otoliths appear, they have sufficient mass to activate the control mechanism(s), presumably through the hair cells, but that a positive feedback is working to produce smaller otoliths at reduced gravity. Centrifuge experiments at the earliest stages have not been reported, so we do not know what the effects would be of rearing at hypergravity during the earliest stages of development.

It is important to note that in all of the systems in which we have demonstrated gravitational effects on otolith growth, the test mass is composed of $CaCO_3$ in the aragonite crystal form (Pedrozo, 1997; Wiederhold, 1994). In mammals, both the utricular and saccular otoliths are constructed of $CaCO_3$ in the calcite form (Carlstrom, 1963). In frogs and newts, the adult saccular otolith is aragonite whereas the utricular otolith is calcite (Wiederhold, 1994). In the newt larvae flown in 1994, we found an increase in the volume of the saccular otolith but not in the utricular otoliths of newts fixed at R+5. Repeated x-ray microfocus imaging (Koike, 1995b) showed a slight increase in the utricular otoliths of one newt at nine months postflight, but a much more pronounced increase in the saccular otoliths. If our findings do apply to calcite otoliths, this will need to be considered before attempts are made to have humans conceived and brought up in space. The size of the saccular otoliths was followed in groups of newts for nine months after a 15-day Shuttle mission. The size of the saccular otoliths was maximal, relative to ground controls, at two to three months postflight and only returned to "normal" size six months postflight. If mammals, including humans, were to be conceived and brought up in microgravity, we do not know whether their otoliths would be abnormally large while in space or whether they would revert to normal size at some time after introduction to one-G. We also do not know if the otolith-related reflexes developed in microgravity would function normally in one-G.

Acknowledgements

The work reported here was supported by NASA grant NAG2-950 and NSF grant IBN-9529136. We would like to thank Dr. Thomas Prihoda for statistical analysis of the data and Jennifer Kwong, our Experiment Support Scientist, for her help before and during the missions.

REFERENCES

A CRYSTALLOGRAPHIC STUDY OF VERTEBRATE OTOLITHS. D. Carlstrom. *Biol. Bull.,* Vol. 125, pages 441–463; 1963.

CRYSTALLOGRAPHY OF *XIPHOPHORUS* OTOLITHS. S. Doty. Personal Communication.

APLYSIA STATOCYST RECEPTOR CELLS: INTRACELLULAR RESPONSES TO PHYSIOLOGIC STIMULI. M. L. Wiederhold. *Brain Res.,* Vol. 78(3), pages 490–494; 1974.

MEMBRANE VOLTAGE NOISE ASSOCIATED WITH CILIARY BEATING IN THE *APLYSIA* STATOCYST. M. L. Wiederhold. *Brain Res.,* Vol. 156, pages 369–374; 1978.

INVESTIGATION OF VESTIBULAR STRUCTURE AND ION COMPOSITION OF SPUR-TOED FROG LARVAE AFTER EXPOSURE TO WEIGHTLESSNESS. D. V. Lychakov and Ye. A. Lavrova. *Kosm. Biol. Aviak. Med.,* Vol. 3, pages 48–52; 1985.

DEVELOPMENT OF CRYSTALLOGRAPHIC FORMS OF OTOCONIA IN NEWT LARVAE. M. L. Wiederhold, P. S. Steyger, H. Steinfenck, R. D. Soloway, H. Koike and I. Kashima. Abstracts of the Seventeenth Midwinter Research Meeting of the Association for Research in Otolaryngology, Vol. 17, page 38; 1994.

EFFECTS OF HYPERGRAVITY ON STATOCYST DEVELOPMENT IN EMBRYONIC *APLYSIA CALIFORNICA*. H. A. Pedrozo and M. L. Wiederhold. *Hear Res.,* Vol. 79, pages 137–146; 1994.

NON-INVASIVE ASSESSMENT OF OTOLITH FORMATION DURING DEVELOPMENT OF THE JAPANESE RED-BELLIED NEWT, *CYNOPS PYRRHOGASTER*. H. Koike, K. Nakamura., K. Nishimura, I. Kashima., M. L. Wiederhold and M. Asashima. *Hear Res.,* Vol. 88, pages 206–214; 1995a.

OTOLITH FORMATION DURING DEVELOPMENTAL PROCESS OF AMPHIBIANS. H. Koike, M. Asashima, S. Kumasaka, K. Nakamura, S. Kobayashi and I. Kashima. *Biol. Sci. in Space* (Tokyo), Vol. 9, pages 164–165; 1995b.

THE EFFECTS OF HYPERGRAVITY ON FUNCTION AND STRUCTURE OF THE PERIPHERAL VESTIBULAR SYSTEM. E. Sondag. Ponsen and Looijen BV, Amsterdam; 1996.

EVIDENCE FOR THE INVOLVEMENT OF CARBONIC ANHYDRASE AND UREASE IN CALCIUM CARBONATE FORMATION IN THE GRAVITY-SENSING ORGAN OF *APLYSIA CALIFORNICA*. H. A. Pedrozo, Z. Schwartz, D. D. Dean, J. L. Harrison, J. W. Campbell, M. L. Wiederhold and B. D. Boyan. *Calcif. Tissue Int.,* Vol. 61, pages 247–255; 1997.

THE STRUCTURE OF THE STATOCYST OF THE FRESHWATER SNAIL *BIOMPHALARIA GLABRATA* (*PULMONATA, BASOMMATOPHORA*). W. Gao, M. L. Wiederhold and R. Hejl. *Hear Res.,* Vol. 109, pages 109–124; 1997a.

DEVELOPMENT OF THE STATOCYST IN THE FRESHWATER SNAIL *BIOMPHALARIA GLABRATA* (*PULMONATA, BASOMMATOPHORA*). W. Gao, M. L. Wiederhold and R. Hejl. *Hear Res.,* Vol. 109, pages 125–134; 1997b.

DEVELOPMENT OF GRAVITY-SENSING ORGANS IN ALTERED GRAVITY. M. L. Wiederhold, W. Y. Gao, J. L. Harrison and R. Hejl. *Grav. Space Biol. Bull.,* Vol. 10(20), pages 91–96; 1997a.

DEVELOPMENT OF GRAVITY-SENSING ORGANS IN ALTERED GRAVITY CONDITIONS: OPPOSITE CONCLUSIONS FROM AN AMPHIBIAN AND A MOLLUSCAN PREPARATION. M. L. Wiederhold, H. A. Pedrozo, J. L. Harrison, R. Hejl and W. Gao. *J. Grav. Physiol.,* Vol. 4(2), pages 51–54; 1997b.

MORPHOMETRY OF FISH INNER EAR OTOLITHS AFTER DEVELOPMENT AT 3G HYPERGRAVITY. R. Anken, T. Kappel and H. Rahmann. *Acta Otolaryngol.,* Vol. 118, pages 534–539; 1998.

THE PROCESS OF OTOCONIA FORMATION IN GUINEA PPIG UTRICULAR SUPPORTING CELLS. Y. Harada, S. Kasuga and N. Mori. *Acta Otolaryngol.* (Stockholm), Vol. 118, pages 74–79; 1998.

THE C.E.B.A.S.-MINIMODULE — DEVELOPMENT, REALIZATION AND PERSPECTIVES. K. Slenzka. Int. Conf. Environ. Systems 99, SAE 1999-01-1988, pages 1–3; 1999.

The Development of an Insect Gravity Sensory System in Space (Crickets in Space)

Authors

Eberhard R. Horn, Günter Kämper, Jürgen Neubert

Experiment Team

Principal Investigator: **Eberhard R. Horn[1]**

Co-Investigators: **Günter Kämper[1], Jürgen Neubert[2]**

Technical Assistant: **Marlene Kuppinger[1]**

Co-workers: **Susanne Förster[1], Pascal Riewe[1], Claudia Sebastian[1], Hans Agricola[3]**

[1]University of Ulm, Ulm, Germany
[2]German Aerospace Establishment, Köln, Germany
[3]University of Jena, Jena, Germany

ABSTRACT

Our Neurolab experiment CRISP (**Cr**ickets **in Sp**ace) examined the balance between genetics and gravity in the development of a gravity-sensing system. Crickets are an excellent model to study this balance since they possess easily accessible external gravity receptors and a simple nervous system. Information from the gravity sensors travels through a pair of neurons called position-sensitive interneurons (PSIs), and is used to keep the cricket's head stable when its body rolls from side to side. In addition, injured gravity sensors can regenerate if they are damaged during development. It is not known, however, if they would regenerate normally if gravity were not present. Crickets at four developmental stages (eggs, first-, fourth-, and sixth-stage larvae) were sent into space for 16 days to explore the possibility of an age-related sensitivity to microgravity. Another group of crickets was exposed to high gravity levels (three-G) for the same amount of time. One gravity sensor was removed from the sixth-stage larvae so that regeneration could be studied. In intact larvae and larvae with a removed sensor, post-flight experiments revealed that the behavioral response (maintenance of head position with body tilt) had a low susceptibility to microgravity. In contrast, the activity of the PSIs revealed a significant sensitivity to microgravity. Nerve firing was significantly reduced in microgravity-exposed larvae, but this recovered after 14 days of re-adaptation to one-G. Parts of the brain where gravitational information is integrated with other senses were not modified by weightlessness exposure. Crickets exposed to three-G revealed no effects on behavior but showed a significant developmental retardation of PSI physiology during one-G re-adaptation. Absence of prominent behavioral effects either early after weightlessness or after three-G exposure may be caused by a fast one-G re-adaptation supported by inputs from other sensors such as position sense and vision. The results demonstrate that the development of the gravity-sensing system is governed not only by genetic programs but also by gravitational experience.

INTRODUCTION

During early life, transient alterations of environmental conditions can cause irreversible anatomical and physiological changes in sensory and motor systems. This was first demonstrated in the visual system of kittens. If one eye was covered for a short period during development, the brain area responsible for vision (the visual cortex) developed abnormally. The physiological responses of neurons within the visual part of the brain and the anatomical organization of this brain area revealed significant modifications in the visually deprived animals compared to their normally reared siblings. Surprisingly, not all periods of life after birth revealed this sensitivity to sensory deprivation. This observation prompted the concept of *critical periods* (Wiesel, 1982); i.e., periods during development when the presence of a particular factor (vision in the example above) is critical for normal development. Detailed knowledge about such periods is important for human medicine since this can help avoid abnormal development in children produced by atypical sensory inputs or movement restrictions during early development. Thus, understanding how environmental alterations can cause changes in the nervous system is important.

One approach for studying the impact of environment on sensory, neuronal, and motor development is *deprivation*. Covering eyes or ears can prevent vision or hearing, respectively; parts of the body can be immobilized to prevent particular movements. Another approach is to use a stimulation above the normal level. For example, exposure to high G-forces is useful to demonstrate the basic susceptibility of a developing sensory system to gravity. However, it cannot answer the question about the unequivocal necessity of gravitational input for normal development. This problem requires the transient, but complete, elimination of gravitational inputs. Exposure to weightlessness is the only way to deprive gravity sensory systems transiently from their specific inputs.

Experimental models are necessary to study basic mechanisms of how modified gravitational conditions affect sensory, neuronal, and motor system development. Animal models help analyze the mechanisms responsible for sensitivities to modified environmental conditions and find ways to overcome abnormal developments. Studies have been conducted on mammals, birds, amphibians, fish, and invertebrates. A survey of progress in this research has been presented by Berthoz and Güell (Berthoz, 1998).

While vertebrates such as fish, amphibians, rats, and mice are preferred for examination, the complexity of their nervous system and the inaccessibility of their gravity sensors make them difficult to study. We had several reasons to choose crickets as a model to study microgravity effects on the development of the nervous system. These insects possess an external gravity sense organ. Its stimulation by postural movements of the crickets (rolling side to side, for example) induces a compensatory head response. In the first stage of larval development, this response can be induced by a single sensory unit (Horn, 1998). In addition, this gravity sense organ regenerates after damage. In parallel with structural recovery, the behavioral response also recovers (Horn, 1996, 2001). Information originating from stimulation of this gravity sense organ is transmitted to higher centers of the nervous system by a single neuron (a PSI) whose activity is unequivocally related to the cricket's posture. A small number of identified neurons located in the higher centers control motor output and, thus, the behavioral response. Both the neurophysiological and behavioral responses use the gravitational vector as the reference direction.

The basic goal of the Neurolab experiment CRISP (Crickets in Space) was to reveal how the development of a gravity sense organ, a behavior related to gravity, and a nerve system sensitive to gravity would be affected by gravity deprivation. These results may have relevance for human medicine. Gravity sense organs are usually closely linked to the muscular system and also have close connections to non-gravity-related sensory systems. What would happen in a human child if during early development the activity of that child's gravity sensors is functionally blocked or at least reduced for some time? This might occur due to a temporary blockade of blood flow through the sense organ leading to a deficit in oxygen and glucose supply. The formation of connections with the visual system or with position sense might be disrupted. These connections are essential for the sensation of up and down, and for normal locomotion. Thus, results from these cricket experiments allow the formulation of hypotheses about neuronal mechanisms relevant to the human nervous system.

METHODS

The crickets were transported into space within the BOTEX incubator. The incubator is equipped with a one-G reference centrifuge for one-G simulation in orbit. During the 16-day Neurolab mission, its temperature was set to 27°C and was maintained in this range with less than a 1°C variation. Crickets were tested behaviorally and physiologically for 16 days after the spacecraft landed.

Behavioral changes (head roll)

Animals roll their eyes or head opposite to an imposed roll to keep vision stable despite the roll. They also move their legs reflexively during body movements to maintain postural stability. These reflexes can be quantified by measuring the angle of compensatory eye or head responses, or by recording the activity of neurons or muscles involved in this reflex machinery. House crickets (*Acheta domesticus*) from the Neurolab experiment CRISP were rolled around their longitudinal axis by 360 degrees in 30-degree steps. Their roll-induced compensatory head response (rCHR) was videotaped from the frontal view and analyzed by a frame-processing system. The standard rCHR characteristic is sine-like (Figure 1, left); its maximum deflection (rCHR amplitude) depends on the developmental stage (Horn, 1998).

The postural information originating in the gravity sense organ, located on a structure called a cercus, is projected to the PSI. The body and branches of these position-sensitive neurons are located in the terminal ganglion of the ventral cord (cercal ganglion). The PSI is part of the neuronal network responsible

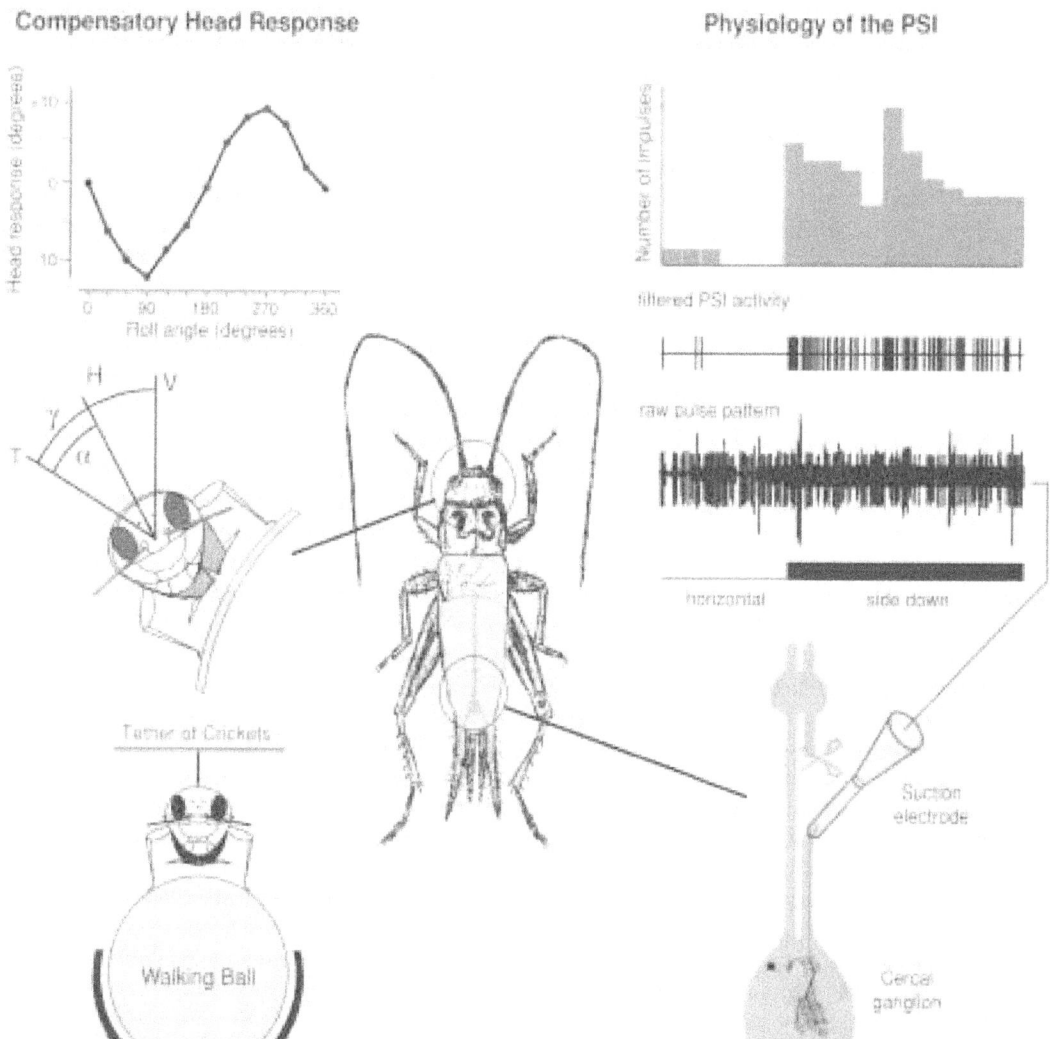

Compensatory Head Response

Physiology of the PSI

Figure 1. Schematic presentation of recording techniques for the compensatory head response (left) and for the extracellular recording of the PSI activity (right). Left: For the behavioral analysis, crickets could move a walking ball while they were tethered to a holder to exclude movements of the body except the head. A lateral roll of the cricket by the roll angle γ (angle between T and V) elicited a compensatory head roll α (angle between T and H). The response characteristic is sine-like. V, spatial vertical, i.e., direction of the gravitational vector; T, front to back axis of the cricket's thorax; H, front to back axis of the cricket's head. Right: For the neurophysiological recordings, the connection between the terminal ganglion and the one next to it were cut and the end of the connective was placed into the recording electrode using suction. The activity pattern recorded during changes of the cricket's posture consists of impulses from position- and wind-sensitive neurons (raw pulse pattern). The PSI activity was filtered out by specific features of the individual pulses; the histogram presents the time-course of the activity during a lateral roll.

for transmitting spatial information from the gravity sense organs to the motor centers of the brain. This, in turn, activates the neck muscles. One property of the PSI is that its activity rate increases during roll to the side of the cricket where most of the branches for that particular neuron are located. Roll to the opposite side, away from the branches, has no significant effect on the discharge frequency (Sakaguchi, 1983). In intact larvae, we recorded the spiking activity by suction electrodes from the

connections on one side of the ganglion (Figure 1, right). In crickets with a regenerated gravity-sensitive system, recordings were taken from both sides to compare the activity between a PSI receiving its input from a normally developed sense organ to one activated by a regenerated sensor.

Eggs and first-, fourth-, and sixth-stage larvae were sent into space to explore the possibility of an age-related sensitivity to microgravity. Additionally, a cercus containing gravity sensors was removed in sixth-stage larvae to study regeneration. Eggs were chosen because the first-stage larvae hatching in space from these eggs would have had no rCHR-experience prior to launch. First-stage larvae were chosen because they possess only a single sensory cell on each cercus; i.e., in this stage, gravity experience before launch can be regarded as the minimum possible under Earthbound conditions (Horn, 1998). Older stages were chosen because of the increasing connectivity within their nervous system that might affect the extent of adaptation to microgravity.

High–G (hypergravity) study

Hypergravity exposures were performed following the identical time schedule used for the Neurolab mission, also using the BOTEX incubator. We chose three-G conditions because we had learned from previous experiments on compensatory eye movements during lateral roll in amphibian tadpoles and young fish that a nine-day, three-G exposure induces significant changes (Sebastian, 1996; Horn, 1999).

Formation of nerve networks

Information transfer from one neuron to another and information processing within nerve networks is based on chemical neurotransmitters such as gamma-aminobutyric acid (GABA), glutamate, noradrenalin, and about 100 other neuropeptides. In some instances, different neurons have different neurotransmitters associated with them. In other neurons, many neurotransmitters are collocated. Thus, evolution has created a large variability of identified neurochemical networks, particularly in insects (Agricola, 1995; Agricola, 1997). The fact that insect neurons can be considered as chemically identified individuals with well-defined shapes (phenotype) offers the opportunity to analyze not only the mechanisms responsible for this diversity, but also to use the insect brain as a model to study the interference of environmental factors with genetic programs in the formation of neuronal networks. Thus, looking at the particular transmitters associated with neurons gives a sense of how gravity may affect the development of the brain and, in particular, of individual neurons.

In the cricket's brain, neurons linked to the motor and sensory system (sensorimotor centers) use neuropeptides such as allatostatin, perisulfakinin, proctolin, or leucokinin to process and modulate neuronal information. Neurons that use these neuropeptides to transmit information are called peptidergic neurons. Some of these neurons connect brain sites with the highest level of integration to lower centers of the nervous system. Other neurons possess connections only with neurons within the brain. Besides their ability to modulate neuronal activity, these neuropeptides are also involved in information transfer from one neuron to another and are responsible for maintaining the structural and functional integrity of muscles (myotrophic function). We studied neurons from the front of the brain, the pars intercerebralis. These neurons are closely connected to one of the main association centers in the brain and control the output to neurons responsible for locomotion.

Representative types of peptidergic neurons from the pars intercerebralis of the larval cricket brain were used to explore the effects of gravity deprivation on the formation of neuronal patterns during post-embryonal development. Studies were performed with developmental stages that had either hatched during the first days of the mission, or had developed to stage one or four at onset of microgravity. Two sets of one-G controls were available; they included crickets exposed to simulated one-G in space, or crickets continuously reared under one-G conditions on Earth. The peptidergic neurons were visualized using fluorescent antibodies that attached to the neuropeptides. In this way, the structure of a neuron with a specific neuropeptide can be revealed. In addition, the intensity of the fluorescence (immunoreactivity) points to the amount of the specific neuropeptide. We used primary antibodies to insect neuropeptides dip-allatostatin II (n=14 stainings), CCAP (n=31), corazonin (n=6), leucokinin (n=6), locustatachykinin II (n=13), perisulfakinin (n=20), and proctolin (n=12). One hundred and two stainings were performed within 68 brains; in some brains, double antibody stainings were used.

RESULTS

The eggs hatched and the larvae molted successfully in orbit. They developed the peripheral and central structures related to the sense of gravity, and were able to regenerate gravity sense organs in the absence of the usual environmental conditions. Immediately after the spacecraft landed, the larvae revealed no sign of disturbed walking behavior.

Behavioral development and regeneration

The compensatory head response during rCHR was studied for 16 days in all four stages exposed to microgravity. The mean rCHR characteristics of microgravity-exposed crickets recorded between postflight days one to eight revealed no significant differences compared to those recorded from ground-reared one-G controls (Figure 2). Structural and functional regeneration occurred in larvae in which one cercus with its gravity sense organ was amputated two to three days before launch. As in the ground controls, the regenerated gravity sense organ possessed fewer sensory units than the normal one, and the rCHR was weaker in crickets with a regenerated cercus. But neither for the regenerated nor for the intact larvae were significant behavioral differences obtained between flight and ground animals (Figure 3, upper plots). In two of the fourth-stage samples, two specific effects became obvious:

- a retardation of reflex development in slowly developing animals, and

- a retardation of one-G re-adaptation in larvae hatched in orbit.

The retardation of reflex development in slowly developing larvae was observed for the sixth-stage larvae. Generally, insects such as crickets, grasshoppers, or cockroaches grow by molting; i.e., the age of these insects can be defined by their

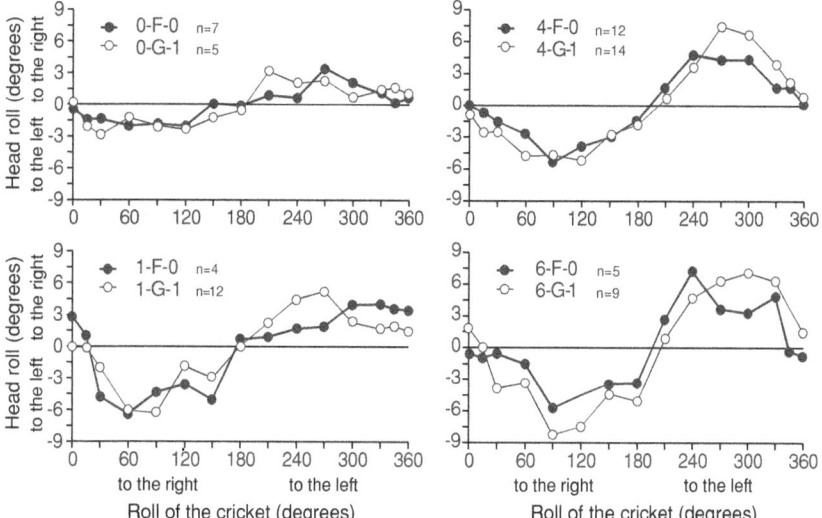

Figure 2. Characteristics of the rCHR for cricket larvae exposed to microgravity for 16 days compared to their ground controls. Samples were defined by the stage at microgravity onset. X-F-0=microgravity-exposed larvae launched at stage X=1, 4, or 6, or as eggs (X=0); X-G-1=the respective one-G ground controls; n=number of larvae. Note the absence of a significant effect on the response characteristics. Recordings were obtained between postflight days one and four. During this period, crickets were kept at 13°C to avoid further molting and development.

number of molts. Developmental acceleration causes a shorter, and developmental retardation a longer, time span between two molts. During the 16-day flight the sixth-stage larvae developed to seventh-, eighth- or ninth-stage larvae; i.e., they performed one, two, or three molts as their ground controls did. From the larvae with a regenerated cercus, the microgravity group that had molted only twice showed a significantly weaker rCHR than their stage-matched, ground-reared siblings. No difference existed between larvae that had molted three

times in orbit and on the ground. In the intact controls, there was a tendency (p<0.1) for a similar stage-related susceptibility (Figure 3, lower plots).

Crickets that hatched during the period of microgravity revealed impaired one-G re-adaptation capabilities compared to larvae that had one-G experience prior to the onset of microgravity (Figure 4, upper plots). In particular, during the two-week postflight period, larvae that hatched in space did not increase their rCHR significantly while their siblings hatched on

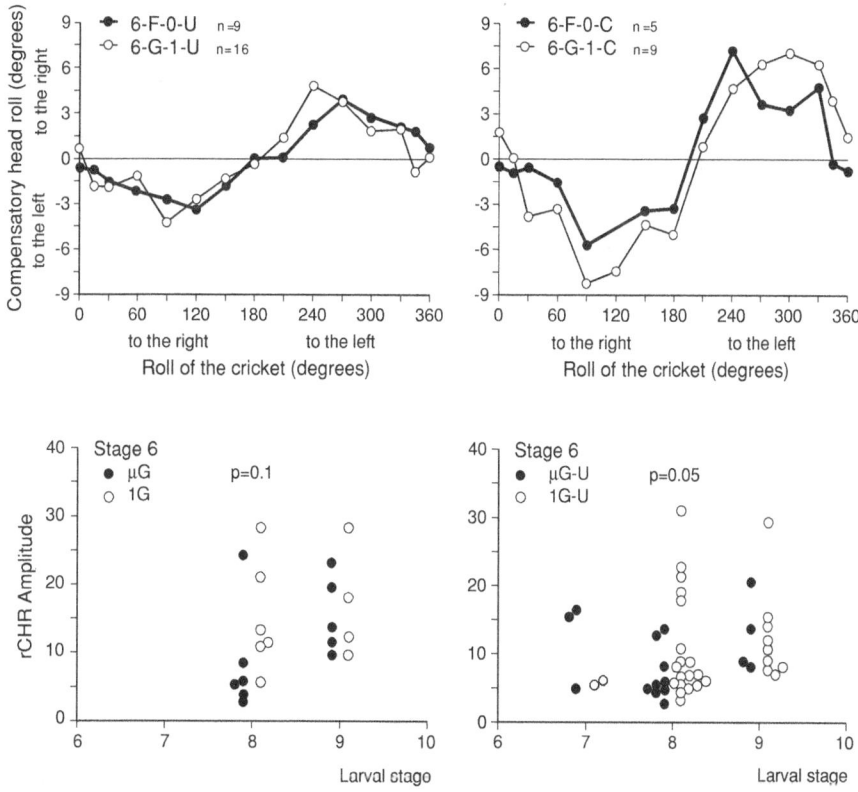

Figure 3. Effect of microgravity on the rCHR in larvae with one regenerated cercal gravity sense organ compared to intact controls. Larvae were launched at stage six. **Upper plots**: rCHR-characteristics, which are sine-like in all samples. Samples were defined by the stage at µg-onset. 6-F-0-U and 6-F-0-C=microgravity-exposed larvae launched at stage six with an injured or intact gravity sensory system, respectively; 6-G-1-U and 6-G-1-C=the respective one-G-ground controls. **Lower plots**: The maximal left-right head movement during a 360-degree lateral roll (rCHR amplitude) plotted in relation to the developmental stages. Note the significant decrease of the rCHR amplitude in larvae that hatched only twice in space (stage eight). Intact controls: left; injured larvae with a regenerated gravity sense organ: right. Each dot (microgravity-larvae) or circle (one-G larvae) represents one cricket.

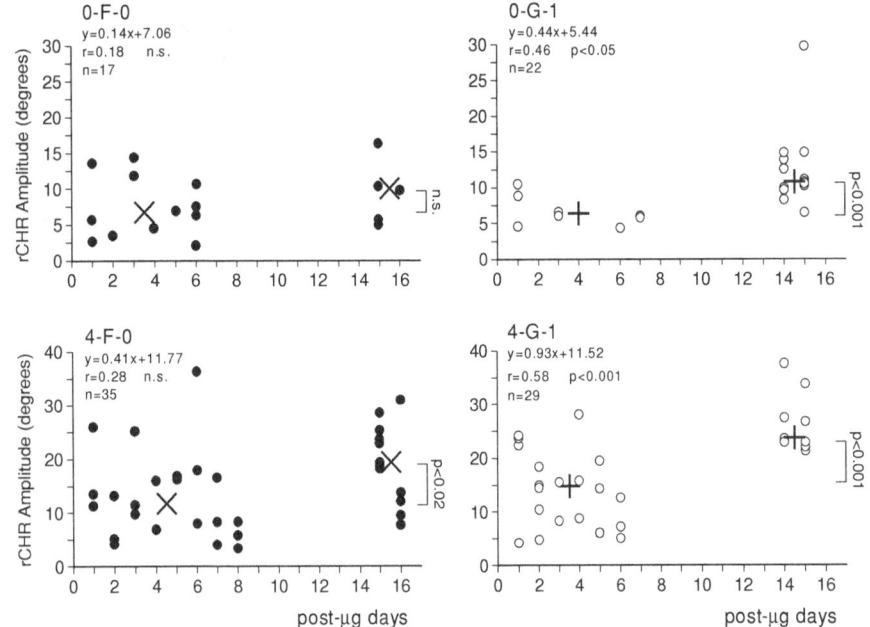

Figure 4. Recordings of the rCHR in cricket larvae during the period of one-G re-adaptation. Each dot and circle represents one animal, + and x represent median values calculated for the first and second recording periods (postflight days one to eight, and 15 to 16, respectively). In addition to the statistical evaluation from the u-test, the correlation coefficient r and the linear regression function y(x) are also given. X-F-0 and X-G-1 with X=0 (**upper plots**) or 4 (**lower plots**); see Figure 2 for definitions.

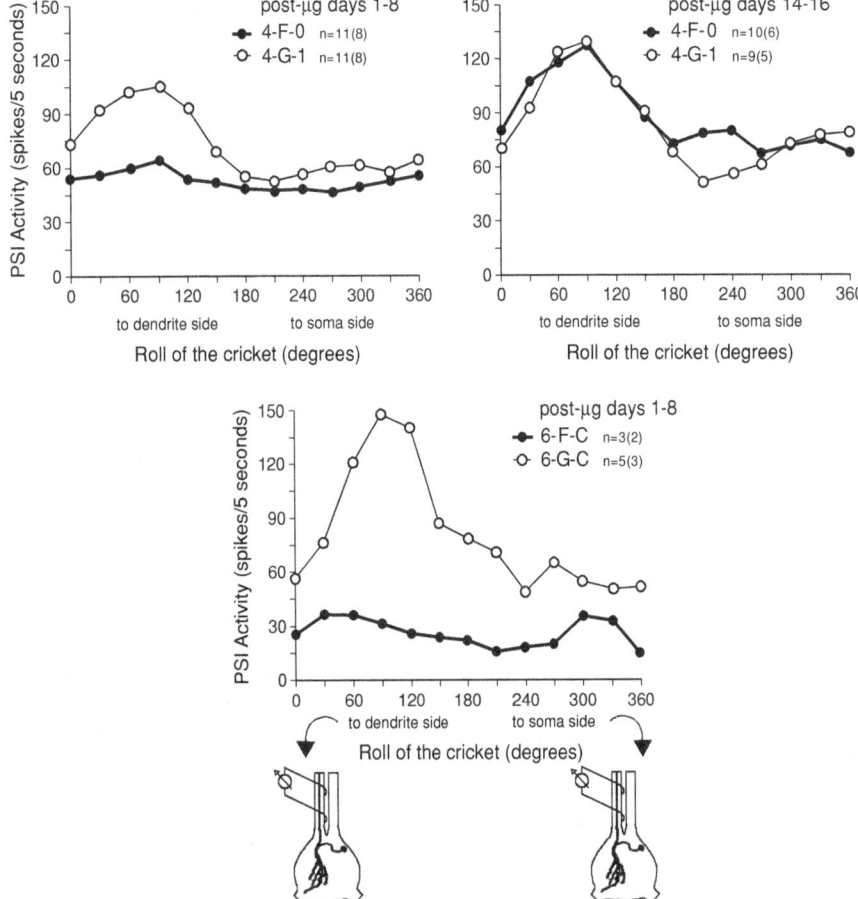

Figure 5. Effects of microgravity on the sensitivity of the PSI. At onset of microgravity, the cricket larvae were at stage four (upper plots) or stage six (lower plot). In stage four, PSI responsiveness was decreased during postflight days one to eight, but recovered to normal development at days 15 and 16. X-F-0 and X-G-1 with X=4 and 6 (see Figure 2 for definitions).

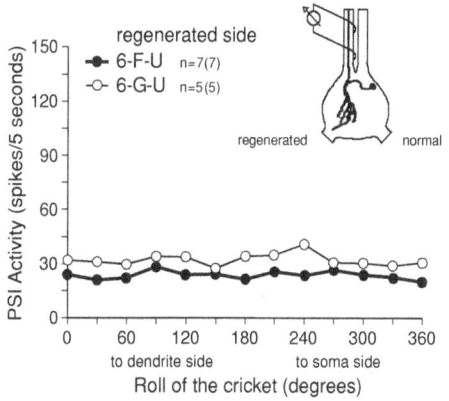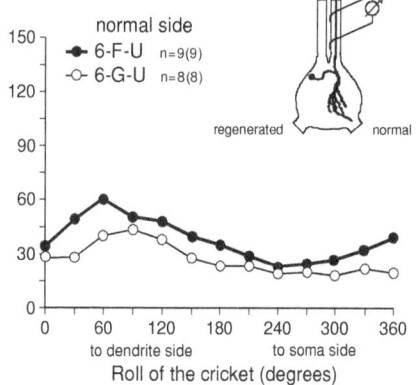

Figure 6. PSI activity in cricket larvae with a regenerated cercal gravity sensory system. Amputation of one cercus was performed at stage six, two days before microgravity onset. Recordings from the PSI whose axon runs along the regenerated side of the nervous system (left) or on the intact side (right). n(N), n recordings from N larvae. Recording period was between postflight days one to eight. Crickets were kept at 13°C during that period.

the ground during the same period did. Crickets investigated during postflight days one to eight were kept at 13°C to block further development, while crickets observed during postflight days 15 to 16 were kept at 27°C to allow further development. For comparison, observations from fourth-stage larvae are included in this figure. Both flight and ground larvae increased their rCHR significantly after flight (Figure 4, lower plots).

Physiological development and regeneration

The PSIs revealed a significant sensitivity to microgravity. These observations contrasted with the behavioral ones. In particular, in the stage-four group the modulation of the activity during roll of the cricket to the sensitive side of the PSI was significantly reduced between postflight days one and eight with respect to the ground controls. Normalization of the discharge frequency modulation to the level of standard development was observed on postflight days 15 and 16. In the stage-six group, the sensitivity of the PSI from microgravity-exposed larvae was also decreased with respect to ground controls. Due to the low number of larvae, the difference was not significant (Figure 5).

Recordings from the PSI were also taken in crickets with a regenerated cercal gravity sense organ. Comparisons were made between (1) flight and ground samples and (2) recordings taken from the PSI receiving its input from the regenerated or the normally developed cercus. Activity modulation was very low or even absent in the PSI linked to the regenerated cercal gravity sense organ; it was significantly larger for the PSI transmitting positional information from the intact side to the cerebral centers (Figure 6).

Hypergravity effects on the rCHR and the PSI

Studies on hypergravity exposures were conducted according to the time schedule of the Neurolab mission. In particular, behavioral studies were performed with all four sets of developmental stages, while neurophysiological studies were restricted to the stage-four and stage-six groups. In general, the behavioral observations were similar to those obtained

from the microgravity-exposed crickets. Under three-G conditions, first eggs hatched and larvae molted at the same ratio as the one-G controls. In addition, the rCHR-characteristics were mainly unaffected by the 16-day, three-G exposure. Similar to the Neurolab observations, the neurophysiological recordings from the PSI revealed differences between the three-G and one-G reared larvae. But while microgravity-exposed larvae had a lower sensitivity after flight and recovered completely during one-G re-adaptation, three-G exposed larvae showed the same frequency modulation of the PSI during the first four post-three-G days as their one-G controls. However, during further development and one-G re-adaptation, the response modulation in the animals with three-G experience was maintained without any change while their one-G controls increased the extent of impulse frequency modulation according to normal development (Figure 7). This observation resembled results obtained from hypergravity-exposed tadpoles of the amphibian *Xenopus laevis* whose abilities for complete recovery during one-G re-adaptation were also blocked (Horn, 1996).

Effects of gravity on the formation of nerve networks in the brain

The results demonstrated that cell patterns from larvae with microgravity experience stained with antibodies for dip-allato-statin II, CCAP, leucokinin, locustatachykinin II, perisulfakin, and proctolin resembled the patterns obtained from one-G controls. Differences between microgravity-exposed crickets and their one-G controls were obtained for leucokinin neurons within the pars intercerebralis from larvae that were at stage one at launch. In one-G controls but not in microgravity-exposed crickets, these neurons were intensively stained by antibodies for leucokinin. Staining of neurons from the sensorimotor centers was not obtained for the corazonin antibody. This result is in contrast to observations in cockroaches. The intensity of neuropeptide expression determined from the extent of fluorescence was similar for most of the antibodies except for locustatachykinin-IR structures. Neurons from crickets with microgravity experience revealed a much higher intensity of immunoreactivity than those from their one-G-reared controls.

DISCUSSION

These studies have shown that the physiology of an identified gravity-sensitive neuron (the PSI) is strongly affected by gravity deprivation. The behavior of the larvae and the anatomy of the neurons from cerebral sensorimotor centers are less or not affected. The lack of coincidence between the results from the behavioral and neurophysiological studies suggest that an integratively acting system exists for normalizing the developmental changes that occurred in weightlessness. We propose that crickets use their multichannel gravity sensory system (Horn, 1983; Horn, 1985; Horn, 1986) to normalize and stabilize the development of the gravity sensory system as fast as possible after gravity deprivation.

If the above explanation, which connects the contradictory behavioral and physiological results, is correct, crickets may offer advantages for further study. They would offer the possibility to study not only microgravity effects at the cellular level, but also the interaction between several sensory channels when one or more channel is deprived of its naturally occurring input during development.

In addition, the experiments demonstrated that the development of neuropeptidergic cerebral structures is mainly independent of gravitational conditions. Several factors, however, prevent us from generalizing these results:

- Lack of significant effects of gravity deprivation on peptidergic neurons of the cricket's brain might be caused by their low or absent participation in gravity processing. Neurons connected with the gravity sensory system will probably be the most affected by microgravity.

- The time window of microgravity exposure may be critical. At onset of microgravity, the full set of neurons had already developed, even in those crickets that hatched in orbit. For Neurolab, eggs from the last quarter of embryonic development were selected according to our experiment-specific criterion that eggs that hatched in orbit should grow as long as possible under microgravity conditions. However, cell proliferation takes place during the first 30% to 50% of embryonic development (Edwards, 1979).

- The effects may be small, so that larger numbers of specimens are necessary for statistical evaluation. The observations of an increased immunoreactivity and the failure of specific antibody stainings in some crickets with microgravity experience support this postulation.

For the future, it is necessary to extend the studies on identified peptidergic neurons, gravity-related behavior, and the

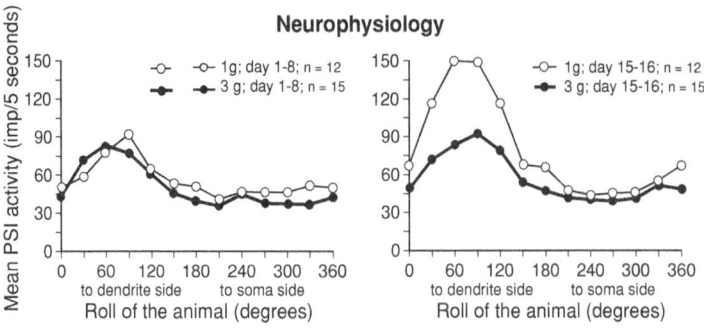

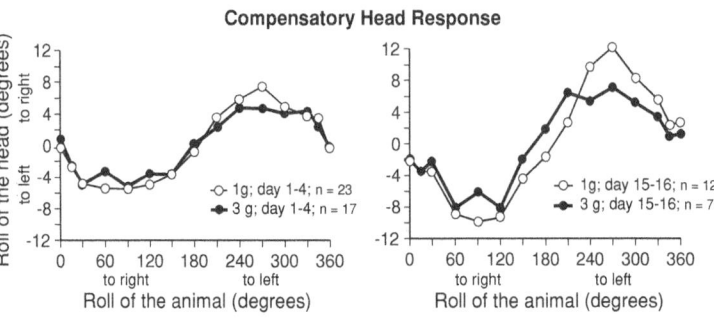

Figure 7. Effects of three-G exposures on the development of the rCHR (**lower plots**) and on the physiology of the PSI (**upper plots**). At onset of the three-G period, these crickets were at developmental stage four; they molted twice or three times during three-G exposure. Note the low or absent effect on behavior and the developmental retardation of the PSI activity during the period of one-G re-adaptation.

physiology of gravity-sensitive neurons to the period of development during which proliferating neurons are exposed to microgravity. This is possible if fertilization and at least the first 30% to 50% of embryonic development occurs under microgravity conditions.

General considerations and perspectives

In this Neurolab study, we studied crickets; but in our previous spaceflight experiments, we also studied fish and amphibians. In all three species, we studied the effects of microgravity on compensatory head and eye movements induced by lateral roll. Conclusions from this comparative approach are useful (1) to finding out the basic mechanisms of how animals adapt to the microgravity environment, and (2) to develop hypotheses of how people can adapt to the space environment and, in particular, what might happen to people if they live in space and their complete development from the fertilized egg to an adult occurs in the absence of gravity. The basic observations from our neurophysiological and comparative behavioral studies have been:

- Gravity is necessary for normal development; deprivation causes disturbances in neuronal activity (Figure 5) or behavior (Figure 8). Even when behavioral modifications are not seen, this does not necessarily mean independence of development from gravity (for crickets, compare Figure 8, left, with Figure 5).

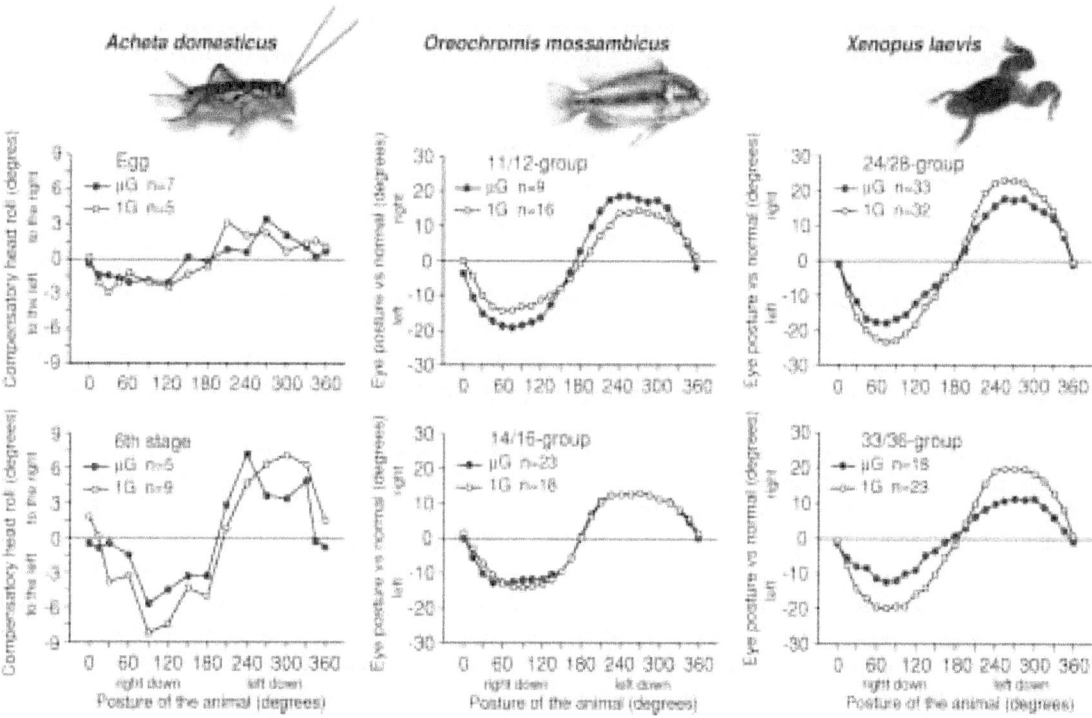

Figure 8. Effects of microgravity on the development of compensatory responses of the eyes or head in developing animals. For each species, mean response characteristics from two different ages are presented. Ages are defined by the animals' developmental stages at µg-onset. Note the lack of µg-effects in crickets (*Acheta*), a stage-dependent effect in fish (*Oreochromis*), and significant effects in both tadpole stages (*Xenopus*). It is important to note that crickets possess three different gravity-sensitive pathways that contribute to the rCHR; only the cercal was stimulated in the postflight tests. Fish control body and eye postures by their otolith organs and by their eyes, which induce the dorsal light response. *Xenopus* is considered as an animal whose equilibrium during swimming is mainly controlled by the vestibular system. *Acheta*: unpubl. data from Neurolab; *Oreochromis*: modified from Horn, 1999, and Sebastian, 2001; *Xenopus*: modified from Sebastian, 1996, and Sebastian, 1998.

• Both microgravity and hypergravity activate adaptive mechanisms, which were observed after terminating the altered gravitational conditions. They revealed either a higher or a lower sensitivity to roll stimulus (Figure 8, middle and right columns).

Most of these studies revealed normalization of development after termination of the altered gravitational conditions. This observation does not exclude the persistence of microgravity-induced effects, because the exposure time was restricted to short periods during development. In fact, observations such as the development of malformations of the body during a space-flight (albeit reversible during one-G re-adaptation) and the high susceptibility to micro- and hypergravity in slowly growing animals as well as in animals that hatched during microgravity or hypergravity exposure (Sebastian, 1996; Horn, 1999) make the occurrence of significant residual, pathological effects likely. These observations prompted several hypotheses.

• Re-adaptation to normal one-G conditions after termination of micro- or hypergravity is faster when more parallel channels exist that can control and affect the specific behavioral response. Comparison between the species used in our studies on compensatory head or

eye movements revealed that crickets can use at least three, fish two, and *Xenopus* one independent gravity-sensitive channel(s) to normalize the development of their sense of gravity. Adaptation was fastest in crickets and took the longest time in *Xenopus*; i.e., the redundancy of multichannel systems is advantageous for normalizating development after terminating environmental changes.

• A high susceptibility exists before and during the period of cell proliferation in the neuronal network underlying behavior (Sebastian, 1998). All exposures were performed before or during the period of neuronal proliferation of the nuclei involved in the central processing of gravity.

The search for mechanisms that allow for life to adapt to orbit is the basic aim of gravitational biological research. For the future, three aspects of this adaptation have to be considered and studied in proper animal models.

• The effects of long-term exposures, under both hyper- and microgravity, that allow the animal to develop from an egg into an adult.

- An analysis of the adaptive mechanisms activated during micro- or hypergravity exposure that enable the organism to accommodate the altered environment.

- The consequences of fertilization in space (or during hypergravity), and whether these organisms would be able to adapt to one-G.

The fertilization studies have high importance. Each experiment on the effects of long-term exposures, allowing complete development under microgravity, has to make clear that the basic step of fertilization is possible in orbit. Successful orbital fertilizations were performed artificially in an amphibian (*Xenopus laevis*) (Souza, 1995; Ubbels, 1997) or naturally in fish (Japanese *medaka oryzias* latipes) (Ijiri, 1997). We observed in our laboratory successful mating under hypergravity in the crickets. The conduct of subsequent studies in cricket larvae that hatch from eggs fertilized in space and develop under these conditions to adults will give more insight into how the nervous system responds to weightlessness. Results obtained from these studies will allow for hypotheses to be formed that can subsequently be tested in lower vertebrates (such as fish and amphibians) and ultimately in mammals.

REFERENCES

EMBRYONIC DEVELOPMENT OF AN INSECT SENSORY SYSTEM, THE ABDOMINAL CERCI OF *ACHETA DOMESTICUS*. J.S. Edwards, S.W. Chen, *Wilh Roux's Arch.*, Vol. 186, pages 151–178; 1979.

POSTNATAL DEVELOPMENT OF THE VISUAL CORTEX AND INFLUENCE OF ENVIRONMENT. T.N. Wiesel, *Nature,* Vol. 299, pages 583–591; 1982.

GRAVITY RECEPTION IN CRICKETS: THE INFLUENCE OF CERCAL AND ANTENNAL AFFERENCES ON THE HEAD POSITION. E. Horn, H.J. Bischof, *J. Comp. Physiol.,* Vol. 150, pages 93–98; 1983.

THE EQUILIBRIUM DETECTING SYSTEM OF THE CRICKET: PHYSIOLOGY AND MORPHOLOGY OF AN IDENTIFIED INTERNEURON. D.S. Sakaguchi, R.K. Murphey, *J. Comp. Physiol.,* Vol. 150, pages 141–152; 1983.

TONIC AND MODULATORY SUBSYSTEMS OF THE COMPLEX GRAVITY RECEPTOR SYSTEM IN CRICKETS. E. Horn, W. Föller, *J. Insect Physiol.* Vol. 31, pages 937-946; 1985.

GRAVITY. E. Horn, In *Comprehensive Insect Physiology, Biochemistry and Pharmacology*, Vol. 6, Nervous System: Sensory (Kerkut GA, Gilbert LI, eds), Pergamon Press, Frankfurt, pages 557–576; 1986.

AMPHIBIAN DEVELOPMENT IN THE VIRTUAL ABSENCE OF GRAVITY. K.A. Souza, S.D. Black, R.J. Wassersug, *Proc. Natl. Acad. Sci.* Vol. 92, pages 1975–1978; 1995.

COMPARATIVE ASPECTS OF PEPTIDERGIC SIGNALING PATHWAYS IN THE NERVOUS SYSTEM OF ARTHROPODS. H.J. Agricola, P. Bräunig, In *The Nervous Systems of Invertebrates: An Evolutionary and Comparative Approach* (Breidbach O, Kutsch W, eds), Birkhäuser, Berlin, pages 303–327; 1995.

A HYPERGRAVITY RELATED SENSITIVE PERIOD DURING THE DEVELOPMENT OF THE ROLL INDUCED VESTIBULOOCULAR REFLEX IN AN AMPHIBIAN (*XENOPUS LAEVIS*). E. Horn, C. Sebastian, *Neurosci. Lett., Vol.* 216, pages 25–28; 1996.

ALTERED GRAVITATIONAL EXPERIENCE DURING EARLY PERIODS OF LIFE AFFECTS THE STATIC VESTIBULO-OCULAR REFLEX OF TADPOLES OF THE SOUTHERN CLAWED TOAD, *XENOPUS LAEVIS*. C. Sebastian, K. Eßeling, E. Horn, *Exp. Brain Res., Vol.* 112, pages 213–222; 1996.

DEVELOPMENT OF AN INSECT GRAVITY SENSORY SYSTEM IN SPACE – A PROJECT FOR THE NEUROLAB MISSION. E. Horn, C. Sebastian, J. Neubert, G. Kämper, ESA SP-390, pages 267–272; 1996.

EXPLANATIONS FOR A VIDEO VERSION OF THE FIRST VERERBRATE MATING IN SPACE – A FISH STORY. K. Ijiri, *Biol. Sci. in Space.,* Vol. 11, pages 153–167; 1997.

FERTILISATION AND DEVELOPMENT OF *XENOPUS* EGGS ON SOUNDING ROCKETS AND IN A CLINOSTATE. G.A. Ubbles, ESA SP-1206, pages 125–136; 1997.

KOLOKALISATION VON ALLATOSTATIN UND ANDEREN NEURO-MEDIATOREN IM NERVENSYSTEM VON INVERTEBRATEN. H.J. Agricola, *Habilitation Thesis*, University of Jena, Jena, 156 pages; 1997.

INDUCTION OF A GRAVITY-RELATED RESPONSE BY A SINGLE RECEPTOR CELL IN AN INSECT. E. Horn, W. Föller, *Naturwissenschaften*, Vol. 85, pages 121–123; 1998.

SPACE NEUROSCIENCE RESEARCH. A. Berthoz, A. Güell, (eds) *Brain Res. Rev.* (Special Issue), Vol. 28, pages 1–233; 1998.

THE MINIMUM DURATION OF MICROGRAVITY EXPERIENCE DURING SPACE FLIGHT WHICH AFFECTS THE DEVELOPMENT OF THE ROLL INDUCED VESTIBULOOCULAR REFLEX IN AN AMPHIBIAN (*XENOPUS LAEVIS*). C. Sebastian, E. Horn, *Neurosci. Lett.,* Vol. 253, pages 171–174; 1998.

A COMPARISON OF NORMAL VESTIBULO-OCULAR REFLEX DEVELOPMENT UNDER GRAVITY AND IN THE ABSENCE OF GRAVITY. E. Horn, C. Sebastian, ESA SP-1222, pages 127–138; 1999.

ALTERED GRAVITATIONAL FORCES AFFECT THE DEVELOPMENT OF THE STATIC VESTIBULOOCULAR REFLEX IN FISH (*OREOCHROMIS MOSSAMBICUS*). C. Sebastian, K. Esseling, E. Horn, *J. Neurobiol.,* Vol. 46, pages 59–72; 2001.

FUNCTIONAL REGENERATION OF A GRAVITY SENSORY SYSTEM DURING DEVELOPMENT IN AN INSECT (*GRYLLUS BIMACULATUS*). E. Horn, W. Föller, *Neuroreport,* Vol. 12, pages 2685–2691; 2001.

Development of the Vestibular System in Microgravity

Authors

Jacqueline Raymond, Danielle Demêmes,
Emmanuelle Blanc, Claude J. Dechesne

Experiment Team

Principal Investigator: **Jacqueline Raymond**

Co-Investigators: **Claude J. Dechesne,
Danielle Demêmes,
Emmanuelle Blanc**

Technical Assistants: **Stéphanie Ventéo,
Florence Gaven,
Patrice Bideaux**

Université Montpellier II, Montpellier, France

ABSTRACT

We investigated the effects of weightlessness on both the development of a gravity sensor (the utricle in the inner ear) and on the relay of information from this sensor to the brain. Since gravity provides the principal sensory stimulus for the utricle, exposure to spaceflight markedly reduces sensory inputs to this organ. If this were to occur in the developing nervous system, the lack of sensory input might result in changes in utricle organization and, consequently, in the wiring in brain areas where the utricle sends its information. To test this hypothesis, we exposed developing rats (eight days old at launch) to weightlessness for 16 days on the Neurolab mission. In the utricle, we observed no effect of microgravity on sensory cells and neuron connections. In the brain, however, neuron development in the vestibular nuclei of the flight rats was affected. The most striking effect was a lack of connections into the vestibular nuclei from the part of the brain concerned with balance and coordination of movement (the cerebellum). These observations suggest that the lack of gravity did have an effect on the development of nerve connections in the brain, but not in the gravity sensor itself. The areas most affected were the vestibular nuclei and the cerebellum. These results suggest that gravity may be important for the normal development of the balance system.

INTRODUCTION

The mammalian gravity detectors are located in the inner ear and consist of two receptors, the utricle and the saccule. These sensors detect accelerations caused by head movements, and also provide information about up and down. This sensory information is used by the central nervous system to control eye, head, and body position (Figure 1 and Figure 2A).

During growth and development, sensory systems change their structure. In the visual and auditory systems, sensory experience (i.e., seeing and hearing) can profoundly affect the developing central nervous system and control its final organization. In particular, there are critical time periods during development when sensory experience is necessary for sensory circuits to mature normally. For example, if an eye is covered during a particular time in development, the part of the brain serving that eye will not develop properly. If the eye were covered later in development, this effect would not be seen. While this is known to be true for vision, it has not been firmly established if gravity is essential for normal development. This is particularly important for the balance (vestibular) system and the connections it makes.

During weightlessness, the utricle and saccule are no longer exposed to a constant gravitational force and this affects the messages they send to the central nervous system. This unique sensory deprivation is likely to change the organization of vestibular sensory circuits if it occurs during a critical period in their development and maturation. It may

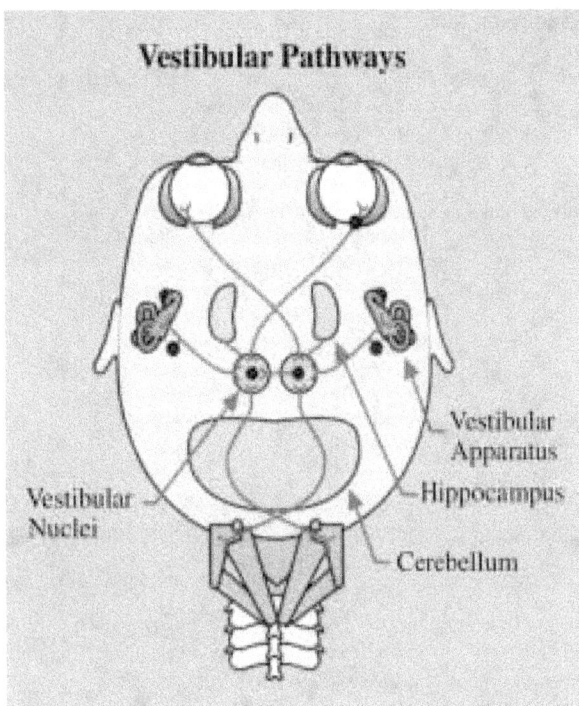

Figure 1. Schematic drawing showing the main pathways integrating vestibular information. Information from the inner ear (vestibular apparatus), which includes the utricle and saccule, travels to the vestibular nuclei. This information is used to control movement and eye movements through connections with the cerebellum and eye muscles.

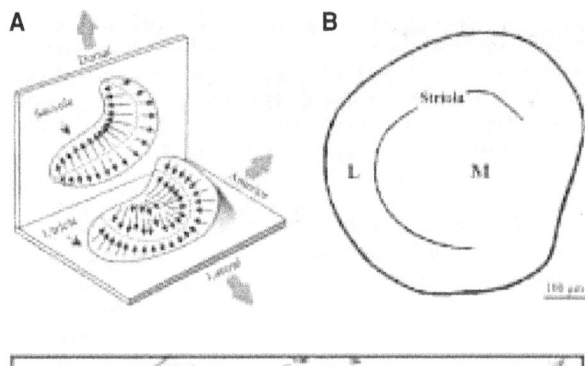

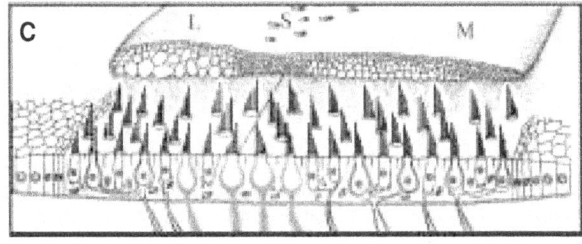

Figure 2 Organization of the macular receptors. A—Planes of orientation of the saccule (vertical) and utricle (horizontal). B—Distribution of the three different zones present in the utricle: lateral (L), striolar (S), and medial (M). C—Transverse section of the utricle, illustrating the organization of the sensory epithelium, which is covered with a thick layer of crystals: the otoliths. The striola (S) divides the population of sensory hair cells into two oppositely polarized groups. Calretinin, a calcium-binding protein, is present in the afferent nerve calyces (red).

affect not only the gravity sensors themselves, but also the central nervous pathways that process and integrate the vestibular messages. These brain pathways start in the vestibular nuclei.

To test whether the gravity sensors and central pathways would be affected by weightlessness, developing rats (eight days old at launch) were exposed to spaceflight for 16 days on the Neurolab mission. We investigated the potential plasticity (the ability to make new connections) of the vestibular system, in both structural and biochemical terms, at four different anatomical levels:

- The gravity sensory organ—the utricle. The utricle contains a sensory layer with hair cells for detecting gravity and accelerations. The messages are transmitted to the brain by afferent neuron fibers (Figure 2C, Figure 3), which connect into the vestibular nuclei. These structures and their connections mature principally during the period extending from just before birth until the first week after birth (Figure 4) (Dechesne, 1992; Desmadryl, 1992).

- The vestibular ganglion containing the cell bodies of the nerve fibers traveling to the central nervous system.

- The efferent system, which is the system that sends messages from the brain back to the utricle. The movements detected by the sensory hair cells and transmitted to the brain by the afferent neurons are controlled by this efferent system. This system contacts the sensory cells and afferent fibers in the sensory layer (Figure 3). The nerve cells of the efferent system originate in the brain stem. The maturation of efferent control occurs mostly between birth and the second week after birth (Figure 4), so this system would be developing at the time the rats in this study were in space.

- The vestibular nuclei—the first step in the integration of vestibular messages. We focused mainly on the vestibular neurons and two of the main pathways entering the vestibular nuclei. One pathway originates from the vestibular sensory organs and the other from the cerebellum. Vestibular neuron development begins just before birth and occurs mainly during the first two weeks after birth. Development of the connections from the vestibular organs begins soon after birth and continues for three weeks. Cerebellar control develops between the first and fourth postnatal weeks. These pathways would be developing while the rats in this study were in space (Figure 7).

We used chemical markers to identify specific sensory cells in the utricle, distinct neurons connecting to the utricle, and particular neurons in the vestibular nuclei. These markers belonged to three categories:

- *Calcium-binding proteins.* These proteins are involved in calcium regulation in sensory cells and neurons. We used parvalbumin, calbindin, and calretinin because their ability to identify specific structures has been well described in vestibular circuits. Parvalbumin is present in the Type I hair cells situated in the striolar area (Figure 3) (Demêmes, 1993). Pathways originating from the vestibular organs are also identified within the brain by detecting parvalbumin. Calretinin identifies the nerve calyces surrounding Type I hair cells containing parvalbumin, and is also present in some Type II hair cells, situated in peripheral areas (Figure 3) (Dechesne, 1996). In addition, the vestibular nuclei present a subpopulation of neurons that produces calretinin (Kevetter, 1996). Pathways originating from the cerebellum are identified by detecting calbindin (De Zeeuw, 1995; Kevetter, 1996).

- *Neurotransmitter proteins.* To identify neurons in the efferent system, we used the neurotransmitter calcitonin gene-related peptide.

- *Synaptic proteins.* To visualize the synapses and their development, we used two markers: an intrinsic membrane protein of synaptic vesicles, synaptophysin, and a presynaptic plasma membrane protein, SNAP-25.

METHODS

We studied rats (Sprague-Dawley rats from Taconic Farms, Germantown, NY) in microgravity and compared them to corresponding ground control rats. Liftoff of the Neurolab STS-90 Shuttle took place when the rats were eight postnatal (P) days

Organization of the neuronal circuits in the maculae

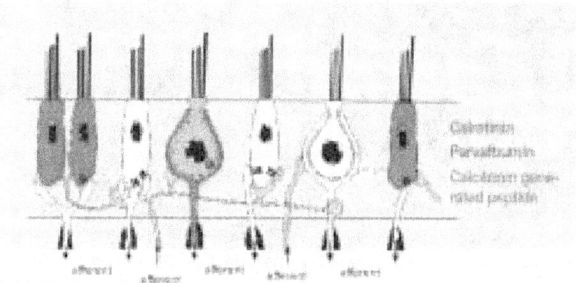

Figure 3 Schematic diagram of the various circuits integrating and controlling vestibular information in the utricular sensory epithelium. Neurochemical markers are used to identify different compartments. Parvalbumin (green) is present in the sensory cells of the striolar zone. Calretinin (red) identifies the afferent nerve calyces surrounding the sensory cells containing parvalbumin. Calretinin (red) is also present in some sensory cells of the peripheral zones. Efferent fibers and their synaptic endings containing the calcitonin gene-related peptide (gray) innervate the Type II hair cells and the afferent calyces and fibers. Afferent fibers take information to the central nervous system; efferent fibers carry information from the central nervous system.

Development of the rat utricle

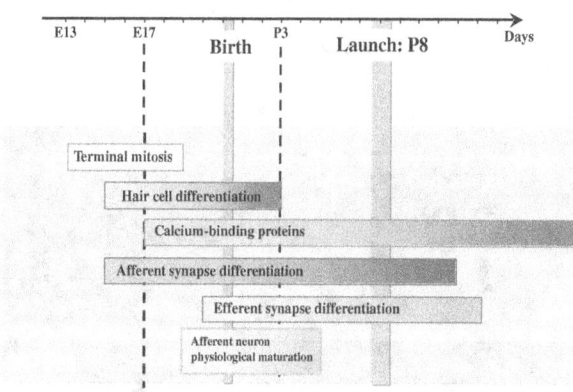

Figure 4 Sequences of events characteristic of the development of the rat utricle. At launch time on postnatal day 8, the structural differentiation of the sensory hair cells and their afferent connections is almost complete. Efferent connections and function mature later.

old and tissues were collected from control animals at this stage (basal, P8). Samples were taken on flight day eight (FD8), corresponding to postnatal day 16 (P16); flight day 15 (FD15) corresponding to postnatal day 23 (P23); and eight hours after landing corresponding to postnatal day 25 (P25). Two types of ground control animals were used. Vivarium (VIV) controls were housed in standard cages and cared for according to usual husbandry practices. Asynchronous ground controls (AGCs) were housed in animal enclosure modules (AEMs).

Each cage, whether in flight or on the ground, contained one mother and six young rats. Our team obtained six rats at FD8, six at FD15, and five on landing. The various experimental procedures for obtaining rat tissues (anesthesia, fixation, and dissection) were performed by the astronauts inflight and by us, on the ground, eight hours after landing, at Kennedy Space Center.

Microscopic analyses were carried out with a BioRad-MRC 1024 laser scanning confocal microscope.

RESULTS

Organization of the utricular sensory epithelium and its connections

Different areas in the otolithic receptors have specific functional properties (Sans, 2001) due to differences in the distribution of sensory hair cell types (Type I/flask-shaped and Type II/cylindrical, Figure 3), types of connections (calyces on Type I hair cells and boutons on Type II hair cells, Figure 3) and the presence of specific proteins (Figure 3). In the saccule and utricle, three different zones are present: the striolar zone, bounded on either side by the peripheral medial and lateral zones (Figures 2B, 2C). Their calcium-binding protein content can identify the various sensory cell types and afferent fibers. Parvalbumin is present in the Type I hair cells situated in the striolar area (Figure 3). Calretinin identifies the nerve calyces that surround

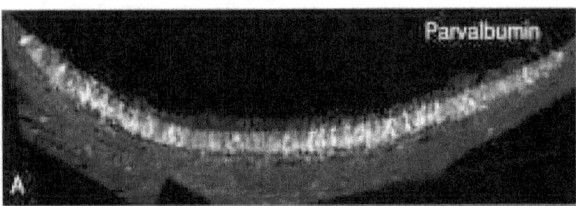

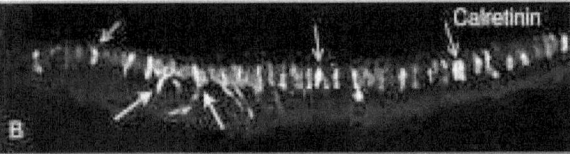

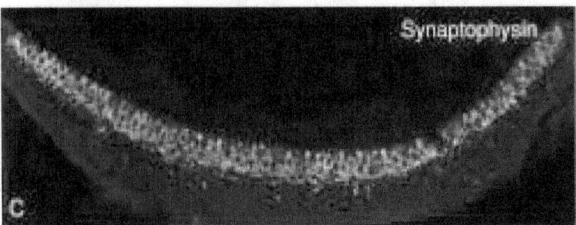

Figure 5 Organization of the utricle on landing, after 16 days of flight. A. Distribution of the sensory cells containing parvalbumin. B. Distribution of calretinin in the sensory cells (thin arrows) and in some afferent calyces (thick arrows). C. Distribution of synaptophysin in synaptic boutons and in afferent calyces. This is a normal distribution and does not differ from the controls.

the Type I hair cells containing parvalbumin, and is also present in some Type II hair cells, situated in peripheral areas (Figure 3). The efferent fibers and endings contain the calcitonin gene-related peptide (Figure 3).

In the flight rats, we observed no anatomical reorganization of the three zones of the utricle in comparison with the controls. In addition, the biochemical markers showed that the distributions of sensory cells containing parvalbumin (Figure 5A) and of the calyx nerve fibers containing calretinin (Figure 5B) were similar in flight and control rats at all stages. Similarly, the distribution of synapses, identified on the basis of their synaptophysin content (Figure 5C), was normal in the flight rats at all stages.

The development and distribution of the efferent fiber network and its synaptic connections, visualized with the calcitonin gene-related peptide, were also similar in microgravity and on the ground, with gradual maturation and the increasing formation of synapses (Figure 6).

Integration and cerebellar control of vestibular information in the vestibular nuclei

In the vestibular nuclei, parts of the developmental sequences for neuron growth (the selective formation of afferent pathways and synaptic maturation) take place in the period during which the rats in this study were exposed to weightlessness (Figure 7). Vestibular neurons and pathways of vestibular and cerebellar origin can be identified both anatomically and by

their specific expression of various calcium-binding proteins. The vestibular nuclei present a subpopulation of neurons that produces calretinin. Detecting parvalbumin identifies the pathways originating from the vestibular organs and detecting calbindin identifies those originating from the cerebellum.

In the rats that developed in space, we observed changes in the neuronal development of the vestibular nuclei. These differences were observed in flight rats after eight days and 15 days of flight and on landing. In flight rats, neuronal cell bodies were markedly smaller and their dendrites displayed less growth and branching (Figure 8B) than those of the controls (Figure 8A). The most striking difference between flight rats (Figure 8D) and controls (Figure 8C) was the lack of development and paucity of cerebellar projections in parts of the vestibular nuclei at various developmental stages in the flight animals.

Effects on synaptic protein distribution in the central nervous system

In mature neurons, synaptic proteins are highly concentrated in axon endings, where they help regulate neurotransmitter release. Their distribution during nervous system development follows a pattern closely linked to axon growth and synaptic maturation. At very early stages, they are distributed diffusely throughout the neuron and all its extensions. As functional synaptic networks mature, synaptic proteins are mostly restricted to the axon endings. We followed the maturation of synapses in the vestibular nuclei. To provide a control, we also

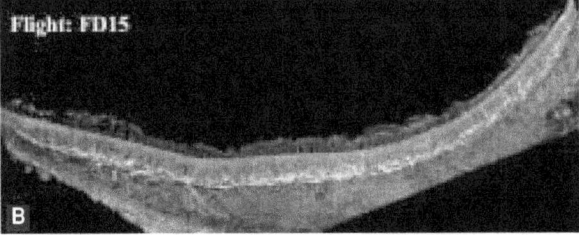

Figure 6 Distribution of the calcitonin-gene related peptide in the sensory epithelium at launch on postnatal day eight (P8) (A) and after 15 days of flight, FD15 (B). A. At the time of launch, the development of the efferent fiber network was incomplete. Note the differences in density of the efferent fibers and synapses in the three zones: lateral (L), striolar (S) and medial (M). B. After 15 days of flight, the density of calcitonin-related peptide has increased in the three zones, in a similar manner to that observed in the controls. (From Demêmes, 2001, with permission; reproduced from *Developmental Brain Research*.)

examined structures not specifically involved in gravity detection—the cochlear nuclei. We used two markers of synapses: synaptophysin and SNAP-25.

During and after flight, the cellular distributions of synaptophysin and SNAP-25 in the vestibular (Figure 9) and cochlear nuclei differed significantly from those in the controls. In controls, they were undetectable in neuron somas and axons, and concentrated in axon endings where they appeared as tiny particles (Figure 9A, 9C). In contrast, in flight rats, synaptic proteins were readily detected throughout most neuron expansions and appeared as dense clusters (Figure 9B, 9D).

DISCUSSION

Development of a gravity sensor, the utricle, in weightlessness

We observed no reorganization of the afferent and efferent connections in the utricle. The afferent system was almost completely mature at the time of launch (Figure 4) and no

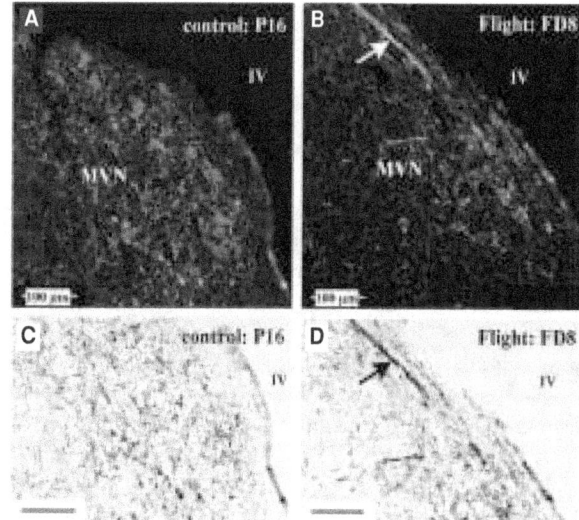

Figure 8 Identification of vestibular neurons (containing calretinin, in red) and cerebellar fibers and endings (containing calbindin, in green) in control (P16, **A**) and in rats after eight days of flight (FD8, **B**) in the medial vestibular nucleus (MVN). **C** and **D** show only the cerebellar fibers and endings (in black), in the same conditions. **A**, **C**—In controls, the distribution of the cerebellar fibers is uniformly dense around the calretinin-containing neurons. **B**, **D**—In contrast, in the flight animals, the density of cerebellar endings is very low. The cerebellar fibers also display an abnormal organization into bundles (arrows). IV: Ventricle IV.

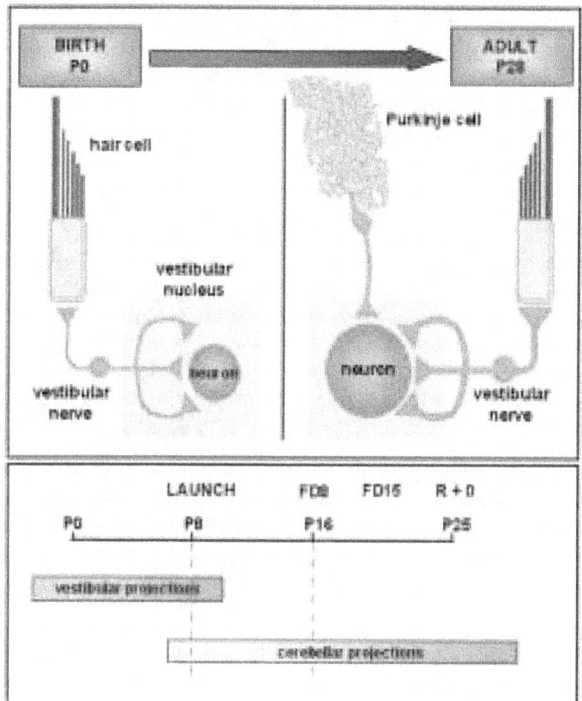

Figure 7 Time sequences for the establishment and postnatal development of vestibular (orange) and cerebellar (green) afferent connections in the vestibular nuclei. The hair cells are in the utricle and the Purkinje cells are in the cerebellum. Both of these structures connect to the vestibular nucleus. Note that at launch, on postnatal day eight, the vestibular nerve connections are almost complete and their endings have an almost mature distribution. In contrast, afferent cerebellar connections (i.e., from the Purkinje cells), accompanying the maturation of vestibular reflexes, occur later during the period of exposure to weightlessness. This suggests that these connections may be more susceptible to weightlessness.

significant changes in organization occurred during spaceflight. We conclude that microgravity did not affect the late development of afferent neurons in the utricle. However, other ultrastructural studies on adult rats after spaceflight have shown an increase in the number of synapses in some hair cells (Ross, 1998). In contrast, the efferent fiber network was still immature at the time of launch (Figure 4). During the period corresponding to the period of flight, the number of efferent endings increases and these endings are relocated to the base of the sensory cells. These modifications occurred similarly in flight and on the ground (Demêmes, 2001).

The lack of effect of microgravity on the organization of the utricle suggests that, for the factors studied, the otolithic receptors and their afferent and efferent networks are insensitive to gravitational changes during the period corresponding to the exposure of these animals to weightlessness. Thus, for these organs, the critical periods of development probably occur before eight days of age (P8) (Figure 4). However, this hypothesis cannot account for the lack of difference in development of the efferent system since much of this system maturation took place mainly during spaceflight. The efferent system integrates information from many different sources not directly linked to the detection and transmission of gravitational signals. Modifying only the messages associated with gravity may not have a drastic effect on the organization of this system.

Development of connections to the vestibular nuclei and cerebellum

Our observations indicate that spaceflight profoundly impaired the development of vestibular neurons and cerebellar branching in the vestibular nuclei. These findings are interesting since, following spaceflight, behavioral changes related to the major central vestibular pathways have been reported (changes in postural balance and vestibulo-ocular reflexes) and provide evidence for sensory-motor rearrangements during exposure to weightlessness in adult animals. Our findings suggest that the development of this system may be affected as well. Other developmental studies carried out during the Neurolab mission examined abnormal neural development in other areas. In the neuromuscular system, spaceflight reduces the complexity of motor nerve terminals (see science report by Riley et al., in this publication), indicating a delay in motor neuron maturation. Also, rats that developed in space displayed behavioral defects. On landing, they had not developed normal mature righting behavior, and they had not developed such behavior even 110 days after landing, demonstrating a permanent effect on righting strategy (see science report by Walton et al., in this publication).

All these findings suggest a critical period exists for the development of normal spinal and vestibular structure and function. Overall, our data suggest that the maturation of cerebellar inputs is critical for mediating effects involved in integrating movement and sensation in the vestibular nuclei. However, it is unclear whether this major modification is simply a delay in development, or a definitive failure to establish the correct connections. In this Neurolab experiment, the number of rats in our investigations was too small for testing a set of rats after a period of recovery on Earth. This would have revealed if the changes could have been modified. Also, factors other than weightlessness during a critical period of development, such as stress or fluid shifts in the Space Shuttle environment, may be involved in these changes. For example, we observed that most of the flight rats were underdeveloped on landing, with weight gains 40% lower than controls.

Further experiments are required to determine whether terrestrial re-adaptation might reverse the structural changes observed in the maturation of the vestibular neurons and their cerebellar connections.

Changes in synaptic proteins

During and after flight, the distributions of synaptic proteins in the vestibular and cochlear nuclei differed significantly from those in the controls. Various studies in adult animals have shown an increase in synaptic proteins after mechanically or chemically induced damage to the central nervous system and in various pathological conditions (anoxia, poisoning). Few data are available concerning the effects of spaceflight on the central nervous system. Following spaceflight, synaptic changes have been reported in the cerebellum of adult rats. These show immature synapses similar to those present during early development and those observed in this study. In contrast, hypergravity environment studies have reported changes in synaptic morphology, with a higher packing density of synaptic vesicles. The transport

Synaptophysin   SNAP-25

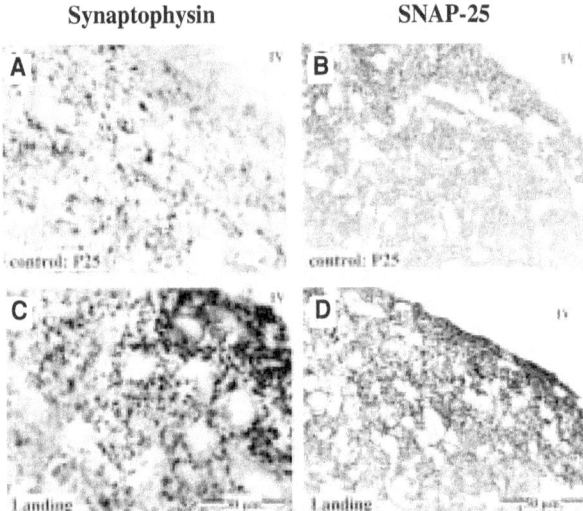

Figure 9 Synaptophysin (**A**, **B**) and SNAP-25 (**C**, **D**) detection in the medial vestibular nucleus of controls (P25) and rats after landing. The presence of synaptophysin and SNAP-25 is characterized by particles (in black). Synaptophysin particles are more densely packed and more intensely stained (**B**) in the vestibular nucleus of flight rats than in control rats (**A**). For SNAP-25, a very striking difference was observed between controls (**C**) with sparse particles and flight rats (**D**) with a very dense distribution of particles. IV: ventricle IV

of synaptic vesicles to their final location at neuron synapses depends on microtubules. Microtubule organization is gravity-dependent: it does not occur or is greatly impaired under low-gravity conditions. The abnormal patterns that we observed may therefore be linked to modifications in functional microtubule organization in microgravity, resulting in the up-regulation and relocation of synaptic proteins. In young rats, these modifications may have been particularly drastic because they occurred during a period of axonal growth.

The observed reorganization of synaptic protein distribution after exposure to weightlessness raises questions about the role of gravity in nerve terminal modeling during development and in the maintenance of mature synaptic contacts. However, factors associated with spaceflight other than microgravity may also have affected synaptic development in the central nervous system. Neuron developmental processes are known to be affected by multiple indirect effects. The influences of these factors should be tested aboard the Space Station by similar analyses of equivalent tissues from rats or mice. Complementary studies should include other aspects of transport and storage of the molecular components of the active synaptic zone to identify the biochemical pathways involved in these modifications.

CONCLUSION

We observed two main effects on the rats subjected to weightlessness. Vestibular neuron growth was decreased and cerebellar endings in the vestibular nuclei were underdeveloped. Such effects may have occurred because these connections developed during the time period while the rats where in weightlessness. The second type of effect was nonspecific and concerned the abnormal distribution of synaptic proteins in the central nervous system of the flight rats. Further studies are required to confirm these observations.

The findings suggest that gravity may be important for the normal development of brain pathways related to balance and posture.

Acknowledgements

Supported by CNES grants 98-793 and 99-793.

REFERENCES

RECENT ASPECTS OF DEVELOPMENT OF THE VESTIBULAR SENSE ORGANS AND THEIR INNERVATION. G. Desmadryl, C.J. Dechesne and J. Raymond. In: R. Romand (Ed.) *Development of Auditory and Vestibular Systems 2*. Amsterdam: Elsevier, pages 461–487; 1992.

THE DEVELOPMENT OF VESTIBULAR SENSORY ORGANS IN HUMAN. C. J. Dechesne. In R. Romand (Ed.) *Development of Auditory and Vestibular Systems 2*, Amsterdam: Elsevier, pages 419–447; 1992.

CELLULAR DISTRIBUTION OF PARVALBUMIN IMMUNOREACTIVITY IN THE PERIPHERAL VESTIBULAR SYSTEM OF THREE RODENTS. D. Demêmes, M. Eybalin and N. Renard. *Cell and Tissue Research*, Vol. 274, pages 487–492; 1993.

POSTSYNAPTIC TARGETS OF PURKINJE CELL TERMINALS IN THE CEREBELLAR AND VESTIBULAR NUCLEI OF THE RAT. C.I. De Zeeuw and A. S. Berrebi. *Eur. J. Neurosci.*, Vol. 7, pages 2322–2333; 1995.

DEVELOPMENT OF CALRETININ IMMUNOREACTIVITY IN THE MOUSE INNER EAR. C.J. Dechesne, D. Rabejac and G. Desmadryl. *J. Com. Neurol.*, Vol. 346, pages 517–529; 1996.

PATTERN OF SELECTED CALCIUM-BINDING PROTEINS IN THE VESTIBULAR NUCLEAR COMPLEX OF TWO RODENT SPECIES. G.A. Kevetter. *J. Com. Neurol.*, Vol. 365, pages 573–584, 1996.

EFFECTS OF GRAVITY ON VESTIBULAR NEURAL DEVELOPMENT. M.D. Ross and D.L. Tomko. *Brain Res. Rev.* Vol. 28, pages 44–51; 1998.

DEVELOPMENT OF THE EFFERENT VESTIBULAR SYSTEM IN GROUND AND IN MICROGRAVITY DURING A 17 DAYS SPACEFLIGHT. D. Demêmes, C.J. Dechesne, S. Ventéo, F. Gaven and J. Raymond. *Dev. Brain Res.*, Vol. 128, pages 35–44; 2001.

THE MAMMALIAN OTOLITHIC RECEPTORS: A COMPLEX MORPHOLOGICAL AND BIOCHEMICAL ORGANIZATION. A. Sans, C.J. Dechesne and D. Demêmes. In: P. Tran Ban Huy and M. Toupet (Ed.) *Advances in Oto-Rhino-Laryngology,* Karger, Basel, Vol. 58, pages 1–14, 2001.

Development of the Aortic Baroreflex in Microgravity

Authors

Tsuyoshi Shimizu, Masao Yamasaki, Hidefumi Waki,
Shin-ichiro Katsuda, Hirotaka Oishi, Kiyoaki Katahira,
Tadanori Nagayama, Masao Miyake, Yukako Miyamoto

Experiment Team

Principal Investigator: **Tsuyoshi Shimizu[1]**

Co-Investigators: **Masao Yamasaki[1], Hidefumi Waki[1], Shin-ichiro Katsuda[1], Hirotaka Oishi[1], Kiyoaki Katahira[1], Tadanori Nagayama[1], Masao Miyake[2], Yukako Miyamoto[1], Michiyo Kaneko[3], Shigeji Matsumoto[4], Syunji Nagaoko[5], Chiaki Mukai[6]**

Technical Assistants: **Haruyuki Wago[1], Toshiyasu Okouchi[1]**

[1]Fukushima Medical University School of Medicine, Fukushima, Japan
[2]The University of Tokyo, Tokyo, Japan
[3]Tokyo Kaseigakuin Junior College, Tokyo, Japan
[4]Nippon Dental University, Tokyo, Japan
[5]Fujita Health University, Aichi, Japan
[6]National Space Development Agency of Japan, Ibaraki, Japan

ABSTRACT

Baroreceptors sense pressure in blood vessels and send this information to the brain. The primary baroreceptors are located in the main blood vessel leaving the heart (the aorta) and in the arteries in the neck (the carotid arteries). The brain uses information from the baroreceptors to determine whether blood pressure should be raised or lowered. These reflex responses are called baroreflexes.

Changing position within a gravity field (i.e., moving from lying to sitting or standing) powerfully stimulates the baroreflexes. In weightlessness, the amount of stimuli that the baroreflexes receive is dramatically reduced. If this reduction occurs when the pathways that control the baroreflexes are being formed, it is possible that either the structure or function of the baroreceptors may be permanently changed.

To study the effect of microgravity on structural and functional development of the aortic baroreflex system, we studied young rats (eight days old at launch) that flew on the Space Shuttle *Columbia* for 16 days. Six rats were studied on landing day; another six were studied after re-adapting to Earth's gravity for 30 days. On both landing day and 30 days after landing, we tested the sensitivity of the rats' baroreflex response. While the rats were anaesthetized, we recorded their arterial pressure, heart rate, and aortic nerve activity. After the tissues were preserved with perfusion fixation, we also examined the baroreflex structures.

On landing day, we found that, compared to the controls, the flight rats had:

- fewer unmyelinated nerve fibers in their aortic nerves
- lower baroreflex sensitivity
- significantly lower contraction ability and wall tension of the aorta
- a reduced number of smooth muscle cells in the aorta.

In the 30-day recovery group, the sensitivity of the baroreflex showed no difference between the flight rats and the control groups, although the unmyelinated fibers of the aortic nerve remained reduced in the flight rats.

The results show that spaceflight does affect the development of the aortic baroreflex. The sensitivity of the reflex may be suppressed; however, the function of the blood pressure control system can re-adapt to Earth's gravity if the rats return before maturation. The structural differences in the input pathway of the reflex (i.e., the reduction in nerve fibers) may remain permanently.

INTRODUCTION

Gravity profoundly affects blood circulation. Without the blood pressure control system, our blood pressure would increase when we were lying down and decrease whenever we stood up. Fortunately, our pressure control system allows us to adjust rapidly to changes in position.

In weightlessness, the cardiovascular system does not have to adjust to changes in body position. If animals were to grow up in space and not be exposed to these stimuli, it is not known how their cardiovascular systems would develop. Animals that develop in microgravity would not experience the usual changes that gravity produces in the cardiovascular system, and this stimulation may be important for normal development. In our experiment, we studied the effect of microgravity on the development of an important part of the blood pressure control system (Gootman, 2000)—the aortic baroreflex.

The aortic baroreflex is a negative feedback mechanism that maintains blood pressure at a constant level by a reflex (Figure 1). In general, a reflex arc is composed of: (a) sensors, (b) afferent nerves that carry the sensory information to a control center, (c) a control center, (d) efferent nerves that carry the appropriate responses back to the tissues, and (e) effectors.

In the aortic baroreflex, the sensors are free endings of the aortic nerve distributed in the outer and middle layers of the wall of the aortic arch and the right subclavian artery.

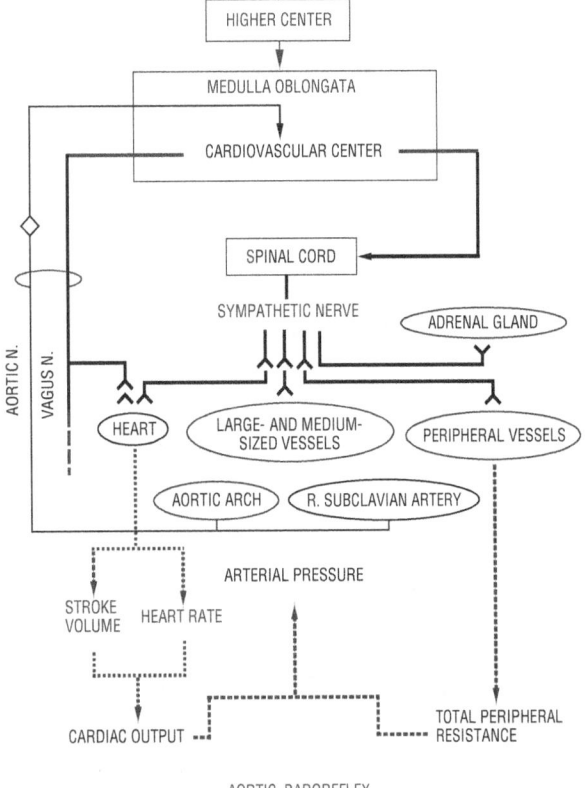

Figure 1. Schematic represention of the aortic baroreflex system (see Text). N: nerve

The afferent nerves (carry signals to the brain) are the left and right aortic nerves. The control center is located in the medulla oblongata of the brain. The efferent nerves (carry signals from the brain) are both the sympathetic and parasympathetic nerves of the autonomic nervous system that distribute to the effectors—the heart and blood vessels. The sensory receptors and afferent fibers are stimulated by stretching the vessel's wall when blood flow increases and blood pressure is elevated. A decrease in blood pressure removes this stimulation. The control center in the brain stem is stimulated or inhibited by receiving this sensory information. The output from the center controls the output of the cardiac parasympathetic nerve (the vagus nerve) and of the sympathetic nerves that originate in the spinal cord and innervate both the blood vessels and the heart. These impulses return blood pressure to normal by affecting the heart's contractile force, the number of heartbeats per minute, and the diameter of blood vessels.

In rats, the baroreflex system grows and develops to maturity during the first eight to 12 weeks after birth. We have previously found the following facts about the development of the aortic baroreflex in the rabbit (Shimizu, 1990) and rat (Yamasaki, 1996):

1. The aortic nerve is composed of myelinated nerve fibers (myelin is a protective sheath on the nerve), which have a low threshold for excitation, and unmyelinated nerve fibers, which have a high threshold and irregular discharges.

2. After birth, the myelinated fibers continue to increase in number, myelin thickness, and axon diameter.

3. In the rabbit, the increase in the number of myelinated fibers correlates lineally with age-related increases in mean arterial blood pressure.

4. The response of the aortic nerve to increases in arterial pressure is smaller in the young rat and rabbit than in the mature rat and rabbit.

5. The bradycardiac response of the aortic baroreflex (i.e., decrease in the heart rate in response to an increase of blood pressure) is smaller in the young rabbit than in the mature rabbit.

6. At around eight to 12 weeks of age, the composition of the aortic nerve in both rat and rabbit and the bradycardiac response in the rabbit become the adult type.

Ground-based simulations of weightlessness have shown changes in the aortic baroreflex. To simulate the headward shift of body fluid that occurs when a human enters space (Charles, 1994), we used head-down tilt (HDT) studies. With HDT, blood flow and arterial pressure increase in the ascending aorta. When rabbits aged three to four weeks were raised in a special cage where the animals experienced HDT for 34–36 days, the number of unmyelinated fibers of the aortic nerve was significantly reduced compared with that of the control groups. The proportion of unmyelinated fibers to all nerve fibers in the aortic nerve was smaller in the HDT group than in the control groups, although there was no difference in the number of myelinated fibers among three groups. Aortic nerve activity and the bradycardiac response of the aortic baroreflex were less in the HDT group than in the control groups (Yamasaki, 2002).

We proposed the following hypotheses based on these ground-based experiments. In the developing rat exposed to microgravity in space:

1. the mechanical behavior of the baroreceptor region of the aortic arch and its branches would be modified due to the redistribution of blood flow that occurs under conditions of microgravity,

2. the composition of the aortic nerve would be changed,

3. there would be a modification of the baroreceptor reflex function that would affect the nervous control of blood circulation.

The purpose of the present space experiment was to substantiate our hypotheses by observing and analyzing the effect of microgravity on baroreflex responsiveness, the quantity of aortic nerve fibers, and the structure and function of the various tissues related to the baroreflex.

METHODS

General Procedure

Sprague-Dawley rats, aged eight days, were flown on the Space Shuttle *Columbia* for 16 days. Two types of control groups were prepared on the ground. One was the asynchronous ground control (AGC) group, where the rats were raised in flight-like cages, and the other was the vivarium (VIV) control group, where the rats were raised in commercial cages.

Animal Care and Welfare

The experiments were performed according to the guidelines of the NASA-Ames Research Center (ARC) *Animal Care and Use Handbook* and were reviewed by the Institutional Animal Care and Use Committee (IACUC) at NASA-ARC and the NASA-Kennedy Space Center (KSC). We also followed the Guidelines for Animal Experiments at Fukushima Medical University, and the Guiding Principles for the Care and Use of Animals in the Field of Physiological Sciences declared by the Physiological Society of Japan.

Assignment and Anesthesia of the Animals

On landing day (recovery day zero or R+0), six flight rats aged 24 days were studied and six more rats were reserved for 30 more days (recovery day 30 or R+30) on the ground. Anesthesia was provided using a 25% urethane solution (1.2 to 1.5g/kg in female rats, 1.5 to 2.0g/kg in male rats) injected into the intraperitoneal cavity. The dose of urethane was determined according to the body weight and sex of each rat on the basis of our previous examinations (unpublished data). The control groups were treated using the same protocol as the flight groups.

Functional Examinations

Tests for baroreflex responsiveness and afferent sensitivity – Arterial pressure, heart rate, aortic nerve activity, and respiratory movements were measured under anesthesia. These parameters were recorded simultaneously on magnetic tape, computer disk, and penrecorder chart; and changes in these parameters accompanying the changes in blood pressure were observed. Arterial pressure was increased or decreased by the venous injection of phenylephrine or sodium nitroprusside, respectively. Arterial baroreflex responsiveness was expressed by dividing the percentage change in heart rate by the percentage change in mean blood pressure (percent change of heart rate (ΔHR%)/percent change of mean blood pressure (ΔMBP%)). Sensitivity of the aortic nerve was determined by calculating the percentage change in aortic nerve activity for a change in mean blood pressure (percent change of the aortic nerve activity (ΔANA%)/percent change of mean blood pressure (ΔMBP%)).

Tests for mechanical properties of the aortic wall – The mechanical properties of the thoracic aorta wall were examined in six rats prior to fixation. The aorta was divided into two portions. One portion was gradually frozen to $-70°C$ and transported to Japan to test its mechanical properties (elasticity, extensibility, etc.) (Patel, 1972). The other portion was used for contraction-relaxation tests at the landing site.

Tensile tests – To examine the mechanical properties of the aortic wall, we did a specific tensile test with a tensile testing instrument (TDM-30J, Miebea, Inc., Japan). This examination quantifies the elasticity, plasticity, and extensibility of the vessel wall.

Contraction-relaxation test – The lower half of the thoracic aorta was divided into three to four ring-shaped samples (approximately 2.0–3.0 mm in width and 0.5–1.0 mm in diameter). The contraction and relaxation were elicited by application of drugs that change excitation of the smooth muscle in the vessel wall; i.e., phenylephrine for producing contraction and acetylcholine for producing relaxation. We also studied the role of the cells lining the inside of the vessel by applying L-NAME to inhibit production of nitric oxide in the endothelium cell or by removing the endothelium mechanically. These experiments were performed under indomethacin application to inhibit production of vasoactive prostanoids, which are also produced by endothelium cells.

Histological Examinations

Fixation and dissection of tissues – After the baroreflex tests, tissues were preserved with chemical fixatives. A modified Kalnowski solution containing 1% paraformaldehyde and 1% glutaraldehyde was used.

Electron microscopic examination of the aortic nerve – Electron microscopic montages of transverse sections of the aortic nerve trunks were printed out at 13,400 times magnification. The unmyelinated and myelinated nerve fibers were counted on the montages. The long and short axes (a and b) of the nerve fibers and the myelin thickness (T) were measured with calipers.

Examination of gene expression in various tissues – Different genes in the tissues can be turned on or expressed in response to various stimuli, and this can produce important growth factors and proteins. In our study, we analyzed the expression of epidermal growth factor (EGF), proliferating cell nuclear antigen (PCNA), lysozyme, and vitamin D_3 receptor (VDR). Total RNA was isolated from the liver, spleen, intestine, and thymus, and RNA concentration was quantified by spectrophotometry.

Statistical analysis – The data were statistically analyzed with one-way analysis of variance (ANOVA) or non-paired student's t-test and followed by Scheff's F test or Bonferroni/ Dunn test. The values were expressed as mean±SD, and statistical significance was set at $p < 0.05$.

RESULTS

General Observations

The flight rats walked with difficulty after landing. Approximately four hours after landing at NASA-KSC, they moved slowly and squatted frequently. By about six hours after landing, they were able to move quickly and smoothly. Initially, the rats' fur was slightly brown and wet. In most flight rats, the tips of the tails were slightly necrotic. The body weight of the flight rats on R+0 was reduced compared to the ground control groups. An average value of body weight (n=6) was 53.4±4.8g inflight, 78.9± 7.1g in AGC, and 84.3±4.2g in VIV. The animals in the recovery group (R+30) inflight grew well and did not differ in weight from the AGC and VIV groups (average body weight (n=6) was 214.2±46.0g inflight, 203.0±13.4g in AGC, and 221.1±9.3g in VIV).

Functional Observations

Basal blood pressure, heart rate, and respiratory movements – Before the baroreflex tests, we measured blood pressure, heart rate, and respiratory movements under anesthesia (Figure 2). Among the six rats in the flight group used on R+0, two animals showed very low blood pressure (less than 40 mmHg of mean blood pressure (MBP)) and had no response to the drugs. The body weight of these rats was 47.4g and 47.7g, which was markedly lighter than the body weight of the other four rats (53.1±61.2g). Their heart rates were 378 beats per minute (bpm) and 361 bpm, which was less than the heart rate in the other four rats (462–520 bpm). Consequently, we did not record aortic nerve activity on these two rats, and we excluded these two rats from the analysis of the cardiovascular variables.

The baseline values for heart rate, blood pressure, and respiratory rate are shown in Table 1. On landing day, blood pressure in the flight rats was reduced compared to both in AGC and VIV, although this was statistically significant only between the flight rats and AGC. On R+30 there were no significant differences between the flight rats and the control groups. Blood pressure in the flight rats on R+30 was significantly higher than that measured on R+0.

On landing day, heart rates in the flight rats were higher than in each control group, although this was statistically significant only between the flight rats and VIV controls. On R+30, all groups showed lower heart rates than those on R+0, but there was no significant difference between the groups on R+30.

Respiratory rate decreased between landing day and R+30, but there was no difference among three groups at R+30.

Table 1. Baseline values for heart rate, blood pressure, and respiratory rate on landing day and 30 days after landing.

	Landing Day	30 days postlanding
HR Flight	492±24	343±35
HR AGC	452±33	386±32
HR VIV	428±41	415±36
RR Flight	122±16	104±32
RR AGC	133±7	105±17
RR VIV	119±8	106±13
MBP Flight	80.1±3.6	102.6±17.9
MBP AGC	96.9±5.7	98±15.2
MBP VIV	89.5±6.8	107.6±10.6
SBP Flight	90.4±10.3	120.8±14.5
SBP AGC	106.1±4.9	114.8±10.5
SBP VIV	99.6±9.9	121.1±9.3
DBP Flight	67.5±4.0	83.5±18.3
DBP AGC	84.5±6.7	98.0±15.2
DBP VIV	79.7±7.1	91.0±12.0

HR = heart rate (bpm), RR = respiratory rate (rpm), MBP = mean blood pressure (mmHg), SBP = systolic blood pressure (mmHg), DBP = diastolic blood pressure (mmHg). Flight = flight rats, AGC = asynchronous ground control group, VIV = vivarium control group.

Responses of arterial baroreflex – On R+0, the maximum decrease in heart rate produced by phenylephrine was less in the flight animals than in either control group. Also, the increase in blood pressure for a given dose of phenylephrine was significantly lower for flight rats. These differences were gone by R+30.

The index of baroreceptor reflex sensitivity (ΔHR%/ ΔMBP%) was calculated from the percentage changes from baseline values in mean blood pressure and heart rate. On R+0, the flight rats showed the lowest value (−0.19±0.08). There was a significant difference between flight and AGC groups (−0.47±0.14). When barorereflex sensitivity was calculated using absolute values rather than percentage change, no significant differences were found among all groups. The flight rats, however, showed the lowest values. The baroreflex sensitivity index was widely dispersed on R+30, and no significant differences were observed among the three groups (Figure 3A).

Sensitivity of the aortic nerve as the input pathway in the aortic baroreflex – Figure 2 shows typical responses of aortic nerve activity accompanying blood pressure changes due to phenylephrine injection. These data were obtained at R+0 in flight rats.

In all the flight rats, the integrated aortic nerve activity (IANA) showed qualitatively similar responses to those

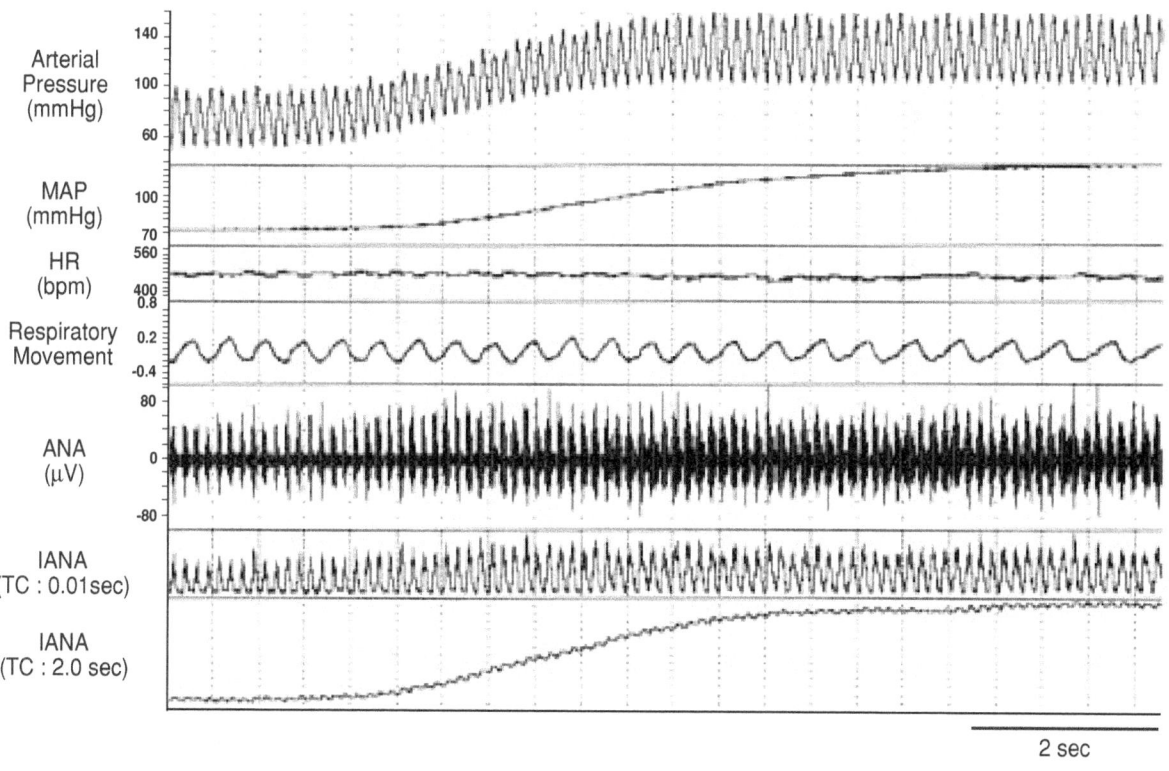

Arterial Pressure (mmHg)

MAP (mmHg)

HR (bpm)

Respiratory Movement

ANA (µV)

IANA (TC : 0.01sec)

IANA (TC : 2.0 sec)

2 sec

FLT#3, Male, 61.2g

Figure 2. Responses of physiological parameters to the injection of phenylephrine recorded in a flight rat on landing day. MAP: mean aortic pressure (mmHg) HR: heart rate (bpm) ANA: aortic nerve activity (µV) IANA: integrated aortic nerve activity (µV) TC: time constant

observed in the ground controls; i.e., when blood pressure increased, IANA also increased, and when mean aortic pressure (MAP) decreased, IANA also decreased. However, the sigmoid curves of the MAP-IANA relationship in the flight rats shifted to the left compared to those of the VIV and AGC groups.

The afferent sensitivity of the aortic baroreceptor reflex in response to an increase of blood pressure on R+0 was lower in flight rats than in the controls. Two flight rats showed the lowest sensitivity (1.76 and 1.54) among all the rats tested (range 1.90~4.40) (Figure 3B), although the differences between each group were not statistically different. The afferent sensitivity on R+30 in the flight rats was significantly higher than that on R+0. No significant differences in the afferent sensitivity between the groups on R+30 were observed.

Functional properties of the thoracic aorta

Mechanical characteristics – The tension produced by strain in the flight rats was significantly smaller than that in either the AGC or VIV groups (Figure 4A). The contour of the stress-strain curve derived from the corresponding tension-strain curve was almost similar among the three groups. The elastic moduli and the relaxation strength showed no significant difference among the three groups. Plastic deformation of the strip of the aortic wall (0.1±0.01mm : mean±SE) was observed in all samples in the flight group five minutes after the stress-strain test. It was not observed at all in the control groups.

Vasocontraction and vasorelaxation – Phenylephrine caused dose-dependent contraction in the rings of the aorta. The sensitivity of the aortic wall with endothelium to phenylephrine (10^{-9}M to 3×10^{-6}M), as a percent of the maximal response, averaged 6.4 for the three flight rats tested. This was smaller than the sensitivity in AGC (7.3±0.1) and in VIV (7.1±0.1), although we could not evaluate this statistically because some aortic rings in the flight group did not respond to the drugs. The maximum contraction in response to phenylephrine in the AGC and VIV groups was stronger than that for the flight group. Even in the presence of L-NAME, these curves showed similar relationships to the results of the above test (Figure 4B).

Acetylcholine (10^{-3}M to 3×10^{-6}M) caused dose-dependent relaxation in the aorta rings with endothelium, and the relaxation showed no difference among the flight, AGC, and VIV groups. The relaxation was almost abolished in both situations of application of L-NAME (3×10^{-4}M) to the ring with endothelium and the ring without endothelium.

Histological Observations

Fine structure of the aortic nerve – The aortic nerve samples available for electron microscopic analysis were extremely limited because of structural and technical problems (Figures 5A and 6). Careful examination of high-magnification montages of the transverse section in the left aortic nerves in five rats in each

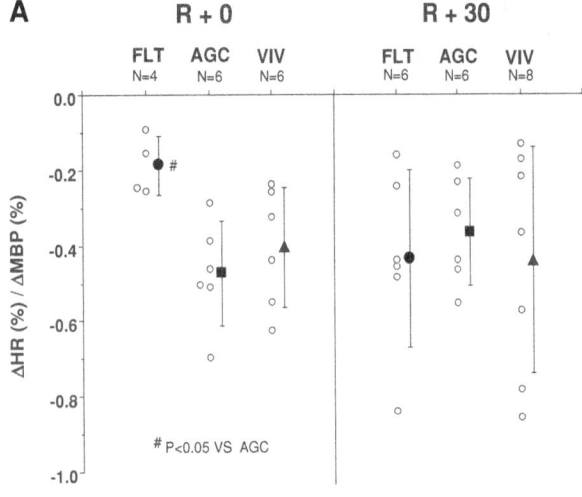

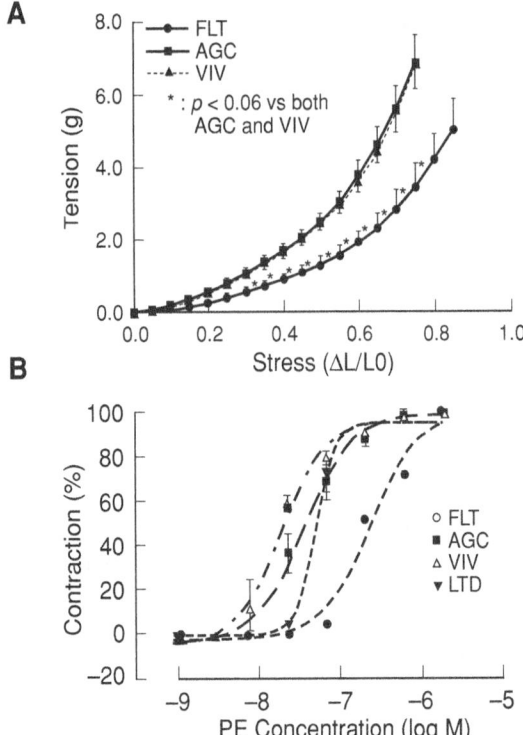

Figure 4. Tension production against stress and contraction elicited by phenylephrine in the wall of the thoracic aorta. FLT: flight group, AGC: asynchronous control group, VIV: vivarium control group, LTD: feeding-limited nursing group

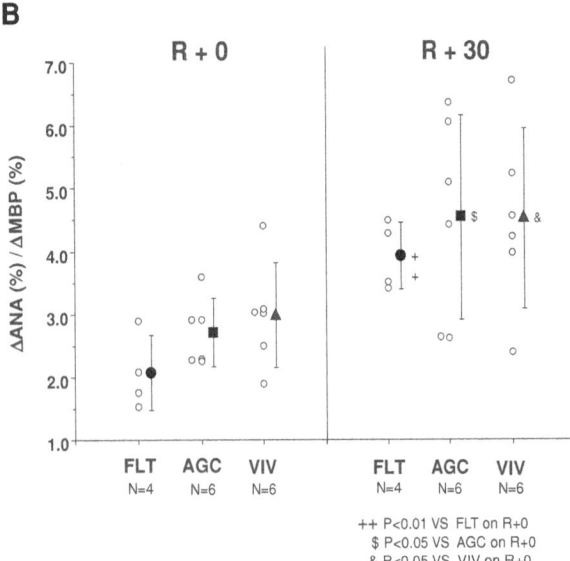

Figure 3. Sensitivity of the baroreflex (A) and the aortic nerve. (B) ΔMBP%, ΔHR% and ΔANA%: maximum percent changes of mean blood pressure, heart rate, and integrated aortic nerve activity calculated with a time constant of 2.0 seconds in three experimental groups on the landing day (R+0) and on the 30th day of the recovery (R+30).

group at R+0 revealed that the number of unmyelinated fibers and the ratio of unmyelinated fibers to all fibers in aortic nerve fascicle were significantly less in the flight rats (139±37, 70.0±3.3%) than those in either AGC (207±36, 77.6±4.8%) or VIV (283±121, 75.2±9.9%) rats. There were no significant differences in the number of myelinated fibers among the three groups. The axon diameter (1.44±0.18 μm inflight, 1.3±0.24 μm in AGC, and 1.43±0.19 μm in VIV) and thickness (0.27±0.03 μm in light, 0.26±0.04 μm in AGC, and 0.26±0.02 μm in VIV) of myelin in each aortic nerve showed no significant difference among the three groups.

On R+30 groups, the number of unmyelinated fibers was 130±71 in flight (n=4), 290±127 in AGC (n=5), and 255±135 in VIV (n=3); and the ratio of unmyelinated fibers to all fibers was 64.7±7.7%, 77.3±6.1%, and 72.3±10.0%, respectively. Both the number of aortic unmyelinated fibers and the ratio of myelinated to unmyelinated to all fibers remained significantly reduced in the flight rats at R+30.

Light microscopic observation of the aortic wall and the intestinal epithelium – The thickness of the aortic wall in the flight rats was about 70% of that in the AGC and VIV rats (Figure 5B). The amount of smooth muscle cells was significantly less in the flight group. There were no significant differences among the three groups in amount of elastin and collagen fibers. The length of the crypt and Paneth cells in the small intestinal epithelium, which are important structures for digestion and absorption of nutrients, was not different among three groups.

Relations of functional and structural parameters to the body weight – The body weight of the flight rats four hours after landing was about 65% that of the AGC and VIV groups. To determine whether some of the changes we noted could be due to differences in body weight, we examined the correlation between the parameters we measured and the body weight in each group. There were no correlations between the parameters examined and the body weight, except for a relationship between mean blood pressure and body weight in the AGC group of R+30 recovery rats.

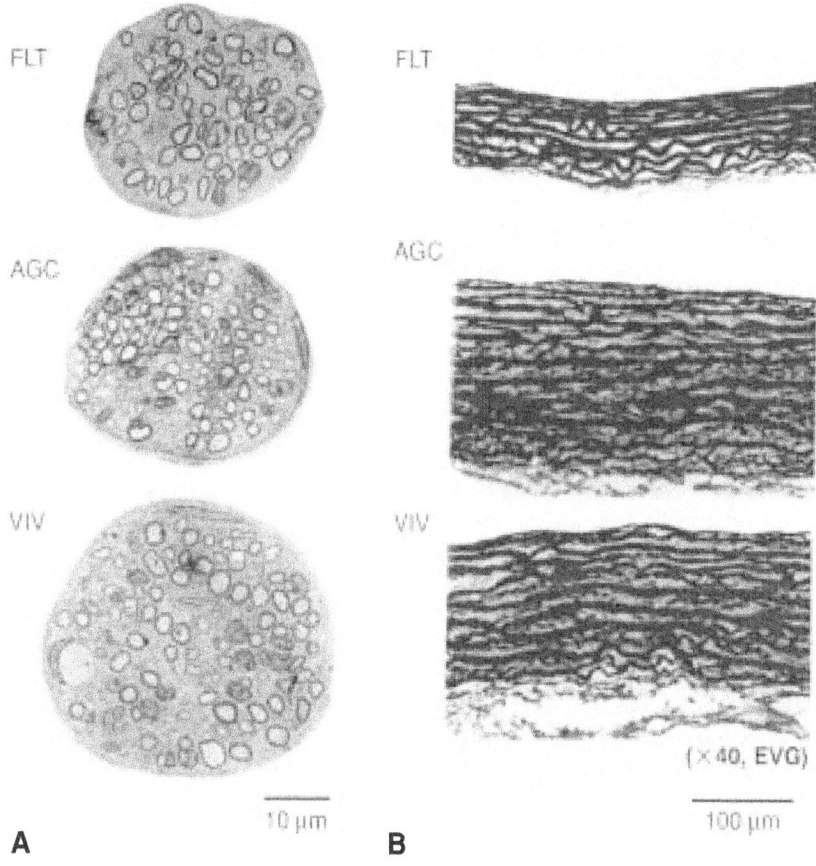

Figure 5. Electron microscopic photos of cross-sections of the aortic nerve (A) and light microscopic photos of the circumferential sections of the thoracic aorta (B).

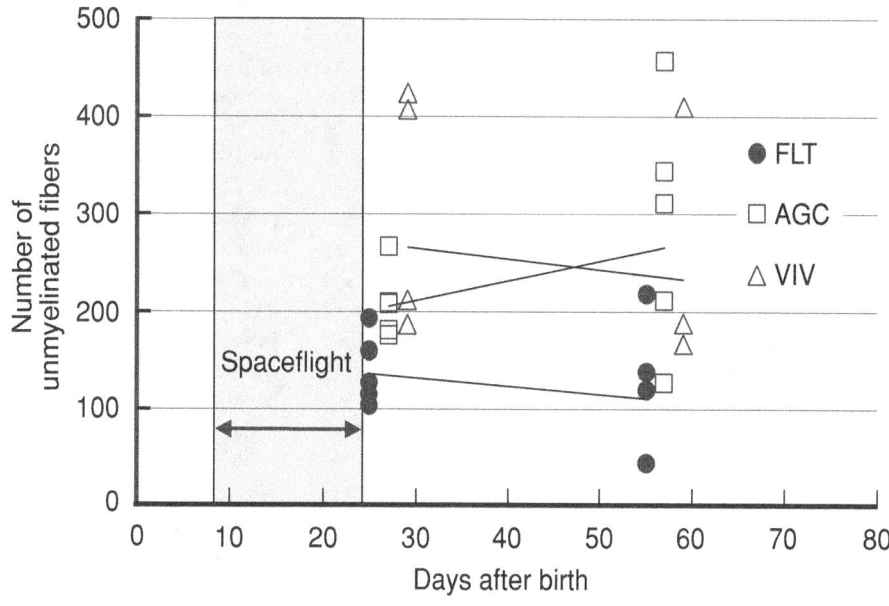

Figure 6. Numbers of unmyelinated fibers of the left aortic nerve counted on landing day (R+0, age 24 days) and the 30th day of recovery (R+30, age 54 days). There were five in each group on R+0, and four in the flight group, five in the AGC group, and three in the VIV group on R+30.

Gene expression of various organs – Expression of VDR in the small intestine did not differ between the flight and ground control rats. Expression of platelet-derived growth factor (PDGF-b), which promotes the proliferation of vascular smooth muscle cells, and nerve growth factor (NGF 1B), which affects the growth and maintenance of nerve cells, was very low in the spleen of the flight rats.

DISCUSSION

The results of our space experiments clearly demonstrated the following changes in the flight rats compared to ground controls:

- In the aortic nerve, the number of unmyelinated fibers (the high-threshold and irregularly discharging nerve fibers) was reduced, although the number and myelin thickness of myelinated fibers (the low-threshold and phasic firing nerve fibers) showed no difference from the controls.

- The sensitivity of both the reflex and the afferent nerve in the aortic baroreflex was reduced.

- The aortic wall also showed significant changes in structure and function. The wall thickness was thin with fewer smooth muscle cells present.

- The tension development against a strain and contraction force produced by a constricting drug (phenylephrine) was weak.

These findings support our hypotheses that microgravity affects the development of the aortic baroreflex. This provides a new understanding of the effect of gravity on nervous system development.

The experiment did have some limitations. The body weights of the flight rats were low compared to the body weights of ground control rats. This raises the question of whether the low body weight was caused by malnutrition (e.g., difficulties of nursing, malabsorption of nutrients, or changes in feeding behavior of the mothers), which influenced the results. The structural and functional differences observed in the flight rats might be due to malnutrition rather than weightlessness. There was, however, no correlation between the cardiovascular parameters we measured and the body weight at the same age in all experimental groups. Histological study and gene expression examinations of the small intestinal epithelium also showed that the epithelial structure of the small intestine developed normally for the absorption of nutrients (including calcium). The weight of the small intestinal tract and liver in the flight rats was not different from that in the controls. Furthermore, the myelination of the aortic nerve was quite normal. While it is reported that myelination is inhibited in malnutrition (Krigman, 1976), in a group of rats that was raised by feeding-limited nursing and that had a low body weight comparable to the 24-day-old rats in the flight group at the same age, the contraction properties of the aorta produced by application of phenylephrine showed no difference from the control rats. Summarizing this evidence, the changes found in the flight rats are most likely not a consequence of malnutrition but rather are due to the space environment.

The study also provided important information on the re-adaptation back to one-G. The functional examination in the flight rats on R+0 started about six hours after landing and lasted for almost 12 hours. When the flight rats were inspected approximately three hours after landing, their behavior was similar to that observed under hypergravity conditions produced during parabolic flight. The rats gradually were able to move actively, and their walking behavior re-adapted to the ground within three to six hours after landing. In our experiments, the attenuation of the aortic baroreflex was still observed in rats that were tested on R+0 more than 10 hours after landing. From these observations, it appears that the regulation of the cardiovascular system seems not to re-adapt rapidly to the ground, and this could be due to structural changes that develop in space. Therefore, the time consumed before we performed the functional tests in this series of experiments on R+0 did not seriously affect the results obtained.

Based on these data, we propose the following mechanism for the changes observed in the flight rats. Following launch, body fluid redistributes and blood vessels located above the heart should receive more blood as compared to one-G. This increases the amount of blood the heart pumps, which increases blood flow in the ascending aorta, aortic arch, and arteries that perfuse the head. This elevates blood pressure in these vessels and expands these vessels' walls at the same time. The stretch of the aortic, subclavian, and carotid arteries excites the baroreceptors and elicits the baroreflex. As a result of the baroreflex, elevation of blood pressure is suppressed by relaxation of peripheral vessels and a slowing of the cardiac rhythm, thereby decreasing overall blood flow. These incidents have been partially demonstrated by us and others using HDT and parabolic flight methods (Shimizu, 1992; Yamasaki, 2002). The puffy face, nasal congestion, thickening of the eyelids, and reduction of leg girth reported by astronauts during real spaceflight have also demonstrated these effects (Charles, 1994).

If the stay in space continues, these acute phenomena become more stable, and the increase in body fluid and muscle cells that should occur during development slows down compared to what would occur in one-G. Not only the antigravity muscles but also other skeletal muscles work less and so are smaller. As a result, the body remains small-sized; i.e., a light body weight as compared to the ground controls of the same age. In space, blood is provided to every portion of the body with a constant pressure gradient. The distribution of blood flow may be affected more in space by the control of vessels to particular organs rather than by the increase of total blood flow in the aorta (Rowell, 1993). Active contraction of the aortic wall during the filling period of the heart is not essential. The absolute value of cardiac output (the amount of blood the heart pumps per minute) also becomes smaller compared to one-G because of a decrease in total body fluid. As a result, episodes elevating blood pressure in space are reduced as compared to on Earth. Therefore, the growth of the muscular component of the aortic wall is slower, which suppresses the development of contraction forces. Fewer afferent nerve fibers are needed to respond to high blood pressure. Baroreflex sensitivity to blood pressure changes at high pressure levels is also reduced. This suppression of the baroreflex may be due to the structural properties of the afferent pathway (the aortic nerve) and effector organs (vessels). Also, it could be due partly to the effects of microgravity on the cardiovascular control center or on efferent pathways of the baroreflex.

After the growing rat returns to Earth from space, the function of the baroreflex system to control blood pressure can develop normally and adapt to the one-G Earth environment—if the rat returns while it is still developing. There is a possibility, however, that structural differences in the afferent pathway (aortic nerve) may remain permanently.

In conclusion, the space environment affects the development of the aortic baroreflex system and the regulation of blood circulation.

Acknowledgements

We thank NASA, NASDA, JSUP, JSF, and particularly the supporting teams of these organizations for giving us the opportunity to perform these experiments in space and for helping us complete them successfully. We thank Ms. Erica Wagner, the experiment support scientist for our team, for her devotion and excellent contributions to this study. We also thank Ms. Chieko Nagayama, Mutsumi Sato, Kumiko Takahashi, and Masae Yamasaki for their secretarial support and data analysis. This work was also financially supported by Grant A: 08407004 and Fund for Basic Experiments Oriented to Space Utilization from the Ministry of Education, Science, Sports and Culture of Japan.

REFERENCES

THE RHEOLOGY OF LARGE BLOOD VESSELS. D.J. Patel in *Cardiovascular Fluid Dynamics.* Volume 2. Ed. by D.H. Bergal, Academic Press London & NY, pages 1–64; 1972.

UNDERNUTRITION IN THE DEVELOPING RAT: EFFECT UPON MYELINATION. M.R. Krigman, E.L. Hogan. *Brain Res.*, Vol. 107, pages 239–255, 1976.

FUNCTIONAL, HISTOLOGICAL AND DEVELOPMENTAL CHARACTERISTICS OF BRADYCARDIAC MECHANISMS OF THE AORTIC NERVE BAROREFLEX IN THE RABBIT. Symposium: Brain and blood pressure control. T. Shimizu, S. Matsumoto, M. Yamasaki, T. Kanno, T. Nagayama. *Therap. Res.*, Vol. 11, No. 12, pages 3871–3878; 1990.

CHANGES IN CARDIAC OUTPUT AND BLOOD DISTRIBUTION IN THE INITIAL STAGE OF EXPOSURE TO THE MICROGRAVITATIONAL ENVIRONMENT UNDER EXPERIMENTALLY SIMULATED CONDITIONS IN THE RABBIT. T. Shimizu, M. Yamasaki, T. Kanno, T. Nagayama, S. Matumoto. in *Proceedings of 3rd International Symposium on Space Medicine in Nagoya 1992 (ISSM '92).* Edited by N. Matui, H. Seo. Nagoya University Press, pages 221–229; 1992.

HUMAN CARDIOVASCULAR CONTROL. L.B. Rowell. Oxford University Press; 1993.

CARDIOPULMONARY FUNCTION. J.B. Charles, M.W. Bungo G.W. Fortner. in *Space Physiology and Medicine.* Edited by A.E. Nicogossian, C.L. Hantoon, S.L. Pool, pages 286–304. Lea & Febiger; 1994.

POSTNATAL CHANGES OF THE NUMBER OF MYELINATED FIBERS IN THE RAT AORTIC NERVE. M. Yamasaki, T. Shimizu, T. Kannno, T. Nagayama, H. Waki, M. Tanno, K. Katahira, C. Mukai, S. Nagaoka, H. Sato. *Jpn. J. Physiol.*, Vol. 46, Supplement, page S197; 1996.

POSTNATAL DEVELOPMENT OF AUTONOMIC REGULATION OF CARDIOVASCULAR FUNCTION. Symposium: Development of the autonomic nervous system. P.M. Gootman in *Auton. Neurosci.: Basic & Clinical.* Vol. 82, Pages 1–3; 2000.

EFFECTS OF HEAD-DOWN TILT ON POSTNATAL DEVELOPMENT OF THE AORTIC BAROREFLEX IN THE RABBIT. M. Yamasaki, T. Shimizu. *Jpn. J. Physiol.*, Vol. 52 No. 2, pages 149–161; 2002.

Neural Development Under Conditions of Spaceflight

Authors

Meredith D. Temple, Maria J. Denslow,
Kenneth S. Kosik, Oswald Steward

Experiment Team

Principal Investigator: **Kenneth S. Kosik**[1]

Co-Principal Investigator: **Oswald Steward**[2]

Co-Investigators: **Meredith D. Temple**[3]
Maria J. Denslow[4]

Technical Assistant: **Justin Harder**[1]

[1]Harvard Medical School, Brigham and Women's Hospital,
Boston, USA
[2]Reeve-Irvine Research Center, University of California
at Irvine, Irvine, USA
[3]National Institute of Biomedical Imaging and Bioengineering,
Bethesda, USA
[4]Harvard Medical School and Beth Israel Deaconess Medical
Center, Boston, USA

ABSTRACT

One of the key tasks the developing brain must learn is how to navigate within the environment. This skill depends on the brain's ability to establish memories of places and things in the environment so that it can form "cognitive maps." Earth's gravity defines the plane of orientation of the spatial environment in which animals navigate, and cognitive maps are based on this plane of orientation. Given that experience during early development plays a key role in the development of other aspects of brain function, experience in a gravitational environment is likely to be essential for the proper organization of brain regions mediating learning and memory of spatial information. Since the hippocampus is the brain region responsible for cognitive mapping abilities, this study evaluated the development of hippocampal structure and function in rats that spent part of their early development in microgravity. Litters of male and female Sprague-Dawley rats were launched into space aboard the Space Shuttle *Columbia* on either postnatal day eight (P8) or 14 (P14) and remained in space for 16 days. Upon return to Earth, the rats were tested for their ability to remember spatial information and navigate using a variety of tests (the Morris water maze, a modified radial arm maze, and an open field apparatus). These rats were then tested physiologically to determine whether they exhibited normal synaptic plasticity in the hippocampus. In a separate group of rats (flight and controls), the hippocampus was analyzed using anatomical, molecular biological, and biochemical techniques immediately postlanding. There were remarkably few differences between the flight groups and their Earth-bound controls in either the navigation and spatial memory tasks or activity-induced synaptic plasticity. Microscopic and immunocytochemical analyses of the brain also did not reveal differences between flight animals and ground-based controls. These data suggest that, within the developmental window studied, microgravity has minimal long-term impact on cognitive mapping function and cellular substrates important for this function. Any differences due to development in microgravity were transient and returned to normal soon after return to Earth.

INTRODUCTION

The Neurolab mission provided a unique opportunity to evaluate how microgravity affects brain and cognitive development. An important aspect of brain function is the ability to form memories of the environment, referred to as "cognitive maps" (O'Keefe, 1978). The processing and recall of spatial information enabling cognitive mapping depends on a part of the brain called the hippocampus and its related structures (O'Keefe, 1971; O'Keefe, 1979; O'Keefe, 1978; O'Keefe, 1980). Forming mental maps is an essential part of our existence. Imagine what it would be like not to have the capability to go to the store and successfully return home. Indeed, it is this very fundamental ability that is lost in neurological disorders such as Alzheimer's disease.

Although the ability to successfully navigate one's environment can be lost for any number of reasons, its proper development depends on early experiences (Paylor, 1992). For example, experience in a complex ("enriched") environment during early development can improve later cognitive performance in a number of tasks (Paylor, 1992). Animals that experience complex environments during early development have higher brain weight, increased cortical thickness, and increases in the number of synapses on cortical neurons. This indicates that experience actually affects the development of neural circuitry (Ferchmin, 1970; Walsh, 1981; Green, 1983).

Earth's gravitational field defines the primary planes of orientation of the spatial environment in which animals navigate, and the cognitive maps we develop are based on this plane of orientation. For example, humans think of floors, walls, and ceilings and "map" elements of the environment based on this planar orientation. Experience in a gravitational environment is likely essential for the proper development of brain regions that mediate learning and memory of spatial information. Normal development not only requires experience, a concept referred to as activity-dependence, but it also requires that the experience occurs within a certain window of time during development. A well-known example is the child who may lack sight due to premature cataracts or who has diminished sight in one eye due to a misalignment of the eyes. If these problems are not surgically corrected within the first few years of life, the brain will develop abnormally. After a certain age, the critical period to fix these problems will pass and the visual problems will be permanent. Sight is required during development to establish a normal visual system.

What is not known is whether the absence of gravitational cues would affect the normal development of the spatial cognitive navigation system. Studies involving the recording of individual cells within the hippocampus reveal that certain cells, now referred to as "place cells," fire preferentially when an animal is located at a particular position in its environment (Knierim, 2000; also see science report by Knierim et al. in this publication). Data suggest that the pattern of activity of place cells represents the location of the animal relative to particular environmental cues and that these cells and their connections may be critical for forming mental maps. Without gravity, it is possible that these connections could be altered.

In fact, a series of experiments was also conducted as part of the Neurolab mission to study how living in microgravity affects place cell firing (Knierim, 2000; see also science report by Knierim in this publication).

The Neurolab Space Shuttle mission (April 1998) provided an unprecedented opportunity to learn the extent to which animals depend on gravity for the normal development of neural systems that store and analyze spatial information. Until now, it has been impossible to assess whether the absence of gravity would affect the development of spatial memory function. Hence, the purpose of the present study was to determine whether development in an environment lacking gravity would alter rats' navigational abilities, cognitive performance, hippocampal anatomy, and hippocampal cellular function upon return to Earth.

Because the Neurolab mission provided a novel opportunity to study the role of gravity in neural development, it was difficult to hypothesize what the specific impact of maturing during spaceflight and without gravity would have on proper neurological development. The absence of gravity could represent an "enriching" experience similar to that of a "complex environment," by allowing animals to experience a multidimensional rather than a two-dimensional environment. Alternatively, development in an environment without gravity could be deleterious, depriving animals of experience with the planar environment that is defined by Earth's gravity.

Our team evaluated the anatomy, neurophysiology, and biochemistry of the hippocampus, and also performed behavioral tests that require a normally functioning hippocampus in rats that had developed in microgravity. The results were compared to their Earth-bound counterparts.

METHODS

Flight paradigm

Rats of two different ages were launched into space on board the Space Shuttle *Columbia*. One group was eight days old (postnatal day eight (P8)) at the time of launch, and the other group was 14 days old (P14). The younger rats had not yet opened their eyes. At this age, rats spend essentially all of their time near their mother (a rodent pre-toddler stage of development). The older animals were at a stage of development three to four days following eye opening. At this age, rats are beginning to venture away from their mothers to explore their local environment.

Prior to launch, the young rats were evenly distributed among the birthing mothers and were designated the experimental flight (FLT) groups. Sixty-four hours before the launch of the Orbiter *Columbia*, the cross-fostered litters were loaded into Research Animal Holding Facility (RAHF) cages (P8 litters) or animal enclosure modules (AEMs) (P14 litters) with each cage holding one mother and eight young. Both the RAHF and AEM cages provided water (lixit devices) and food (spring-loaded food bars). Modifications were made so that the water and food could be delivered to the animals on orbit without the aid of gravity. The same procedure (pooling, redistribution, and loading of rats) was performed on the control groups (also P8 and P14) designated vivarium (VIV) and asynchronous ground control (AGC), respectively. The VIV groups lived in standard animal vivarium housing, while the AGC groups lived in housing identical to that aboard the Space Shuttle *Columbia*. The AGC and VIV groups for both litters were run on four-day and eight-day delays with their timeline running exactly parallel to the FLT group.

The flight lasted 16 days, during which time food and water consumption in each cage was monitored regularly. Because rats mature rapidly over this time interval, both groups of rats reached young adulthood in space. Approximately eight hours after landing, the animals were removed from the Orbiter, examined by a veterinarian, and distributed to the individual labs for study. All animal care and experimental procedures conformed to National Institutes of Health (NIH) guidelines and were approved by the Institutional Animal Care and Use Committees at NASA Ames Research Center, California, the University of Virginia, and/or Harvard University, Massachusetts.

Behavioral analysis of flight animals

Upon return to Earth (R+0), a total of 43 rats from FLT litters of both age groups (P8 and P14 at launch) and their respective control groups were tested behaviorally over a one-month period. Three behavioral testing paradigms were used. One task (the Morris water maze) evaluates a rat's ability to learn and remember the location of a submerged platform in a tank filled with opaque water. There were constant cues in the room to provide spatial information about the location of the submerged platform. This task could be likened to the navigation problems faced by rats in the wild that live in semiaquatic environments (an urban sewer for example). Rats were tested as they learned to find the hidden platform to assess learning ability, and they were retested on R+25 (postlanding) to determine whether they remembered the location of the platform. Initial testing began three days after landing (R+3) for the P14 litters and nine days postlanding (R+9) for the P8 litters. During the initial testing, animals had four trials per day (one from each start location) for four consecutive days.

Another task evaluated the rats' ability to navigate on an eight-arm radial maze baited with highly attractive food substances. This task assesses rats' foraging strategies and is similar to the problems faced by rats seeking food in the wild. The optimal foraging strategy is one in which places are visited without repetition—that is, to avoid recently visited sites where food has already been found and removed. Hence, this task assesses memory of places recently explored. Animals were tested for five (nonconsecutive) days (two trials/day) on R+16, 19, 22, 25, and 27. Finally, we also assessed the rats' exploratory behavior in an open field. This free exploration task provided information regarding whether the rats would recognize walls as barriers. In space, walls are not barriers; instead, they are simply alternative surfaces upon which to navigate. The ability to recognize a wall as a barrier is likely to require experience with that boundary. All rats received one trial per day on days R+2, 3, 4, 5, and 7.

Electrophysiological analysis of hippocampal function

At the end of the behavioral testing, 22 rats from the P8 litters were assessed neurophysiologically to evaluate a form of synaptic plasticity in the hippocampus that is thought to play a role in the memory of place (long-term potentiation (LTP)). For this experiment, rats were anesthetized, and stimulating and recording electrodes were placed to activate key hippocampal pathways. We focused on the pathway called the "perforant path," which carries information from the cerebral cortex to the hippocampus. This is the pathway in which LTP was first discovered. To explore the capacity for LTP, synaptic responses generated by the perforant pathway were evaluated before and after the delivery of patterns of stimulation that are optimal for inducing LTP. The key issue was whether the rats that developed in space had the same capacity for LTP as rats raised on Earth. The stimulation to induce LTP involved the delivery of three "bouts" of stimulation, in which each bout consisted of 10 eight-pulse trains delivered at 400 Hz.

Histological analyses of flight hippocampus

On the day of recovery from the Space Shuttle *Columbia* (R+0), the brains of six rats from the P8 FLT group were studied. The tissue was fixed by microwave irradiation for five minutes while submerged in P.L.P. (2% paraformaldehyde, l-lysine, sodium periodate). The brains were then post fixed in P.L.P. for 18 hours and transferred on R+1 to a 15% glycerol solution in phosphate-buffered saline to cryoprotect the tissue. The same harvesting procedure was completed for the VIV (n=6) and AGC (n=6) control groups on the four-day and eight-day delay, respectively. Forty-micron coronal sections were taken at the level of the hippocampus. The tissue was collected in serial sections so that each group of sections contained a representative section of the hippocampus at ~800µ intervals. One group of serial sections was stained with a Nissl stain to mark the cell bodies. The remaining serial sections were then immunoctyochemically stained using a free-floating technique. Antibodies against the following markers were used: phosphorylated neurofilament (Sternberger Monoclonals Incorporated (SMI) (Lutherville, MD) #31), non-phosphorylated neurofilament (SMI #312), synaptophysin (Sigma #SVP 38) (Sigma Chemicals Pty Ltd., Perth, Western Australia) and SNAP-25 (SMI #81), mGluR1 (Upstate Biotechnology, Inc., Lake Placid, NY), GluR 2/3 (Upstate Biotechnology, Inc., Lake Placid, NY), mGluR5, NMDA R1 (Chemicon International, Inc., Temecula, CA), NMDA R2A/B (Chemicon), and microtubule associated protein (Kosik Lab 5F9). All immunocytochemical stainings were labeled with a Cy 3 secondary fluorescent antibody and imaged on a confocal microscope. Z-series images were taken at 40× and 63× where each image of the Z-series represents 0.50 (of the section). Images were then transferred to the IPLab Imaging System (IPLab, Stockholm, Sweden) software and analyzed. For neurofilament staining, the optimal level was determined and the surrounding frames were averaged. The composite frame was then thresholded to highlight the immunopositive areas of the section. The level of thresholding was set to equal the mean intensity of each region analyzed. The immunopositive areas were then calculated as a percent area of the total measured region, and neurofilament densities were calculated.

RESULTS

Flight rats exhibit normal learning and memory in spatial tasks

Despite the fact that the FLT rats spent much of their early development in a weightless environment, they exhibited no lasting behavioral abnormalities. The FLT animals in both age groups (P8 and P14) exhibited essentially normal learning during the initial Morris water maze testing (Figure 1) and normal memory during the retest (Figure 2). The same was true of the radial arm maze (data not shown). Rats also exhibited normal exploratory behavior in the open field (including normal responses to walls) (data not shown).

The only exception involved the FLT group launched at 14 days of age. These rats began testing in the Morris water maze on the third day after landing (R+3) and took longer to find the hidden platform and swam faster during testing that day. However, by the second day of testing, the P14 FLT rats' performance was comparable to the control groups.

Flight rats, regardless of age, used different search strategies in the early phases of Morris water maze testing. By the third day of maze testing, both FLT groups were using the same strategy as control animals. For complete data analysis, see Temple et al. (Temple, 2002).

Flight rats exhibit a normal capacity for hippocampal synaptic plasticity

Long-term potentiation (LTP) is a form of synaptic plasticity that is thought to be important for the establishment of memories. LTP is seen when particular pathways are activated by brief high-frequency trains of stimuli (similar to the patterns of activity exhibited by neurons in the hippocampus and related structures). The standard paradigm for assessing LTP is to deliver single-pulse stimulation to the perforant path while recording synaptic responses from a structure called the dentate gyrus in the hippocampus. In this study, we first determined "baseline" response amplitude, and then delivered three bouts of high-frequency stimulation to induce LTP. P8 FLT rats exhibited similar baseline responses, and exhibited the same capacity for LTP as their respective controls (Figure 3). These results demonstrate that by one month after returning to Earth, animals that developed in space exhibited no detectable physiological abnormalities. It will be of considerable interest to determine whether

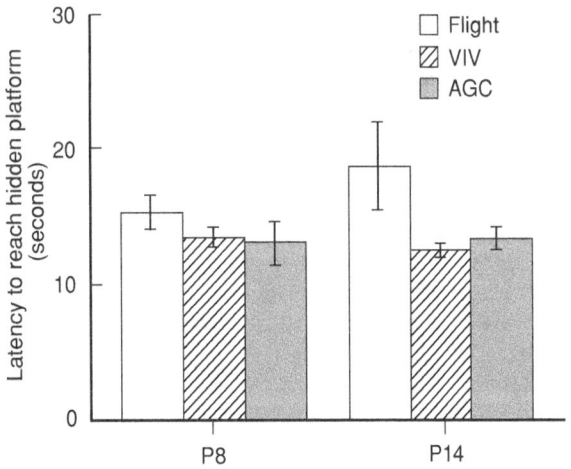

Figure 1. Cognitive performance was assessed using a hidden platform version of the Morris water maze task. During the initial testing phase, rats were given four trials/day for four consecutive days. They had a maximum of 60 seconds to locate the hidden platform. This figure represents a collapsed analysis of latency measures for each group across the four days of testing. On the first day of testing, the P14 Flight group took longer to locate the hidden platform. However, statistical analyses revealed no significant differences among the FLT, VIV, and AGC groups within each age group (P8 or P14).

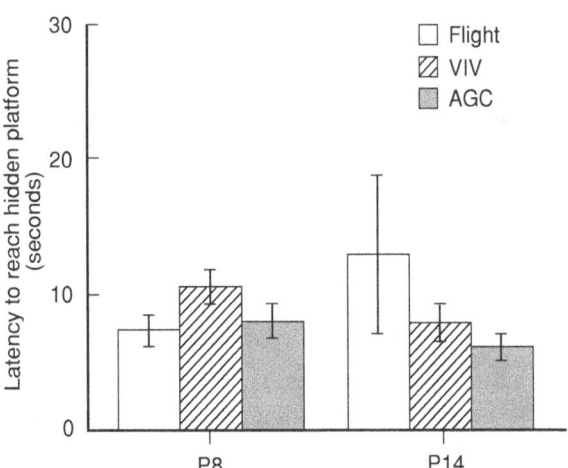

Figure 2. On R+25 postlanding, rats were tested again in the Morris water maze for retention of the task (i.e., memory of the location of the hidden platform). Rats were tested one day (four trials) with a maximum of 60 seconds to locate the hidden platform. This figure represents the average latency for each group during the one day of retesting (i.e., four trials). Analysis of the latency measure for each litter indicated no significant differences among the FLT, VIV, and AGC groups within each age group (P8 or P14).

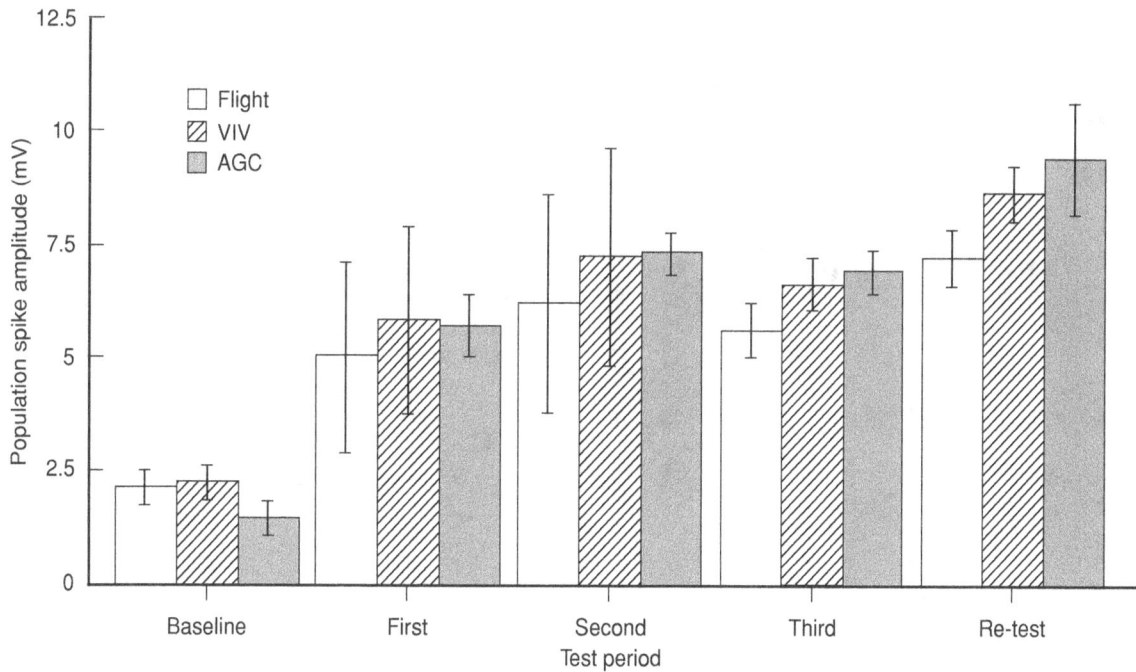

Figure 3. Electrophysiology studies were conducted in a group of P8 FLT, VIV, and AGC animals at R+ 30 postlanding. The bar graph illustrates the degree of LTP (long-term potentiation) induced by high-frequency stimulation of the perforant path projections to the dentate gyrus. Baseline = average response amplitude prior to the delivery of high-frequency stimulation. The three sets of bars indicate average response after separate bouts of stimulation (each bout consisting of 10 trains of eight pulses at 400 Hz delivered once every 10 seconds). Three bouts of stimulation were delivered to induce LTP, and then 10 test pulses were delivered to assess response amplitude. After the final bout, response amplitude was tested for two hours. Then, a final bout of stimulation was delivered to determine the degree of saturation of response amplitude (final set of bars). There was no significant difference among groups in the extent of the initial change in response amplitude or in the duration of LTP.

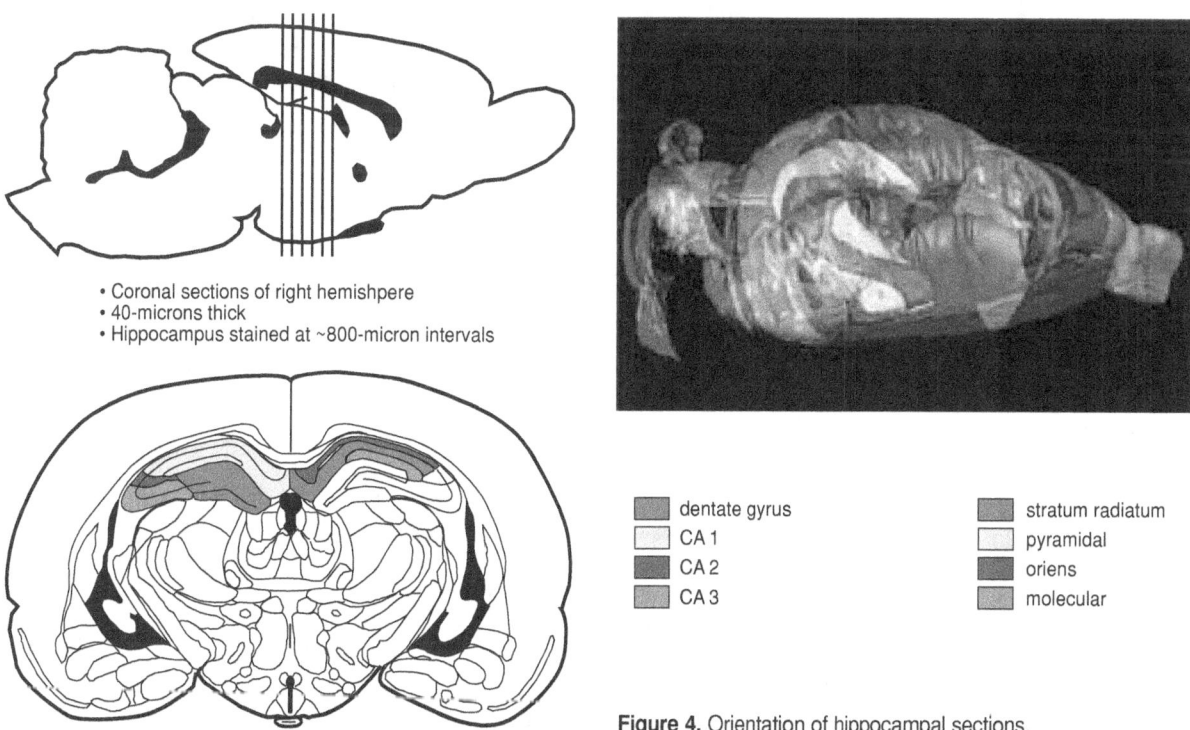

- Coronal sections of right hemishpere
- 40-microns thick
- Hippocampus stained at ~800-micron intervals

dentate gyrus		stratum radiatum
CA 1		pyramidal
CA 2		oriens
CA 3		molecular

Figure 4. Orientation of hippocampal sections.

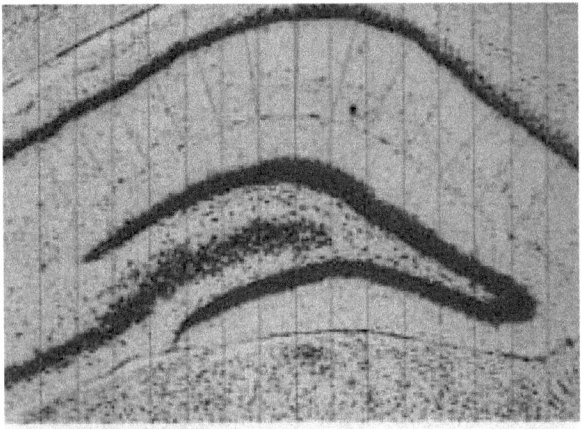

- Grid overlaid on Nissl
- One dimensional measurement of CA 1 field
  (measured at grid intersection with hippocampal fissure)

Figure 5. Grid for measurement of intra-hippocampal distances to estimate changes in CA1.

there may be differences if the rats are younger at the time of spaceflight and/or if the rats are tested just after returning from a microgravity environment.

Flight rats do not differ from controls in hippocampal anatomy or by antibody detection

Corresponding to the absence of behavioral and electrophysiological findings after spaceflight was the absence of any detectable changes in protein staining (Figure 4-9). The hippocampus appeared to contain a normal number of neurons, normal appearing dendritic arbors, normal axons, normal synapses, and normal receptors. This was true of the group of rats that was launched when eight days old and evaluated immediately upon returning to Earth.

DISCUSSION

Taken together, our findings suggest that development in an environment lacking gravity does not produce any permanent changes in an animal's ability to use spatial information and form memories of place within its environment. The very subtle behavioral differences observed in the P14 FLT litter in the first few days after reentry were transient. There were also no detectable differences in hippocampal anatomy, biochemistry, and a key measure of synaptic function (the capacity for LTP).

Because of sharing paradigms among investigators, we were unable to test both age groups in all three sets of experiments (behavior, electrophysiology, and hippocampal anatomy/antibody detection). In addition, there is no way to distinguish the experience of microgravity from the experience of launch/landing, which produces rapid changes in pressure gradients. We consider that launch/landing and microgravity are both components of any spaceflight experience.

The NASA Neurolab mission provided an unprecedented opportunity to study how spaceflight and microgravity impact development. This is the first study to evaluate brain development and cognitive function of animals that have spent much of their early life in space. Our data strongly suggest that early experience in an environment lacking gravity does not appear to affect adversely the development of an important aspect of the neural learning and memory system. Given the sensitivity of neural systems to early environment, it is quite surprising that the development of the neural systems that mediate spatial cognitive functions are so impervious to what is certainly an extremely unusual experience during the critical period of development—when animals first explore their environment. Apparently, this key memory system can rapidly re-adapt.

These findings bear importantly on the question of how prolonged spaceflight might affect the brain. Our results indicate that the systems that are critical for spatial cognitive function can readily re-adapt to a gravitational field even when most

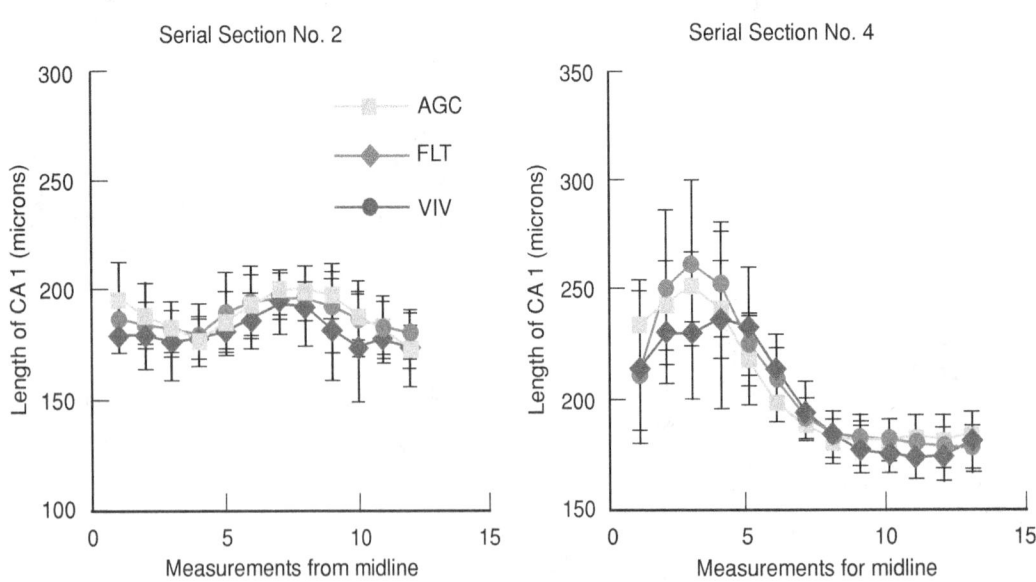

Figure 6. Hippocampal morphometry by Nissl stain reveals no differences in flight rats.

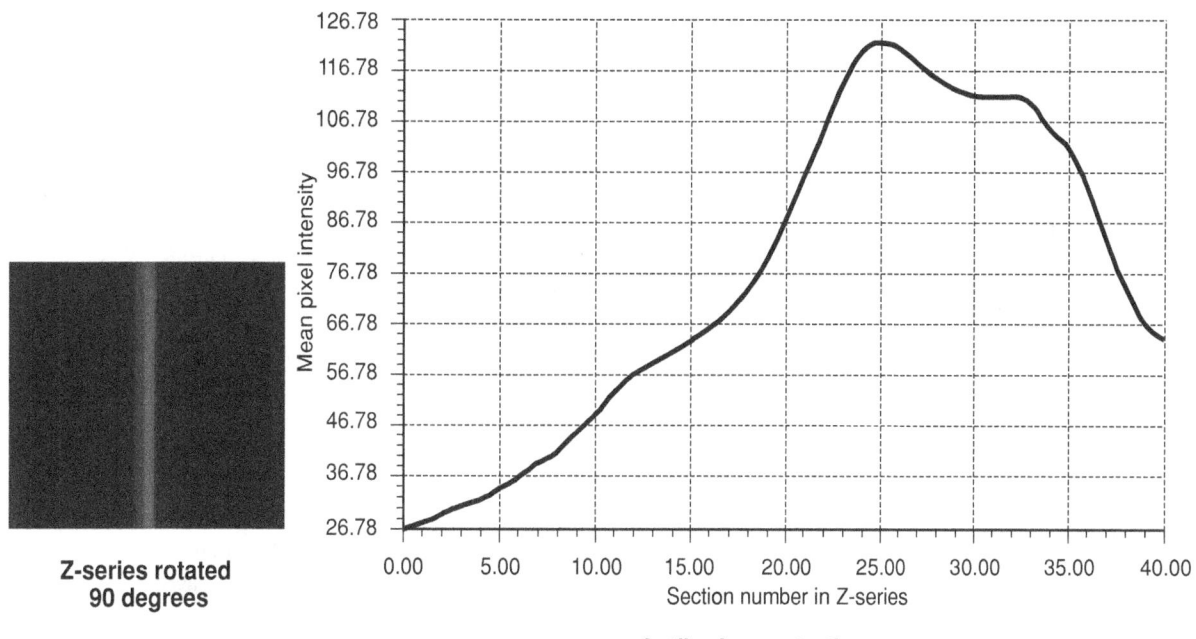

**Z-series rotated
90 degrees**

Antibody penetration curve

Figure 7. Penetration of the representative (synaptophysin) antibody in the Z-axis.

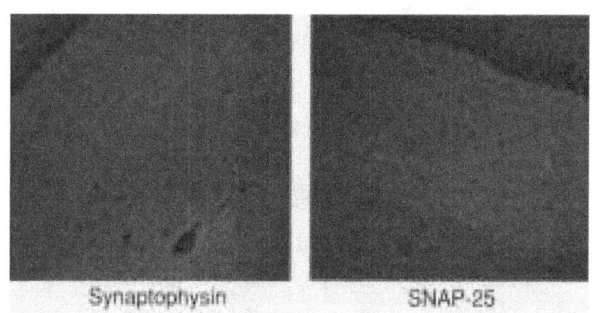

Synaptophysin   SNAP-25

Figure 8. Diffuse antibody labeling of hippocampus from flight rat. For "diffusely" staining antibodies (i.e., receptor and synaptic antibodies), the average pixel intensity of each frame of each Z-series was measured to determine the optimal level of fluorescent expression. The five frames at the optimal level were then averaged into one frame, and the mean pixel intensity for the various regions was measured.

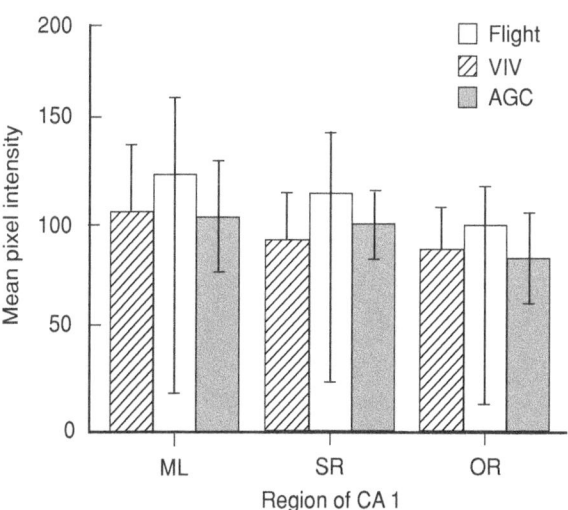

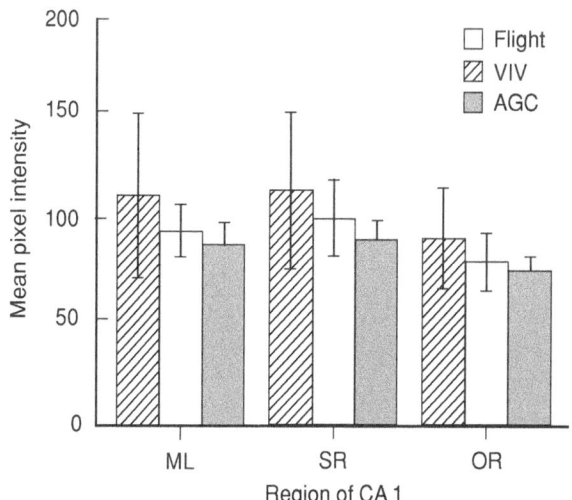

Figure 9. Quantification of the synaptophysin label in flight vs. control rats. No differences were found between the flight and the control rats.

Figure 10. Quantification of the SNAP-25 label in flight vs. control animals. No differences were found between the flight and the control rats.

early development is spent in a gravity-free environment. It is indeed crucial to know that it will be possible to live in space for prolonged periods and even grow up in space without irreversibly altering the memory systems that mediate our basic ability to navigate within the Earth environment.

Acknowledgements

Supported by NASA grant NAG2-964. The authors gratefully acknowledge all of those individuals involved in the NASA Neurolab mission, particularly our team's Experimental Support Scientists (ESSs) Erica Wagner and Tracie Hughes. Special thanks also to the crew of STS-90 for their extraordinary efforts to make the animal experiments a success. We also thank Justin Harder for his technical assistance and Dr. Leonard Jarrard and Dr. Robert Hamm for their advice in designing the behavioral testing paradigm.

REFERENCES

STUDIES OF BRAIN WEIGHT AND RNA CONTENT AFTER SHORT PERIODS OF EXPOSURE TO ENVIRONMENTAL COMPLEXITY. P. A. Ferchmin, V. A. Eterovic, and R. Caputto. *Brain Res.*, Vol. 20, pages 49–57; 1970.

THE HIPPOCAMPUS AS A SPATIAL MAP: PRELIMINARY EVIDENCE FROM UNIT ACTIVITY IN THE FREELY-MOVING RAT. J. O'Keefe and J. Dostrovsky. *Brain Res.*, Vol. 34, pages 171–175; 1971.

THE HIPPOCAMPUS AS A COGNITIVE MAP. J. O'Keefe and L. Nadel. Clarendon Press, Oxford, 1978.

A REVIEW OF THE HIPPOCAMPAL PLACE CELLS. J. O'Keefe. *Prog. Neurobiol.,* Vol. 13, pages 419–439; 1979.

ON THE TRAIL OF THE HIPPOCAMPAL ENGRAM. J. O'Keefe and D. H. Conway. *Physiol. Psychol.*, Vol. 8, pages 229–238; 1980.

EFFECTS OF ENVIRONMENTAL COMPLEXITY AND DEPRIVATION ON BRAIN ANATOMY AND HISTOLOGY: A REVIEW. R.N. Walsh. *Int. J. Neurosci.*, Vol. 12, pages 33–51; 1981.

EFFECTS OF COMPLEX OR ISOLATED ENVIRONMENTS ON CORTICAL DENDRITES OF MIDDLE-AGED RATS. E.J. Green, W.T. Greenough, and B.E. Schlumpf. *Brain Res.,* Vol. 264, pages 233–240; 1983.

BRIEF EXPOSURE TO AN ENRICHED ENVIRONMENT IMPROVES PERFORMANCE ON THE MORRIS WATER TASK AND INCREASED HIPPOCAMPAL CYTOSOLIC PROTEIN KINASE C ACTIVITY IN YOUNG RATS. R. Paylor, S.K. Morrison, J.W. Rudy, L. T. Waltrip, and J.M. Wehner. *Behav. Brain Res.,* Vol. 52, pages 49–59; 1992.

THREE-DIMENSIONAL SPATIAL SELECTIVITY OF HIPPOCAMPAL NEURONS DURING SPACE FLIGHT. J.J. Knierim, B.L. McNaughton, and G.R. Poe. *Nat. Neurosci.*, Vol. 3, pages 209–210; 2000.

SPATIAL LEARNING AND MEMORY IS PRESERVED IN RATS AFTER EARLY DEVELOPMENT IN A MICROGRAVITY ENVIRONMENT. M. D. Temple, K. S. Kosik, and O. Steward. *Neurobiol. Learning Mem.*, Vol. 78, pages 199-216; 2002.

The Effect of Weightlessness on the Developing Nervous System

Authors

Richard S. Nowakowski, Nancy L. Hayes

Experiment Team

Principal Investigator: **Richard S. Nowakowski**

Co-Investigator: **Nancy L. Hayes**

University of Medicine and Dentistry of New Jersey,
Robert Wood Johnson Medical School, Piscataway, USA

ABSTRACT

As the brain develops, cells proliferate, and then their progeny migrate to the locations they will occupy in the adult brain where they differentiate. These events involve interactions both between cells and between organelles within cells. The buoyancy of the cells and organelles can affect these interactions. On Earth, gravity is responsible for maintaining the buoyancy of organelles and cells, and in the weightless condition of space these differences are lost. We tested the hypothesis that the loss of these differences will disrupt cell proliferation and migration within a particular part of the brain—the neocortex.

To study these developmental processes, the developing brains of prenatal mice were labeled during space-flight with two markers of DNA synthesis (tritiated thymidine and bromodeoxyuridine). These brains were preserved in fixative, stored in the cold, and returned to Earth for study after the flight. The postflight analysis showed there were significant differences in the development of the brains of fetal mice in space vs. those maintained in identical conditions at the Kennedy Space Center. Detailed work to determine precisely how brain development is modified continues as these specimens are being further analyzed. This experiment demonstrated that complex studies on brain development can be performed successfully in space.

Section 4 **Blood Pressure Control**

Standing after spaceflight: The effects of weightlessness on blood pressure

INTRODUCTION

Many people occasionally feel lightheaded when they stand up. This lightheadness is sometimes caused by dehydration or illness, and it can progress to fainting. What the lightheaded sensation signals is that not enough blood is reaching the brain. Since the head is above the heart when people stand or sit, the cardiovascular system must work against gravity to maintain brain blood flow. A person whose cardiovascular system is having trouble providing this upward flow is said to have orthostatic intolerance. This can be a daily problem for people with certain diseases, but it is also a problem for astronauts after spaceflights.

Experiments on previous life science Spacelab missions showed that all crewmembers could easily stand quietly for 10 minutes before flight. After flight, however, 60% needed to sit down before the 10 minutes were up. Some showed significant decreases in blood pressure, while others needed to sit despite an apparently normal, though low, blood pressure. Overall, the results suggested that the autonomic nervous system—the part of the nervous system that automatically controls blood pressure—might have changed in weightlessness.

To regulate blood pressure, the autonomic nervous system has two main tools. It can either change the amount of blood flowing by increasing or decreasing the pumping of the heart, or it can change the resistance to blood flowing in the arteries. This is analogous to increasing the pressure in a garden hose by either opening the faucet or squeezing on the hose to narrow the diameter. To accomplish this seemingly simple task, however, a complex and redundant system has evolved. The cardiovascular system has high- and low-pressure receptors to sense pressure in various areas. The brain has a cardiovascular center to integrate all the information it receives. The brain sends out signals to alter the function of the heart and blood vessels to maintain normal blood pressure. The results are then sensed and the necessary corrections are made. In fact, the system shows continuous slight oscillations as the pressure sensors and brain work together to constantly adjust blood pressure.

On Neurolab, an integrated series of experiments studied the autonomic nervous system to determine how the blood pressure control system changed while crewmembers were weightless.

THE BLOOD PESSURE CONTROL SYSTEM

The bottom left diagram in Plate 5 shows how the blood pressure control system works. Pressure receptors (called baroreceptors) in the heart, aortic arch and blood vessels serving the brain (carotid arteries) sense the pressure within the cardiovascular system. This information travels to the brain areas (blue lines) where the information is interpreted. If blood pressure is low, signals from the brain are sent to speed the heart (red lines to sinoatrial (SA) and atrioventricular (AV) nodes) and to constrict blood vessels (red lines to smooth muscle in arteries). When blood pressure is high, the heart is slowed and the blood vessels are dilated. The parasympathetic system mainly controls the heart rate responses, and the sympathetic system mainly controls the signals to blood vessels.

The diagram in the upper right of Plate 5 shows the approach that can used to study the function of this system. Blood pressure can be measured using cuffs on the arm and finger. Heart rate can be obtained from the electrocardiogram (ECG). The information sent to blood vessels can be measured by a technique called microneurography. With this technique, a very thin needle (the size of an acupuncture needle) is attached to an amplifier. The needle is then advanced into the peroneal nerve just below the knee, and the signals traveling to the blood vessels in the leg muscles (muscle sympathetic nerve activity) can be recorded.

Blood flow to the brain can be estimated using the noninvasive transcranial Doppler technique. With this device, high-frequency sound waves can be used to show how blood flow to the brain is regulated. These measurements are important because not all people who have orthostatic intolerance after spaceflight experience a drop in blood pressure. Instead, some crewmembers may have a problem with the control of brain blood flow.

The autonomic nervous system can be challenged by a variety of tests. The diagram in the upper right shows the lower body negative pressure device, which surrounds the lower body and is connected to a vacuum hose. The vacuum creates a low pressure (or suction) inside the bag. This causes blood to move from the chest into the legs, just as it does when an individual stands up in Earth's gravity.

The cold pressor glove provides a cold stimulus to the hand. This activates the blood pressure control system and raises blood pressure, but not by stimulating baroreceptors. The lower body impedance leads measure fluid distribution in the body. These measurements help to study how fluid distribution may contribute to problems in blood pressure control.

The diagram in the upper left shows the results from another test—the Valsalva maneuver. With the Valsalva maneuver, a person takes a breath and then bears down (or strains). The straining stimulates the baroreceptors. The diagram shows that as finger blood pressure drops during the straining maneuver, the nerve activity going to blood vessels (filtered and integrated neurogram) increases dramatically. The results from the different tests of the blood pressure control system (lower body negative pressure, cold pressor, Valsalva, and others) when evaluated together reveal which part of the autonomic system may be functioning improperly.

THE NEUROLAB AUTONOMIC EXPERIMENTS

Dr. Blomqvist and his team showed that the blood pressure control system, rather than having a defect, was working normally after the flight. During the flight, the control of brain blood flow and the responses to the cold pressor test did not change compared to preflight measurements. The team headed by Dr. Eckberg showed that sympathetic nervous system responses to the Valsalva maneuver also were unchanged. Contrary to preflight expectation, Dr. Robertson and his laboratory showed that the astronauts had mildly elevated, not decreased, resting sympathetic nervous system activity in space. Sympathetic nervous system responses to stresses that simulated the cardiovascular effects of standing (lower body negative pressure) were brisk both in space and after the flight. Dr. Baisch and his team showed the changes in fluid distribution that accompanied these findings.

Taken together, these experiments defy predictions that the autonomic nervous system would change dramatically during short-term spaceflights. The data suggest that the reduction in blood volume that occurs in space is a key contributor to blood pressure control problems after spaceflight. Another contributor may be the reduced responsiveness of blood vessels to the signals they receive. These findings add to our understanding of the complex causes of orthostatic intolerance—both on Earth and in space.

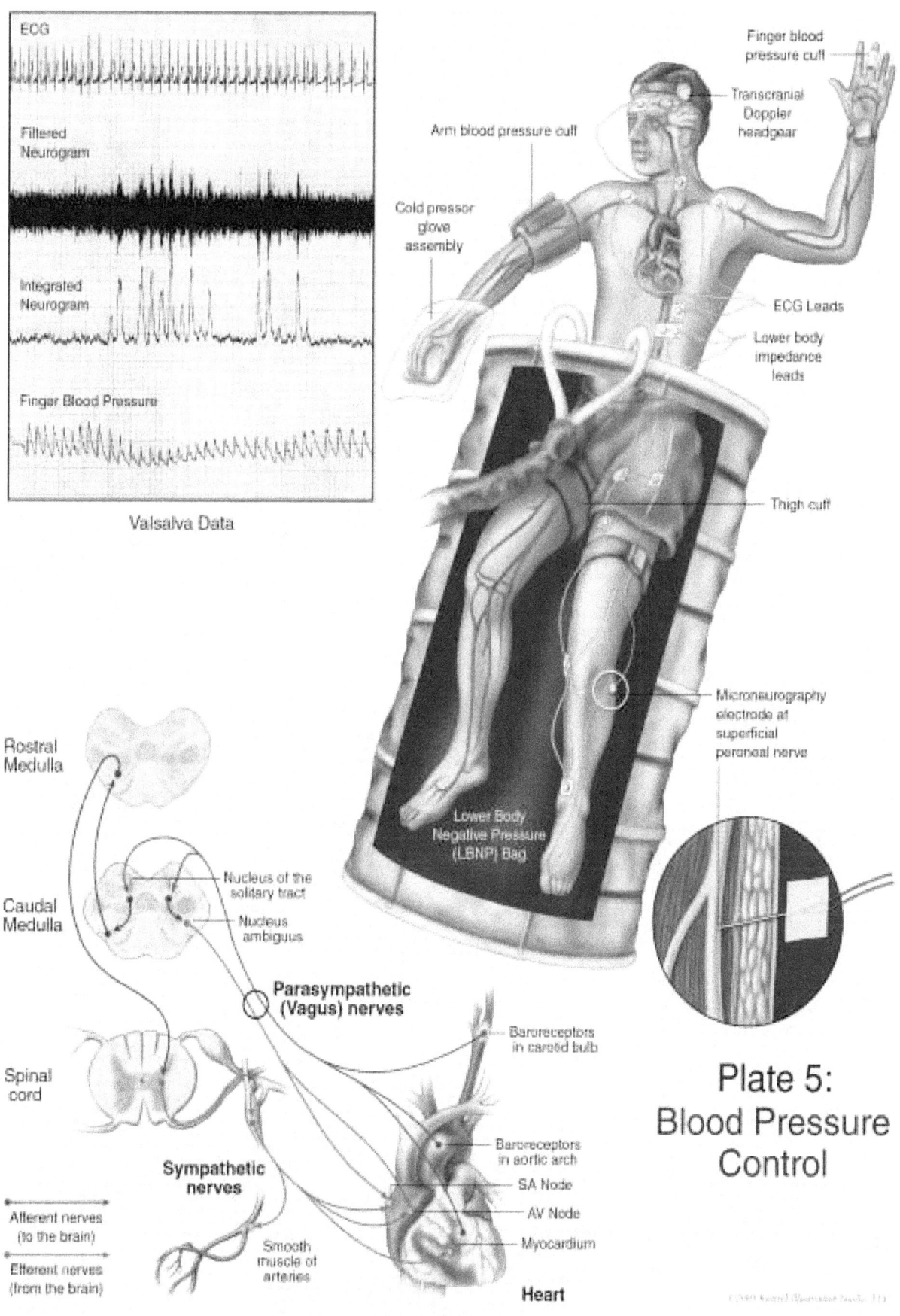

ECG

Filtered
Neurogram

Integrated
Neurogram

Finger Blood Pressure

Valsalva Data

Finger blood
pressure cuff

Transcranial
Doppler
headgear

Arm blood pressure cuff

Cold pressor
glove
assembly

ECG Leads

Lower body
impedance
leads

Thigh cuff

Microneurography
electrode at
superficial
peroneal nerve

Lower Body
Negative Pressure
(LBNP) Bag

Rostral
Medulla

Caudal
Medulla

Nucleus of the
solitary tract

Nucleus
ambiguus

Parasympathetic
(Vagus) nerves

Baroreceptors
in carotid bulb

Spinal
cord

Baroreceptors
in aortic arch

SA Node

AV Node

Myocardium

Sympathetic
nerves

Afferent nerves
(to the brain)

Efferent nerves
(from the brain)

Smooth
muscle of
arteries

Heart

Plate 5:
Blood Pressure
Control

173

Neural Control of the Cardiovascular System in Space

Authors

Benjamin D. Levine, James A. Pawelczyk,
Julie H. Zuckerman, Rong Zhang, Qi Fu,
Kenichi Iwasaki, Chet Ray, C. Gunnar Blomqvist

Experiment Team

Principal Investigator: **C. Gunnar Blomqvist[1]**

Co-Investigators: **Benjamin D. Levine[1]**
James A. Pawelczyk[2]
Lynda D. Lane[3]
Cole A. Giller[1]
F. Andrew Gaffney[3]

[1]University of Texas Southwestern Medical Center, Dallas, USA
[2]The Pennsylvania State University, University Park, USA
[3]Vanderbilt University, Nashville, USA

ABSTRACT

During the acute transition from lying supine to standing upright, a large volume of blood suddenly moves from the chest into the legs. To prevent fainting, the blood pressure control system senses this change immediately, and rapidly adjusts flow (by increasing heart rate) and resistance to flow (by constricting the blood vessels) to restore blood pressure and maintain brain blood flow. If this system is inadequate, the brain has a "backup plan." Blood vessels in the brain can adjust their diameter to keep blood flow constant. If blood pressure drops, the brain blood vessels dilate; if blood pressure increases, the brain blood vessels constrict. This process, which is called "autoregulation," allows the brain to maintain a steady stream of oxygen, even when blood pressure changes. We examined what changes in the blood pressure control system or cerebral autoregulation contribute to the blood pressure control problems seen after spaceflight. We asked: (1) does the adaptation to spaceflight cause an adaptation in the blood pressure control system that impairs the ability of the system to constrict blood vessels on return to Earth?; (2) if such a defect exists, could we pinpoint the neural pathways involved?; and (3) does cerebral autoregulation become abnormal during spaceflight, impairing the body's ability to maintain constant brain blood flow when standing upright on Earth? We stressed the blood pressure control system using lower body negative pressure, upright tilt, handgrip exercise, and cold stimulation of the hand. Standard cardiovascular parameters were measured along with sympathetic nerve activity (the nerve activity causing blood vessels to constrict) and brain blood flow.

We confirmed that the primary cardiovascular effect of spaceflight was a postflight reduction in upright stroke volume (the amount of blood the heart pumps per beat). Heart rate increased appropriately for the reduction in stroke volume, thereby showing that changes in heart rate regulation alone cannot be responsible for orthostatic hypotension after spaceflight. All of the astronauts in our study had an increase in sympathetic nerve activity during upright tilting on Earth postflight. This increase was well calibrated for the reduction in stroke volume induced by the upright posture. The results obtained from stimulating the sympathetic nervous system using handgrip exercise or cold stress were also entirely normal during and after spaceflight. No astronaut had reduced cerebral blood flow during upright tilt, and cerebral autoregulation was normal or even enhanced inflight. These experiments show that the cardiovascular adaptation to spaceflight does not lead to a defect in the regulation of blood vessel constriction via sympathetic nerve activity. In addition, cerebral autoregulation is well-maintained. It is possible that despite the increased sympathetic nerve activity, blood vessels did not respond with a greater degree of constriction than occurred preflight, possibly uncovering a limit of "vasoconstrictor reserve."

INTRODUCTION

The Challenge of Upright Posture

One of the unique features distinguishing humans from quadrupeds is the requirement for standing upright. Teleologically, this upright posture frees the hands to perform complex tasks and, therefore, is considered an essential component of human evolution. By placing the brain above the heart, however, the upright posture presents a complex problem for the blood pressure control system.

During the acute transition from supine to upright posture, a large volume of blood suddenly moves out of the chest and into the lower body. Large gravitationally induced pressure gradients (hundreds of mmHg) force blood away from the brain. Within seconds, the cardiovascular control system must sense this change, and then rapidly adjust flow (by increasing heart rate) and resistance (by constricting the blood vessels) within the circulation to restore an adequate blood pressure and maintain blood flow to the brain. If this process fails, fainting or "syncope" will occur—a medical problem affecting millions of Americans, and accounting for nearly 2% of all emergency room visits in the United States.

The Control System—The Autonomic Nervous System

A complex network of nerves called the autonomic nervous system mediates this rapid response. This system contains two main branches—the sympathetic and the parasympathetic pathways. A good analogy for how this system works would be the environmental control system in a room. A thermostat set at a specific temperature serves as the target. If the temperature drops too low, the heater comes on and raises the temperature; similarly if it gets too hot, the air conditioner kicks in and brings the temperature down. If all components are working well (sensors, heater, air conditioner), the temperature stays relatively constant. A similar process occurs in the human body. Pressure sensors, called "baroreceptors," are located in key areas such as the carotid arteries at the base of the brain, the aortic arch, and the heart itself. If the pressure gets too low, the "heater" or sympathetic nervous system speeds the heart and constricts the blood vessels; if the pressure gets too high, the "heater" turns off, relaxing the blood vessels. The parasympathetic nervous system or "air conditioner" then acts to slow the heart rate and reduce the blood flow.

Cerebral Autoregulation

If this system is not working perfectly, the brain has a "backup plan." Over a relatively large range of pressures, the blood vessels of the brain can adjust their own diameter to keep blood flow constant. If blood pressure drops, the blood vessels dilate; if blood pressure increases, the blood vessels constrict. This process, which is called "autoregulation," allows the brain to maintain a steady stream of oxygen and nutrients, even when blood pressure changes during the activities of daily life.

Cerebral autoregulation is a dynamic process that is frequency dependent. In other words, the ability of local vascular control mechanisms to buffer changes in arterial pressure and keep cerebral blood flow (CBF) constant may be more or less effective, depending on the time period or frequency at which changes in blood pressure occur. Previous work in our laboratory has described the cerebral circulation as a "high pass filter." In other words, when blood pressure changes rapidly, or with "high" frequency (>0.20 Hz, or faster than every five seconds), the autoregulatory process can't keep up and blood flow changes along with pressure. In contrast, at lower frequencies (<0.07 Hz, or slower than every 13 seconds), autoregulation is more effective and the changes in blood pressure can be buffered or filtered. This allows the maintenance of stable brain blood flow despite changes in blood pressure.

The brain has such a high metabolic rate that if its blood flow is interrupted even for a few seconds, normal neuronal function is disrupted and syncope will occur. Ultimately, syncope occurs because of a reduction in CBF sufficient to cause loss of consciousness. Usually, the fall in CBF is secondary to a dramatic drop in blood pressure, which overwhelms the capacity of cerebral autoregulation to maintain a constant flow. In some patients with recurrent syncope, however, cerebral autoregulation may be compromised. This may result in a fall in CBF during standing or sitting, even with relatively minor changes in blood pressure.

The Problem: Syncope after Spaceflight

Because gravity plays such a critical role in determining the pressure and distribution of blood flow within the circulation, the absence of gravity, such as occurs in spaceflight, affords a unique environment in which to examine these control systems. Moreover, in the earliest days of the crewed space program, the clinical importance of this issue became evident. When astronauts returned to Earth and tried to stand up, it was found that many of them couldn't—they experienced lightheadedness, dizziness, and fainting. This problem is called "orthostatic intolerance."

This problem was a topic of intense investigation during the first American and European dedicated space life sciences (SLS) missions: SLS-1, SLS-2, and Spacelab-D2. In the SLS studies, our research group made a number of important observations (Buckey, 1996): (a) about two-thirds of astronauts could not stand for 10 minutes after returning from even a short (one–two weeks) spaceflight; (b) the hearts of all of the astronauts pumped less blood in the upright position (reduced "stroke volume") after spaceflight; (c) all of the astronauts had an increase in heart rate on standing, to compensate for the reduced stroke volume; (d) in contrast to this augmented heart rate, those astronauts with the least ability to tolerate standing after spaceflight were unable to constrict their blood vessels to a greater degree than they were before flight; and, thus, they were unable to compensate for the reduced stroke volume. A similar finding was made subsequently by NASA investigators (Fritsch-Yelle, 1996), who confirmed this problem in a larger number of astronauts. They also found that the concentrations in the blood of norepinephrine (the neurotransmitter released by the sympathetic nervous system) were lower in those astronauts with the most impaired orthostatic tolerance. These data suggested that spaceflight could be leading to a possible defect in the sympathetic nervous system that impaired the ability to constrict the blood vessels. Finally, (e) some astronauts could not remain standing, even though their blood pressure appeared to be normal. This finding raised the possibility that spaceflight could also impair autoregulation in the brain, leading to abnormal cerebral blood flow, even with adequate autonomic neural control of blood pressure.

These then were the questions we wanted to address with our Neurolab experiments: (1) does the adaptation to spaceflight cause a unique adaptation in the autonomic nervous system that would impair the ability to increase sympathetic activity and constrict blood vessels when gravitational gradients are restored on return to Earth?; (2) if such a defect could be found, could we pinpoint the specific afferent (the sensors sending signals from the body to the brain) or efferent (the sympathetic signals to the blood vessels themselves) neural pathways that were involved?; and (3) is there an abnormality of cerebral autoregulation that develops during spaceflight and impairs the ability to maintain CBF constantly in the upright position on Earth, even if blood pressure is well maintained?

METHODS

To answer these questions, we needed to be able to quantify precisely the sympathetic responses of the autonomic nervous system, as well as systemic and cerebral blood flow, in the context of an experimental protocol that perturbed multiple components of the blood pressure control system. Our experimental setup is shown in Figure 1.

The subject lay supine in a clear plastic box, which allowed the application of LBNP (see science report by Ertl et

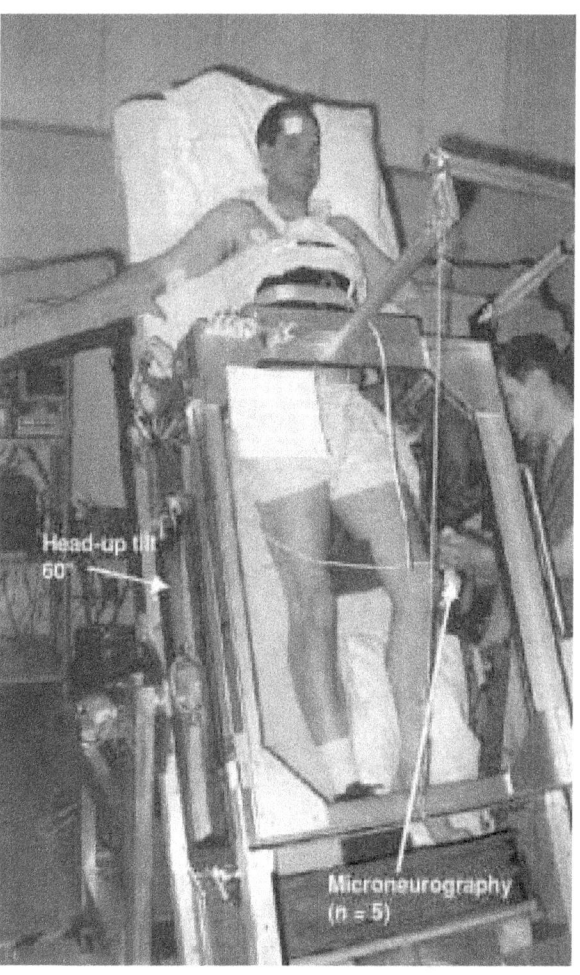

Figure 2. This figure shows the test apparatus in Figure 1 tilted upright for the 60-degree upright tilt measurements. Microneurography is being performed through the side window in the LBNP device.

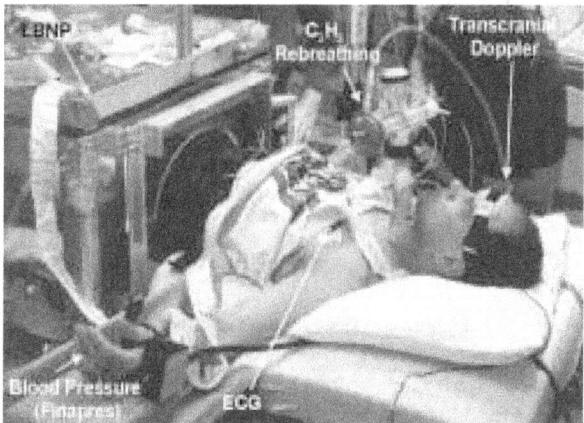

Figure 1. Alternate payload specialist Alexander Dunlap is shown during a preflight lower body negative pressure (LBNP) testing session. The head strap holds the transcranial Doppler device that makes the brain blood flow measurements in place. Acetylene (C_2H_2) rebreathing is used to measure cardiac output. The finger blood pressure device and electrocardiogram (ECG) electrodes are also shown.

al. in this publication). This procedure moves blood from the chest to the lower body, as standing does, and stresses the cardiovascular control system. Beat-by-beat, arterial pressure was measured in the finger using photoplethysmography (Finapres, Ohmeda, Madison, WI); it also was measured intermittently in the brachial artery using a blood pressure cuff (SunTech, Raleigh, NC). Heart rate was measured with an ECG, and cardiac output was measured with the foreign gas C_2H_2 rebreathing technique. Sympathetic nerve activity was measured directly with high resolution, using microelectrodes placed in the efferent sympathetic nerves as the nerves passed by the knee using the microneurographic technique. Ertl et al. provide details of this technique in a technical report (see technical report by Ertl et. al. in this publication). A small, removable window in the LBNP box allowed access for microneurography, both supine and during upright tilt as shown in Figure 2.

The study of CBF in humans aboard the Space Shuttle required a technique that was safe, noninvasive, and allowed repeatable estimates of changes in flow on a beat-to-beat

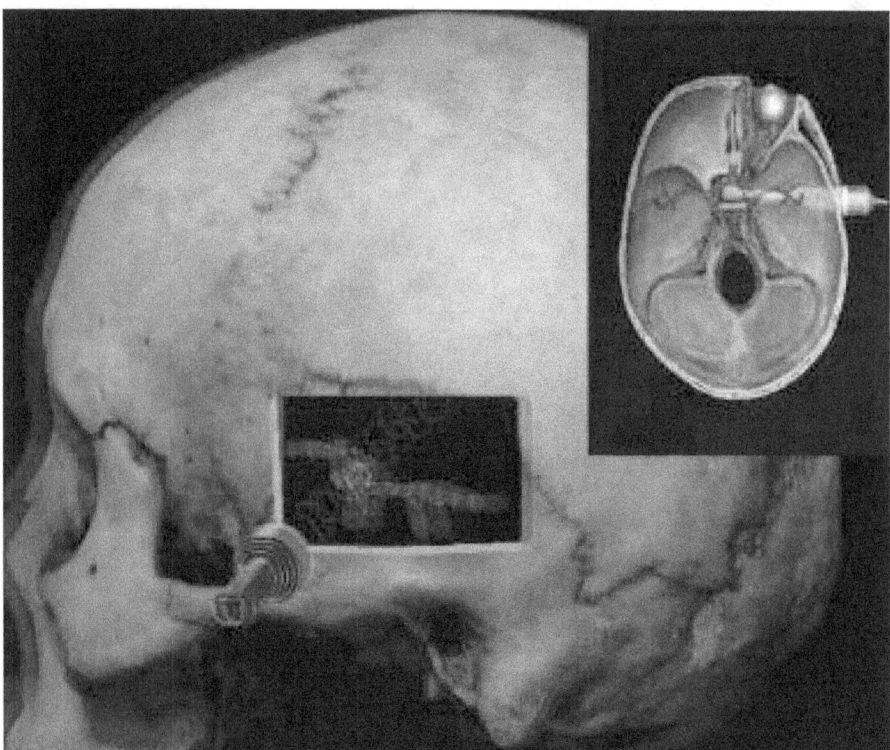

Figure 3. Transcranial Doppler. The middle cerebral arteries head toward the ears. The Doppler ultrasound travels through the skull and is partially reflected back by the blood flowing in the artery. The velocity of blood flow in the middle cerebral artery can be calculated from the reflected ultrasound. Used courtesy of Rune Aaslid.

basis. To meet these requirements, we used the transcranial Doppler technique that takes advantage of the ability of ultrasound to penetrate the skull at relatively low frequencies (2 mHz). This technique measures flow in the middle cerebral artery (Figure 3).

Because of its safety, ease of use, and ability to monitor rapid changes in CBF from velocity, transcranial Doppler has become a standard clinical tool for evaluating diseases of the cerebral circulation. By recording dynamic changes in cerebral blood flow velocity simultaneously with beat-by-beat changes in arterial blood pressure, we were able to use special mathematical techniques, called "spectral analysis," to evaluate the dynamic relationship between pressure and flow, known as "dynamic cerebral autoregulation."

By calculating the mathematical relationship between pressure and velocity, called the "transfer function," we were able to measure the strength of association (coherence) and the magnitude of association (gain) between blood pressure and cerebral blood flow at multiple frequencies. These served as indices of cerebral autoregulation. By these criteria, a high coherence means that pressure and velocity vary together very closely, implying relatively weak autoregulation. If coherence is low, the signals are independent of each other, suggesting that autoregulation is working well. In addition, when coherence is sufficiently high, the gain can be calculated. Under such circumstances, even with a high coherence, if gain is small, large changes in blood pressure lead to only small changes in CBF, implying effective autoregulation. Similarly, a large gain implies that large changes in pressure lead to similarly large changes in flow, and to relatively poor autoregulation.

An additional problem had to be overcome to measure CBF reliably—how to place the Doppler probe in exactly the same position both inflight and on the ground, with a technique that could be learned easily by nonexpert astronaut operators. To solve this problem, we used a polymer employed by dentists to take impressions of teeth. From this we made a mold of the lateral portion of the skull and ear that could hold the probe.

Protocol

We developed a protocol involving various stimuli that stressed different aspects of the blood pressure control system. The details of some of these, such as LBNP and the Valsalva maneuver, are described in detail in the science reports by Ertl et al. and Cox et al. in this publication. In this section, we will focus on three specific challenges: orthostatic stress, handgrip exercise, and cold stimulation.

To examine autonomic neural control in the upright posture, we studied our six astronaut subjects first lying flat, and then tilted upright to 60 degrees. The head-to-foot gravitational stress is equivalent to the sine of the tilt angle (90 degrees or standing upright=100% of Gz, the gravitational force directed toward the feet; 180 degrees or lying flat=0% of Gz). Therefore, 60 degrees allowed us to achieve nearly 90% of the Gz force, but still allowed the subjects to be lying quietly for careful study. Tilt studies were performed approximately two months before flight (with microneurography), two weeks before flight (without microneurography), and on landing day (with microneurography). During flight, orthostatic stress was produced using LBNP.

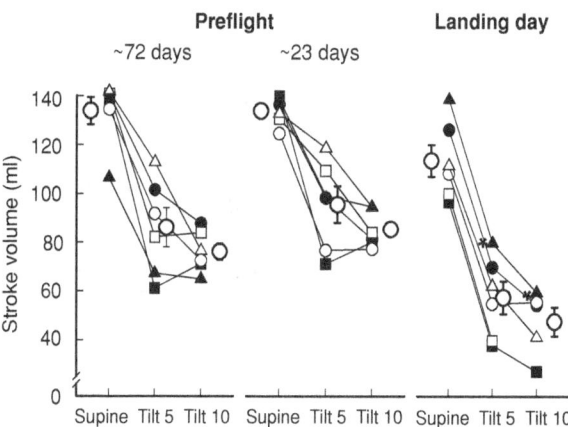

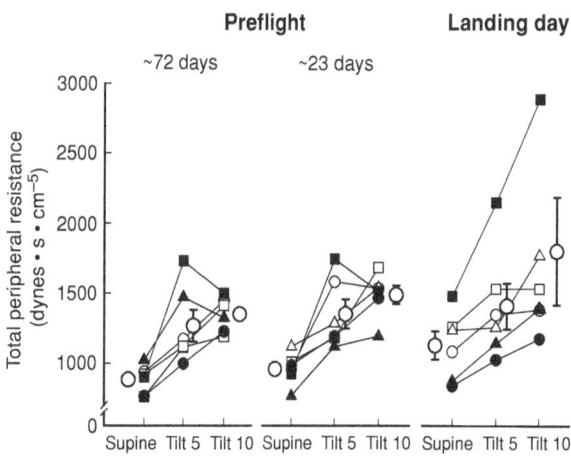

Figure 4. Stroke volume responses to 60-degree upright tilt. The graph shows average and individual stroke volumes in the supine position and after five and 10 minutes of 60-degree head-up tilt. Each astronaut is represented by a different symbol. An asterisk (*) shows a change that was significant (p<0.05) compared to the same time preflight. On landing day, supine stroke volumes were lower, and early and late stroke volume reductions during tilt were greater. (From Levine, 2002, with permission; reproduced from *The Journal of Physiology.*)

Figure 6. Calculated total peripheral resistance responses to 60-degree head-up tilt. Total peripheral resistance increased significantly when the crewmembers were tilted from supine to 60-degree upright positions both pre- and postflight. These increases were not significantly different on landing day. (From Levine, 2002, with permission; reproduced from *The Journal of Physiology.*)

The sympathetic nervous system is the primary mediator by which the cardiovascular response to exercise is controlled. During exercise, heart rate and blood pressure increase, and both cardiac output and muscle blood flow also increase to provide exercising muscle with the fuel for muscle contraction. This process is remarkably tightly regulated, and the study of the "exercise pressor reflex" is one of the classic tools in autonomic physiology. The control of the cardiovascular response to exercise involves the initiation of the reflex

response from higher-order centers in the brain (so-called "central command"), plus feedback signals from muscle, which sends information regarding the fatigue level or metabolic state ("metaboreceptors") and the strength ("mechanoreceptors") of muscle contraction. Because spaceflight also leads to muscle atrophy, some investigators have wondered whether signals generated from skeletal muscle could be impaired, leading to abnormal sympathetic nerve activity.

In 1937, an ingenious strategy was developed to isolate these signals during handgrip exercise. The investigators had their subjects squeeze a handgrip device at 30% of their maximal force, while recording heart rate and blood pressure (and more recently, by other investigators, sympathetic nerve activity). After two minutes, they inflated a cuff on the upper arm to very high levels (300 mmHg) designed to occlude blood flow to the limb and prevent the washout of metabolites within skeletal muscle that were stimulating muscle metaboreceptors. This is also called "post exercise circulatory arrest." After the cuff was inflated, the subjects stopped exercising. At that point, central command had ceased (there was no more effort to contract) and the muscle was relaxed, but the metaboreceptors were still being stimulated. This approach demonstrated that most of the increase in heart rate during such exercise was mediated by central command from the brain (McCloskey, 1972). However, virtually all of the increase in sympathetic nerve activity, and about half of the increase in blood pressure, was mediated from peripheral metaboreceptors.

In the Neurolab investigations, we used this protocol to determine: (a) if the pathways from the brain through the sympathetic nervous system were intact; (b) if they could be activated appropriately by exercising skeletal muscle; and (c)

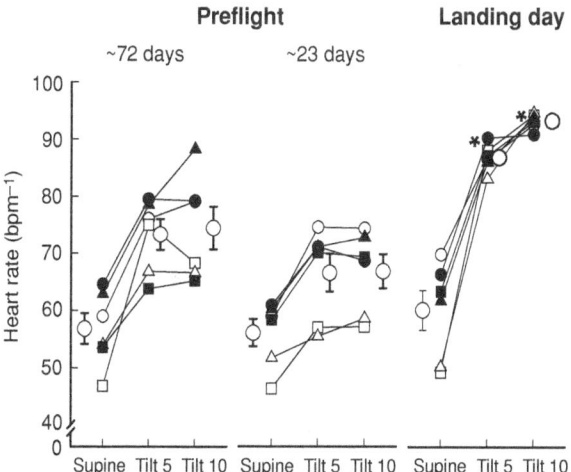

Figure 5. Heart rate responses to 60-degree head-up tilt. Heart rate increased to a significantly greater level during tilting after landing as compared to preflight. (From Levine, 2002, with permission; reproduced from *The Journal of Physiology.*)

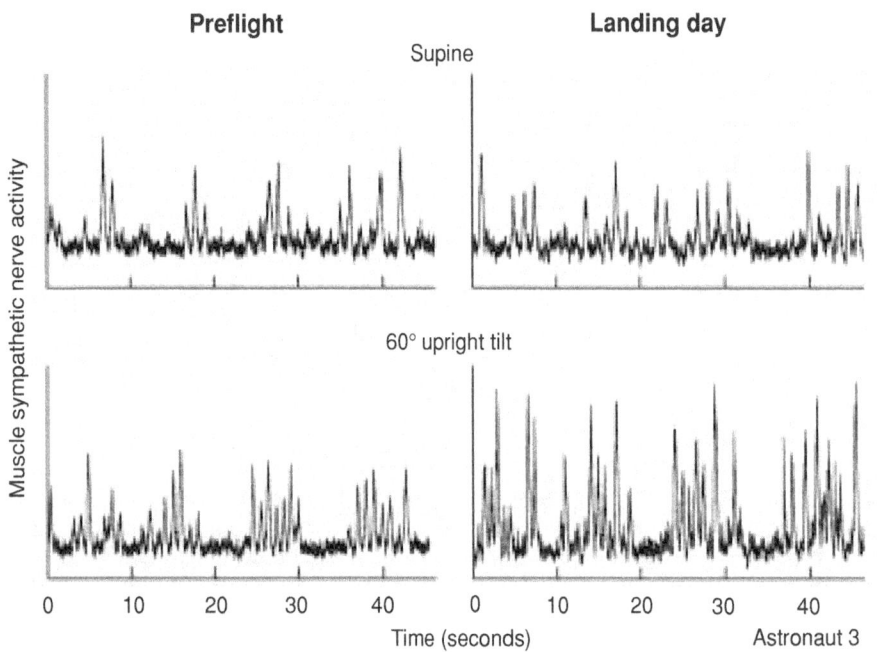

Preflight

Supine

Muscle sympathetic nerve activity

60° upright tilt

0   10   20   30   40

Landing day

0   10   20   30   40

Time (seconds)   Astronaut 3

Figure 7. Muscle sympathetic nerve responses of one astronaut. Sympathetic nerve activity increased from the supine to upright-tilt positions both pre- and postflight. The activity was greater postflight compared to preflight levels in both the supine and upright-tilt positions. (From Levine, 2002, with permission; reproduced from *The Journal of Physiology.*)

---

if they would raise the blood pressure similar to before spaceflight. We modified the classic protocol by continuing the handgrip exercise to fatigue to try to get the maximal possible stimulus to sympathetic activation possible, and to control for possible changes in muscle strength during spaceflight (Seals 1993). Handgrip studies were performed approximately two months before flight (with microneurography), two weeks before flight (without microneurography), towards the end of the flight on day 12 or 13, and on landing day (both with microneurography).

Finally, to make sure that the sympathetic nervous system could be activated by a stimulus that we didn't think would change during spaceflight, we chose to use another classic test, the "cold pressor" test, where the subject's hand is placed in ice water for two minutes. This test stimulates peripheral "nociceptors" and increases sympathetic nerve activity (thereby raising blood pressure) by a completely different set of afferent pathways. The test thus served as a control to make sure that the central and efferent pathways were intact and functioning normally. If orthostatic or handgrip responses were abnormal but cold pressor responses were normal, we could isolate the defect to the afferent, or sensing, side of the control system. If all three were abnormal, the defect would more likely be in the brain or the efferent sympathetic neural pathways. Since a bucket of ice would not be practical to use on the Space Shuttle, we used a specially designed mitt filled with a gel that allowed us to deliver a cold stimulus to the hand. Cold pressor tests were performed preflight and inflight along with the handgrip studies, but could not be performed on landing day because of time restrictions.

RESULTS

Hemodynamic measurements were performed on all six crewmembers during all pre- and postflight data collection sessions. Microneurographic recordings were performed on five of these six subjects. During flight, neither the commander nor the pilot was available for study, so only four subjects completed all the inflight experiments.

Stroke volumes (the amount of blood pumped with each heartbeat, Figure 4) in the supine position and stroke volume

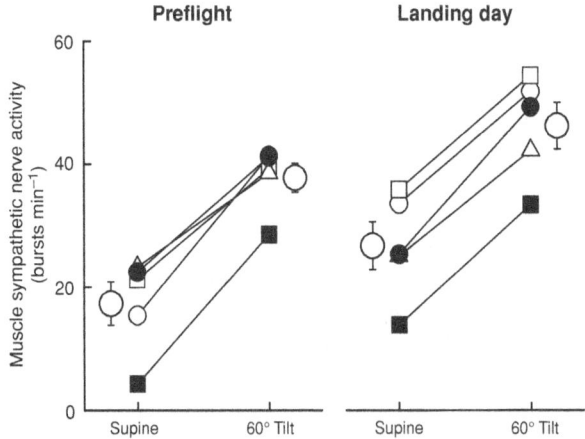

Preflight   **Landing day**

Muscle sympathetic nerve activity (bursts min⁻¹)

60

40

20

0

Supine   60° Tilt   Supine   60° Tilt

Figure 8. Muscle sympathetic nerve burst frequencies for all astronauts. Supine sympathetic nerve activity was significantly greater postflight. The increase in sympathetic nerve activity with tilting was comparable pre- and postflight, but carried sympathetic nerve activity to higher levels postflight. (From Levine, 2002, with permission; reproduced from *The Journal of Physiology.*)

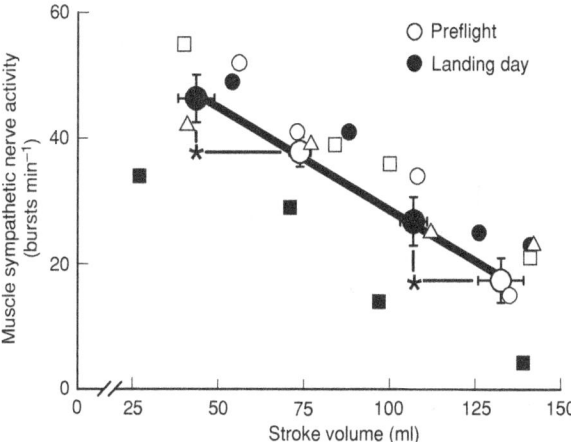

Figure 9. Muscle sympathetic nerve burst frequency plotted as a function of left ventricular stroke volume. The relationship between stroke volume and sympathetic nerve activity was similar both pre- and postflight. (From Levine, 2002, with permission; reproduced from the *Journal of Physiology.*)

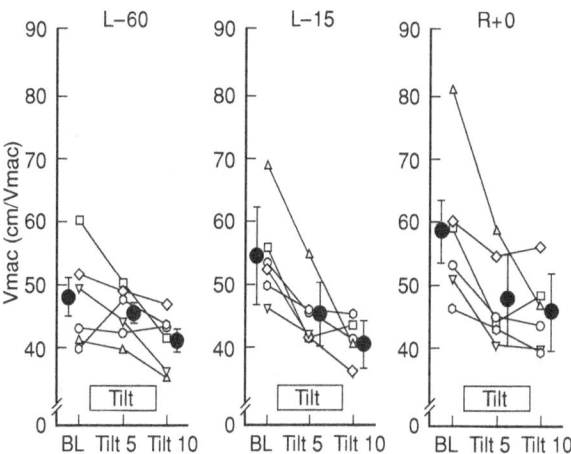

Figure 10. Cerebral blood flow responses to tilting. CBF fell slightly with tilt, but the change did not differ between pre- and postflight testing.

reductions during 60-degree upright tilt were remarkably similar during the two preflight sessions. Preflight stroke volumes decreased by about 40% after five minutes, and by 50% after 10 minutes of 60-degree upright tilt. On landing day, stroke volumes during supine rest were lower (p<0.05), and early and late stroke volume reductions during 60-degree upright tilt were greater than on either preflight day. Stroke volumes during postflight tilting were lower than those measured during either preflight session in all subjects (p<0.01).

Heart rate (Figure 5) during supine rest and heart rate increases during 60-degree upright tilt were similar in the two preflight sessions. The greater decreases of stroke volume registered during postflight tilting (Figure 4) were associated with greater increases of heart rate (p<0.01, compared with preflight levels). As a result, cardiac outputs, or the amount of blood pumped per minute (heart rate × stroke volume), during tilting were comparable during pre- and postflight sessions. Pre- and postflight cardiac outputs averaged 7.8+0.3 vs. 7.3+0.6 (p=0.54) in the supine position, and 5.6+0.3 vs. 4.9+0.6 (p=0.37) at 10-minute tilt.

Total peripheral resistance during supine rest and total peripheral resistance increases during 60-degree upright tilt (Figure 6) were similar during the preflight sessions. Peripheral resistance in the supine position was insignificantly greater on landing day than during preflight sessions. In contrast to the much greater increases of heart rate that occurred during tilting on landing day (Figure 5), increases of total peripheral resistance with tilting were not greater than those measured preflight (p=0.32). Thus, on landing day, greater stroke volume reductions were not matched by greater increases of total peripheral resistance. On landing day, five of the six astronauts had total peripheral resistance increases during tilting similar to those measured preflight. (In two astronauts, peripheral resistance increases were slightly greater; and in three astronauts, peripheral

resistance increases were slightly smaller.) Only one subject, a high-performance jet pilot, experienced a substantially greater increase of total peripheral resistance during tilting on landing day than during the preflight sessions (Figure 6, right panel, filled squares).

The net result of reduced stroke volumes, increased heart rates, and unchanged total peripheral resistances was that arterial pressures were preserved. There were no significant differences between supine and tilting measurements, or among measurements made during preflight and landing day sessions. Diastolic pressure increased significantly from supine levels during tilting

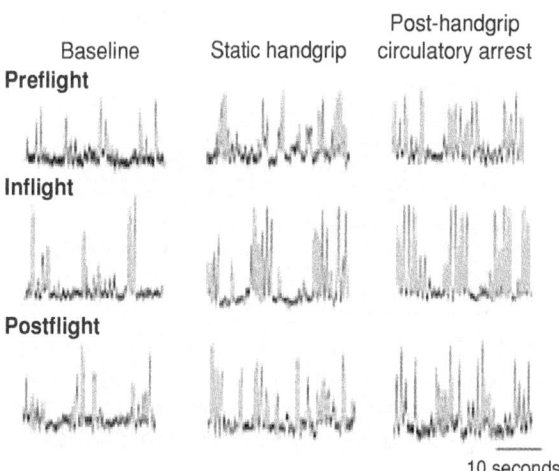

Figure 11. Muscle sympathetic nerve activity (MSNA) during handgrip. Sympathetic nerve activity was higher postflight in all subjects before static handgrip. Handgrip exercise produced the same peak increases in MSNA both during and after spaceflight. (From Fu, 2002, with permission; reproduced from *The Journal of Physiology.*)

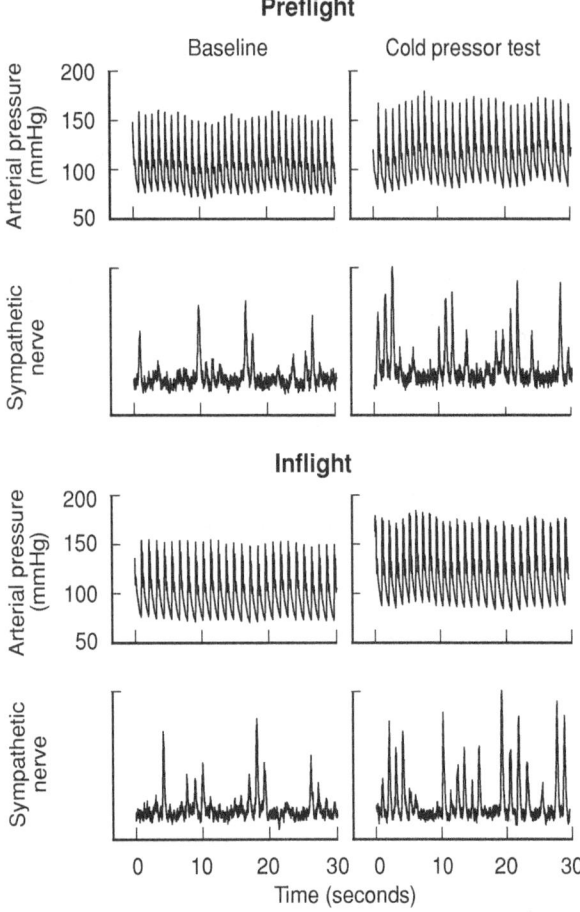

Preflight

Baseline | Cold pressor test

Inflight

Time (seconds)

Figure 12. Sympathetic nerve recordings from one crewmember during the cold pressor test. All subjects showed similar increases in MSNA and blood pressure during the cold pressor test preflight and inflight. This confirms the integrity of reflexes that increase in sympathetic nerve activity in response to cold stimulation. (From Fu, 2002, with permission; reproduced from *the Journal of Physiology.*)

($p<0.05$) in both preflight and landing day trials, with no significant differences among them. Postflight, most individual recordings showed large respiratory oscillations of arterial pressure during tilting that were not observed preflight. Systolic pressure standard deviations, used as indexes of this variability, did not change significantly from supine to tilting preflight ($7.3+1.2$ vs. $8.2+1.6$ mmHg, $p=0.11$), but they increased significantly, by >50%, on landing day ($6.7+0.8$ vs. $10.1+2.1$ mmHg, $p<0.01$). Increases of systolic pressure standard deviations during tilting on landing day were significantly greater than those measured during preflight sessions ($p<0.001$).

Sympathetic neural responses – Figure 7 shows muscle sympathetic nerve activity of a representative crewmember. Preflight recordings (left) document typical pulse-synchronous sympathetic bursting during supine rest and increased burst frequency during tilting. The postflight supine recording (upper right panel) is scaled so that its mean burst height is

equivalent to the mean preflight burst height. The vertical scale is the same in postflight supine and upright recordings, and faithfully indicates the changes of burst amplitude that occurred with tilting. Figure 7 shows increases of sympathetic nerve activity when the subject changed from supine to upright-tilt positions, both pre- and postflight, and higher levels of sympathetic nerve activity postflight compared to preflight levels in both supine and upright-tilt positions.

Figure 8 depicts individual and mean measurements of muscle sympathetic nerve burst frequency from all subjects. Increases of muscle sympathetic nerve activity provoked by 60-degree upright tilt were remarkably consistent among astronauts during both pre- and postflight sessions. However, supine muscle sympathetic nerve burst frequency was significantly greater in postflight than in preflight sessions ($p<0.05$). Therefore, increases of muscle sympathetic nerve activity with tilting, which were comparable both pre- and postflight, carried muscle sympathetic nerve burst frequency to higher levels postflight.

As noted above, steady-state arterial pressures were similar in supine and tilted positions, both before and after spaceflight. Therefore, arterial pressure measurements provided no evidence for the changes of baroreceptor activity, which must have occurred during pre- and postflight tilting, and postflight hypovolemia (dehydration). Left ventricular stroke volumes, on the other hand, did change as expected. Figure 9 depicts average and individual pre- and postflight muscle sympathetic nerve activity as a function of left ventricular stroke volume. Linear regression coefficients were extremely high, both for individual astronauts ($R^2=0.91–1.00$) and mean data (heavy line, $R^2=0.99$, $p<0.01$).

Stroke volumes were largest, and muscle sympathetic nerve activities were smallest, during preflight supine rest (Figure 9, extreme right). Stroke volumes and muscle sympathetic nerve activities during postflight supine rest fell about halfway between preflight supine and upright values. Stroke volumes were lowest, and muscle sympathetic nerve activity was highest, during postflight tilting (Figure 9, extreme left).

CBF responses to tilting are shown in Figure 10. There was a small, but statistically significant decrease in CBF in the tilted position that did not differ during both preflight sessions. Inflight, CBF was well maintained during low-level LBNP. In fact, small decreases in CBF velocity, which had been observed during LBNP of –30 mmHg during both preflight sessions, were not observed inflight. On landing day, this response was similar to preflight. No astronaut had a reduction in CBF despite maintenance of arterial pressure. In keeping with this preservation of CBF velocity during LBNP or tilting, spectral analysis showed that coherence between blood pressure and CBF velocity was unchanged at all frequencies. Moreover, the gain between pressure and flow at low frequencies, where autoregulation is most effective, was decreased significantly by 26% (FD6–FD7), 23% (FD12–FD13), and 27% (on landing day) as compared with the preflight value of 1.11 cm/s/mmHg ($p<0.05$). In other words, smaller oscillations in CBF occurred for a given change in blood pressure, which is suggestive of improved rather than impaired autoregulation.

Clinical outcome in response to tilt – All subjects completed the entire protocol, and were able to remain upright at 60 degrees for 10 minutes both before and after spaceflight.

Handgrip responses – Microneurograms representing sympathetic nerve activity at baseline, during fatiguing handgrip exercise, and during post-handgrip circulatory arrest, from one astronaut are shown in Figure 11.

For all subjects, the contraction-induced rises in heart rate were similar among preflight, inflight, and postflight conditions. MSNA was higher postflight in all subjects before static handgrip (26±4 postflight vs. 15±4 bursts min^{-1} preflight, p=0.017). The contraction-evoked peak increases in MSNA were not different before, during, or after spaceflight (41±4, 38±5, and 46±6 bursts min^{-1}, respectively, all p>0.05). MSNA during post-handgrip circulatory arrest was higher postflight than it was preflight or inflight (41±1 postflight vs. 33±3 preflight and 30±5 bursts min^{-1} inflight, p=0.038 and 0.036, respectively). These data demonstrate that fatiguing handgrip exercise elicits the same peak increases in MSNA, blood pressure, and heart rate during and after 16 days in space, with no evidence for impairment in central command or reflex stimulation of peripheral mechano- or metaboreceptors.

Cold pressor test – Sympathetic nerve recordings from one representative subject before and during spaceflight are shown in Figure 12. For all subjects, similar increases in MSNA and blood pressure during the cold pressor test were observed preflight and inflight, thereby confirming the integrity of peripherally stimulated reflex increases in sympathetic nerve activity and vasomotor responsiveness. Together, both handgrip and cold pressor data confirm that stimulation of muscle or other nociceptive afferent receptors and their reflex responses are not impaired by short-duration spaceflight.

DISCUSSION

Blood Pressure Control in the Upright Position: A Problem with Stroke Volume

This experiment demonstrated a number of important findings —some expected, and others unexpected. As has been demonstrated previously, we confirmed that the key cardiovascular adaptation to spaceflight was a reduction in upright stroke volume. The astronauts lost blood volume while in space (i.e., they became effectively "dehydrated"), and their hearts may have become smaller and less distensible (Levine, 1997; Perhonen, 2001a; Perhonen, 2001b). So when crewmembers stood up after being in space, less blood was left in their hearts to pump compared to preflight. In this case, the heart behaves somewhat like a rubber band—the more it stretches, the more it snaps back during pumping. Similarly, the less it stretches, the less blood it can pump. This reduction in upright stroke volume appears to be the primary specific effect of gravity (or its prolonged absence) on the cardiovascular system.

When thinking about the processes regulating blood pressure, it may be helpful to consider the circulation as an electrical circuit that obeys a form of Ohm's law (V=IR, or voltage = current×resistance), where blood pressure is the "voltage,"

blood flow is the "current," and total peripheral resistance (TPR) is the "resistance." In this model, BP=Qc×TPR, or blood pressure = cardiac output (blood flow in liters/min)×total peripheral resistance. Since cardiac output=heart rate (beats/min)×stroke volume (mL/beat), the blood pressure can be thought of as the "triple product" of heart rate×stroke volume×total peripheral resistance (Levine, 1991). During changes in gravitational stress, such as moving from the supine to the upright position, stroke volume changes acutely, with less blood pumped per beat into the arterial tree and less pulsatile distension of arterial baroreceptors. The cardiovascular control system responds to this challenge by increasing the heart rate and/or the TPR. A useful analogy might be a firefighter trying to get water to the roof of a house that is on fire. The problem is that gravity is pulling the water to the ground (akin to a low stroke volume). The firefighter can then either turn the pump up faster (similar to increasing the heart rate), or the firefighter can place a finger over the end of the hose (similar to constricting the blood vessels and raising the peripheral resistance).

Reflex Responses to Upright Posture

If after spaceflight stroke volume was reduced by gravity pulling blood into the feet, how did the cardiovascular control system respond to this challenge? Although some investigators have demonstrated alterations in the baroreflex control of heart rate during or after spaceflight (see science report by Cox et al., in this publication), heart rate increased appropriately for the reduction in stroke volume in every astronaut. Thus, a low heart rate, by itself, cannot be responsible for the orthostatic hypotension observed after spaceflight.

Previous work has focused attention on the sympathetic nervous system and regulation of the vascular resistance as a potential mediator of postflight orthostatic intolerance. During SLS-1, SLS-2, and Spacelab-D2, Buckey and colleagues (Buckey, 1996) demonstrated that approximately two-thirds of the astronauts studied could not stand quietly for 10 minutes following spaceflight of one to two weeks, whereas they could all complete the test before flight. Afterwards, those astronauts with the best orthostatic tolerance were able to raise their vascular resistance to greater levels than they were preflight, thus compensating well for the reduction in stroke volume. However, the astronauts with the poorest orthostatic tolerance couldn't increase the vascular resistance during standing more than they did preflight.

Subsequently, Fritsch-Yelle et al. (Fritsch-Yelle, 1996) studied a larger number of astronauts and showed that in this series, about 25% of the astronauts couldn't stand quietly for 10 minutes. These investigators also measured plasma levels of norepinephrine, the neurotransmitter released from the sympathetic nerve endings that causes blood vessels to constrict. Fritsch-Yelle et al. found that the astronauts with the worst orthostatic tolerance had lower levels of norepinephrine in their blood than those who tolerated standing best. However, there were two important caveats to these studies. First, the more tolerant astronauts had their blood drawn while they were still standing, and the sympathetic nervous system was still highly activated. Because they were feeling faint, the less-tolerant astronauts had their blood drawn only after they were placed in a supine position. Since the autonomic nervous system responds very rapidly to changes in body position, this protocol could have biased the results toward lower values in the less-tolerant astronauts. Second, most of the time people who faint do so because the cardiovascular reflexes "make a mistake"—that is, the sympathetic nervous system withdraws, so that the blood vessels are no longer constricted (similar to taking the finger off the end of the hose in the firefighter example). Such a sympathetic withdrawal or fainting reaction could reduce the levels of norepinephrine and the peripheral resistance in orthostatically intolerant astronauts, even if it had increased appropriately at the beginning of the stand.

These experiments set the stage for the Neurolab studies of the sympathetic nervous system in the upright posture after spaceflight. Both studies raised the possibility that the cardiovascular adaptation to spaceflight might result in a defect in the sympathetic nervous system that would prevent astronauts from increasing sympathetic activity in the upright position, impairing their ability to raise vascular resistance and compensate for the reduction in upright stroke volume. However, the results from Neurolab were surprising, yet convincing. All of the astronauts studied in these experiments had an increase in sympathetic nerve activity during upright tilting on Earth after two weeks of spaceflight. Moreover, this increase in sympathetic nerve activity was perfectly well calibrated for the reduction in stroke volume induced by the upright posture (Figure 9). Finally, the results from stimulating the sympathetic nervous system by other pathways—handgrip exercise (stimulating muscle metaboreceptors) or cold stress (stimulating peripheral nociceptors)—were also entirely normal during and after spaceflight. From these experiments, we can say convincingly that the cardiovascular adaptation to spaceflight does NOT lead to a defect in the autonomic regulation of sympathetic nerve activity.

Cerebral Autoregulation

But, what about the brain? Even though blood pressure was well-regulated in these studies, was there evidence that cerebral autoregulation was impaired? Ground-based experiments in support of these studies suggested that this might be the case, and animal studies have confirmed changes in the cerebral blood vessels with simulated microgravity. In the Neurolab experiments, however, CBF was precisely regulated under all conditions. There was no evidence of reduced CBF during upright tilt in any astronaut. Moreover, during the inflight experiments using LBNP, the greater orthostatic stress inflight was accompanied by normal or even enhanced cerebral autoregulation, thereby preserving CBF. Using sophisticated frequency domain measures of autoregulation, we were unable to demonstrate impaired cerebral autoregulation at any frequency. Rather at the lower frequencies, cerebral autoregulation appeared to be improved, not impaired, during and immediately following spaceflight. The mechanism of this improvement is speculative, but it may result from hypertrophy of cerebral blood vessels due to a persistent central fluid shift.

Orthostatic Intolerance After Spaceflight

So, if the neural control system is not impaired and cerebral autoregulation is preserved, why then do some astronauts faint after spaceflight? Unfortunately, we cannot answer that question definitively. One of the problems with interpreting our experiment in this regard was that no astronaut had frank orthostatic intolerance during testing on landing day. We suspect that at least some of this problem may have been related to the specific experimental conditions. Astronauts were only tilted to 60 degrees upright to allow microneurography, instead of standing fully upright as in previous experiments. Moreover, our subject numbers were small, raising the possibility of sampling bias. However, all evidence points to the fact that the Neurolab crew behaved similarly and had a similar cardiovascular adaptation to spaceflight as did other, previously evaluated astronauts. Their upright stroke volumes were very low, and their heart rates were high. Their mean blood pressures were stable when tilted upright, but we observed prominent oscillations in blood pressure around this mean. This is typical of other astronauts, and indicative of individuals whose blood pressure control system is being severely challenged. Thus we suspect, but cannot prove that, even if we had studied a much larger group of astronauts, the results would have been the same. Failure to augment sympathetic nerve activity properly is not likely to be an important mechanism in this condition.

Some additional insight may be gained from examining the changes in total peripheral resistance after flight. Despite the increased sympathetic nerve activity, most astronauts did not increase their vascular resistance to a greater degree than they did preflight. This observation suggests that these may be a limited "vasomotor reserve" in many individuals, such that more sympathetic nerve activity does not necessarily result in greater vascular resistance. To continue our firefighter analogy—the finger may be already compressed over the edge of the hose as tightly as it can be, and squeezing tighter won't help anymore. Some investigators have suggested that such a limited vasomotor reserve may be responsible for some patients who have orthostatic intolerance in everyday life. Thus, the examination of astronauts who lose the ability to tolerate gravity, after brief periods without gravity in space, may provide important insights into patients with orthostatic intolerance on Earth.

Acknowledgements

In a complex spaceflight experiment such as this one, there are so many individuals who contributed substantively to the success of the project that it is extremely difficult to acknowledge them all. At a minimum, the crew of STS-90 must be acknowledged for an outstanding effort as subjects and experimentalists inflight. Special acknowledgement must also be made of the efforts of Mike Grande and Stuart Johnston from NASA/Lockheed Martin, and Matt Morrow, Troy Todd, and Dak Quarles from Presbyterian Hospital of Dallas who put in effort above and beyond the call of duty to ensure the success of the project. In addition, Mike Smith, Satoshi Iwase, Mitsuro

Saito, and Yoshiki Sugiyama provided their expertise with microneurography for the postflight experiments. This project was supported by grants from NASA (NAS9-19540) and the NIH (HL53206).

REFERENCES

REFLEX CARDIOVASCULAR AND RESPIRATORY RESPONSES ORIGINATING IN EXERCISING MUSCLE. D.I. McCloskey and J.H. Mitchell. *J. Physiol.,* Vol. 224, pages 173–186; 1972.

PHYSICAL FITNESS AND CARDIOVASCULAR REGULATION: MECHANISMS OF ORTHOSTATIC INTOLERANCE. B.D. Levine, J.C. Buckey, J.M. Fritsch, C.W. Yancy, Jr., D.E. Watenpaugh, P.G. Snell, L.D. Lane, D. L. Eckberg, C.G. Blomqvist. *J. Appl. Physiol.,* Vol. 70, pages 112–122; 1991.

INFLUENCE OF FORCE ON MUSCLE AND SKIN SYMPATHETIC NERVE ACTIVITY DURING SUSTAINED ISOMETRIC CONTRACTIONS IN HUMANS. D. R. Seals. *J.. Physiol.,* Vol. 462, pages 147–159; 1993.

ORTHOSTATIC INTOLERANCE AFTER SPACEFLIGHT. J.C. Buckey, Jr., L.D. Lane, B.D. Levine, D.E. Watenpaugh, S.J. Wright, W.E. Moore, F.A. Gaffney, and C.G. Blomqvist. *J. Appl. Physiol.,* Vol. 81, pages 7–18; 1996.

SUBNORMAL NOREPINEPHRINE RELEASE RELATES TO PRESYNCOPE IN ASTRONAUTS AFTER SPACEFLIGHT. J.M. Fritsch-Yelle, P.A. Whitson, R.L. Bondar, and T.E. Brown. *J. Appl. Physiol.,* Vol. 81, pages 2134–2141; 1996.

CARDIAC ATROPHY AFTER BED-REST DECONDITIONING: A NONNEURAL MECHANISM FOR ORTHOSTATIC INTOLERANCE. B.D. Levine, J.H. Zuckerman, J.A. Pawelcyzk. *Circulation,* Vol. 96, pages 517–525; 1997.

DETERIORATION OF LEFT VENTRICULAR CHAMBER PERFORMANCE AFTER BED REST: "CARDIOVASCULAR DECONDITIONING" OR HYPOVOLEMIA? M.A. Perhonen, J.H. Zuckerman, B.D. Levine, *Circulation,* Vol. 103(14), pages 1851–1857; 2001a.

CARDIAC ATROPHY AFTER BED REST AND SPACEFLIGHT. M.A. Perhonen, F. Franco, L.D. Lane, J.C. Buckey, Jr., C.G. Blomqvist, J.E. Zerwekh, R.M. Peshock, P.T. Weatherall, B.D. Levine. *J. Appl. Physiol.,* Vol. 91(2), pages 645–653; 2001b.

HUMAN MUSCLE SYMPATHETIC NEURAL AND HAEMODYNAMIC RESPONSES TO TILT FOLLOWING SPACEFLIGHT. B.D. Levine, J.A. Pawelczyk, A.C. Ertl, J.F. Cox, J.H. Zuckerman, A. Diedrich, I. Biaggioni, C.A. Ray, M.L. Smith, S. Iwase, M. Saito, Y. Sugiyama, T. Mano, R. Zhang, K. Iwasaki, L.D. Lane, J.C. Buckey, Jr., W.H. Cooke, F.J. Baisch, D. Robertson, D.L. Eckberg, and C.G. Blomqvist. *J. Physiol.,* Vol. 538, pages 331–340; 2002.

CARDIOVASCULAR AND SYMPATHETIC NEURAL RESPONSES TO HANDGRIP AND COLD PRESSOR STIMULI BEFORE, AND AFTER SPACEFLIGHT. Q. Fu, B. D. Levine, J. A. Pawelczyk, A.C. Ertl, A. Diedrich, J. F. Cox, J. H. Zuckerman, C. A. Ray, M. L. Smith, S. Iwase, M. Saito, Y. Sugiyama, T. Mano, R. Zhang, K. Iwasaki, L. D. Lane, J. C. Buckey Jr, W.H. Cooke, R. M. Robertson, F. J. Baisch, C. G. Blomqvist, D. L. Eckberg, D. Robertson and I. Biaggioni. *J. Physiol.;* Vol. 544.2, pages 653-664, 2002.

Influence of Microgravity on Arterial Baroreflex Responses Triggered by Valsalva's Maneuver

Authors

James F. Cox, Kari U.O. Tahvanainen,
Tom A. Kuusela, William H. Cooke,
Jimey E. Ames, Dwain L. Eckberg

Experiment Team

Principal Investigator: **Dwain L. Eckberg**[1]

Co-Investigators: **James F. Cox,**[1] **Kari U.O. Tahvanainen,**[2] **Tom A. Kuusela,**[3] **William H. Cooke,**[4] **Jimey E. Ames**[1]

[1]Hunter Holmes McGuire Department of Veterans Affairs and Medical College of Virginia at Virginia Commonwealth University, Richmond, USA
[2]Kuopio University Hospital, Kuopio, Finland
[3]University of Turku, Turku, Finland
[4]Michigan Technological University, Houghton, USA

ABSTRACT

When astronauts return to Earth and stand upright, their heart rates may speed inordinately, their blood pressures may fall, and some returning astronauts may even faint. Since physiological adjustments to standing are mediated importantly by pressure-regulating reflexes (*baroreflexes*), we studied involuntary (or *autonomic*) nerve and blood pressure responses of astronauts to four, 15-second periods of 15- and 30-mmHg straining (*Valsalva's maneuver*). We measured the electrocardiogram, finger blood pressure, respiration, and muscle sympathetic nerve activity in four healthy male astronauts before and during the 16-day Neurolab Space Shuttle mission. We found that although microgravity provoked major autonomic changes, no astronaut experienced fainting symptoms after the mission. Blood pressure fell more during straining in space than on Earth (the average reduction of systolic pressure with 30-mmHg straining was 49 mmHg during and 27 mmHg before the mission). However, the increases of muscle sympathetic nerve activity that were triggered by straining were also larger in space than on Earth. As a result, the *gain* of the sympathetic baroreflex, taken as the total sympathetic nerve response divided by the maximum pressure reduction during straining, was the same in space as on Earth. In contrast, heart rate changes, which are mediated by changes of *vagus* nerve activity, were smaller in space. This and earlier research suggest that exposure to microgravity augments blood pressure and sympathetic adjustments to Valsalva straining and differentially reduces vagal, but not sympathetic baroreflex responsiveness. The changes that we documented can be explained economically as a consequence of the blood volume reduction that occurs in space.

INTRODUCTION

All terrestrial life forms have to contend with gravity. Some of gravity's effects are readily observed. Others are not apparent but, nonetheless, important. For example, in animals, gravity pulls blood toward the Earth. Such effects are posture-specific: they are maximal during upright standing and minimal during recumbency. When humans stand upright, the shift of blood to the lower body could cause blood pressure to fall markedly. People are usually unaware of gravitational pull because their blood pressure-regulating reflexes (*baroreflexes*) respond to the challenge by increasing their heart rates and blood vessel tone. Such adjustments are involuntary, and are made by the *autonomic nervous system*.

At the beginning of the space exploration era, scientists were concerned that microgravity would overwhelm the circulatory system. Their concern proved unfounded. Cardiovascular function is normal in space despite the major gravitational changes that occur. Interestingly, the most dramatic consequences of spaceflight are observed *after*, not *during* space missions, when astronauts stand upright. In many astronauts, standing causes their hearts to beat very rapidly and their blood pressures to fall. In some returning astronauts, these changes are quite pronounced—some feel lightheaded, as if they are going to pass out, a condition known as *pre-syncope*; and some actually faint (*syncope*). Eventually (within days), symptoms during standing disappear, and astronauts re-adapt to living on Earth. Presumably, these symptoms have their origins in physiological changes occurring during spaceflight.

Mechanisms responsible for the cardiovascular changes that take place in microgravity are not understood fully (Blomqvist, 1983; Fortney, 1996). However, many sophisticated measurements have been made in astronauts, and much is known. Published research indicates clearly that human cardiovascular function in space is not fixed, and that measurements made during the earliest days of space missions are very different than those made later. For example, there is an immediate shift of fluids from the lowest parts of the body toward the head. These changes are responsible for astronauts' "moon faces" and "bird legs" in space. Associated with the early headward fluid and blood shifts, heart chamber sizes increase. However, after the first few days of spaceflight, blood volume declines by about 10% from measurements made on Earth. Heart chambers shrink and are smaller than they were (in the supine position) before space travel. As might be expected, the body takes note of these changes and makes adjustments, by altering autonomic nerve traffic to vital organs.

The autonomic nervous system

Autonomic nerve traffic from the central nervous system is carried by the left and right *vagus nerves* and by many *sympathetic nerves*. The principal cardiovascular target of the vagus (or *parasympathetic*) branch of the autonomic nervous system is the pacemaker of the heart, the sinus node. Increases of vagus nerve traffic increase concentrations of the neurotransmitter, *acetylcholine*, at the sinus node and slow the heart

rate. Conversely, reductions of vagus nerve traffic allow more acetylcholine to be taken back into the vagus nerve endings, and the heart rate speeds. The vagus nerve does not control blood pressure directly, but it does modulate blood pressure by changing heart rate.

Sympathetic nerves directly influence both the cardiac pacemaker and the blood pressure. Sympathetic nerve traffic tends to fluctuate reciprocally with vagus nerve traffic; and the effects of the sympathetic neurotransmitter, *norepinephrine*, tend to be opposite those of the parasympathetic neurotransmitter, acetylcholine. Increases of sympathetic nerve traffic to the sinus node speed the heart; and increases of sympathetic nerve traffic to the small resistance blood vessels, the *arterioles*, constrict their muscular walls and increase blood pressure.

Traffic carried over the two branches of the autonomic nervous system is modulated according to signals coming to the brain from sensors located throughout the body. Arterial *baroreceptors* are particularly important mediators of responses to standing. Baroreceptors are nerves with stretch-sensitive endings located in the walls of the *carotid arteries* in the neck, and in the aorta, (the largest artery in the body) in the chest. Consider how the baroreflex responds to a simple challenge—standing. Within a few seconds of standing, blood pressure falls. Reductions of pressure within arteries reduce the stretch of the artery walls, and the arteries shrink. Shrinkage of the carotid arteries and aorta reduces the stretch on baroreceptive nerves, and they reduce their level of firing. Reduced baroreceptor traffic is sensed in an area of the brain stem known as the *solitary tract nucleus*. The information that arrives in the solitary tract nucleus—that baroreceptor traffic has fallen off—is carried by a bucket brigade of intermediary neurons that connect, or *synapse*, with each other, and eventually arrives at the *motor neurons* that are responsible for vagus and sympathetic nerve firing.

The reduction of baroreceptor nerve traffic is interpreted, appropriately enough, as a reduction of blood pressure. The vagus and sympathetic motor neurons respond in opposite ways to the change of baroreceptor traffic, which they both see. Vagus firing declines, less acetylcholine is released at the sinus node, and the heart rate speeds. This faster heart rate translates into more strokes of blood being ejected into the circulation per minute, and this helps to restore blood pressure toward usual levels. Sympathetic motor neurons take the opposite tack. Sympathetic nerve traffic increases, and this speeds the heart and constricts the blood vessels. The sympathetic changes complement the faster heart rate that results from reduced vagus nerve traffic. Both sets of responses tend to restore blood pressure toward usual levels. The two systems work together seamlessly, and the person who stands may not even be aware that the circulation has been challenged.

Autonomic responses to space travel

Prior to the Neurolab mission, research in astronauts focused on the vagus limb of baroreflex responses. There was a practical reason for this: changes of vagus nerve traffic to the sinus node can be inferred from changes of heart rate measured from the electrocardiogram, which can be recorded safely anywhere.

In the earlier research, baroreceptor input to the brain was changed by a mechanical device that tricked the baroreflex. As we mentioned, the carotid arteries have neurons embedded in their walls that sense the degree of distension. When pressure inside a carotid artery increases, the artery stretches, the level of neuron firing increases, and a reflex adjustment is made. As would be predicted, the carotid arteries stretch when pressures inside them increase. The carotid arteries also stretch when pressures *outside* them *decrease*. Astronauts' baroreflexes were tricked by applying a vacuum to the neck collars that they were wearing. Thus, the usual pressure inside the carotid arteries pushed against a vacuum outside the arteries. This caused the carotid arteries to stretch, the baroreceptor nerves to increase firing, and the heart rate to slow. Thus, in earlier research, scientists merely recorded the degree of heart rate slowing with the electrocardiogram, and related the slowing to the neck pressure of the moment. The conclusion from this research was that in space, *vagal* baroreflex function is impaired. For the same degree of neck pressure reduction, the heart rate slowing is less during (Fritsch, 1992b) or immediately after spaceflight (Fritsch, 1992a; Fritsch-Yelle, 1994) than it was before spaceflight.

Does this mean that since vagal baroreflexes are impaired in space, *sympathetic* baroreflex responses are also impaired? This was the question asked and answered during the Neurolab mission. The challenge was to measure sympathetic nerve traffic. As mentioned, changes of vagus nerve traffic can be inferred from changes of the period between heartbeats on the electrocardiogram. However, there is no comparable indirect way to measure sympathetic nerve traffic. To have an indication of the heartbeat-by-heartbeat changes of sympathetic nerve traffic, it is necessary to measure sympathetic nerve electrical activity *directly*. Sympathetic *microneurography* was developed in the 1970's by two Swedish neurologists, Hagbarth and Vallbo, who fashioned minute needles, inserted the needles through the skin into their own nerves, and led off and amplified the electrical activity that was present in those nerves. This method is described in the technical report in this volume by Ertl et al.

For Neurolab research, we perturbed the baroreflex system simply, by asking astronauts to perform *Valsalva's maneuver* (Smith, 1996). The Italian scientist, Antonio Maria Valsalva, published a book on anatomy in 1704, *The Human Ear*. Buried in that Latin text is one phrase that has, for almost three centuries, maintained Valsalva's position in the firmament of medical science: *"... inflation of the middle ear by closing the mouth and nostrils and blowing, so as to puff out the cheeks."* Valsalva's maneuver is nothing more than straining against a closed airway, as would occur when lifting a heavy book. This is done, however, in a highly controlled manner.

METHODS

Valsalva's maneuver

Astronauts' workdays during the Neurolab mission were filled to overflowing. Therefore, it was necessary that each task be presented simply and clearly. As with any kind of scientific

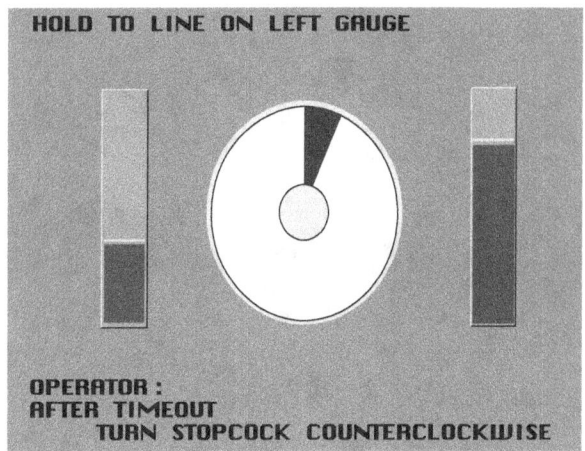

Figure 1. Laptop computer screen used to guide astronauts during Valsalva straining.

experiment, it was important that whatever the astronauts were asked to do, they must do it precisely and *reproducibly*. Therefore, we developed a computer program that guided astronauts through their Valsalva maneuvers. One screen that the astronauts saw (on laptop computers positioned in front of them) is shown in Figure 1.

Over a period of eight minutes, astronauts strained four times in precisely controlled ways. Before each period of straining, they breathed through a mouthpiece, in synchrony with a tone that sounded every four seconds, to a depth shown by the "gas gauge" displayed in Figure 1, right. Then, on cue from the computer, they were asked to strain and to increase the pressure in their mouthpieces to match the height of the gauge on the left. They performed four Valsalva maneuvers and strained to a level of either 15 or 30 millimeters of mercury (mmHg) in a random order that was chosen by the computer. The "clock," shown in the center of Figure 1, appeared as soon as astronauts began straining. As they strained, they watched a "second hand," shown as the blue wedge, fill the circle, rotating clockwise. The second hand made one complete revolution in 15 seconds. As soon as the second hand had reached 12 o'clock, the astronaut stopped straining and resumed breathing as before.

Thus, before and during the Neurolab mission, each of four astronauts strained four times to intensities of either 15 or 30 mmHg (two each), for 15 seconds, in random order. Figure 2 shows one astronaut holding a mouthpiece as he prepares to perform Valsalva maneuvers. This astronaut is barely discernible because of the wealth of equipment surrounding him. A second astronaut, who is readily seen, is watching over the experiment. The astronaut who is to perform the Valsalva maneuvers is lying supine, with his lower body enclosed in an airtight chamber that was used for a different experiment. The laptop computer that guided him during actual measurements is not shown. Figure 2 gives some indication that although it is a simple matter for astronauts to strain four times for 15 seconds each, preparations for such challenges during the Neurolab mission were complicated indeed.

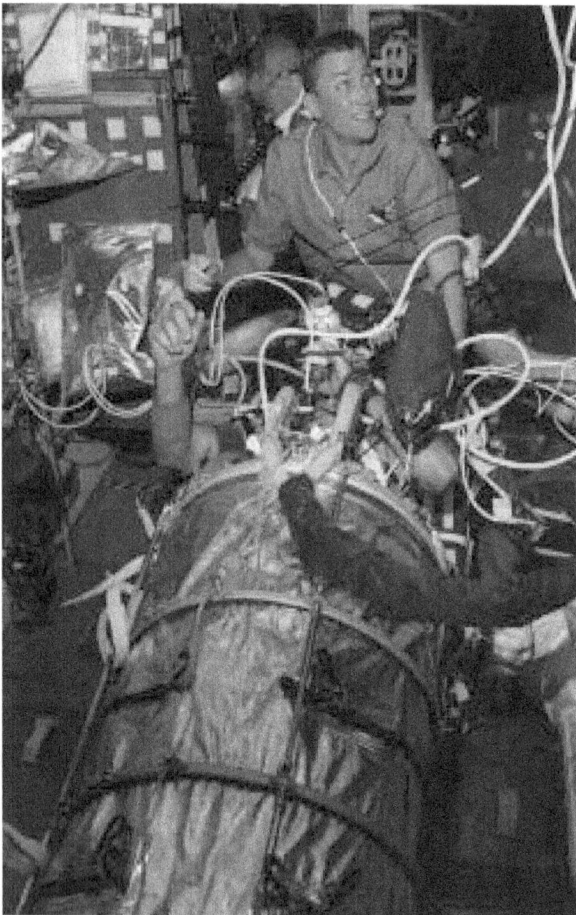

Figure 2. The view from inside the Space Shuttle, as one astronaut prepares to perform Valsalva maneuvers during the Neurolab mission.

Although we studied six astronauts on six occasions before, during, and after the Neurolab mission, this chapter focuses on results from the four astronauts who were studied during the two sessions that featured sympathetic microneurography, about 72 days prior to the Neurolab launch, and on the 12th or 13th day of the mission.

Measurements

Our measurements included electrocardiograms (signals reflecting the heart's electrical activity, recorded from electrodes attached to the skin and amplified), beat-by-beat finger blood pressure,* the depth of each breath, mouthpiece pressure (to indicate the force of Valsalva straining), and muscle sympathetic nerve activity.

Vagal traffic travels to the heart in right and left vagus nerves. These nerves emerge from the *brain stem* and travel in the neck to the chest. The sympathetic nerves are unlike the vagus nerves, in that they do not come from the brain (they emerge from many levels of the spinal cord) and they are not discrete nerves. Rather, sympathetic nerve *fibers* join other nerves and thus become parts of multifunctional superhighways with traffic traveling in both directions—out to body structures, and back to the central nervous system. An example of such a nerve is the *peroneal nerve*, the one used for Neurolab recordings, which travels from behind the knee and down the lateral side of the leg. Within this nerve are motor neurons that cause leg muscles to contract on demand, and several types of sensory nerves that apprise the central nervous system of the state of muscle contraction. Also within the peroneal nerve are sympathetic nerve fibers that travel to those same muscles and regulate their blood flow. Since muscle is over 40% of adult body mass, sympathetic nerves to muscle arterioles are very important.

The technical report by Ertl et al. in this volume describes the technique of sympathetic microneurography in detail. Briefly, electrical shocks are applied just posterior and lateral to a knee to locate the peroneal nerve. When the shock is applied directly over the nerve, the foot twitches. A mark is placed on the skin at this location, and the electrode that was used to deliver the shocks is moved to a different location, where the process is repeated. After several repetitions, a track of dots on the skin traces the path of the nerve. Then, a fine needle, 0.5 mm in diameter, is stuck through the skin and aimed directly toward the nerve. (A ground needle is also inserted nearby, just beneath the skin.) The electrical activity from the nerve electrode is amplified. The astronaut inserting the needle listens to this amplified signal with headphones. When the uninsulated tip of the needle touches a nerve fiber, a highly characteristic barrage of nerve firing is heard (it sounds like a dive bomber). The astronaut then makes very fine adjustments until he or she hears sounds typical of muscle sympathetic nerve activity: episodic *bursts* of firing, at the same frequency as the heart, that wax and wane with breathing.

RESULTS

Valsalva responses

Figure 3 shows the responses of one astronaut to Valsalva maneuvers performed 72 days before, and on day 12 of the Neurolab mission. The classic blood pressure responses to 30-mmHg Valsalva straining (Smith, 1996) are illustrated in the *middle panels*. At the beginning of straining (time zero), blood pressure rises, secondary to displacement of blood from the

---

*The simple and elegant device used to estimate blood pressure was invented by a Czech scientist, Jan Peñáz. It was manufactured commercially by Ohmeda, under the name Finapres—*Finger arterial pressure*—and was modified for use in space by TNO, a biomedical laboratory in Amsterdam University, The Netherlands. A red, light-emitting diode illuminates the middle digit of the middle finger, and a sensor measures the intensity of the light transmitted through the finger. This digit is encircled by a cuff that is connected by a tube to a pressure chamber. Each time the heart beats, the finger fills with blood, and the amount of light transmitted through the finger changes. The pressure in the finger cuff is adjusted continuously, to maintain a constant level of light transmission. A gauge in the pressure chamber measures the amount of pressure required to maintain light transmission constant, and this is recorded as the blood pressure.

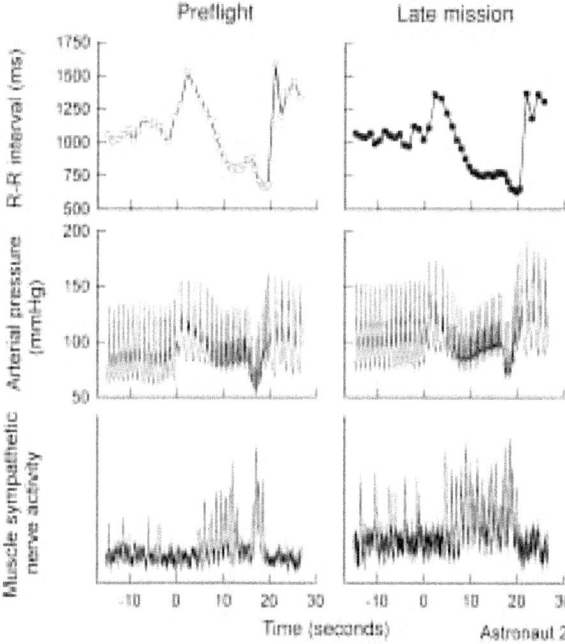

Figure 3. Valsalva maneuvers performed by one astronaut 72 days before the Neurolab and on mission day 12. In this and other astronauts, sympathetic responses to straining were much greater in space (right) than on Earth. (From Cox, 2002, with permission; reproduced from *The Journal of Physiology.*)

abdomen and chest to the rest of the body. Then, as straining continues and pressure increases in the abdomen and chest are maintained, finger blood pressure falls. [Blood moves throughout the circulation simply according to pressure gradients—blood (or fluid in any other hydraulic system) moves from higher to lower pressure areas. During straining, less blood reaches the heart because pressure in the limbs is lower than pressure in the abdomen and chest.] Then, as straining continues, compensatory mechanisms kick in, and pressure begins to return to normal.

After the astronaut stops straining (at 15 seconds), there is an abrupt further reduction of pressure. This occurs because vein pressure has been increasing as blood has dammed up in the limbs during straining. When the astronaut takes his or her first breath (and thereby creates a negative pressure in the chest), increased quantities of blood suddenly move from the limbs to the chest and heart. Finally, as the astronaut resumes normal breathing, blood pressure rises to levels higher than those measured before straining (middle panels, right).

As our discussion suggests, some of the blood pressure transients shown in the middle panels of Figure 3 result simply from hydraulic forces—they are *mechanical.* However, other of the blood pressure changes result from neural activity. The bottom panels of Figure 3 show how Valsalva straining influences sympathetic *nerve activity.* Notice that before straining (left of time zero), sympathetic activity increased episodically as small vertical spikes or "bursts." The quantity and quality of the sympathetic bursts changed when

the astronaut began straining. First, sympathetic traffic was turned off, between zero and about five seconds; this is seen particularly well in the tracing made in space (bottom, right). Sympathetic silence reflected simply a baroreflex response to the very brief increase of blood pressure that occurred at the beginning of straining. (When blood pressure rises, sympathetic activity is silenced as baroreflex mechanisms work to return blood pressure to normal.) Then, as blood pressure fell, the size and number of sympathetic bursts increased. This also represents a baroreflex response: sympathetic bursts come out to restore blood pressure to usual levels. Note that the increased sympathetic nerve activity, brought out by straining, reversed the downward blood pressure trend; and, even though straining continued (between about five and 10 seconds), blood pressure began to return toward the usual levels. This is particularly evident in space (right middle panel).

The sudden reduction of blood pressure after the end of Valsalva straining (at about 15 seconds) and the consequent increase of sympathetic nerve activity also illustrate (quite remarkably) the elegance of baroreflex sympathetic responses. This is shown particularly well in the preflight recording (bottom, left). The very brief blood pressure reduction was answered by an equally brief increase of sympathetic nerve activity. Finally, at the very end of the responses, blood pressure rose (middle panels, right). The terminal rise of blood pressure reflects both the constriction of arterioles in the body, secondary to the preceding bursts of sympathetic nerve activity, and the greater return of blood to the heart. After straining, more blood is being pumped by the heart than during straining, and the blood is being pumped into a constricted vascular bed. Baroreflex physiology also can be seen after Valsalva straining; i.e., when pressure is high, sympathetic neurons fall silent.

The top panels of Figure 3 show changes of the period between heartbeats. (The *R wave* of the electrocardiogram is the tall, narrow spike that indicates the spread of electricity through the main pumping chambers of the heart. The interval between each heartbeat, the *R-R interval*, is used as an indirect index of the amount of vagus nerve traffic traveling to the heart.) Thus, at the beginning of straining, blood pressure rises and R-R intervals increase (the heart rate slows) as vagus nerve traffic to the heart increases. The R-R interval responses throughout all stages of the Valsalva maneuver elegantly reflect baroreflex physiology. Notice how changes of R-R intervals move exactly in parallel to changes of blood pressure.

Valsalva responses on Earth and in space

Did spaceflight alter astronauts' responses to Valsalva maneuvers? For one answer to this question, compare recordings made before (left) with those made during the Neurolab mission (right) in Figure 3. It is arguable whether or not the vagally mediated R-R interval changes (top panels) were less in space than on Earth. However, it seems clear that the sympathetic nerve responses (bottom panels) were different in space. Before straining (to the left of zero second), sympathetic bursts were more numerous in space. During straining, sympathetic bursts were also more numerous in space than on Earth.

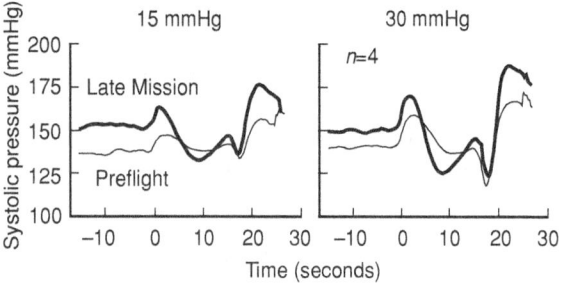

Figure 4. Average systolic pressure changes during 15- and 30-mmHg Valsalva maneuvers for all four astronauts. Systolic blood pressure changes were greater during the late mission session (dark lines) than during the first preflight session. Diastolic pressure changes (not shown) paralleled systolic pressure changes. (From Cox, 2002, with permission; reproduced from *The Journal of Physiology*.)

Why were the sympathetic responses in this and the other astronauts so much greater in space than on Earth? Figure 3 provides a clue: the reductions of blood pressure that triggered the increases of sympathetic activity (middle panels) also may have been greater. To test this possibility, we averaged all blood pressure responses to both levels of Valsalva straining in all four astronauts. Figure 4 shows that systolic pressure (the upper number given with blood pressure readings) reductions during, and increases after, straining were *greater* in space (heavy lines) than on Earth. One might conclude from this that the greater sympathetic outpouring in response to Valsalva straining meant simply that the baroreflex was doing its job: the blood pressure challenge was greater and, therefore, the sympathetic response also was greater.

As mentioned, astronauts performed Valsalva maneuvers six times, before, during, and after the Neurolab mission. Average measurements from the four astronauts give an excellent notion of how space travel alters blood pressure responses to straining. Figure 5 shows a box plot of changes (Δ) of systolic pressure during all recording sessions. (The top, center, and bottom lines of each box indicate the 75th, 50th, and 25th percentiles of pressure changes from baseline values before straining.) The reduction of pressure during straining was clearly greatest during the second inflight session. Our study was not designed to explore *why* pressure falls more during straining in space than on Earth. No doubt it was due importantly to the reduction of blood volume that astronauts experience when they travel into space.

Figure 6 shows total nerve activity for all four astronauts, before (zero straining pressure) and during 15- and 30-mmHg Valsalva maneuvers. (Note that the scale is larger for Astronaut 4 than for the other astronauts; this astronaut had particularly large sympathetic responses to Valsalva straining.) We did not perform statistical analyses because there were only four subjects. However, the fact that all astronauts changed in the same way suggests strongly that sympathetic responses to blood pressure reductions are greater in space than on Earth.

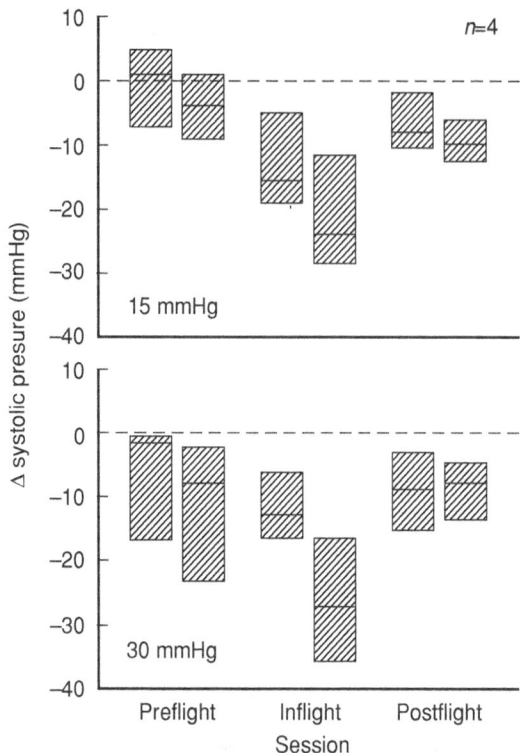

Figure 5. Box plot of systolic pressure changes for all recording sessions for all four astronauts. The top, center, and bottom lines of each box indicate the 75th, 50th, and 25th percentiles of pressure changes from baseline values before straining. (From Cox, 2002, with permission; reproduced from *The Journal of Physiology*.)

Did spaceflight really change sympathetic nerve responses? Or, was the sympathetic response greater simply because the pressure changes were greater? To answer these questions, we related astronauts' responses to the stimuli that provoked those

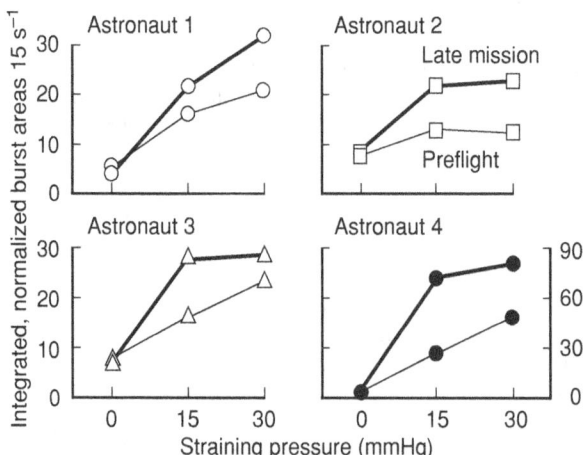

Figure 6. Total muscle sympathetic nerve activity over the 15-second period of Valsalva straining. (From Cox, 2002, with permission; reproduced from *The Journal of Physiology*.)

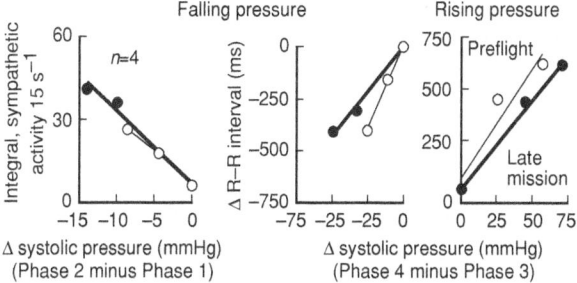

Figure 7. Average sympathetic and vagal responses plotted as functions of arterial pressure changes. Heavy lines depict measurements made on flight days 12 or 13, and light lines depict measurements made ~72 days preflight on Earth. In each panel, measurements made during the one-minute period before Valsalva straining are plotted at zero pressure change. The next symbol denotes responses to 15-mmHg straining, and the second symbol denotes responses to 30-mmHg straining. The two left panels depict responses to straining, and the right panel depicts responses after release of straining. (From Cox, 2002, with permission; reproduced from *The Journal of Physiology*.)

responses and derived *stimulus-response relations*. We divided R-R interval and sympathetic responses by blood pressure changes during or after straining, and we plotted three pairs of data for each measurement. We reasoned that if the changes we recorded in space meant simply that baroreflex responses were normal, responses divided by the blood pressure changes that provoked them should be unchanged in space.

Figure 7 shows average integrated (or total) sympathetic nerve activity plotted as functions of changes of diastolic pressure reductions during straining (*left panel*). (We plotted diastolic pressure, the lower number given in blood pressure measurements, because diastolic pressure regulates sympathetic nerve activity, and systolic pressure does not.) In the left panel of Figure 7, measurements made in space are shown as filled circles and heavy lines, and measurements made on Earth are shown as open circles and light lines. Each linear relation has three points: no straining, 15-mmHg straining, and 30-mmHg straining. It seems obvious, from the left panel in Figure 7, that sympathetic nerve responses to pressure reductions in space are *normal*. The stimulus-response relations are nearly superimposable—the sympathetic baroreflex is simply doing its job.

Figure 7 also shows vagal R-R interval changes during (middle panel) and after straining (right panel). In both cases, the slopes of the relations made on Earth (light lines) seem to be steeper than those made in space.

DISCUSSION

When astronauts return to Earth from space, they may have difficulty standing upright. Their hearts may beat much more rapidly than before they went into space, and their blood pressures may fall. Some astronauts may even faint (Buckey, 1996). We studied the pressure-regulating reflexes that enable people

to stand without fainting on Earth and in space. Since our earlier research showed clearly that space travel impairs the vagal, or heart rate, portion of the baroreflex (Fritsch, 1992a, 1992b; Fritsch-Yelle, 1994), we tested the hypothesis that space travel also impairs the sympathetic, or blood pressure, portion of the baroreflex. The data in this chapter reject our hypothesis: sympathetic baroreflex responses are normal in space. Therefore, it appears that microgravity *differentially* affects human baroreflexes—vagal baroreflex responses are impaired, and sympathetic baroreflex responses are preserved.

Sympathetic baroreflexes in space

In our study, we focused on autonomic *transients* provoked by Valsalva maneuvers. Our principal new finding is that although Valsalva straining triggers much greater increases of muscle sympathetic nerve activity in space than on Earth, the *stimulus*, falling blood pressure, is also greater. As a result, reflex gain (the sympathetic response plotted as a function of the diastolic pressure reduction, Figure 7, left panel) is normal. Ours is one of three studies involving sympathetic microneurograms obtained from Neurolab astronauts. The results of the three studies, which employed different research provocations and tested different hypotheses, are remarkably congruent. Ertl and coworkers (see science report by Ertl et al. in this publication) showed that muscle sympathetic nerve and plasma norepinephrine spillover responses to lower body suction (used to simulate the effects of standing in space) are similar in space and on Earth. Levine and his colleagues (see science report by Levine et al. in this publication) showed that sympathetic nerve responses to actual upright tilt on landing day after the Neurolab mission were the same as before the mission.

Potential mechanisms

Autonomic responses to Valsalva straining are determined importantly by blood volume. Both acute reductions of central blood volume, such as those provoked by standing on Earth, and those that occur after prolonged head-down bed rest increase arterial pressure and heart rate responses to Valsalva straining. Therefore, the blood volume reduction that occurs in space leads acutely to greater arterial pressure reductions and increases of muscle sympathetic nerve activity during Valsalva straining than in the supine position on Earth.

As mentioned earlier in this chapter, reflex adjustments occur as responses to changes of information carried to the brain by sensors distributed throughout the body. We are not certain which sensors mediated these changes we report. However, as we suggest, we strongly suspect that arterial baroreflexes were involved. Blombery and Korner (Blombery, 1982) studied what they called "Valsalva-like" maneuvers in conscious rabbits, and showed that nearly all vascular and heart rate responses are abolished by cutting the nerves leading from baroreceptive arteries.

Absence of changes of baseline arterial pressures and R-R intervals (we measured similar baseline blood pressures and R-R intervals in space and on Earth) does not discount arterial baroreceptor participation in the adjustments made by astronauts to

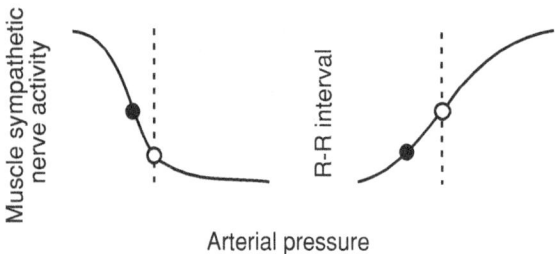

Figure 8. One possible mechanism to explain baroreflex changes in space. These figures show average reverse sigmoid carotid baroreceptor—muscle sympathetic nerve—and sigmoid—R-R interval relations—for healthy young volunteers studied earlier (Rea, 1987). The resting position of these subjects is indicated by the open circles and vertical dashed lines. Microgravity may shift subjects leftward on both relations (closed circles). (From Cox, 2002, with permission; reproduced from *The Journal of Physiology*.)

microgravity. Experimental changes—which should, but do not, change arterial pressure, including low levels (<20 mmHg) of lower body suction—reduce the dimensions of the important baroreceptive artery in the chest, the aorta, and increase muscle sympathetic nerve activity. We speculate that in space, reduced blood volume is compensated for by baroreflex-mediated increases of sympathetic nerve activity and that this, in turn, maintains arterial pressure at normal levels. The baroreflex adjustment is so effective that without the direct recordings of sympathetic nerve activity we made (or norepinephrine spillover, an indirect index of whole body sympathetic nerve activity), there would be no evidence of arterial baroreflex engagement. We do not exclude other possibilities, including changes in sensory inputs from antigravity muscles or heart muscle, both of which atrophy in actual or simulated microgravity (Levine, 1996).

It is possible that microgravity simply shifts astronauts to the left on their stimulus-response relations. Figure 8 shows sympathetic and R-R interval responses of healthy volunteers to brief, graded neck suction or pressure on Earth (Rea, 1987). These data suggest that healthy young adults work near the middle of the linear region of their vagal blood pressure—R-R interval relations (right)—and close to the threshold of their blood pressure—sympathetic relations (both positions are designated by open circles).

Limitations

The principal limitation of our study is the small number of subjects. Because we studied only four astronauts, we present descriptive results rather than statistical analyses. A second limitation of our study is the Valsalva maneuver itself, which is deceptively simple. Although Valsalva's maneuver is time-honored (Eckberg, 1992), *the actual baroreflex stimulus is not measured*. Arterial baroreceptors sense distortion, not pressure, and we did not measure baroreceptive artery dimensions. However, it is likely that the changes of finger blood *pressure*

we measured paralleled the changes of carotid artery pressure and *dimensions*.

CONCLUSIONS

We measured the sympathetic and vagal transients that result when astronauts strain, before and during the Neurolab Space Shuttle mission. One likely explanation for our findings is that space travel chronically shifts astronauts' positions on their arterial pressure-sympathetic and -vagal response relations (Figure 8) to the left. We do not exclude other possibilities, including atrophy of antigravity and cardiac muscle. However, the changes that we documented can be explained economically as a consequence of the blood volume reduction that occurs in space, itself a chronic adjustment to space travel.

Acknowledgements

We owe profound thanks to the astronauts who gave unstintingly of their time and emotional energy to make this research project successful. We also thank James E. Ames, Virginia Commonwealth University, and a large cadre of scientists and engineers, including particularly Hasan Rahman and Suzanne McCollum at Lyndon B. Johnson Space Center. This research was supported by National Aeronautics and Space Administration contracts NAS0-19541 and NAG2-48, and a grant from the National Heart, Lung, and Blood Institute, UO1HL-56417. The research was also supported by a grant from Helsingin Sanomat, Finland.

REFERENCES

ROLE OF AORTIC AND CAROTID SINUS BARORECEPTORS ON VALSALVA-LIKE VASOCONSTRICTOR AND HEART RATE REFLEXES IN THE CONSCIOUS RABBIT. P.A. Blombery, and P.I. Korner. *J. Autonom. Nerv. Syst.*, Vol. 5, pages 303–315; 1982.

CARDIOVASCULAR ADJUSTMENTS TO GRAVITATIONAL STRESS. C.G. Blomqvist, H.L. Stone, In *Handbook of Physiology. The Cardiovascular System. Peripheral Circulation and Organ Blood Flow*, Part 2. (JT Shepherd, FM Abboud, eds), American Physiological Society, Bethesda, Maryland, pages 1025–1063; 1983.

CAROTID BARORECEPTOR-MUSCLE SYMPATHETIC RELATION IN HUMANS. R.F. Rea and D.L. Eckberg. *Am. J. Physiol.*, Vol. 253, pages R929–R934; 1987.

HUMAN BAROREFLEXES IN HEALTH AND DISEASE. D.L. Eckberg and P. Sleight. Clarendon Press Oxford, pages 61–77; 1992.

SHORT-DURATION SPACEFLIGHT IMPAIRS HUMAN CAROTID BARORECEPTOR-CARDIAC REFLEX RESPONSES. J.M. Fritsch, J.B. Charles, B.S. Bennett, M.M. Jones, and D.L. Eckberg. *J. Appl. Physiol.* Vol. 73, pages 664–671; 1992a.

EFFECTS OF WEIGHTLESSNESS ON HUMAN BAROREFLEX FUNCTION. J.M. Fritsch, and D.L. Eckberg. *Aviat. Space Environ. Med.*, Vol. 63, page 439; 1992b (Abstract).

Spaceflight Alters Autonomic Regulation of Arterial Pressure In Humans. J.M. Fritsch-Yelle, J.B. Charles, M.M. Jones, L.A. Beightol, and D.L. Eckberg. *J. Appl. Physiol.,* Vol. 77, pages 1776–1783; 1994.

Orthostatic Intolerance After Spaceflight. J.C. Buckey, Jr., L.D. Lane, B.D. Levine, D.E. Watenpaugh, S.J. Wright, W.E. Moore, F.A. Gaffney, and C.G. Blomqvist. *J. Appl. Physiol.,* Vol. 81, pages 7–18; 1996.

The Physiology of Bed Rest. S.M. Fortney, V.S. Schneider, J.E. Greenleaf. In *Handbook of Physiology*, Section 4, *Environmental Physiology*, Vol. II. (M.J. Fregly, C.M. Blatteis, eds), Oxford University Press, Oxford, pages 889–939; 1996.

Valsalva's Maneuver Revisited: A Quantitative Method Yielding Insights into Human Autonomic Control. M.L. Smith, L.A. Beightol, J.M. Fritsch-Yelle, K.A. Ellenbogen, T.R. Porter, and D.L. Eckberg. *Am. J. Physiol.,* Vol. 271, pages H1240–H1249; 1996.

Influence of Microgravity on Sympathetic And Vagal Responses to Valsalva's Manoeuvre. J.F. Cox, K.U.O. Tahvanainen, T.A. Kuusela, B.D. Levine, W.H. Cooke, T. Mano, S. Iwase, M. Saito, Y. Sugiyama, A.C. Ertl, I. Biaggioni, A. Diedrich, R.M. Robertson, J.H. Zuckerman, L.D. Lane, C.A. Ray, R.J. White, J.A. Pawelczyk, J.C. Buckey, Jr., F.J. Baisch, C.G. Blomqvist, D. Robertson, and D.L. Eckberg. *J. Physiol. Lond.,* Vol. 538, pages 309–322; 2002.

The Human Sympathetic Nervous System Response to Spaceflight

Authors

Andrew C. Ertl, André Diedrich, Sachin Y. Paranjape, Italo Biaggioni, Rose Marie Robertson, Lynda D. Lane, Richard Shiavi, David Robertson

Experiment Team

Principal Investigator: **David Robertson**

Co-Investigators: **Italo Biaggioni**
Andrew C. Ertl
Rose Marie Robertson
André Diedrich

Technical Assistant: **Sachin Y. Paranjape**

Engineer: **Richard Schiavi**

Vanderbilt University, Nashville, USA

ABSTRACT

The sympathetic nervous system is an important part of the autonomic (or automatic) nervous system. When an individual stands up, the sympathetic nervous system speeds the heart and constricts blood vessels to prevent a drop in blood pressure. A significant number of astronauts experience a drop in blood pressure when standing for prolonged periods after they return from spaceflight. Difficulty maintaining blood pressure with standing is also a daily problem for many patients. Indirect evidence available before the Neurolab mission suggested the problem in astronauts while in space might be due partially to reduced sympathetic nervous system activity. The purpose of this experiment was to identify whether sympathetic activity was reduced during spaceflight. Sympathetic nervous system activity can be determined in part by measuring heart rate, nerve activity going to blood vessels, and the release of the hormone norepinephrine into the blood. Norepinephrine is a neurotransmitter discharged from active sympathetic nerve terminals, so its rate of release can serve as a marker of sympathetic nervous system action. In addition to standard cardiovascular measurements (heart rate, blood pressure), we determined sympathetic nerve activity as well as norepinephrine release and clearance on four crewmembers on the Neurolab mission. Contrary to our expectation, the results demonstrated that the astronauts had mildly elevated resting sympathetic nervous system activity in space. Sympathetic nervous system responses to stresses that simulated the cardiovascular effects of standing (lower body negative pressure) were brisk both during and after spaceflight. We concluded that, in the astronauts tested, the activity and response of the sympathetic nervous system to cardiovascular stresses appeared intact and mildly elevated both during and after spaceflight. These changes returned to normal within a few days.

INTRODUCTION

Many human physiological responses are controlled by the nervous system without any conscious effort by the individual. For example, when people are startled, their skin prickles. When they are excited, their eyes dilate; and when they are frightened, blood is directed to their muscles and shunted away from the digestive tract to prepare for fighting or running away. These examples are part of the autonomic (or automatic) nervous system's almost immediate response and adjustment to everyday occurrences. The two branches of the autonomic nervous system that control these adjustments are the sympathetic nervous system, which usually controls alerting responses, and the parasympathetic nervous system, which usually controls resting and relaxing responses.

Automatic autonomic nervous system adjustments are essential for standing. During standing, gravity causes about 700 mL of blood and body fluids to move from the chest into the legs and trunk. Autonomic reflexes increase heart rate and constrict blood vessels in many areas, thereby increasing resistance to blood flow in less essential regions. Blood pressure would fall if these autonomic adjustments did not occur. Instead, blood pressure is maintained, allowing humans to remain conscious while standing upright. These adjustments also keep blood pressure relatively constant under varied everyday conditions.

Blood pressure is sensed by pressure receptors in the blood vessels of the carotid sinus in the neck, in the aortic arch, and in the heart. The pressure receptors are called baroreceptors. When blood pressure is high, the baroreceptors inhibit sympathetic nerve activity from the brain, which allows a relative relaxation of blood vessels. When these receptors sense low blood pressure, the outgoing (or efferent), sympathetic activity increases and causes blood vessels to constrict. These reflex reactions are called baroreflexes. When the sympathetic nerves are activated, they release the hormone norepinephrine into the blood. Norepinephrine is a neurotransmitter discharged from active sympathetic nerve terminals, so its rate of release can serve as a marker of sympathetic nervous system activity.

When astronauts return from a spaceflight and experience gravitational forces on Earth again, they may have difficulty standing (Buckey, 1996). Their bodies' adaptations to spaceflight (such as a reduction in blood volume) can result in elevated heart rates and the feeling that they are about to faint. If these symptoms were severe during landing or in an emergency after landing, this could present a risk to the astronauts. Therefore, an understanding of this problem is important to NASA.

The inability to tolerate an upright position is called orthostatic intolerance (OI). Spaceflight-induced OI results from factors such as reduced blood volume and possibly autonomic nervous system dysfunction produced by a prolonged reduction of sympathetic nervous system activity while in space (Robertson, 1994).

There are several reasons why microgravity might reduce sympathetic nervous system activity in space. When body fluids are no longer pulled towards the legs by gravity and therefore remain in the central part of the body (Watenpaugh, 1995), this might stretch pressure receptors in the heart and inhibit sympathetic activation. The absence of the regular baroreflex-mediated sympathetic stimulation, which occurs on Earth with frequent body position changes and subsequent fluid shifts, might lead to temporary baroreflex dysfunction in space and exaggerated but insufficient sympathetic responses when astronauts are reexposed to Earth's gravity (Vernikos, 1996).

An early study (Leach, 1983) of sympathetic activity in space suggested that sympathetic nervous system activity was reduced in astronauts. This study showed that norepinephrine levels in blood were decreased inflight compared to preflight measurements. The level of norepinephrine in blood, however, is not always an accurate marker of sympathetic nervous system activity. The blood level of norepinephrine represents a balance between the norepinephrine entering the blood and the clearance of norepinephrine from the blood. A more useful measure of sympathetic nervous system activity therefore is how much norepinephrine is actually entering the bloodstream, not just the blood level.

Subsequent studies of norepinephrine levels in space have shown either no change or an increase in norepinephrine levels. To resolve whether sympathetic nervous system activity was in fact reduced in space, we conducted a study that would measure the actual release (or spillover) of norepinephrine into the bloodstream. In addition, we measured sympathetic nerve activity directly using microneurography (see technical report by Ertl et al. in this publication). Measurement of sympathetic nerve activity with microneurography enables direct on-line monitoring of sympathetic nervous system function, is remarkably stable in individual subjects over time, and yields measurements that correlate well with norepinephrine spillover. With this combination of measurements, we could determine definitively whether reduced or increased sympathetic activity occurred during spaceflight.

METHODS

The experiment required a way to provide a standardized stress to the cardiovascular system both in space and on Earth. With standing, blood leaves the chest and moves to the lower body. Standing, however, could not be used as a cardiovascular stress in space, since the astronauts are experiencing weightlessness. Instead, suction was applied to the lower body of supine astronauts with their hips and legs sealed in a chamber just above the pelvis. The suction moves blood from the chest to the lower body, in a way that is very similar to what occurs with standing. This suction (or lower body negative pressure (LBNP)) was used preflight, inflight, and postflight as the cardiovascular stress.

The LBNP chamber used on Earth was made of rigid plastic, and the LBNP chamber used in space was made of collapsible airtight fabric. Both had windows to allow leg access for microneurography. After seven minutes of resting baseline recording, suction was applied at –15 and –30 mmHg for seven minutes each. The simulated gravitational stress of LBNP at these levels is somewhat less than standing in an upright posture on Earth, and it is considered by NASA to be a safe stress for astronauts in space. Data were collected in space during the third to fifth minute of each seven-minute period, and blood was drawn at the end of each segment. Data were collected by computer and analyzed on Earth. The blood samples were processed in space and analyzed in our laboratory on Earth.

Heart rate was measured by electrocardiogram. Adhesive electrodes were placed on the chest to monitor the heart's pulse and electrical activity. Blood pressure was monitored continuously using finger arterial pressure by placing a small pressure cuff on a finger. Muscle sympathetic nerve activity (MSNA) was recorded by microneurography (see technical report by Ertl et al. in this publication). Briefly, the tip of a 200-μm tungsten microelectrode was introduced into the peroneal nerve behind the knee and placed among C-fibers carrying the sympathetic nervous system signals to the small blood vessels supplying the leg muscles. To provide another measure of sympathetic nervous system activation, plasma norepinephrine spillover and clearance were measured before, during, and after the Neurolab flight. This technique used a tracer dose of radioactive norepinephrine, which was infused into an arm vein. Blood samples were taken from the opposite arm during the experiment. Analysis of these samples provided the kinetics of norepinephrine appearance and disappearance from the circulation. Use of this methodology helps us understand the divergent results from norepinephrine measurements determined in previous spaceflight experiments. The amount of radioactive norepinephrine was so small that it did not change norepinephrine levels in the plasma or affect the astronaut's heart or circulation (Ertl, 2002). These measurements were made at rest in a supine posture and after seven minutes of LBNP at –15 and –30 mmHg on Earth and in space.

RESULTS

We indicated the number of subjects for each measurement on the graphs. Not all astronauts were studied with all protocols because of NASA regulations and technical difficulties.

While the average heart rate of four astronauts at rest in the LBNP was similar preflight and inflight (56±4 beats per minute), heart rate during –30 mmHg LBNP was greater in

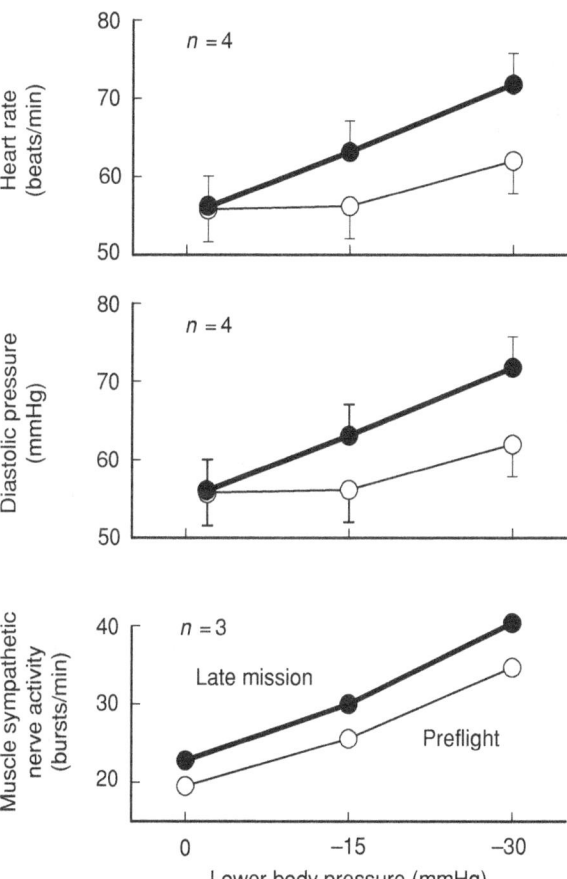

Figure 1. Hemodynamic responses to LBNP. Although average baseline heart rates and diastolic pressures were similar during preflight and inflight sessions, increments during graded LBNP were greater in space. All three astronauts had more muscle sympathetic bursts per minute before and during LBNP in space than on Earth. (From Ertl, 2002, with permission; reproduced from *The Journal of Physiology*.)

space (72±4 beats per minute) than before spaceflight (62±4 beats per minute). In Figure 1, the steeper slope of the heart rate response to LBNP during spaceflight indicates a greater response to LBNP in space compared to Earth. Average blood pressures did not change significantly during LBNP in any session. Figure 2 depicts MSNA in one astronaut, recorded 73 days preflight and 13 days inflight. This subject's MSNA prior to LBNP increased from 15 bursts per minute preflight to 20 beats per minute inflight. The astronaut's MSNA further increased during –15 and –30 mmHg LBNP, and these increases also were greater during inflight than preflight sessions (19 vs. 32, and 39 vs. 48 bursts per minute). Average MSNA, in bursts per minute, before and during spaceflight for three astronauts is depicted in Figure 1, bottom panel. The average activity for the three astronauts was slightly elevated both at rest and during LBNP in space compared to on Earth.

Plasma norepinephrine and norepinephrine kinetics

In the same three astronauts who underwent sympathetic microneurography, baseline plasma norepinephrine concentrations were higher inflight than preflight (range of increases: 35–169%, Figure 3). Average norepinephrine clearance and spillover were also elevated in space from preflight values. In five astronauts, plasma norepinephrine concentrations and whole body norepinephrine spillover and clearance were significantly higher one day after landing as compared to preflight levels. These values all returned to preflight levels by day five or six.

DISCUSSION

Our study provided the first comprehensive analysis of human sympathetic nervous system function during spaceflight. In space, baseline sympathetic neural outflow, whether measured by plasma norepinephrine spillover or MSNA, was slightly increased and sympathetic responses to LBNP were exaggerated. The astronauts we studied responded normally to the simulated orthostatic stress of LBNP, and they also maintained their blood pressures.

Baseline sympathetic outflow

In three astronauts, microgravity moderately increased the number of sympathetic bursts per minute in baseline MSNA. These limited data agree with most studies measuring sympathetic microneurography during the simulated microgravity of head-down bed rest on Earth. The data show, however, that we need to reject our hypothesis that spaceflight reduces sympathetic neural outflow. The finding of increased levels of MSNA during the Neurolab mission suggests that the significantly increased levels of sympathetic nerve activity recorded in five astronauts after the Neurolab mission (see science report by Levine et al. in this publication) is not an artifact resulting from uncontrollable stresses on landing day.

The changes in astronauts' sympathetic nerve activity were mirrored by the norepinephrine data. The same three astronauts had elevated plasma norepinephrine levels as

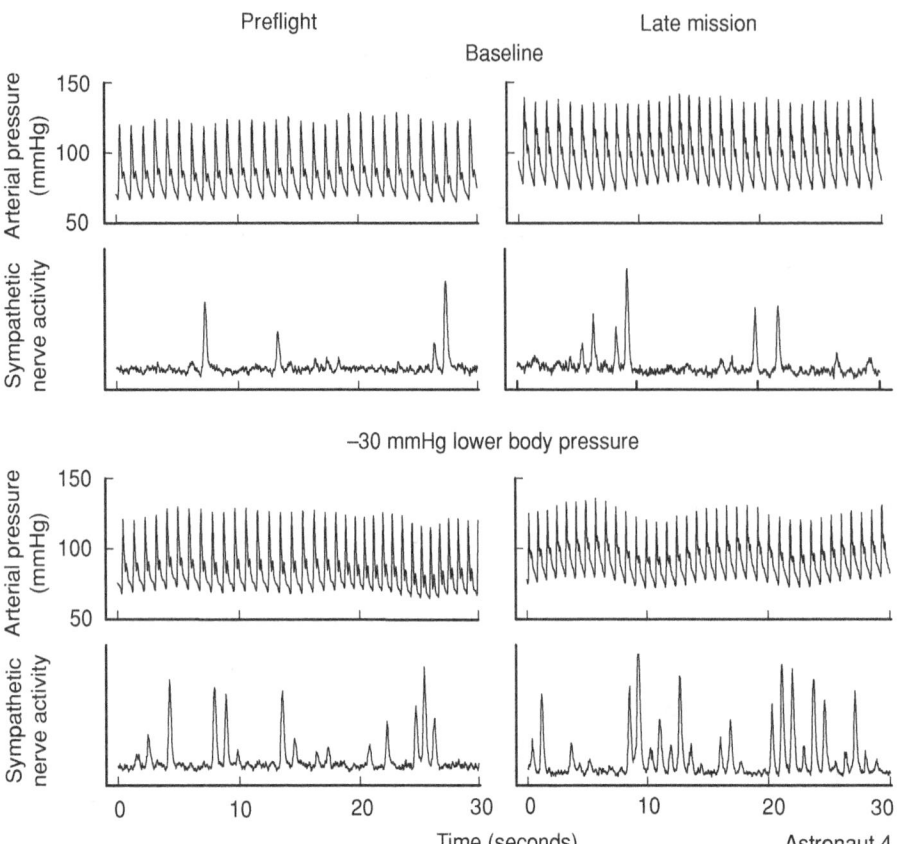

Preflight   Late mission

Baseline

Figure 2. Recordings before and during 30-mmHg LBNP from one astronaut. This and the other two astronauts had greater sympathetic responses to LBNP in space than on Earth. (From Ertl, 2002, with permission; reproduced from *The Journal of Physiology*.)

−30 mmHg lower body pressure

Time (seconds)   Astronaut 4

compared to values obtained in the supine position on Earth (Figure 3). As mentioned earlier, however, elevations of plasma norepinephrine concentrations may reflect reduced clearance as well as increased spillover. We found that norepinephrine clearance is increased in space, a change that should reduce, not increase, plasma norepinephrine levels. One implication of increased norepinephrine clearance is that plasma norepinephrine levels in space underestimate increases of sympathetic nervous activity. In Neurolab astronauts, plasma norepinephrine spillover was increased more than was norepinephrine clearance; and, therefore, plasma norepinephrine levels were higher (Figure 3), not lower, in space.

All of the measurements of sympathetic activity in astronauts—muscle sympathetic nerve activity, plasma norepinephrine concentrations, and norepinephrine spillover—indicate that spaceflight increases baseline sympathetic neural outflow. This finding may explain increases of calf vascular resistance documented in astronauts during brief space missions. Following the Neurolab mission, we continued studies of norepinephrine kinetics in five astronauts, and found that norepinephrine spillover remains elevated for at least one to two days after spaceflight. We were reassured by the agreement between norepinephrine kinetics measurements made five days postflight and those made preflight.

LBNP

LBNP in space further increased MSNA, and these exaggerated increases of MSNA were paralleled by exaggerated increases of norepinephrine spillover. The greater sympathetic responses to LBNP in space than on Earth (where recordings also were made in the supine position) compensated for the LBNP-induced reduction of the already reduced central blood volume and contributed to maintenance of arterial pressure at preflight levels. We also found no support for our hypothesis that sympathetic responses to simulated orthostatic stress in space are impaired. The finding of normal sympathetic responses to steady-state LBNP is consistent with the findings of other Neurolab protocols, which documented normal sympathetic responses to abrupt reductions of baroreceptor input during Valsalva straining in space (see science report by Cox et al. in this publication), and normal sympathetic and hemodynamic responses to upright tilt after spaceflight (see science report by Levine et al. in this publication).

Autonomic mechanisms in space

Cardiovascular measurements confirmed results from earlier studies conducted during simulated or actual spaceflight. We found exaggerated heart rate and arterial pressure increases during LBNP, as has been seen in other studies (Ertl, 2002). Such large responses to LBNP begin after only two days of spaceflight, by which time intravascular volume has declined by 14–17%. Greater increases in heart rate may reflect increased sympathetic stimulation of the heart, increased withdrawal of vagal restraint (Cox, 2002), or a combination of these factors. Exaggerated responses to LBNP might also be mediated by cardiac muscle atrophy (Levine, 2002).

Several mechanisms might explain increased norepinephrine clearance in space. Flow in the vascular bed of the lungs is altered by microgravity, and the lungs are importantly involved in norepinephrine spillover and clearance. Headward redistribution of fluid might also alter blood flow, capillary exchange, and how norepinephrine is disposed in other vascular beds. If such gravitational shifts alter norepinephrine kinetics in space, it is unclear what mechanism sustained the changes during the first postflight measurements.

Limitations

The principal limitations of our study are shared by most research conducted on humans in space: the small numbers of subjects, and the limited time available for individual research protocols. We were able to record MSNA in only three astronauts. However, we also characterized sympathetic activity with plasma norepinephrine concentrations and norepinephrine spillover in the same three astronauts. All indexes of sympathetic function changed in parallel. Further, we were able to obtain complete sets of norepinephrine kinetic data in five subjects before the Neurolab mission, and after landing. These measurements documented elevations of plasma norepinephrine concentrations and norepinephrine spillover one day after landing—measurements that were similar to those recorded about four days earlier in space in three of the five astronauts.

There were severe time constraints for all of the Neurolab protocols, including ours. Nonetheless, we believe that we allowed sufficient time for tritiated norepinephrine to reach a steady-state level for baseline measurements, and that the seven-minute stages of LBNP also were adequate for our purposes. The diet and fluid intakes of Neurolab astronauts were similar to those of astronauts on earlier NASA missions and were not controlled. Since we studied short-duration spaceflight, we do not know if sympathetic mechanisms would have undergone further adaptation had spaceflight been extended. As detailed in the Appendix of Cox et al. (Cox, 2002), our studies followed other protocols.

In conclusion, we studied the influence of spaceflight on sympathetic control mechanisms by measuring plasma norepinephrine concentrations, whole body norepinephrine spillover and clearance, and peroneal nerve muscle sympathetic activity on Earth and in space. In space, baseline sympathetic neural outflow (however it is measured) is increased moderately, and sympathetic responses to LBNP are exaggerated. The consequence of these changes is that in spite of a reduced blood volume, the astronauts we studied responded normally to simulated orthostatic stress (and to upright tilt on landing day), and were able to maintain their blood pressures at normal levels.

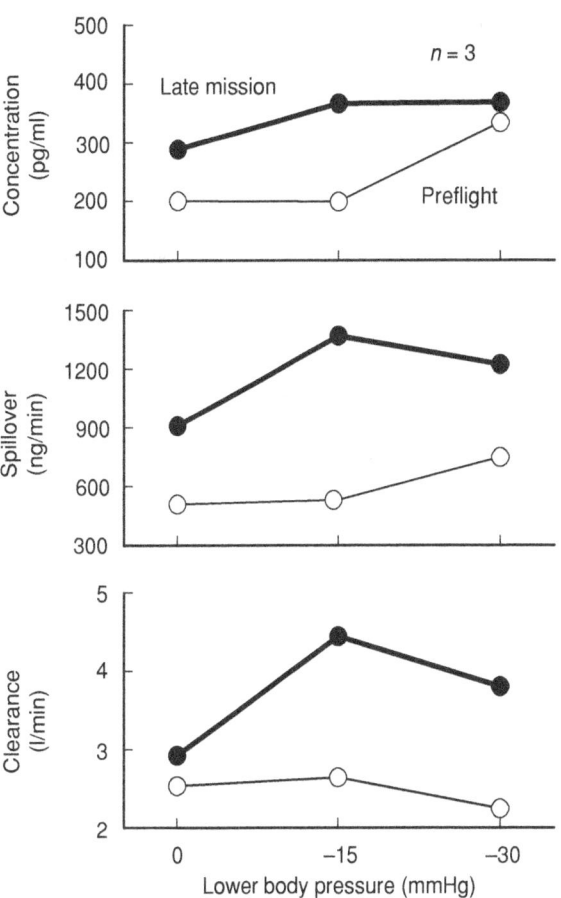

Figure 3. Average norepinephrine data from three astronauts. These data are from the same three subjects whose microneurography responses were depicted in Figure 1, bottom panel. (From Ertl, 2002, with permission; reproduced from *The Journal of Physiology*.)

Acknowledgements

This work was supported in part by NASA grants NAS 9-19483 and NAS 9-19429, and NIH grants M01 RR00095, 5P01 HL56693, 1U01NS33460, 1U01HL53206.

REFERENCES

THE ENDOCRINE AND METABOLIC RESPONSES TO SPACEFLIGHT. C.S. Leach, S. I. Atlchuler, N.M. Cintron-Trevino. *Med. Sci. Sport. Exer.,* Vol. 15, pages 432–440; 1983.

THE SYMPATHETIC NERVOUS SYSTEM AND THE PHYSIOLOGICAL CONSEQUENCES OF SPACEFLIGHT: A HYPOTHESIS. D. Robertson, V.A. Convertino and J. Vernikos. *Am. J. Med. Sci.,* Vol. 308, pages 126–132; 1994.

THE CARDIOVASCULAR SYSTEM IN MICROGRAVITY. D.E. Watenpaugh and A.R. Hargens. In *Handbook of Physiology, Section 4, Environmental Physiology.* M.J. Fregly and C.M. Blatteis eds. Oxford University Press, New York, NY, USA. Vol.1, pages 631–674; 1995.

EFFECT OF STANDING OR WALKING ON PHYSIOLOGICAL CHANGES INDUCED BY HEAD-DOWN BED REST: IMPLICATIONS FOR SPACEFLIGHT. J. Vernikos, D.A. Ludwig, A.C. Ertl, C.E. Wade, L. Keil and D. O'Hara. *Aviat. Space Envir. Med.,* Vol. 67, pages 1069–1079; 1996.

ORTHOSTATIC INTOLERANCE AFTER SPACEFLIGHT. J.C. Buckey, Jr., L.D. Lane, B.D. Levine, D.E. Watenpaugh, S.J. Wright, W.E. Moore, F.A. Gaffney and C.G. Blomqvist. *J. Appl. Physiol.,* Vol. 81, pages 7–18; 1996.

HUMAN MUSCLE SYMPATHETIC NERVE ACTIVITY AND PLASMA NORADRENALINE KINETICS IN SPACE. A.C. Ertl, A. Diedrich, I. Biaggioni, B.D. Levine, R.M. Robertson, J.F. Cox, J.A. Pawelczyk, C.A. Ray, C.G. Blomqvist, D. Eckberg, B. Baisch, J.H. Zuckerman, J.C. Buckey, Jr., L.D. Lane, R. Shiavi, C. Holt, D. Robertson. *J. Physiol.* (London), Vol. 538, pages 321–329; 2002.

HUMAN MUSCLE SYMPATHETIC NEURAL AND HAEMODYNAMNIC RESPONSES TO UPRIGHT TILT FOLLOWING SPACEFLIGHT. B.D. Levine, J.A. Pawelczyk, A.C. Ertl, J.F. Cox, J.H. Zuckerman, A. Diedrich, I. Biaggioni, C.A. Ray, M.L. Smith, S. Iwase, M. Saito, F. Sugiyama, T. Mano, R. Zhang, K. Iwasaki, L.D. Llane, J.C. Buckey, Jr., W.H. Cooke, F.J. Baisch, D. Robertson, D.L. Eckberg and C.G. Blomqvist; (2001). *J. Physiol.* (London), Vol. 538, pages 331–340; 2002.

INFLUENCE OF MICROGRAVITY ON ASTRONAUTS' SYMPATHETIC AND VAGAL RESPONSES TO VALSALVA'S MANOEUVRE. J. F. Cox, K.U.O. Tahvanainen, T.A. Kuusela, B.D. Levine, W.H. Cooke, T. Mano, S. Iwase, M. Saito, Y. Sugiyama, A.C. Ertl, I. Biaggioni, A. Diedrich, R. M. Robertson, J. H. Zuckerman, L.D. Lane, C.A. Ray, R.J. White, J.A. Pawelczyk, J.C. Buckey, Jr., F. J. Baisch, C.G. Blomqvist, D. Robertson and D. L. Eckberg (2001). *J. Physiol.* (London), Vol. 538, pages 309–320; 2002.

Blood Pooling and Plasma Filtration in the Thigh in Microgravity

Experiment Team

Principal Investigator: **Friedhelm J. Baisch**

Co-Investigators: **E.J. Beck, R. Gerzer, K. Moller, K.H. Wesseling, C. Drummer, M. Heer, J.M. Karemaker**

DLR Institute of Aerospace Medicine, Cologne, Germany

Author

Friedhelm J. Baisch

ABSTRACT

When moving from lying down to standing on Earth, some blood moves out of the chest and into the legs. Measurements of leg circumference and leg volume show that they both increase with standing and decrease when lying down. Most of the blood that moves into the leg stays in the blood vessels, but some of the blood plasma will filter out of the blood vessels into the spaces surrounding the muscle cells. This space is called the interstitial space. Filtration of fluid into the interstitial spaces increases when the pressure in blood vessels is high and decreases when pressure in the blood vessels is low. Determining how much of a change in leg volume is due to changes in the amount of blood in blood vessels or to alterations in interstitial fluid (or to other factors) is challenging. This information is important, however, since nerve fibers in the muscles (metaboreceptors) may sense the interstitial volume and signal the nervous system to control interstitial fluid content.

In weightlessness, leg volume decreases markedly. We hypothesized that interstitial fluid volume would also be reduced. We used electrical impedance tomography (EIT) to determine the changes in fluid volume in the thigh before, during, and after spaceflight. Changes were measured at rest and with the application of lower body negative pressure (LBNP), since LBNP changes leg volume. Thigh circumference fell 11.5% (±2.1% SEM) inflight. The pooling of blood in the legs during LBNP (measured in arbitrary EIT units), increased more than twofold at 15 mmHg LBNP (3.9±2.6 preflight and 11.2±2.7 inflight) as well as at 30 mmHg LBNP (9.7±3.1 preflight and 28.0±3.6 inflight) compared to preflight. Postflight measurements did not differ significantly from preflight values. The filtration of plasma from the blood vessels into the interstitial space increased inflight 470%±145% at 15 mmHg LBNP and 278%±76% at 30 mmHg LBNP.

We concluded that microgravity caused a reduction in thigh volume. The increased plasma filtration during LBNP showed that the fluid volume of the interstitial space of the lower limbs was significantly reduced inflight. The measurements support the hypothesis that receptors in the musculature of the lower part of the body may sense a dehydrated state in microgravity.

INTRODUCTION

When moving from lying down to standing on Earth, some blood moves out of the chest and into the legs. Measurements of both calf circumference and leg volume show that they increase with standing and decrease with lying down. Most of the blood that moves into the leg travels in the blood vessels, but some of the blood plasma will filter out of the blood vessels into the spaces surrounding the muscle cells. This space is called the interstitial space. Filtration of fluid into the interstitial spaces increases when pressure in the blood vessels is high and decreases when pressure in the blood vessels is low. Several factors affect the filtration of fluid into the interstitial space including (a) signals from the autonomic nervous system, (b) mechanical properties of vessel walls, and (c) characteristics of the surrounding tissues. Metaboreceptors, which are nerve fibers in the skeletal muscles of the lower limbs, cause an increase in blood pressure and cardiac contractility when activated (Mitchell, 1983). It has been assumed that they tend to distribute cardiac output, and most probably plasma flow, to the skeletal muscles. This in turn controls interstitial fluid content.

In weightlessness, fluid redistribution takes place caused by the absolute loss of gravity-dependent hydrostatic pressures. "Bird legs" (visibly smaller legs) are one of the consequences of the adaptation to microgravity. The loss of plasma volume and reductions in venous pressure in microgravity (Buckey, 1996) may diminish interstitial fluid volume of the lower limbs and thus affect afferent nerve traffic to the autonomic nervous system. We tested the hypothesis that the interstitial fluid volume of skin and muscle of the thigh is reduced in microgravity.

METHODS

Measurements were performed on four astronauts preflight, inflight, and postflight during the Neurolab Spacelab mission. Electrical impedance tomography (EIT) (Kotre, 1997; Hahn,

2001) of a cross section of the thigh was used to measure the dynamics of the fluid volume changes that occurred in both the skin and musculature of the thigh. Circumference measurements in the cross section documented the absolute volume changes of the thigh. A lower body negative pressure (LBNP) stress (Baisch, 2000) with steps of 15 and 30 mmHg was used to vary blood pooling and plasma filtration in the lower limbs. Pooling was assessed by quantification of the EIT changes at the onset of the LBNP steps and plasma filtration by measuring the slope of EIT changes during the LBNP steps (see Figure 1). Preflight measurements were performed 2.5 months prior to the mission. The inflight measurements took place after one week in microgravity. Postflight measurements were performed on days 1 or 2 (R+1) and on days 5 or 6 (R+5) after landing.

RESULTS

The circumference measurements at the thigh showed a volume reduction of 11.5%±2.1% SEM inflight compared to preflight. The pooling of blood, which was measured in arbitrary EIT units and is shown on Figure 2, increased more than twofold at 15 mmHg LBNP (3.9±2.6 preflight and 11.2±2.7 inflight) as well as at 30 mmHg LBNP (9.7±3.1 preflight and 28.0±3.6 inflight) compared to preflight. Postflight measurements did not differ significantly from preflight values. Plasma filtration (Figure 3) changed significantly in microgravity on all four astronauts. It increased 470%±145% at 15 mmHg and 278%±76% at 30 mmHg. Pre- and postflight measurements were not significantly different.

CONCLUSION

The lack of hydrostatic pressure in microgravity caused a reduction in thigh volume. The increase in plasma filtration during LBNP showed that the fluid volume of the interstitial space of

Figure 1. Measurement of pooling and filtration in a cross section of the thigh derived from electrical impedance tomography measurement. An individual tracing from one astronaut is shown during a two-step protocol of lower body negative pressure.

Figure 2. Pooling of blood during two steps (−15 mmHg and −30 mmHg) of lower body negative pressure (LBNP) measured in arbitrary EIT units pre-, in-, and postflight on four astronauts. The first bar of each pair of measurements represents the mean value of 15 mmHg LBNP. The second bar shows 30 mmHg LBNP. Black bars indicate SEM values.

Figure 3. Filtration of plasma during two steps (–15 mmHg and –30 mmHg) of lower body negative pressure (LBNP) measured in arbitrary EIT units per second pre-, in-, and postflight on four astronauts. The first bar of each pair of measurements represents the mean value of 15 mmHg LBNP. The second bar shows 30 mmHg LBNP. Black bars indicate SEM values.

the lower limbs was significantly reduced inflight. Thus the reduction of interstitial fluid volume contributed significantly to the overall reduction in thigh volume. In consequence, it is possible that afferent nerve traffic, e.g., mediated via the metaboreceptors (unencapsulated nerve endings in the interstitial space of the skeletal muscle), may sense fluid volume loss in microgravity. This may trigger increases in blood pressure and the release of fluid-regulating hormones (Norsk, 2000). The measurements support the hypothesis that receptors in the musculature of the lower part of the body may sense a dehydrated state in microgravity, which could increase sympathetic outflow from the autonomic nervous system (Mitchell, 1996).

Acknowledgements

The experiments were supported by the directorate "Raumfahrt" of the German Aerospace Center (DLR), by the former "Deutsche Agentur fuer Raumfahrtangelegenheiten," and by the European Space Agency (ESA).

REFERENCES

CARDIOVASCULAR REFLEX CONTROL BY AFFERENT FIBERS FROM SKELETAL MUSCLE RECEPTORS. IN. *HANDBOOK OF PHYSIOLOGY*. J.H. Mitchell, R.F. Schmidt. The Cardiovascular System Vol. III Bethesda, MD: American Physiological Society, pages 623–58; 1983.

CENTRAL VENOUS PRESSURE IN SPACE. J.C. Buckey, Jr., F.A. Gaffney, L.D. Lane, B.D. Levine, D.E. Watenpaugh, S.J. Wright, W. Yancy, Jr., D.M. Meyer, C.G. Blomqvist. *J. Appl. Physiol.*, pages 19–25; 1996.

NEURAL CONTROL OF THE CARDIOVASCULAR SYSTEM: INSIGHTS FROM MUSCLE SYMPATHETIC Nerve Recordings in Humans. Mitchell J.H., Victor R.G. *Med. Sci. Sport Exer.*, 60-69; 1996.

ELECTRICAL IMPEDANCE TOMOGRAPHY. C.J. Kotre *Brit. J. Radiol.*, pages 200–205; 1997.

CARDIOVASCULAR RESPONSES TO LOWER BODY NEGATIVE PRESSURE STIMULATION BEFORE, DURING, AND AFTER SPACE FLIGHT. F. Baisch, L. Beck, G. Blomqvist, G. Wolfram, J. Drescher, J.-L. Rome, C. Drummer. *Eur. J. Clin. Invest.*, pages 1055–1065; 2000.

RENAL ADJUSTMENT TO MICROGRAVITY. P. Norsk. *Eur. J. Physiol.*, Vol. 441, Suppl: pages R62–R65; 2000.

QUANTITATIVE EVALUATION OF THE PERFORMANCE OF DIFFERENT ELECTRICAL TOMOGRAPHY DEVICES. G. Hahn, F. Thiel, T. Dudykevych, I. Frerichs, E. Gersing, G. Hellige. *Biomed. Tech.*, pages 91–95; 2001.

Section 5 Circadian Rhythms, Sleep, and Respiration

Background *How spaceflight influences circadian rhythms, sleep, and respiration.*

INTRODUCTION

The light and dark cycles of the 24-hour day are fundamental to life on Earth. As much of modern life moves to round-the-clock operations, the limits that circadian (circa—around, dia —the day) rhythms place on people become more apparent. The trucking, shipping, and airline industries have all experienced accidents caused by people trying to stay awake during the time when their bodies wanted them to sleep.

Circadian rhythms also have importance in space. Studies on simple organisms have shown that weightlessness itself can affect circadian rhythms. Changes in circadian rhythms could also be one possible contributor to the poor sleep astronauts often experience on Shuttle missions. Shuttle crewmembers have reported an average sleep period of five to six hours, compared to the typical period of seven to eight hours on Earth. On missions, astronauts may have to become shift workers to handle the myriad tasks related to the Shuttle and the mission. There is little privacy aboard the Shuttle, quarters are confined, and noises or other interruptions may occur. One survey shows that more than 50% of crewmembers use sleeping medication at some point during their missions. Medications like melatonin, which affects circadian rhythms and has few side effects, have been proposed for treating sleep problems in space.

Poor sleep can result from factors in addition to a disruption of circadian rhythms. Studies on Earth have shown a strong relationship between the disruption of breathing patterns and poor sleep. An extreme example is patients with sleep apnea, who awake many times during the night because their breathing has temporarily been obstructed. The obstruction leads to an increase in carbon dioxide in the body, which provides a strong stimulus to wake up. In space, altered sensitivity to levels of carbon dioxide or oxygen in the blood might affect when the crewmembers wake up. On the Neurolab mission, a series of experiments examined the basic effects of weightlessness on circadian rhythms in rats, whether melatonin could be an effective sleep aid for astronauts, and how breathing might affect sleep in weightlessness.

CIRCADIAN RHYTHMS AND SLEEP

The top left diagram in Plate 6 shows how circadian rhythms are regulated. The suprachiasmatic nucleus in the brain is the site of the biological circadian pacemaker. This area receives information on light levels from the eyes, and in turn signals the pineal body to release the hormone melatonin. Levels of melatonin in the blood are strongly associated with the need to sleep. This association of melatonin with sleepiness has led to the suggestion that it might be an effective sleep aid for people who were sleeping poorly because their circadian rhythms were disrupted.

The bottom two diagrams in Plate 6 show the upper airway in cross section. In one-G, gravity moves the tongue and tissues of the neck into the airway. In weightlessness, however, the airway would not experience this compression. As a result, breathing might be improved slightly in space compared to lying supine on Earth.

Studying sleep in space requires significant instrumentation, and this is illustrated in the upper right panel in Plate 6. A crewmember is shown wearing the equipment needed to measure sleep accurately in space. The sleep net headgear contains electrodes to pick up brainwaves (electroencephalogram (EEG)). These brainwaves are used to classify sleep into different stages. The sensors underneath the nose measure airflow. These measurements, when compared with the movements of the chest wall and abdomen, can show if breathing is normal (chest wall and abdominal movements produce airflow) or obstructed (chest wall and abdominal movements don't produce airflow). Other sensors measure heart rate, snoring, and the oxygen level in the blood. This equipment is used for determining how weightlessness affects sleep and whether a countermeasure for sleep loss (like melatonin) would improve sleep.

THE NEUROLAB SLEEP, CIRCADIAN RHYTHM, AND RESPIRATORY EXPERIMENTS

Dr. Charles Fuller and his colleagues measured circadian rhythms in adult rats. Heart rates and body temperature in animals show strong circadian rhythms on Earth. Dr. Fuller's team found that the rats' heart rate and body temperature rhythms changed in weightlessness. This was the first demonstration that microgravity affects the fundamental properties of the circadian timing system in mammals. Dr. Cziesler and his group performed a placebo-controlled, double-blind study of oral melatonin as a countermeasure for sleep disturbances.

No beneficial effects of melatonin were observed. These investigators, however, did note circadian rhythm abnormalities in the crewmembers, along with sleep loss.

The team from Dr. West's laboratory showed that snoring and episodes of airway obstruction were virtually eliminated in space. The sensitivity to carbon dioxide was unaltered. They concluded that sleep disruption in microgravity is not the result of respiratory factors.

The experiments showed that circadian rhythms change in weightlessness. Why exactly these changes occur—and how they affect sleep in zero-G—is still not explained. A better understanding of circadian rhythms is needed to cope with round-the-clock stresses both on Earth and in space.

Plate 6: Circadian Rhythms and Sleep

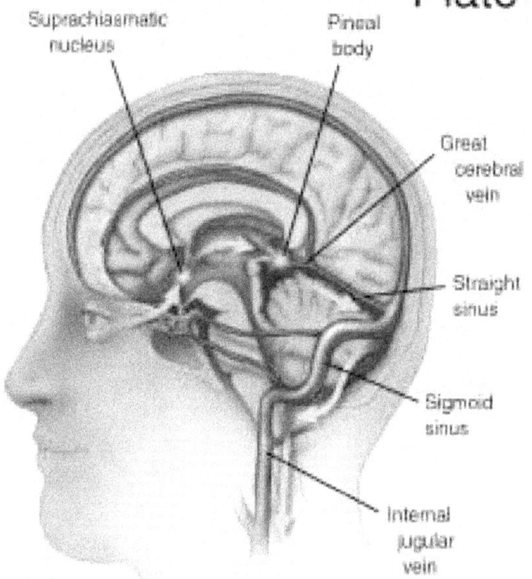

Suprachiasmatic nucleus
Pineal body
Great cerebral vein
Straight sinus
Sigmoid sinus
Internal jugular vein

Release of Melatonin into Circulation

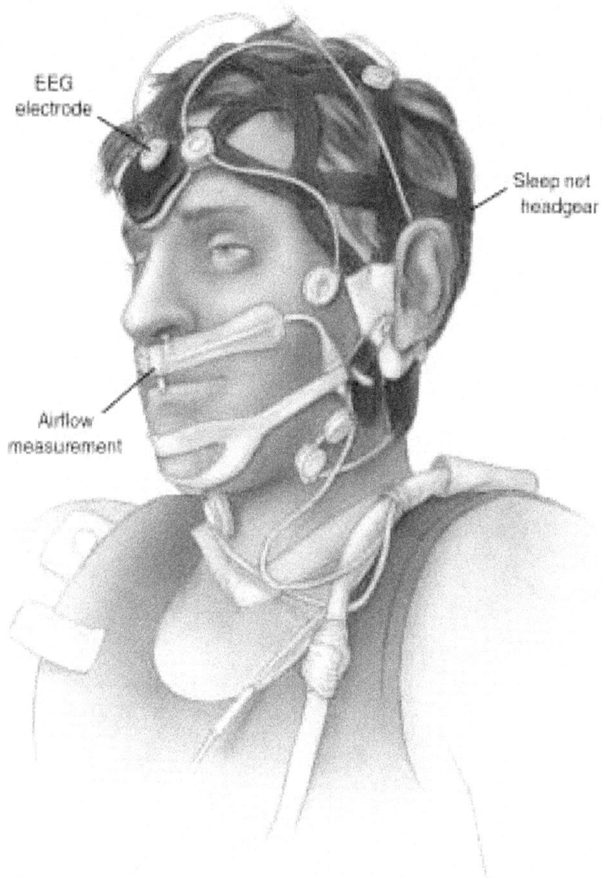

EEG electrode
Sleep net headgear
Airflow measurement

Subject Wearing Sleep Net Headgear

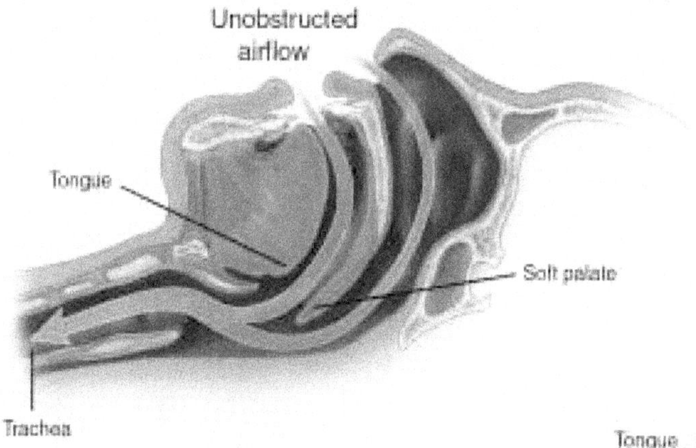

Unobstructed airflow
Tongue
Soft palate
Trachea

Airway at zero-G

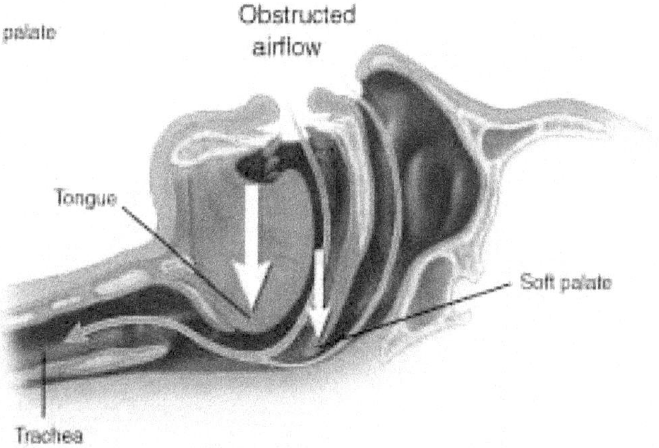

Obstructed airflow
Tongue
Soft palate
Trachea

Airway at one-G

Sleep, Circadian Rhythms, and Performance During Space Shuttle Missions

Authors

Derk-Jan Dijk, David F. Neri, James K. Wyatt,
Joseph M. Ronda, Eymard Riel, Angela Ritz-De Cecco,
Rod J. Hughes, Charles A. Czeisler

Experiment Team

Principal Investigator: **Charles A. Czeisler[1]**

Co-Investigators: **Derk-Jan Dijk,[1] David F. Neri,[1,2] James K. Wyatt,[1] Joseph M. Ronda,[1] Rod J. Hughes[1]**

Technical Assistants: **Eymard Riel,[1] Karen Smith,[1] Angela Ritz-De Cecco[1]**

Engineers: **Jennifer Jackson,[1] Jerzy Krol-Sinclair[1], Alex McCollom[1]**

[1]Harvard Medical School and Brigham and Women's Hospital, Boston, USA
[2]NASA Ames Research Center, Moffett Field, USA

ABSTRACT

Sleep and circadian rhythms may be disturbed during spaceflight, and these disturbances can affect crewmembers' performance during waking hours. The mechanisms underlying sleep and circadian rhythm disturbances in space are not well understood, and effective countermeasures are not yet available. We investigated sleep, circadian rhythms, cognitive performance, and light-dark cycles in five astronauts prior to, during, and after the 16-day STS-90 mission and the 10-day STS-95 mission. The efficacy of low-dose, alternative-night, oral melatonin administration as a countermeasure for sleep disturbances was evaluated. During these missions, scheduled rest-activity cycles were 20–35 minutes shorter than 24 hours. Light levels on the middeck and in the Spacelab were very low; whereas on the flight deck (which has several windows), they were highly variable. Circadian rhythm abnormalities were observed. During the second half of the missions, the rhythm of urinary cortisol appeared to be delayed relative to the sleep-wake schedule. Performance during wakefulness was impaired. Astronauts slept only about 6.5 hours per day, and subjective sleep quality was lower in space. No beneficial effects of melatonin (0.3 mg administered prior to sleep episodes on alternate nights) were observed. A surprising finding was a marked increase in rapid eye movement (REM) sleep upon return to Earth. We conclude that these Space Shuttle missions were associated with circadian rhythm disturbances, sleep loss, decrements in neurobehavioral performance, and alterations in REM sleep homeostasis. Shorter than 24-hour rest-activity schedules and exposure to light-dark cycles inadequate for optimal circadian synchronization may have contributed to these disturbances.

INTRODUCTION

Travel is often associated with disturbed sleep, and spaceflight is no exception. In fact, spaceflight combines elements of jetlag and shift work—both of which are associated with disturbed sleep. In space, crewmembers may experience high workloads, anxiety, excitement, space motion sickness symptoms, and a noisy and often uncomfortably cold or warm sleeping environment. Even now, more than 40 years after the first human spaceflight, space travel is still an adventure, and not without risk, anxiety, and stress. The Space Shuttle, although much more comfortable than early Apollo cabins, is certainly not comparable to a hotel room. The pilot and commander often sleep in their chairs on the flight deck. Fortunate astronauts may have their own private sleep "cabinet." More commonly, they must prepare for sleep by attaching to the wall of the middeck or Spacelab with Velcro. Astronauts are often scheduled to rise earlier every day, advancing their bed and wake times by five hours or more during the course of a mission; this is similar to the time zone change imposed by an eastbound trip from the United States to Europe. At the same time, either a sunrise or a sunset occurs every 45 minutes while the Space Shuttle is in low Earth orbit, sending potentially disruptive signals to the circadian pacemaker in those exposed to these 90-minute "days."

Biological Rhythms and Sleep Regulation

Earth is a highly periodic planet, and its environments are characterized by tidal, daily, monthly (lunar), and annual cycles. Natural selection has favored organisms equipped with internal biological clocks, particularly daily (circadian) clocks. Such innate clocks allow these living systems to anticipate the periodic environmental changes produced as Earth rotates around its axis. In mammals, circadian oscillations in nearly every aspect of physiology and behavior are driven by a pacemaker, the biological master clock that is located in the suprachiasmatic nuclei (SCN) of the anterior hypothalamus in the brain. The SCN drives circadian rhythmicity via both nerves and hormones. Circadian oscillations are generated by feedback loops of clock genes and their gene products. Indeed, Earth's rotations around its axis are engraved in the genome of nearly all living systems, including humans. Even when humans leave Earth, the circadian system travels along.

In the absence of a periodic environment, the human circadian clock oscillates at its intrinsic period—on average 24 hours and 11 minutes (Czeisler, 1999) in sighted people. When living on Earth, these circadian oscillations are synchronized to the 24-hour day (and resynchronised after humans travel through time zones) by the light-dark cycle. During low Earth orbit, the 24-hour natural light-dark cycle is absent and is replaced by a 90-minute external light-dark cycle related to the orbit of the Space Shuttle around the Earth. This interacts with the near-24-hour light-dark cycle associated with the activity-rest schedules produced by electric lamps and window shades aboard the spacecraft. This complex light-dark cycle may be inadequate to synchronize (or entrain) the human circadian clock to the scheduled rest activity, which often has an average period that is somewhat shorter than 24 hours.

In humans, the circadian pacemaker plays a pivotal role in sleep-wake regulation, and it is important to sleep and wake up at the appropriate phase of the circadian cycle (Dijk, 1995). The circadian clock enables people to stay awake and perform well for a full 16-hour waking day and then sleep well at night. The clock does this by providing a wake-promoting signal that becomes progressively stronger throughout the normal waking day. This signal suddenly dissipates at around 22–23:00, and is replaced by a sleep-promoting signal. Ground-based studies, in which sleep and wakefulness were scheduled to occur at all phases of the internal circadian cycle, have demonstrated that this circadian rhythm of sleep propensity is closely associated with the rhythm of plasma melatonin. This rhythm, which reflects variation in the synthesis of melatonin by the pineal gland, is driven by the SCN. The nocturnal rise of melatonin may help people fall asleep at night by quieting the output of the SCN, thereby silencing the wake-promoting signal. When individuals attempt to sleep outside this phase of melatonin secretion, sleep is disrupted. Ground-based research has shown that administration of melatonin when sleep is attempted at circadian phases at which the body's own melatonin is absent increases sleepiness, improves sleep consolidation, and may facilitate the synchronization of the circadian clock to the desired rest-activity cycle.

Circadian Rhythms and Spaceflight

Over the years, NASA has developed regulations designed to reduce the magnitude of daily shifts in scheduled sleep and to protect scheduled crew rest time. (Although a procedure called "slam shifting," which involves abrupt shifts of up to 12 hours, is now used to align the sleep-wake schedules of Space Shuttle and International Space Station crews upon docking.) Moreover, astronauts are often highly motivated to complete necessary repairs or payload activities after hours. They may be called upon to deal with "off-nominal" situations at all times of day or night, and may stay up later than the scheduled bedtime. Nonetheless, in most cases, they will be awakened at the scheduled wake time regardless of how late they retired.

Data on medication use in space supports the conclusion that sleep is disturbed during Space Shuttle missions. NASA analysis of 219 records of the use of medication during low Earth orbit Space Shuttle missions found that sleeping pills are the most commonly used medication. Sleep medications were reportedly used by astronauts throughout many missions, in contrast to the motion-sickness remedies that are used primarily during the first few days of such missions. The frequent use of sleeping pills is all the more remarkable given that astronauts carry a high homeostatic sleep pressure (i.e., have a strong need to sleep) during the mission, since they sleep on average only six to 6.5 hours per day in space (Monk, 1998; Santy, 1988).

On Earth, sleep disturbance and chronic sleep restriction lead to decrements in daytime performance that jeopardize productivity and safety in the workplace. There is no reason to believe that this would be different in space. Hypnotics that bind to the GABAa-benzodiazepine receptor complex (the most commonly prescribed sleeping pills) have been shown to impair

The Neurolab Spacelab Mission: Neuroscience Research in Space

daytime performance and the ability to respond quickly and adequately when awakened from hypnotic-induced sleep.

It remains unknown whether there are aspects specific to space travel and the space environment that disrupt sleep. Is the timing of sleep disrupted because astronauts' biological clocks are no longer exposed to the Earth's 24-hour day-night cycle? Is the sleep stage composition of sleep need altered in space because floating in microgravity is less fatiguing than walking upright on Earth?

To answer these questions, we investigated sleep, performance, and circadian rhythms in five astronauts (STS-90: four; STS-95: one) and studied the light-dark cycles to which crewmembers were exposed during these two Space Shuttle missions. The efficacy of melatonin as a countermeasure for sleep disturbances during spaceflight was investigated as well. To assess the impact of spaceflight, extensive baseline measurements were obtained prior to the flights. Astronauts agreed to refrain from the use of hypnotics so that the effects of spaceflight on unmedicated sleep could be evaluated. Effects of re-adaptation to the Earth environment were evaluated by recording sleep, performance, and circadian rhythms immediately after return. The results of our sleep experiment and the experiment of our team members on respiration during sleep in space have been reported in full elsewhere (Dijk, 2001; Elliott, 2001). Here we present a brief overview of the background to this research, summarize the main results, and compare our results to those of other researchers who have investigated sleep in space and on the ground.

METHODS

Work-Rest, Rest-Activity Schedules during STS-90 and STS-95

The work-rest schedules of astronauts are tailored to the scheduled time of launch and reentry. The crew of both STS-90 and STS-95 were stationed at the Johnson Space Center (JSC) in Texas and lived on the local time zone (Central Daylight Time (CDT)). Three to four days prior to scheduled launch they left for the Kennedy Space Center (KSC) in Florida, travelling eastward through one time zone (Eastern Time zone).

The launch windows for STS-90 and STS-95 were both located conveniently in the afternoon. In fact, the launch schedules of STS-90 and STS-95 were specifically designed to avoid the necessity of prelaunch circadian phase shifting. To achieve this, launch occurred at 14:19 Eastern Daylight Time for STS-90 and 14:20 Eastern Standard Time for STS-95. Crewmembers woke about seven hours before launch, and the first eight-hour sleep episode in space was scheduled to begin at 01:00 hours. Figure 1 is a raster plot of the mission schedule; it illustrates that the sleep-wake schedules in space were not identical to a normal 24-hour cycle. Astronauts were to rise and retire earlier every day by about 20 (STS-90) and 35 minutes (STS-95). This was done to assure that on the day of reentry, sleep-wake schedules were timed appropriately. Bed and wake times advanced by as much as five hours in the course of these missions.

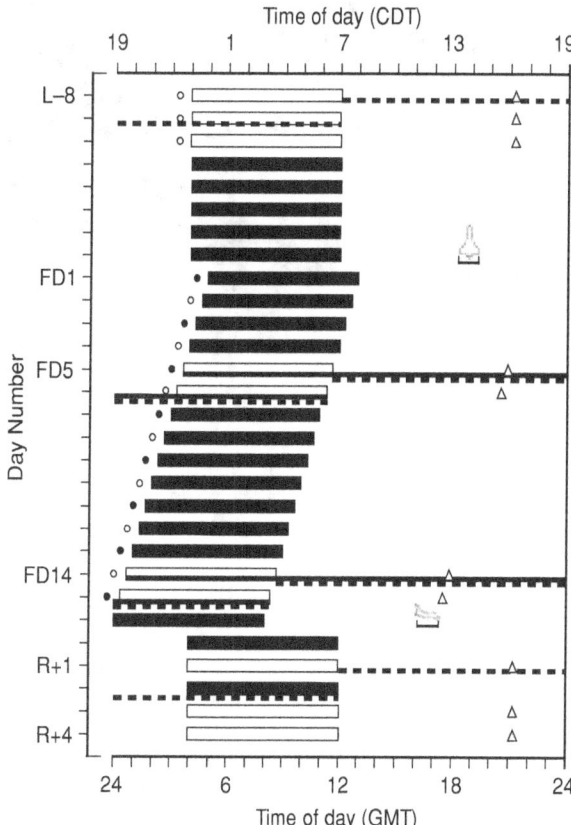

Figure 1. Experimental schedule and scheduled time of the rest-activity cycles for the Neurolab mission (STS-90). This raster plot illustrates when measurements were made shortly prior to (L–), during (FD), and after (R) the mission. Additional baseline measurements were made two months and one month prior to flight. In this figure, flight days (FDs) are numbered such that FD1 starts at launch and ends at the end of the first inflight sleep episode. Timing of scheduled polysomnographic recordings is indicated by open horizontal bars; neurobehavioral performance tests by triangles; urine collection sessions by dashed horizontal lines; and body temperature recordings by a solid horizontal line. Placebo (open circles) and melatonin (closed circles) administration prior to scheduled sleep episodes is indicated for one treatment group (see Dijk et al. (Dijk, 2001) for a description of differences between treatment groups). Time of day is indicated as Central Daylight Time (CDT: upper horizontal axis) and Greenwich Mean Time (GMT: lower horizontal axis). (From Dijk, 2001, with permission; reproduced from *The American Journal of Physiology.*)

Measuring Sleep and Wakefulness

To determine how well the astronauts could adhere to the rest-activity schedules, we used actigraphy—a continuous recording of activity of the nondominant wrist (Figure 2). Ground-based and flight-based studies have shown that such actigraphic recordings can be analyzed to obtain a reliable estimate of total sleep time and other sleep parameters. Figure 3 illustrates the rest-activity cycle in one crewmember of STS-90.

Figure 2. Long-term recording of rest-activity cycles can be accomplished by actigraphy. In this photograph taken during the STS-95 mission, Chiaki Mukai is wearing an actigraph on her nondominant wrist.

Polysomnographic Recording of Sleep

To investigate how sleep structure (i.e., the different stages of sleep from light to deep) was affected, we recorded brainwaves (electroencephalogram (EEG)), submental muscle tone (electromyogram (EMG)), and eye movements (electro-oculogram (EOG)) during sleep episodes in the first half and second half of the missions (Figure 1). This kind of recording is called polysomnography. These data were compared to similar recordings obtained during three sessions prior to (two months, one month, and one week beforehand) and immediately after the mission. The recordings were made using a digital sleep recorder and a specially designed sleep net (Figure 4). (See also technical report by Dijk et al. in this publication.) Human sleep consists of non-rapid eye movement (nREM) sleep and REM sleep. nREM sleep can be further subdivided into stage 1, 2, 3, and 4 sleep. Stages 3 and 4 (also called slow wave sleep (SWS)) are often considered the deepest stages of sleep and are thought to be important for the recovery aspects of sleep (Figure 5). During normal nocturnal sleep on Earth, SWS declines in the course of a sleep episode and REM sleep increases. During a normal night, there will be some wakefulness as well. It takes time to fall asleep and at the end of the night, individuals may wake up several times before finally getting out of bed.

Light-Dark Cycles and Circadian Rhythms

Extensive research on the effects of light exposure on human circadian rhythms has established that the light-dark cycle is a powerful synchronizer of the human circadian pacemaker. Scheduled light exposure can be used to shift the rhythms of astronauts prior to those missions in which the launch window dictates rest-activity cycles to be timed out of synchrony with the normal rest-activity cycle on Earth (Czeisler, 1991). Light levels lower than ordinary room light can modulate the timing of circadian rhythms, and can also exert direct effects on alertness (along with its EEG and EOG correlates) and neuroendocrine variables such as melatonin synthesis. Although the human

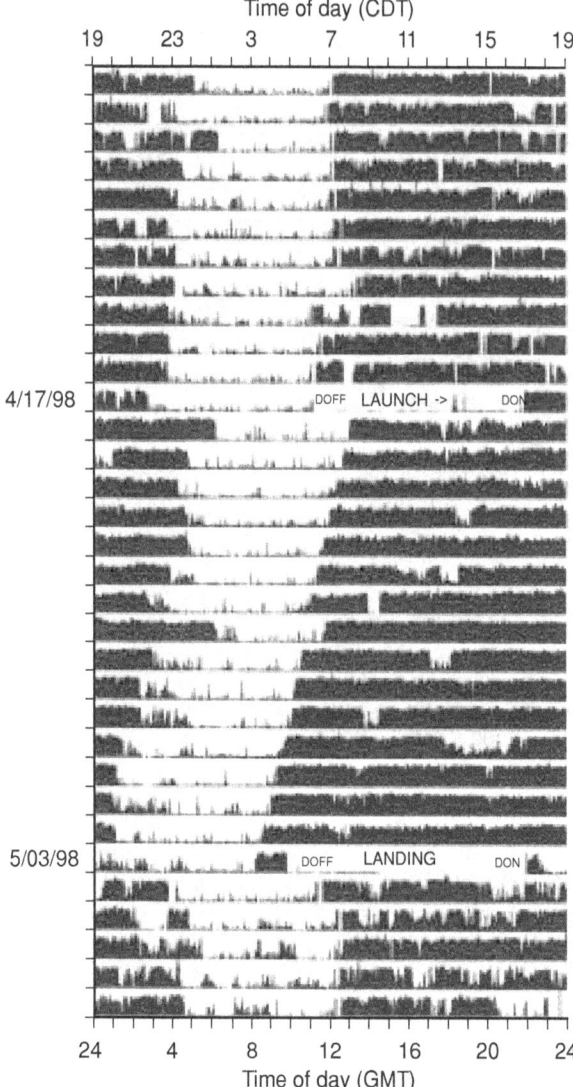

Figure 3. Recording of the rest-activity cycle of an astronaut prior (before 4/17/98), during (4/17/98-5/03/98), and after the Neurolab mission by actigraphy. Consecutive 24-hour segments are plotted below each other. No actigraph was worn at the time of launch or landing. Note the regular advance of activity onset and the variability in the offset of activity during spaceflight, and the acute delay in the rest-activity cycle upon return to Earth. (From Dijk, 2001, with permission; reproduced from *The American Journal of Physiology*.)

circadian pacemaker is most sensitive to light during the biological night—i.e., when melatonin is present in plasma—light exposure during the biological day also exerts effects on circadian phase. This implies that complex light-dark cycles and unscheduled light exposure during spaceflight may have both beneficial and detrimental effects on the synchronization of the human circadian pacemaker. To assess this, light levels were recorded throughout the mission in the flight deck, middeck, and Spacelab using Actillume light recorders (Ambulatory Monitoring, Inc., Ardsley, NY).

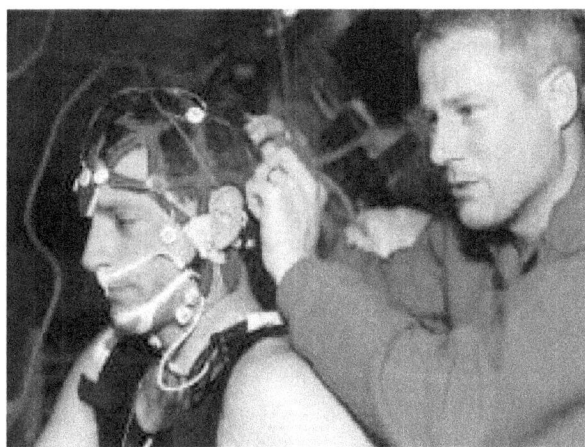

Figure 4. Brainwaves, eye movements, and airflow were recorded by means of a sleep net and additional sensors. This photograph shows payload commander Rick Linnehan being instrumented by payload specialist Jay Buckey during the STS-90 mission.

To determine whether the crewmember's internal circadian rhythms during these missions maintained adequate synchronization with the rest-activity schedules, we measured core body temperature and urinary cortisol rhythms on two occasions during the STS-90 mission (Figure 1) as well as prior to and after the mission. Both body temperature and urinary cortisol have strong circadian rhythms and can be used to follow the changes in circadian rhythms.

Performance and Mood

One important question was whether changes in sleep duration and circadian rhythms would be related to a deterioration of performance during wakefulness. This question is difficult to answer directly because no data exist on performance during spaceflights during which there were no sleep and circadian rhythm disturbances. We could, however, compare performance during spaceflight with performance prior to and after spaceflight. On STS-90 and STS-95, we assessed performance and mood on a specific test battery designed in collaboration

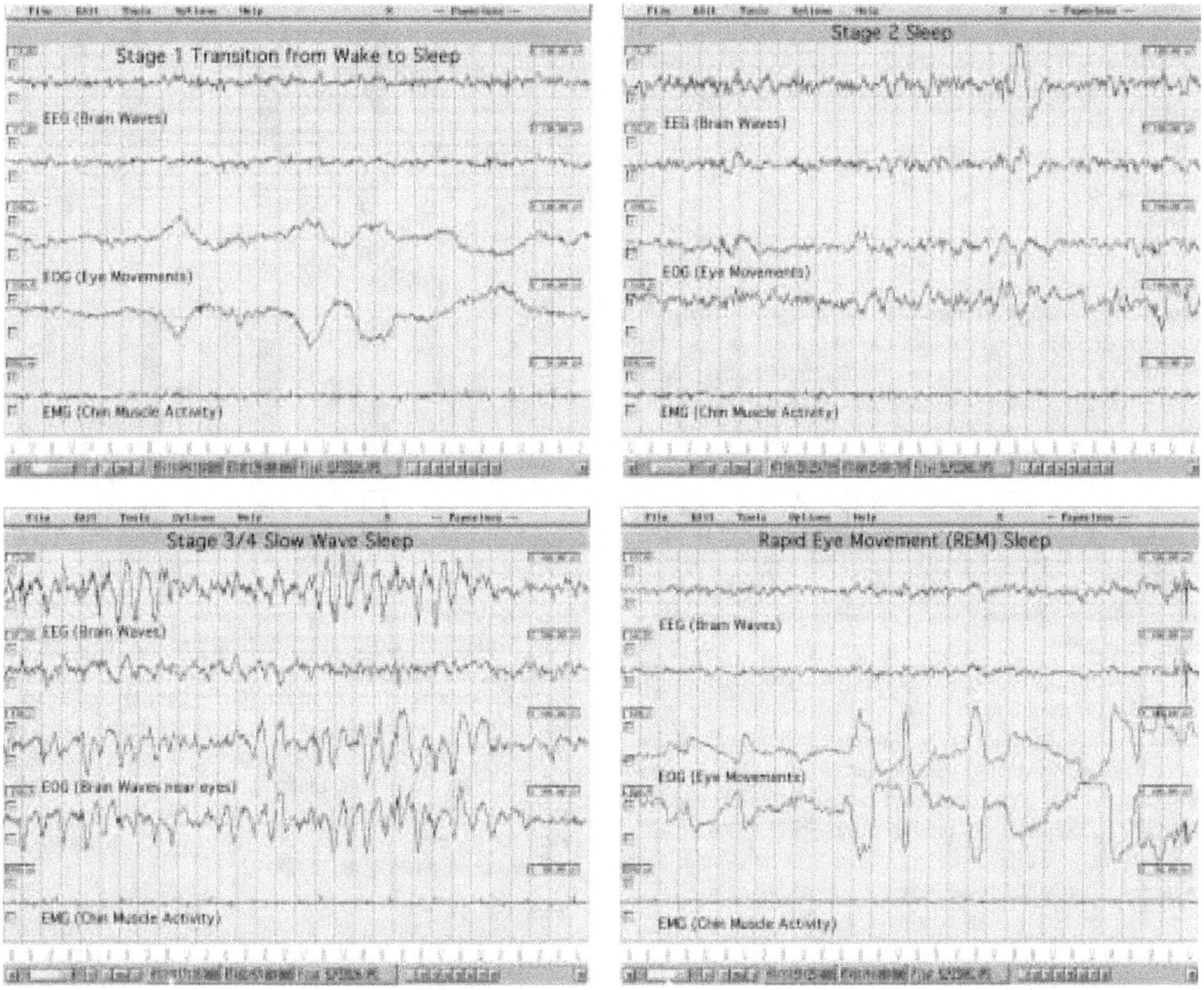

Figure 5. Brainwaves, eye movements, and submental muscle tone during stages of sleep. These traces were recorded during the Neurolab mission and downlinked to JSC immediately after the recording. This allowed researchers on the ground to inspect the quality of sleep recordings during the mission.

with Dr. David F. Dinges of the University of Pennsylvania. This test battery included tests of memory, calculation ability, vigilance, and coordination. In addition, it had several subjective scales to rate sleepiness and mood.

Melatonin

One other objective of our studies was the evaluation of the efficacy of melatonin as a countermeasure for the disturbances of sleep and circadian rhythms during spaceflight. Melatonin is thought to be devoid of some of the side effects of standard hypnotics and could be an alternative to typical sleeping pills. We sought to assess objectively the efficacy of melatonin in space. In preparation for the Neurolab mission, we first evaluated the efficacy of two doses of melatonin (0.3 and five mg) in a ground-based study. In this experiment, the efficacy of melatonin was investigated at all circadian phases by scheduling subjects to a 20-hour sleep-wake routine. During the scheduled wake episodes, light levels were low (<five lux). Our initial analyses of these data indicated that 0.3 mg was as effective as five mg in inducing sleep when sleep occurred outside the phase of endogenous melatonin secretion. With the 0.3-mg dose, elevation of plasma melatonin concentration returned to baseline after the sleep episode. We therefore selected the 0.3-mg dose for evaluation in space in a double-blind, placebo-controlled experiment in which placebo and melatonin nights alternated.

RESULTS

Actigraphy

Figure 3 shows the 24-hour rest-activity cycle prior to flight (and the approximately one-hour phase-advance associated with the travel from JSC to KSC) in one astronaut. Inflight, the progressive phase-advance of wake time associated with the shorter-than-24-hour sleep-wake schedule is clearly visible. Note the major deviation from this schedule on FD8 due to operational demands. It is also interesting to note that the day-to-day variability in the onset of activity (wake time) was much smaller than the day-to-day variability in offset of activity (bedtime). After landing, we can see the abrupt approximately four-hour phase delay of the rest-activity cycle, comparable to flying westward through four time zones. Actigraphic recordings in the other astronauts gave very similar patterns. Quantitative analysis showed that according to these actigraphic recordings, astronauts' average daily sleep period time (i.e., the time from sleep onset to final awakening) was on average only 427.6 (SE: 6.8) minutes; this was approximately 30 to 40 minutes less per night than during sleep episodes prior to and after flight. During these seven hours, a half-hour was spent awake such that total sleep time was approximately 6.5 hours. On some nights, total sleep time was reduced to as short as 3.8 hours.

Polysomnographic Recordings

Figure 6 illustrates the time course of wakefulness, SWS, and REM sleep during sleep episodes recorded in space, prior to spaceflight, and after return to Earth. SWS declines from the first

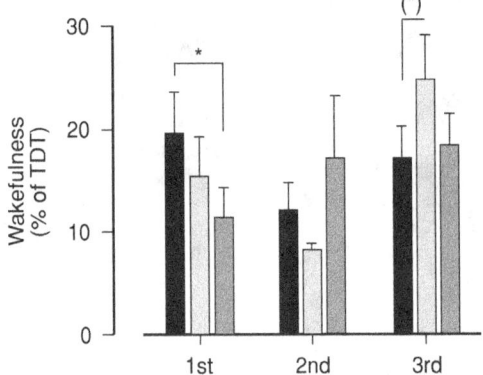
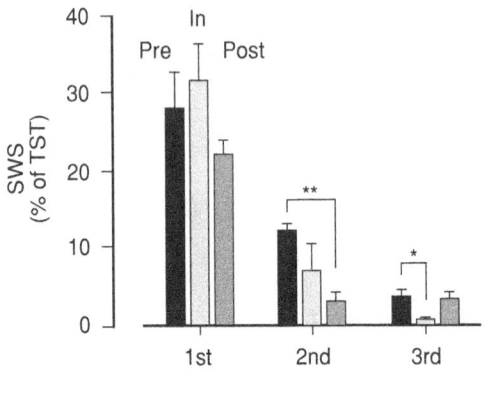
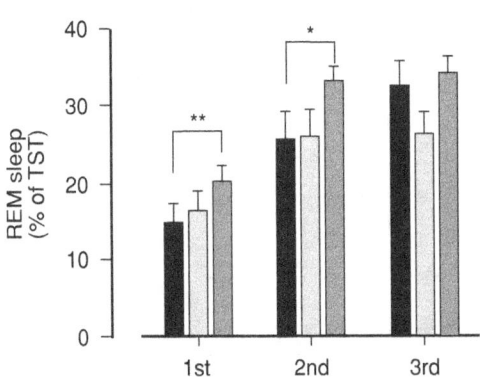

Figure 6. Wakefulness (upper panel), SWS (middle panel), and REM sleep (lower panel), in the first, second, and final third of sleep episodes during the preflight (first solid bar), inflight (second grey bar), and postflight (third solid bar) segments. Wakefulness is expressed as percentage of total dark time (TDT). SWS and REM sleep are expressed as percentage of total sleep time (TST). N=5; Sleep episodes after intake of melatonin are not included. Significance of pair-wise comparisons are indicated: (*)=p<0.1; *=p<0.05; **=p<0.01. Figure based on data published in Dijk et al. (Dijk, 2001).

to the last third of the sleep episode under all three conditions. REM sleep increases from the beginning to the end of sleep. Thus, the overall structure and temporal organization of sleep were not markedly altered during and after spaceflight. It should be mentioned, though, that on average the astronauts rated their

216

sleep as of poorer quality than compared to sleep on Earth. Sleep efficiency (i.e., percentage of time spent asleep/time in bed) was just below 85% inflight, very similar to the preflight and postflight values. Detailed analysis revealed several intriguing alterations in sleep structure in space and upon return to Earth. In the first third of sleep, less wakefulness was present in space and after return than prior to the mission. Thus, astronauts were very well able to fall asleep despite their advanced sleep schedule (inflight) and delayed sleep time (postflight). Interestingly, there was a tendency for more wakefulness in the last third of sleep episodes during the inflight segment. This is surprising because during eastward travel (advance of sleep relative to the endogenous circadian rhythms), it is difficult to fall asleep and difficult to awaken on scheduled local time. SWS was reduced in the last third of inflight sleep episodes.

For REM sleep, the most marked changes occurred after return to Earth. There was more REM sleep in both the first and second third of sleep episodes recorded after landing. Furthermore the latency to REM sleep was significantly reduced from 86.4 (±13.1) minutes prior to flight to only 43.3 (±2.5) minutes postflight. During the second sleep episode after landing (no recording was obtained during the first sleep episode), REM sleep expressed as a percentage of total sleep time was as high as 32%, compared to 24% preflight.

Comparison of the estimates of sleep derived from actigraphy and polysomnography revealed a surprising difference. It appeared that astronauts slept longer when fully instrumented for sleep monitoring. Total sleep time (as derived from actigraphy) during polysomnographically recorded sleep episodes was near seven hours compared to less than 6.5 hours for the nights during which no full polysomnographic recording was obtained.

Light Levels

Light recordings on the flight deck throughout the STS-90 mission revealed very complex and highly variable light-dark cycles. An approximately 90-minute periodicity, associated with the orbit of the spacecraft around the Earth, is superimposed on a slightly shorter than 24-hour oscillation, associated with the scheduled rest-activity cycle (Figure 7). On the flight deck, light levels as high as 79,000 lux were observed. Although on the flight deck the shades were pulled down during scheduled sleep episodes, orbital dawn still entered the flight deck and

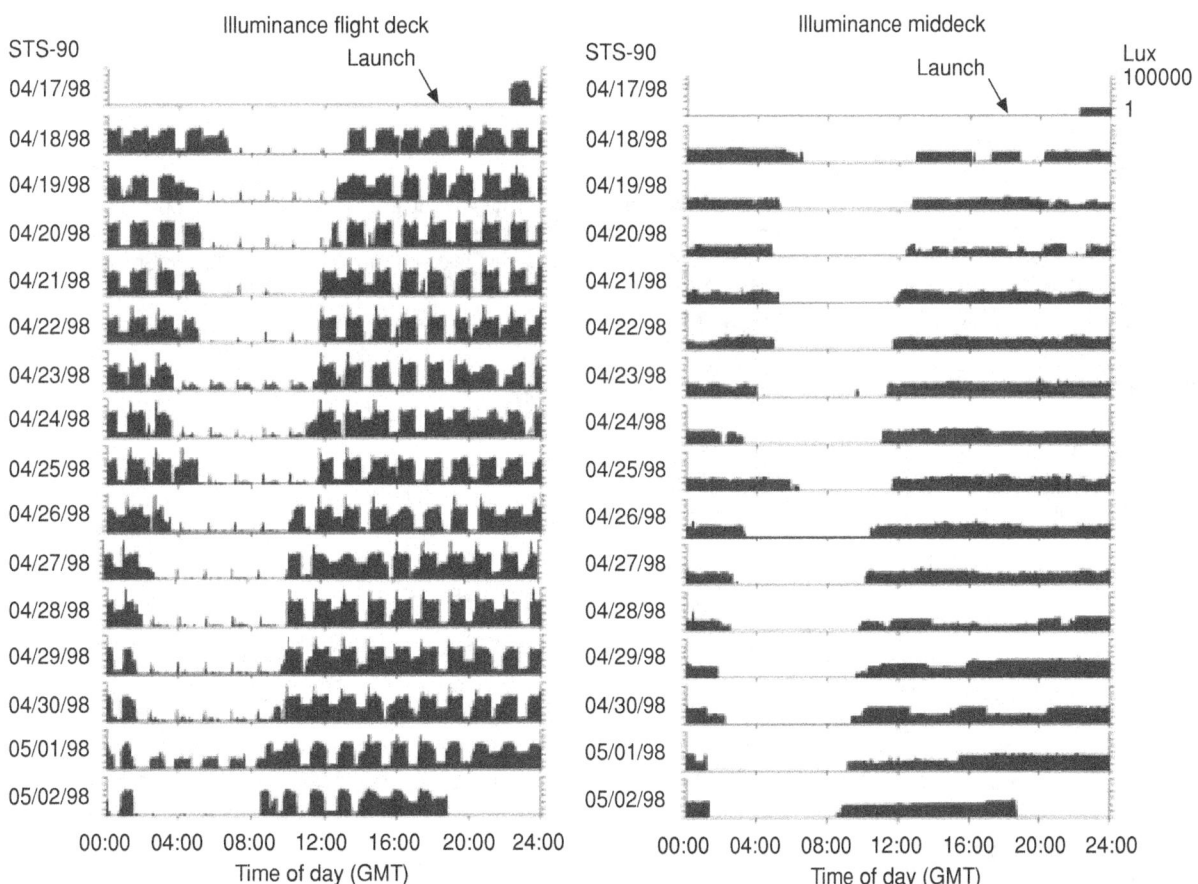

Figure 7. Illuminance on the flight deck and middeck during the Neurolab mission. Days are plotted below each other. Illuminace (lux) is plotted on a logarithmic scale. Please note the 90-minute recurrence of orbital dawn on the flight deck and the shorter-than-24-hour light-dark cycle on the middeck. Figure based on data from Dijk et al. (2001).

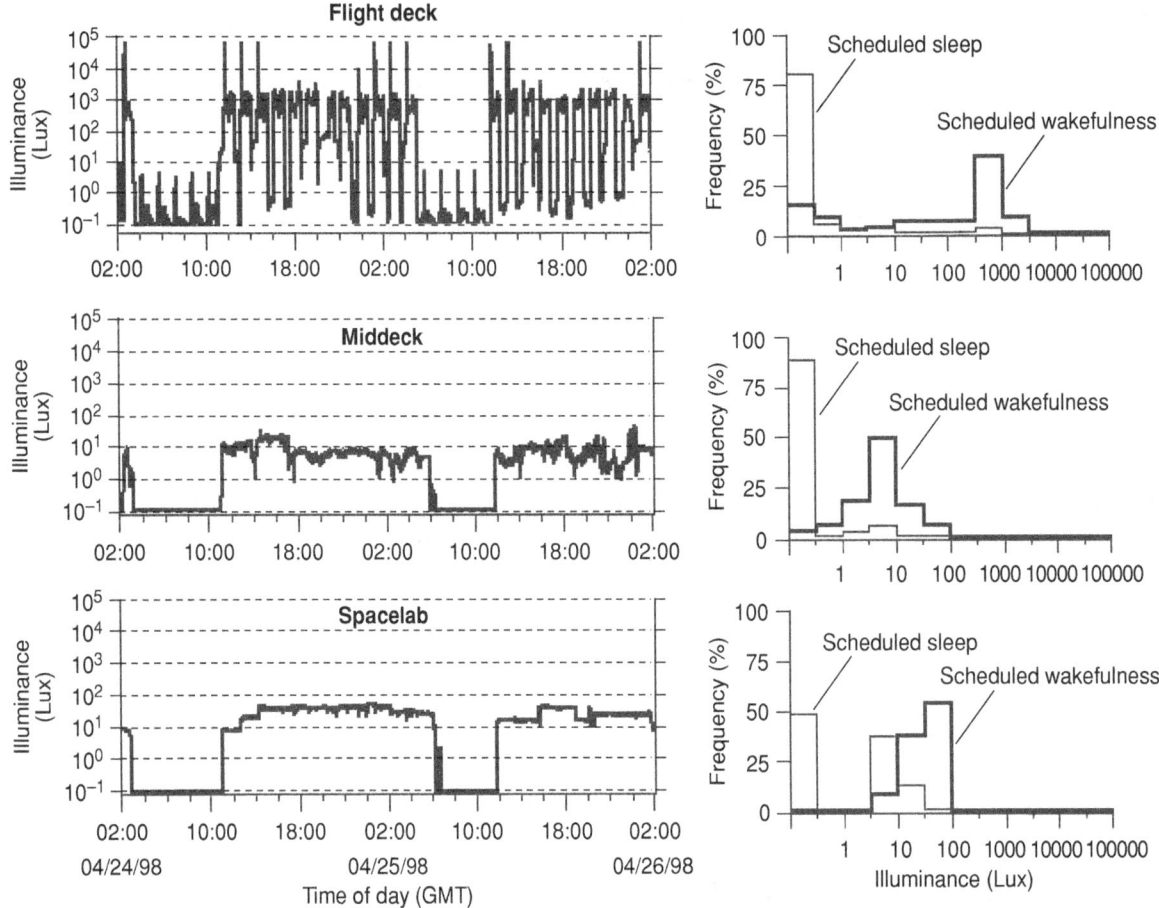

Figure 8. Time course of illuminance (48-hour segment) recorded on the flight deck, on the middeck, and in the Spacehab (left panels). Right panels: Relative frequency distribution of illuminances (bin width 0.5 log units) during scheduled wakefulness (heavy solid line) and scheduled sleep (thin solid line) for the flight deck (upper right panel), middeck (middle right panel), and Spacelab (lower, right panel). Frequency distributions were derived from all data collected during the STS-90 mission. (From Dijk, 2001, with permission; reproduced from *The American Journal of Physiology*.)

the average illuminance during scheduled sleep episodes was 73 lux (Figure 8). On the windowless middeck, the 90-minute periodicity was not present. Only the shorter-than-24-hour light-dark cycle associated with the rest-activity cycle characterized the light environment in this compartment. Light levels on the middeck were very low. The highest illuminance observed was 93 lux, and the mean value during scheduled wake episodes was only nine lux. To put these numbers in perspective: illuminance on the surface of a desk in a well-lit room may be 300 to 500 lux. During a bright sunny day, ambient outdoor light intensity may reach 100,000 lux. The temporal pattern of illuminance in the Spacelab (STS-90) and Spacehab (STS-95) was similar, with average levels also rather low, although slightly higher than on the middeck. However, the Spacelab had a window that provided additional illumination when unshaded. The illuminances we recorded were obtained from light recording devices mounted on the interior walls of the spacecraft and, therefore, do not accurately represent light exposure of individual astronauts who moved around from one compartment to the other.

Circadian Rhythms (Body Temperature and Urinary Cortisol)

Several features of circadian rhythms in space emerged. Interestingly, the onset of the sleep episode was still associated with a drop in core body temperature, despite the absence of "postural" changes. In other words, "masking" of the endogenous circadian core body temperature rhythm by behavioral cycles is still present during spaceflight. Consequently, a simple recording of body temperature may not provide a reliable estimate of the phase of the circadian pacemaker, even in space. Nonetheless, the amplitude of the "masked" temperature rhythm was attenuated in space. Urinary cortisol secretion may be less affected by behavioral cycles and better reflect the status of the circadian pacemaker. We quantified the circadian rhythm of cortisol preflight, early in the flight, late in the flight, and postflight. Preflight and early inflight urinary cortisol reached a peak shortly after scheduled wake time and started to decline within two to four hours. In the second half of the mission, this decline did not occur until six hours after scheduled wake time.

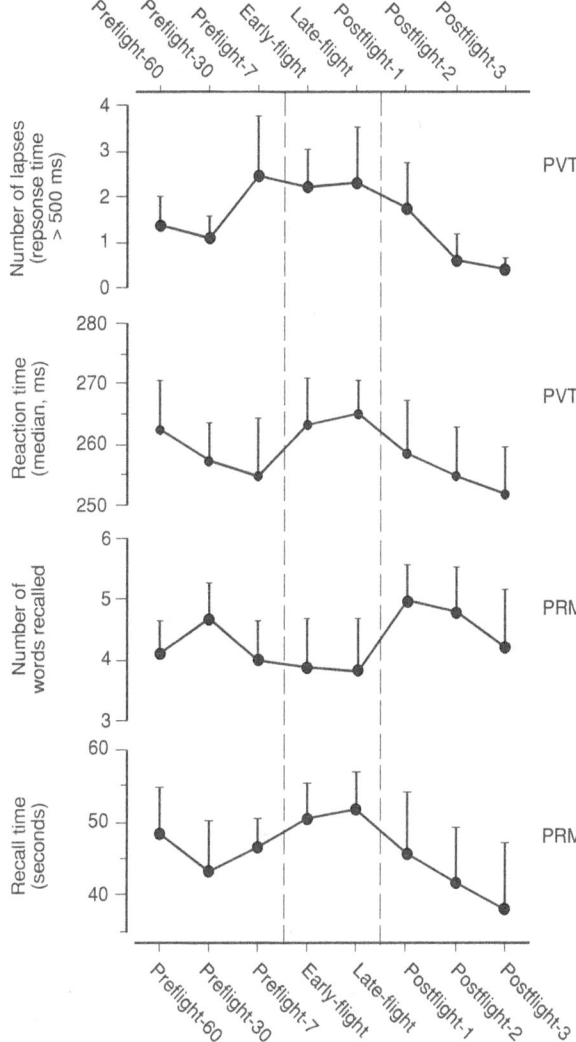

Figure 9. Neurobehavioral performance measures derived from the psychomotor vigilance test (PVT) and the probed recall memory (PRM) test during three preflight segments, two inflight segments, and three postflight segments. n=5 (except for postflight 1 when n=4). (From Dijk, 2001, with permission; reproduced from *The American Journal of Physiology*.)

Performance Measures

In general, several measures of performance indicated better performance on Earth and deterioration in flight. Some of the data are summarized here (Figure 9). Thus, the lowest number of lapses (i.e., reaction times longer than 500 ms) were observed two months and one month prior to flight, and on the third and fourth day after landing. Median reaction times were longer inflight. On a probed recall memory task, time to recall was longest, and fewer words were recalled inflight. After flight, recovery of performance occurred. Similar patterns were observed on other tasks as well as for subjective assessments of mood.

Interestingly, for some of the performance measures, decrements were already observed during the L–7 segment;

i.e., shortly before launch. This may be related to prelaunch apprehension or the increased workload leading to sleep loss prior to launch.

Melatonin Trial

Comparison of sleep after placebo and melatonin, however, did not reveal significant effects of melatonin on sleep in space. This may be related to the specific pattern of melatonin administration (alternating with placebo), the circadian phase of the administration of melatonin on this protocol, or changes in pharmacokinetics of melatonin during spaceflight.

DISCUSSION

Our analyses of sleep in space confirm and extend previous observations (see Stampi, 1994; Gundel, 1997; Monk, 1998; Dijk, 2001; and references therein). They confirm that sleep in spaceflight is shorter than on Earth, and the actigraphic recordings indicate that this problem may be more serious than polysomnographic sleep recording sessions would suggest.

Increase in REM sleep after flight

The effects on REM sleep after the mission are intriguing. Because REM sleep is under control of the circadian clock and sleep times were shifted throughout the mission and then suddenly shifted back upon return, circadian effects may in part explain this increase in REM sleep. However, our analyses of the circadian rhythms of core body temperature and urinary cortisol secretion do not support such an interpretation. Previously, (Frost, 1977) reported an increase in REM sleep after Skylab missions, and he also dismissed a circadian explanation for this phenomenon. An alternative explanation for this REM rebound is that it reflects a homeostatic response to the loss of REM sleep incurred during spaceflight. A more speculative explanation is that this massive increase in REM sleep represents a response to the re-adaptation to one-G. REM sleep has been implicated in learning processes related to sensory-motor tasks in particular. Could it be that "relearning" to walk on Earth is closely related to this REM sleep increase?

Longer sleep times with sleep instrumentation in place

The data show that on nights when the astronauts were wearing the complete sleep ensemble, their sleep was better than on other nights. This result might seem paradoxical since the sleep ensemble involved many electrodes and sensors. Our interpretation of this result relates to the effects of the experimental demands on the astronauts' behavior. The crewmembers who were not wearing the sleep ensemble would instrument those who were, and the non-instrumented crewmembers would try to get the others to sleep on time. On other nights, the crewmembers' adherence to their scheduled sleep-wake cycle was affected by other demands of the flight, whereas they may have seen it as their top priority to sleep at scheduled times during the nights they were fully instrumented. This interpretation was supported by an analysis of the actigraphically determined onset of the sleep episode on nights with and without sleep ensemble. The interval between scheduled bedtime and onset of the sleep episode was 42.1 (SE 14.1, n=5) and 15.9 (9.2, n=5) minutes for the non-monitored sleep and monitored sleep nights, respectively. Analyses of the time course of illuminance at the transitions of scheduled wake to sleep episodes provided further support for this interpretation. The lights on the middeck stayed on until as much as 30 to 40 minutes after scheduled bedtime.

Light Levels Varied Markedly, Perhaps Affecting Circadian Synchronization

The light-dark cycles during these Space Shuttle missions may not have been optimal for circadian synchronization. In fact, if astronauts were to visit the flight deck during their presleep leisure time, they might be exposed to a short light pulse of 60,000 lux or so. Ground-based research has shown that short exposures to bright light can be surprisingly effective at resetting circadian phase. Such exposure to light in the evening would be maladaptive, because bright light exposure at this biological time of day results in a delay of circadian rhythms that are contraindicated for adaptation to a shorter-than-24-hour rest-activity cycle when phase advances are required.

Circadian Rhythm Changes in Space

The results from the urinary cortisol measurements show that the circadian system was unable to keep pace with the advancing sleep-wake schedule on these missions. This is consistent with the results of recent ground-based simulations of human circadian adaptation to a 23.5-hour sleep-wake schedule in similar lighting conditions. Our data contrast with the findings of Dr. Monk and colleagues during the STS-78 mission on which the rest-activity schedules were very similar. This apparent discrepancy may be related to differences in methods of analysis or, alternatively, to the lower light levels on STS-90 compared to STS-78.

CONCLUSIONS AND PERSPECTIVE

Our analyses highlight two specific aspects of the space environment that could contribute to the cumulative sleep loss seen in space: (1) the shorter-than-24-hour rest-activity cycles, and (2) the highly variable and, in some compartments of the spacecraft, very dim light-dark cycles.

Our data indicate that astronauts do not adapt fully to these schedules when exposed to these light-dark schedules. Recently, we investigated the ability of healthy volunteers to adapt to such non-24-hour rest-activity cycles in ground-based studies. The data demonstrated that in humans, the internal circadian oscillations are so robust that in the absence of adequate light-dark cycles, they will not synchronise to either a 23.5-hour rest-activity schedule (similar to Space Shuttle missions) or a 24.6-hour rest-activity schedule (the Martian day).

In this paper and in our full report, we have emphasized group data. Interindividual differences during spaceflight were, however, observed by us and by others (Gundel, 1997). Appreciation and investigation of such differences are warranted. For example, analysis of respiration during sleep in these astronauts revealed that sleep-disordered breathing was attenuated by spaceflight (see science report by Prisk et al. in this publication). For an understanding of sleep disturbances at the individual level, a comprehensive assessment of sleep physiology and circadian physiology may be required. Such integrative approaches may lead to some surprising interpretations of the changes in sleep duration observed during spaceflight (Dinges, 2001).

Circadian phase is a major determinant of sleep duration, structure, and performance; accurate assessment of circadian phase is a prerequisite for a reliable interpretation of sleep and performance data obtained during and after spaceflights. Implementation of nonintrusive methods to assess circadian phase, for example on the basis of salivary melatonin, may be considered. The status of the sleep homeostat is another major determinant of performance. Acute sleep curtailment and cumulative sleep loss will affect the sleep homeostat and performance. New methodologies are being evaluated to assess on line the status of the sleep homeostat by electrophysiologcal and ocular parameters that would allow continuous monitoring of performance capability.

Circadian phase and sleep homeostasis interact in their determination of performance. Attempts to develop biomathematical models, in which these aspects of sleep and performance and the effects of light on the circadian system are integrated, have been published already. Future success and refinement of such predictive models will depend on continued collaboration between biomathematicians and physiologists as well as continued acquisition of more data on sleep, circadian rhythms, and light exposure of astronauts prior to, during, and after short- and long-duration space missions.

Acknowledgements

We thank the crewmembers of the STS-90 and STS-95 missions for their participation and dedication; our "Sleep Team" members from UCSD: Ann Elliott, Kim Prisk, and John West, for guidance, expert advice, and inspiration; the scientific and technical staff at NASA-JSC and Lockheed Martin: Mel Buderer, Suzanne McCollum, Peter Nystrom, Sherry Carter, Floyd Booker, Armando DeLeon, and Carlos Reyes for guidance and support; the technical and administrative staff at the Brigham and Women's Hospital: Karen Smith, Alex McCollom, Jennifer Jackson, Ralph Todesco, Nicole Bruno, Mona Vogel, and Carmella Palmisano; Trevor Cooper and Janelle Fine (both UCSD) for their technical support; Thomas Blackadar, Paul Gaudet (both from Personal Electronic Devices, Wellesley, MA), and Reed W. Hoyt (US Army Research Institute of Environmental Medicine, Natick, MA) for their help with the implementation of the BCTMS; Dennis M. Heher (NASA Ames) and Robin Smith (MIT) for their contribution to the development of the expert sleep system;

Tom Kazlausky (AMI) for advice on the Actillume; and Wim Martens and Carlo Peters (both from Temec Instruments, Kerkrade, the Netherlands) for the modifications and implementation of the Vitaport sleep recording system. We acknowledge the help and advice of David F. Dinges in developing and implementing the neurobehavioral test battery. We are grateful to Dr. Frank M. Sulzman for his contribution to the Neurolab initiative, and to Dr. Jerry Homick for his guidance throughout the Neurolab project. Supported by NASA NAS9-19435 and NIA NAG9-1035.

REFERENCES

EXPERIMENT M133. SLEEP MONITORING ON SKYLAB. In: *Biomedical Results From Skylab* (R.S. Johnston, L.F. Dietlein, eds). J.D. Frost, W.H. Shumate, J.G. Salamy, and C.R. Booher. Washington, D.C.: National Aeronautics and Space Administration, pages 113–126; 1977.

ANALYSIS OF SLEEP ON SHUTTLE MISSIONS. P.A. Santy, H. Kapanka, J.R. Davis, and D.F. Stewart. *Aviat. Space Environ. Med.*, Vol. 59, pages 1094–1097; 1988.

RESEARCH ON SLEEP, CIRCADIAN RHYTHMS AND AGING: APPLICATIONS TO MANNED SPACEFLIGHT. C.A. Czeisler, A.J. Chiasera, and J.F. Duffy. *Exp. Gerontol.*, Vol. 26, pages 217–232; 1991.

SLEEP AND CIRCADIAN RHYTHMS IN SPACE. C. Stampi. *J. Clin. Pharmacol.*, Vol. 34, pages 518–534; 1994.

THE ALTERATION OF HUMAN SLEEP AND CIRCADIAN RHYTHMS DURING SPACEFLIGHT. A. Gundel, V.V. Polyakov, and J. Zulley. *J. Sleep Res.* Vol. 6, pages 1–8; 1997.

SLEEP AND CIRCADIAN RHYTHMS IN FOUR ORBITING ASTRONAUTS. T.H. Monk, D.J. Buysse, B.D. Billy, K.S. Kennedy, and L.M. Willrich. *J. Biol. Rhythms,* Vol. 13. pages 188–201p; 1998.

STABILITY, PRECISION, AND NEAR-24-HOUR PERIOD OF THE HUMAN CIRCADIAN PACEMAKER. C.A. Czeisler, J.F. Duffy, T.L. Shanahan, E.N. Brown, J.F. Mitchell, D.W. Rimmer, J.M. Ronda, E.J. Silva, J.S. Allan, J.S. Emens, D-J Dijk, and R.E. Kronauer. *Science,* Vol. 284, pages 2177–2181; 1999.

MICROGRAVITY REDUCES SLEEP DISORDERED BREATHING IN HUMANS. A.R. Elliott, S.A. Shea, D-J Dijk, J.K. Wyatt, E. Riel, D.F. Neri, C.A. Czeisler, J.B. West, and G.K. Prisk. *Am. J. Respir. Crit. Care Med.,* Vol. 164, pages 478–485; 2001.

SLEEP, PERFORMANCE, CIRCADIAN RHYTHMS, AND LIGHT-DARK CYCLES DURING TWO SPACE SHUTTLE FLIGHTS. D-J Dijk, D.F. Neri, J.K. Wyatt, J.M. Ronda, E. Riel, A. Ritz-De Cecco, R.J. Hughes, A.R. Elliott, G.K. Prisk, J.B. West, and C.A. Czeisler. *Am. J. Physiol. Regul. Integr. Comp. Physiol., Vol.* 281, pages R1647–664; 2001.

SLEEP IN SPACE FLIGHT. BREATH EASY–SLEEP LESS? D.F. Dinges. *Am. J. Respir. Crit. Care Med.*, Vol. 164, pages 337–338; 2001.

Sleep and Respiration in Microgravity

Authors

G. Kim Prisk, Ann R. Elliott,
Manuel Paiva, John B. West

Experiment Team

Principal Investigator: **John B. West**[1]

Co-Investigators: **Ann R. Elliott,**[1] **G. Kim Prisk,**[1] **Manuel Paiva**[2]

Technical Assistants: **Janelle M. Fine,**[1] **Trevor K. Cooper,**[1] **Amy Kissell**[1]

[1]University of California, San Diego, USA
[2]Université Libre de Bruxelles, Brussels, Belgium

ABSTRACT

Sleep is often reported to be of poor quality in microgravity, and studies on the ground have shown a strong relationship between sleep-disordered breathing and sleep disruption. During the 16-day Neurolab mission, we studied the influence of possible changes in respiratory function on sleep by performing comprehensive sleep recordings on the payload crew on four nights during the mission. In addition, we measured the changes in the ventilatory response to low oxygen and high carbon dioxide in the same subjects during the day, hypothesizing that changes in ventilatory control might affect respiration during sleep. Microgravity caused a large reduction in the ventilatory response to reduced oxygen. This is likely the result of an increase in blood pressure at the peripheral chemoreceptors in the neck that occurs when the normally present hydrostatic pressure gradient between the heart and upper body is abolished. This reduction was similar to that seen when the subjects were placed acutely in the supine position in one-G. In sharp contrast to low oxygen, the ventilatory response to elevated carbon dioxide was unaltered by microgravity or the supine position. Because of the similarities of the findings in microgravity and the supine position, it is unlikely that changes in ventilatory control alter respiration during sleep in microgravity. During sleep on the ground, there were a small number of apneas (cessation of breathing) and hypopneas (reduced breathing) in these normal subjects. During sleep in microgravity, there was a reduction in the number of apneas and hypopneas per hour compared to preflight. Obstructive apneas virtually disappeared in microgravity, suggesting that the removal of gravity prevents the collapse of upper airways during sleep. Arousals from sleep were reduced in microgravity compared to preflight, and virtually all of this reduction was as a result of a reduction in the number of arousals from apneas and hypopneas. We conclude that any sleep disruption in microgravity is not the result of respiratory factors.

INTRODUCTION

There are many reasons to expect that the lung will behave quite differently in the microgravity environment of spaceflight than it does here on the Earth. The lung is an unusual organ in that it comprises little actual tissue mass in a relatively large volume. It is an expanded network of air spaces and blood vessels designed to bring gas and blood into close proximity to facilitate efficient gas exchange. As a direct consequence of this architecture, the lung is highly compliant and is markedly deformed by its own weight.

Although there is little, if any, structural difference between the top and bottom of the normal human lung, there are marked functional differences, caused by the effects of gravity. For example, the air spaces at the top of the lung are relatively over-expanded compared to those at the bottom of the lung. As a consequence of this, ventilation (the amount of fresh gas reaching the gas-exchanging region) is higher at the bottom of the lung, because the initial smaller volume there makes the lung more readily able to expand in response to a given breathing effort. There are even larger differences in pulmonary blood flow (perfusion) between the top and bottom of the lung—most of the blood flow goes to the base (bottom) of the upright lung. While both ventilation and pulmonary perfusion increase towards the lower regions of the lung, the differences in perfusion are larger than those in ventilation. As a result, the ventilation-perfusion ratio is higher at the top than at the bottom of the lung. Since it is the ventilation-perfusion ratio that determines gas exchange, regional differences in exhaled gas and effluent blood composition will occur. We have extensively studied the effects of gravity on the lung in previous Space Shuttle missions stretching back to 1991, and on numerous occasions in the NASA KC-135 Microgravity Research Aircraft. While many of the effects of gravity on the lung were predictable, there have been a large number of surprising observations (West, 1997; Prisk, 2000a).

Ventilatory responses

On Neurolab, we turned our attention to a previously unexplored area of pulmonary physiology in microgravity, namely the possible changes in the neural control of ventilation, and the effects of such changes on sleep. Humans have two, largely independent mechanisms that control breathing. The primary control mechanism is a change in the amount of carbon dioxide (CO_2) in the blood. If CO_2 rises (hypercapnia), chemoreceptors in the brain stem cause a marked increase in ventilation. The result of this is to eliminate more CO_2 and return blood CO_2 levels to normal. A separate feedback control system senses a low partial pressure of oxygen (O_2) in the blood (hypoxia) via chemoreceptors located in the carotid bodies in the neck. This also stimulates respiration. Not only are the control paths for O_2 and CO_2 different, the sensors are located in different places. The O_2 sensors are solely peripheral (primarily in the carotid bodies in the neck, with a small component from the aortic bodies in the aortic arch), and the CO_2, sensors are primarily central (brain stem), with only a small peripheral component.

Potential mechanisms of changes in the ventilatory response

It has long been known that cardiovascular changes directly affect respiration. In his 1945 Nobel Prize lecture, Corneille Heymans noted that "… variations in arterial blood pressure exert an effect on the respiratory center … by a reflex mechanism involving the aortic and carotid sinus receptors." In cats, hypotension increased the firing rate of aortic chemoreceptors markedly, and slightly increased the neural output from carotid chemoreceptors. Hypoxia markedly increased the carotid body firing rate. Similarly, in dogs, hypotension increases carotid body activity. This is known to occur via a central pathway through changes in peripheral chemoreceptor activation, as opposed to a direct effect on the peripheral chemoreceptors themselves, because unilateral changes in baroreceptor pressure altered chemoreceptor response on the opposite side (see Prisk, 2000b, for further details). In humans there is less direct evidence for a strong coupling. An increase in blood pressure inhibits the ventilatory response to hypoxia when CO_2 levels are kept constant.

Respiration and sleep

Respiration and sleep are strongly coupled here on Earth, with respiratory disorders being a common cause of sleep disruption. Sleep disruption results from various mechanisms, among them sleep-disordered breathing. Probably the best-known form of sleep-disordered breathing is obstructive sleep apnea (OSA) (Strohl, 1984). In OSA, as the subject is sleeping, the muscles in the upper airway at the back of the throat relax, and in some subjects the result is a closing of the airway. Respiratory effort continues, but is ineffective, so that there is a period in which no fresh air reaches the lungs, lowering O_2 and raising CO_2. Eventually, the changes in O_2 and CO_2 cause an arousal from sleep (although the patient with OSA may not be aware of this) and the airway opens, often accompanied by loud snoring or snorting. The subject then falls asleep again and the cycle repeats. OSA is thought to have potentially serious health consequences, including raised blood pressure and a suspected increase in the risk of heart attack. In addition, the disruption to sleep can result in excessive daytime drowsiness, which can have serious consequences in situations such as driving a car. There are also other forms of sleep-disordered breathing such as central apnea (a cessation of breathing effort altogether) and hypopneas (a reduction in breathing effort that may at times be cyclical).

In spaceflight, sleep disruption is common, and sleeping medications are often used. We hypothesized that alterations in the control of ventilation might be a contributory factor in the sleep disruption seen in spaceflight. The experiment we performed on Neurolab measured the ventilatory responses both to lowered O_2 and increased CO_2 during the day, and, in conjunction with our colleagues from Brigham and Women's Hospital, performed a comprehensive study of sleep quality, quantity, and disruption (see science report by Dijk et al., in this publication).

METHODS

Control of ventilation studies

We used the experimental system that we had previously developed for studying lung function in spaceflight. The astronaut lung function experiment (ALFE) hardware is an automated system that allows the subject to perform numerous lung function tests without the assistance of another crewmember. Gas flow is measured using a flowmeter that is part of the ALFE hardware, and gas concentrations are measured by sampling the gas at the lips of the subject using a mass spectrometer as the subject breathes on the mouthpiece.

The ventilatory response to elevated CO_2 was measured using a rebreathing method (Read, 1967). A bag in the system was filled with a gas mixture of 7% CO_2, 60% O_2, balance nitrogen (N_2) and the subject breathed normally in and out of the bag. Metabolic production by the subject raised the CO_2 in the bag and stimulated breathing until the PCO_2 reached 70 mmHg (~10%), or until four minutes had elapsed, or until the subject was unable to continue. While metabolic consumption lowered the O_2 in the bag, the O_2 never decreased below that in air, eliminating the possibility of any hypoxic stimulus.

The hypoxic response was measured in a similar fashion (Rebuck, 1974). In this case, the bag was filled with a gas mixture of 17% O_2, 7% CO_2, and the balance N_2 and again the subject breathed normally in and out of the bag. Metabolic consumption lowered the O_2 in the bag, stimulating breathing until the inspired PO_2 reached 43 mmHg (~6%), or until four minutes had elapsed, or until the subject was unable to continue. To avoid metabolic production increasing CO_2 in the bag and thereby stimulating breathing, a computer-controlled variable speed fan withdrew some of the gas from the bag and passed the gas through a canister filled with soda lime, which absorbed the CO_2 and thus kept its concentration constant.

The ventilatory responses were calculated from a plot of ventilation against either end-tidal PCO_2 for the hypercapnic response or arterial oxygen saturation (SaO_2) for the hypoxic response. Arterial oxygen saturation was measured with a pulse oximeter on the subject's finger. After a threshold level has been reached, both of these plots show an approximately linear increase in ventilation as CO_2 rises or as SaO_2 falls. The slope and intercept of the fitted lines, as well as ventilation at predetermined levels of CO_2 or SaO_2, provide measures of the ventilatory responses. Figures 1 and 2 show examples of the hypoxic and hypercapnic responses in one of the subjects studied.

As another measure of ventilatory drive, we measured the inspiratory occlusion pressure during air breathing during the early part of the ventilatory response tests described above. The inspiratory occlusion pressure is the pressure generated when the subject begins to inspire against an unexpectedly occluded breathing path. The pressure measured is that 100 milliseconds after the valve is closed, since this is before the subject has time to consciously react to the obstruction. Because the sudden closure of the breathing path was obtrusive when ventilation was stimulated, the occlusion pressure was only measured during the beginning of the ventilatory response tests.

Sleep studies

The sleep system that was developed for this spaceflight experiment consisted of a portable digital sleep recorder, a custom-fitted sleep cap, a body suit with sensors to measure rib cage and abdominal motion, a cable harness, an impedance meter, and a signal quality assessment computer system. This system is fully described in a chapter in this volume by Dijk et al.. Briefly, the sleep recordings included four channels of brainwave activity (EEG), two channels of eye motion (EOG), two channels of muscle tone (EMG), the electrocardiogram (ECG), nasal airflow (thermistor), snoring sounds, light levels (on/off), arterial oxygen saturation SaO_2, and motion of the rib cage and abdomen. Figure 3 shows a subject fully instrumented for sleep during Neurolab.

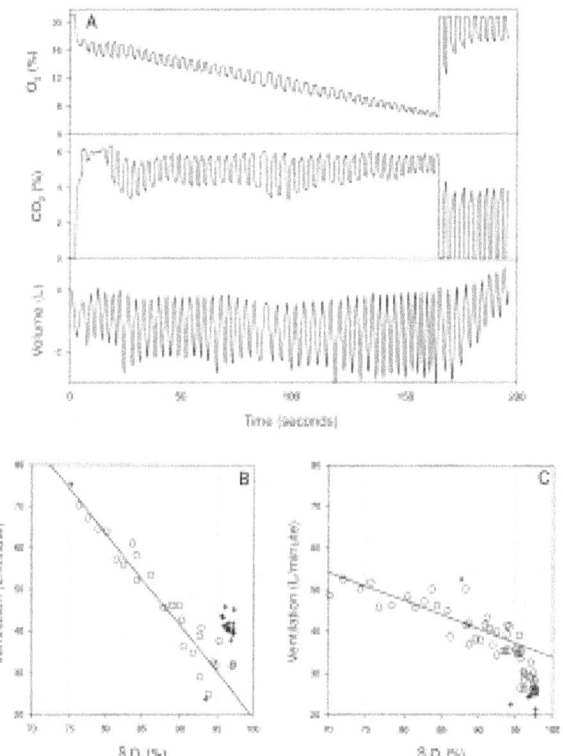

Figure 1. Example of a tracing showing the measurement of the hypoxic ventilatory response (HVR) in one subject standing in one-G (A, B) and in microgravity (C). Subjects rebreathed from a bag for up to four minutes during which time the O_2 in the bag fell and the CO_2 was held constant by a computer-controlled CO_2 removal circuit (A). As O_2 falls, there is a small increase in the respiratory frequency and a substantial increase in the volume of each breath, both of which serve to increase overall ventilation. Also shown is the analysis (B, C) in which breath-by-breath ventilation is plotted as a function of SaO_2. The line is the least-squares best fit to the points lying between 75% and 95% SaO_2, and within two SD of the best fit to all data points in that range. Points lying outside this range are marked by a cross. (From Prisk, 2000b, with permission; reproduced from *Journal of Applied Physiology*.)

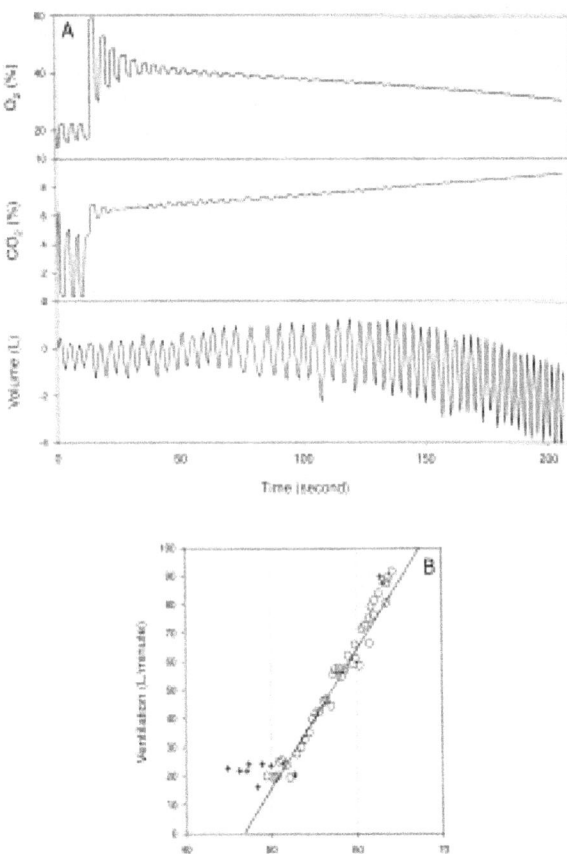

Figure 2. Example of a tracing showing the measurement of the hypercapnic ventilatory response (HCVR) in one subject in microgravity (A). Subjects rebreathed from a bag for up to four minutes during which time the CO_2 in the bag rose. Although O_2 falls, it always remains above 30%, eliminating any contribution from a hypoxic stimulus. The increase in ventilation is more marked than that in the HVR test (Figure 1). Also shown is the analysis in which breath-by-breath ventilation is plotted as a function of PCO_2. The line is the least-squares best fit to the points lying between 50 and 60 mmHg, and within two SD of the best fit to all data points in that range. Points lying outside this are marked by a cross. Note that ventilation only begins to increase above a threshold level of PCO_2, in this case about 50 mmHg.

For the sleep studies, we combined data from Neurolab with data from a fifth subject studied on a later flight, STS-95. A total of 77 polysomnographic recordings were collected preflight, inflight, and postflight on one female and four male subjects (64 on the Neurolab crew, and 13 on STS-95). On each Neurolab subject, there were nine recordings preflight, four inflight, and three postflight. On STS-95, there were six preflight, four inflight, and three postflight recordings. The details of the sleep sessions may be found in the chapter on sleep by Dijk et al. in this publication. This chapter also describes the administration of melatonin, which formed part of the overall Sleep Team experiment design.

The sleep staging of each recording was performed using standard criteria. Obstructive apneas were defined as a cessation of airflow for a minimum of 10 seconds with continued respiratory effort as evidenced by movement of the rib cage or abdomen. Central apneas were characterized by the absence of airflow and respiratory movement for a minimum of 10 seconds. Hypopneas were scored according to standard criteria. Apnea-hypopnea indices ((AHIs) (events/hour)) were determined for the sleep period, the total sleep time, and for the rapid eye movement (REM) and non-REM (nREM) times.

Arousals were determined by the standard criteria outlined by the American Sleep Disorders Association based on changes in the EEG. An arousal was considered to be associated with a respiratory event when it occurred within 15 seconds after the event. Snoring was considered as present if the microphone signal was above a 10% threshold for more than half of a 30-second epoch. For each event, the SaO_2 was averaged over the duration of the event.

Regarding respiration, we saw no difference in the sleep data between the results from nights on which subjects took melatonin and from the nights on which they took a placebo. We therefore combined the data without regard to the presence of melatonin or placebo.

RESULTS

Hypoxic ventilatory response

Microgravity markedly decreased the hypoxic ventilatory response (HVR) (Prisk, 2000b). The slope of the HVR (the increase in ventilation with decreasing arterial oxygen saturation) was approximately halved by microgravity, and became the same as that measured when the subjects were acutely placed in the supine position (Figure 4). Microgravity and the supine position resulted in reductions to approximately $46\pm10\%$ and $53\pm11\%$, respectively, of that in preflight standing control data for the slope of the HVR. There were concomitant reductions in the intercept of the ventilation line, with values being $55\pm7\%$ in microgravity and $59\pm6\%$ for the supine position. Neither the slope nor the intercept measured standing in the postflight period as significantly different from that measured preflight.

The changes in the slope and the intercept both affect the increase in ventilation as arterial oxygen saturation decreases. Spaceflight reduced the ventilation at an SaO_2 of 75% to approximately 65% of that measured standing in the preflight period ($p<0.05$) with a similar reduction when subjects were in the supine position.

There were small increases in the HVR when the subjects were standing in the postflight period (Figure 4), but these failed to reach the level of statistical significance. When we examined the day-to-day changes in the ventilation at an SaO_2 of 75%, we found no consistent changes during the 16 days of the flight. Similarly, there was no change in the response with time either standing or supine during the postflight period.

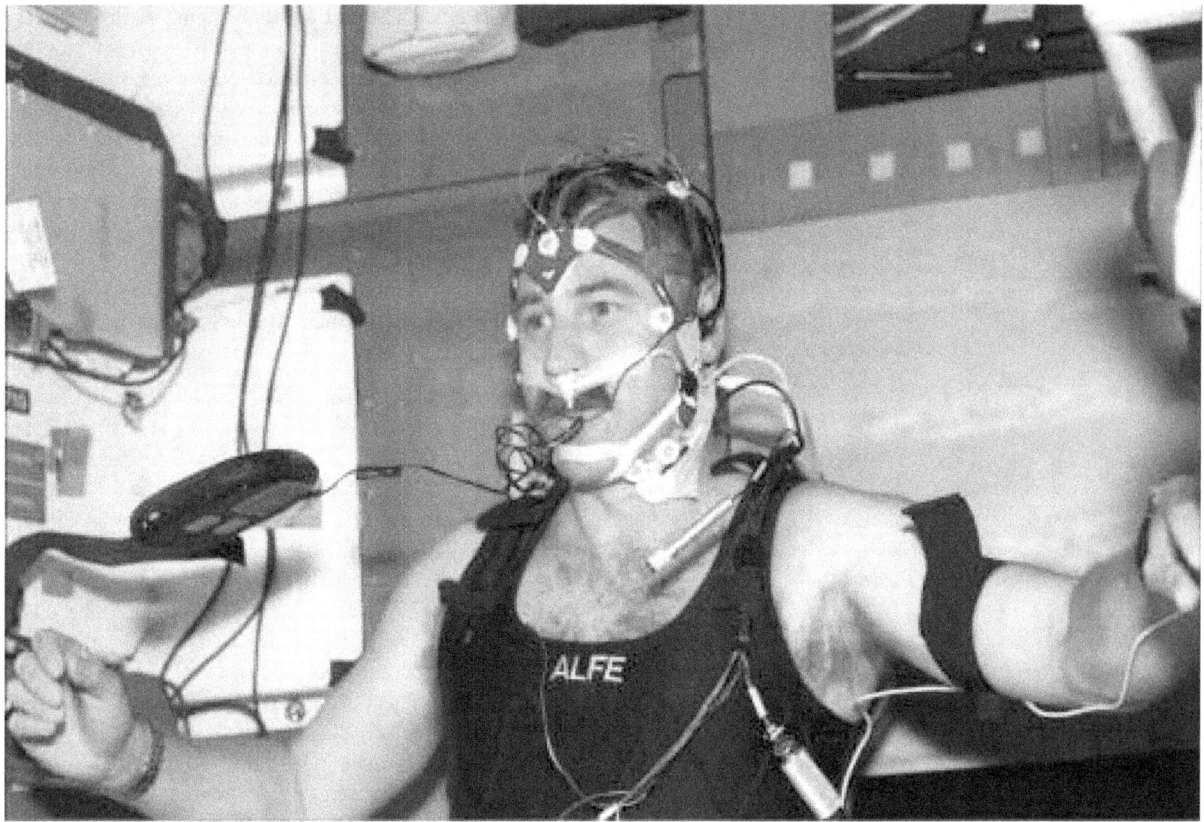

Figure 3. Mission Specialist Dave Williams fully instrumented for sleep during the Neurolab mission. The details of the sleep instrumentation are covered in the report by Dijk et al., in this volume. The CD player is personal equipment. (Photograph NASA 90E5077)

Response to increased CO_2

In sharp contrast to the HVR, microgravity did not significantly alter the hypercapnic ventilatory response (HCVR) (Prisk, 2000b). Figure 5 shows the changes in ventilation resulting from elevated CO_2 by plotting the ventilation calculated from the measured response at a PCO_2 of 60 mmHg. This measurement combines any change in slope with any change in set point for CO_2 above which ventilation begins to increase. As was the case with the HVR, there was no difference in the ventilatory response to CO_2 as the length of time in microgravity increased. Supine, the CO_2 response was slightly reduced, but this did not reach the level of statistical significance preflight.

Overall, the changes we saw in the HCVR suggest a slight steepening of the response with a concomitant shift of the line to the right. The end expiratory PCO_2 measured during quiet breathing rose from 36 mmHg standing preflight to 39 mmHg inflight, and to 41 mmHg supine.

Inspiratory occlusion pressures

The changes in inspiratory occlusion pressures are similar to the overall changes in control of ventilation (above). Figure 6 shows the inspiratory occlusion pressures measured during air breathing and those measured during the early stages of both the HVR

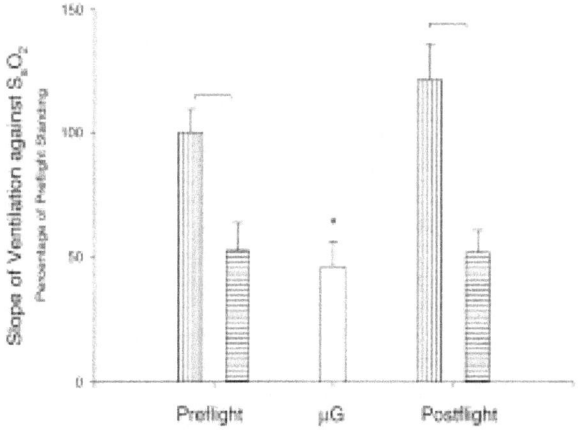

Figure 4. Slope of the hypoxic ventilatory response (HVR) measured standing and supine in one-G, and in microgravity. Both microgravity and the supine position approximately halve the HVR. Markers between adjacent bars indicate $p<0.05$. An * indicates $p<0.05$ compared to preflight standing control. Vertical shading, standing; horizontal shading, supine; open bars, microgravity. (From Prisk, 2000b, with permission; reproduced from *Journal of Applied Physiology*.)

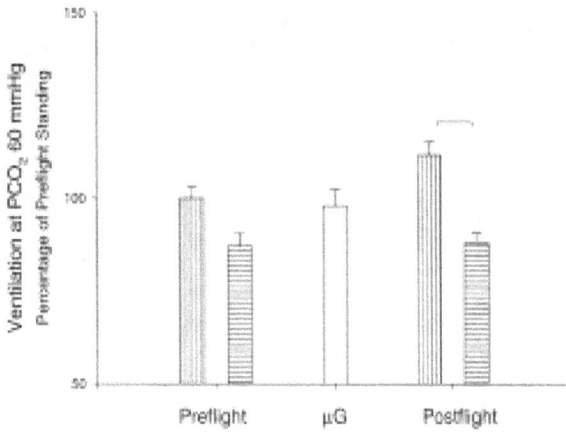

Figure 5. Ventilation measured at a PCO$_2$ of 60 mmHg during the hypercapnic ventilatory response (HCVR) test. Format same as Figure 4. (From Prisk, 2000b, with permission; reproduced from *Journal of Applied Physiology*.)

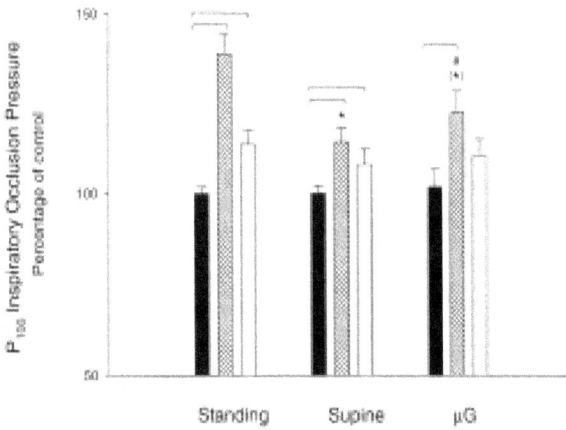

Figure 6. The inspiratory occlusion pressure measured 100 milliseconds after the closure of a valve at the beginning of inspiration (P100). The black bars show the P100 values breathing air. The shaded bars show the P100 measured during the HVR test when the PO$_2$ was between 75 and 85 mmHg. The open bars show the P100 measured during the HCVR test when the PCO$_2$ was between 43 and 50 mmHg.

A marker between adjacent bars indicates p<0.05.

* indicates p<0.05 compared to preflight standing.

\# indicates p<0.05 compared to preflight supine.

Parentheses indicate 0.10<p<0.05.

(From Prisk, 2000b, with permission; reproduced from *Journal of Applied Physiology*.)

and HCVR tests. Occlusion pressures during the hypoxic test (measured during breaths in which the end tidal O$_2$ was between 75 and 85 mmHg) showed a marked increase above air breathing in all cases. However, the increase was significantly less in both the supine position and in microgravity than it was standing. In contrast, occlusion pressures during the measurement of the HCVR (when the PCO$_2$ was between 43 and 50 mmHg) showed a modest increase above air breathing that was not different among the three conditions studied.

Respiration and sleep apneas and hypopneas in microgravity

On the ground, this group of young, normal subjects had a low apnea-hypopnea index (AHI) of only 7.6±1.4 events/hour. Three of the subjects had an AHI below 5.0, one 6.0, and only one had an AHI in the mildly abnormal range at 19.9±2.9. Almost all of these events resulted from hypopneas with an average apnea index of only 0.9±0.3, and a hypopnea index of 6.7±1.2 (Figure 7). The AHI decreased dramatically during microgravity by 52% to 3.4±0.8 events/hour, and almost all of these were hypopneas (3.1±0.8), with almost no apneas (0.3±0.1). Postflight, the AHI increased to 9.4±2.3, which was not statistically different from the preflight values. The AHI preflight was approximately the same in both rapid eye movement (REM) and non-rapid eye movement (nREM) sleep. Inflight, there was a slight increase in the AHI during REM; and postflight, the AHI was higher during REM sleep and was ~50% greater than that preflight (see Elliott, 2001, for details).

Preflight, obstructive apneas accounted for ~21% of the total number of apneas that occurred during a sleep period (1.6±0.7), the rest being central or mixed in origin. Inflight, the number of obstructive apneas decreased to essentially zero (0.1±0.1).

Snoring and arousals associated with respiratory events

Snoring essentially disappeared in microgravity. The percentage of time spent snoring during the preflight sleep periods was 16.8±3.0%, and this was reduced to 0.7±0.5%. Postflight, the snoring returned to preflight levels (Figure 8).

The number of arousals associated with a respiratory event (an apnea or a hypopnea) during the preflight sleep periods was on average 5.5±1.2 arousals/hour within the context of a total number of arousals of 18.0±1.8 arousals/hour (Figure 9). Inflight, the number of respiratory arousals/hour decreased markedly by 70% to 1.8±0.6 arousals/hour while the total number of arousals/hour decreased by only 19% to 13.4±1.5 arousals/hour. Thus, almost all of the decrease in the arousal index was as a result of the reduction in the respiratory arousal index (Figure 9). Postflight, the arousal indices were not different to preflight levels.

The most dramatic decrease in the arousal index was seen in the subject with the highest AHI preflight. That subject's inflight respiratory arousal index dropped from a preflight value of 16.3±2.3 to 6.3±1.1 events/hour. This was within the context of a total arousal index preflight of 35.7 arousals/hour, which was reduced to 23.5±1.5 inflight, again almost completely as a result of the reduction in respiratory arousals.

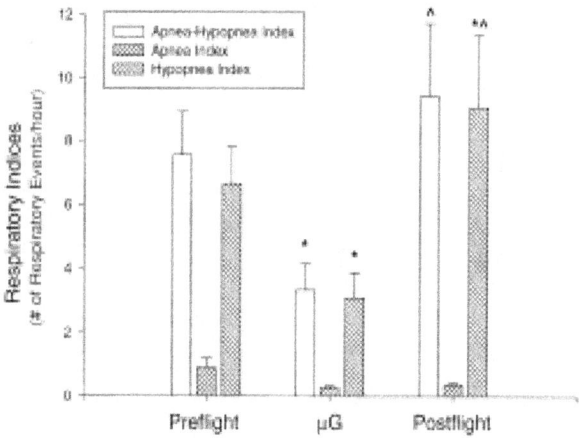

Figure 7. The apnea-hypopnea index and its subdivisons. An * indicates $p<0.05$ compared to preflight and a ^ indicates $p<0.05$ compared to inflight. (From Elliott, 2001, with permission; reproduced from *American Journal of Respiratory and Critical Care Medicine*.)

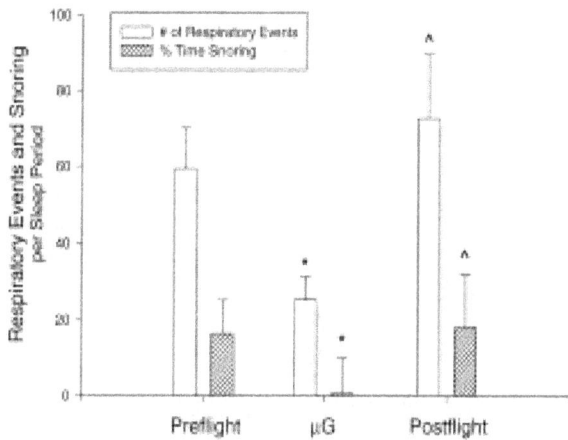

Figure 8. The total number of respiratory events in a sleep period and the percentage of time spent snoring in a sleep period. Significance markings same as Figure 7. (From Elliott, 2001, with permission; reproduced from *American Journal of Respiratory and Critical Care Medicine*.)

DISCUSSION

Changes in control of ventilation in microgravity

The principal finding of this portion of the study is that the hypoxic response in microgravity is only about half of that measured standing in one-G (Figure 4). This is essentially unaltered by the amount of time spent in microgravity up to the 15 days over which we were able to make measurements. Upon return to one-G, the hypoxic response was slightly elevated compared to preflight control data, and this elevation persisted for at least one week. In sharp contrast to this, exposure to microgravity left the ventilatory response to hypercapnia unaltered.

The reduction in the hypoxic ventilatory response resulted from changes in both the slope and the intercept of the ventilatory response, with the slope being reduced slightly more than the intercept. The degree of reduction seen in microgravity in the hypoxic response closely matched that seen in the response measured after the subjects acutely assumed the supine position. Those measurements were generally made within five to 40 minutes of becoming supine.

To a large extent, the differences seen in ventilation at an SaO_2 of 75% result from differences in the increase in the tidal volume, and not from alterations in frequency. Only postflight were there changes in respiratory frequency (a slight increase) that reached the level of significance. These changes match the strategy used to produce a lower ventilation under resting breathing conditions in microgravity (West, 1997; Prisk, 2000) where frequency was largely unaltered and tidal volume decreased to reduce total ventilation compared to standing in one-G.

There was no overall change in the hypercapnic response as measured by the ventilation at a PCO_2 of 60 mmHg caused by exposure to microgravity. However, there was some indication that the slope of the response steepened somewhat both in microgravity and supine, and that this was accompanied by a concomitant increase in the PCO_2 at a calculated ventilation of zero. However, only the zero intercept showed a statistically significant increase. There was also an increase in the end expiratory PCO_2 measured during quiet breathing. The increase from 36 mmHg to 39 mmHg was smaller than that seen between standing and supine (36 to 41 mmHg), but raises the possibility of a shift in the set point of the PCO_2. Measurements made in an environmental chamber study in which the PCO_2

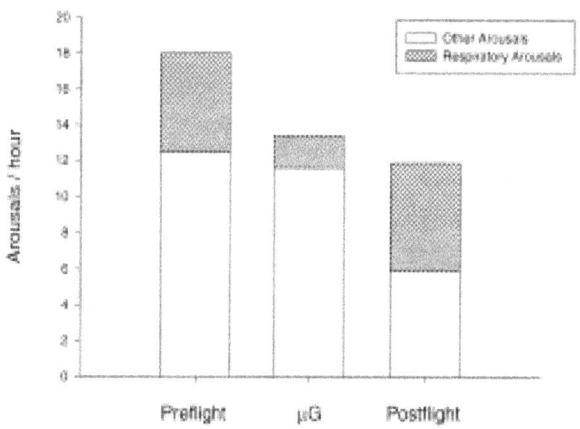

Figure 9. Arousals from respiratory causes (shaded) and non-respiratory causes (open bars). Note that inflight, virtually all of the reduction in arousals is a result of the fewer number of arousals resulting from respiratory causes. (From Elliott, 2001, with permission; reproduced from *American Journal of Respiratory and Critical Care Medicine*.)

was elevated to 1.2% (8.6 mmHg) showed an early increase in the set point (Elliott, 1998) that gradually abated. However, that study failed to show any significant alterations when the environmental PCO_2 was controlled at 5.0 mmHg. In the case of Neurolab, environmental PCO_2 averaged ~2.3 mmHg, a level below that in the chamber studies.

The changes in the ventilatory responses can be explained by changes in neural drive to the respiratory muscles. This conclusion is supported by the inspiratory occlusion pressure measurements (Figure 6). Hypoxia resulted in a substantial rise in the inspiratory occlusion pressure in all cases, although there were marked differences in the magnitude of the increase between the different conditions tested. The greatest response was measured with the subjects standing, where there was an increase of ~40% above that measured breathing air. While there was no change in the inspiratory occlusion pressure measured during air breathing in either the supine position or in microgravity, the increase seen during hypoxia in the supine position and in microgravity was significantly less than the increase measured standing. The reductions suggest that the hypoxic drive is approximately halved by either the supine position or by microgravity, an observation consistent with the ~50% reduction seen in the slope of the ventilatory response (Figure 4). In contrast, hypercapnia (a PCO_2 between 43 and 50 mmHg) resulted in the same increase in occlusion pressures regardless of the condition in which it was measured, consistent with the lack of a significant change in the response to inhaled CO_2.

Changes in the hypoxic ventilatory response resulting from blood pressure changes have been seen before. For example, an increase in carotid level blood pressure of ~10 mmHg results in a 33% smaller increase in ventilation elicited by a hypoxic challenge of breathing 10% O_2. Other studies have shown that the hypoxic response is reduced in the supine position by ~43% compared to the upright, a reduction of similar magnitude to that which we observed in our preflight control data (Figure 4). It therefore seems likely that the changes we observed in the HVR result from changes in blood pressure at the carotid chemoreceptors.

The blood pressure measured at the level of the heart changes only slightly between standing and supine rising only ~3mm Hg. However, the transition from the standing position to supine in one-G abolishes the hydrostatic difference in pressure between heart level and carotid level. Thus, when lying down we would expect an increase in carotid level blood pressure of 15–20 mmHg due to hydrostatic effects. Adding these two effects suggests that overall there is an increase in carotid level pressure of ~20 mmHg. This increase in pressure likely explains the decrease in hypoxic ventilatory response we observed in the supine position.

Microgravity results in only modest changes in heart level blood pressure, with a decrease of ~4 mmHg. However, in the absence of gravity, the hydrostatic gradient between heart level and the carotid region is also absent. Thus, despite the slight decrease in heart level blood pressure, it is likely that carotid level blood pressure in microgravity is considerably above that in the standing posture in one-G. Such an increase is consistent with the physical changes such as facial puffiness

associated with the early period in microgravity. Again, this increase in pressure likely explains the decrease in hypoxic ventilatory response we observed in microgravity.

Blood pressure falls in the upright position in the period immediately following flight to a variable degree. Certainly in these subjects, there was a persisting reduction in cardiac stroke volume of ~10% for the week immediately following flight, although a concomitant tachycardia maintained cardiac output. This suggests that carotid systolic pressure was slightly reduced compared to preflight, and this may have contributed to the increase in HVR we observed in the standing posture postflight (Figure 4).

Most of the hypercapnic ventilatory response can be ascribed to the central chemoreceptors. However, a significant component of the response comes from the carotid chemoreceptors. We reasoned that while there might be no change in the central chemoreceptor response to CO_2 as a result of exposure to microgravity, there may well be some change in the peripheral component of this response. Our data show that this is not the case. These results are consistent with those of other studies that show a reduction in the HVR, but not the HCVR in the supine position compared to the upright position (see Prisk, 2000b, for details).

It might be argued that the reduction in HVR was as a result of some mechanical disadvantage of the respiratory muscles supine and in microgravity. However, our data do not support a reduction in the HVR due to mechanical factors. As Figure 6 shows, inspiratory occlusion pressure breathing air was unaltered by the supine posture in one-G or by microgravity, and was similar in the three conditions during hypercapnia. Similarly, there was no significant change in the HCVR caused by microgravity (Figure 5). These results would not be expected if the cause of the reduced HVR supine and in microgravity was mechanical in origin.

Respiration and sleep in microgravity

Sleep studies during spaceflight have shown a decrease in the total amount of sleep obtained by the astronauts and a reduction in deep restorative (delta) sleep (see see science report by Dijk et al. in this publication). However, the important message from this portion of the studies is that not only are respiratory events not increased in microgravity, they are significantly decreased, and they are clearly not the cause of the sleep disruption in microgravity. This is evidenced by the dramatic reduction in the number of sleep-disordered breathing events, amount of time spent snoring, and the number of arousals associated with these respiratory-related events during microgravity (Figures 8 and 9).

The respiratory system is greatly influenced by the force of gravity. The changes the system goes through by just moving from the standing to the supine posture are significant. In the supine posture, functional residual capacity (the normal resting volume of the lung), expiratory reserve, and tidal volumes are all reduced. Functional residual capacity and expiratory reserve volumes are reduced in space, but to a lesser degree when compared to the supine posture; and tidal volume is reduced to a greater extent in space when compared to supine (West, 1997).

In the supine posture, gravity also works to reduce upper airway size and increase upper airway resistance by causing the tongue, soft palate, uvula, and epiglottis to move back toward the posterior pharyngeal wall. The tongue cross-sectional area, uvular width, and soft palate thickness all increase in the supine position resulting in a reduction in the oropharyngeal cross-sectional area. All these anatomical changes caused by moving from the upright to supine posture result in an increase in upper airway resistance. It is widely believed that it is the gravitational forces acting on the upper airway structures that cause an increase in upper airway resistance that results in an increased probability that snoring, hypopneas, and frank obstructive sleep apnea will occur during sleep.

All five subjects in this study showed some degree of snoring from mild to moderate ranging from 8.7 to 32.6% of the sleep period. In microgravity, snoring was almost completely eliminated in all subjects. The change in snoring habits correlate with the changes in the number of respiratory events during sleep period (Figure 8).

Even though the majority of the subjects in this study showed respiratory disturbances as determined from the AHI within the normal range (<5.0/hour), four out of the five subjects showed a reduction in their AHI in microgravity. The greatest reduction in the AHI of almost 60% was seen in the subject with the largest AHI preflight (19.9/hour).

On average, the reduction in AHI was greatest in the nREM periods (a 68% reduction) as opposed to during REM (a 30% reduction). During REM sleep, there is an increase in upper airway resistance when compared to nREM sleep attributed to a loss of upper airway dilator muscle tone. There is also an inhibition of other respiratory muscles, including intercostal muscles, which alters the configuration of the rib cage and its contribution to tidal breathing. Both effects are independent of gravity per se. Though these two mechanisms play a role in the generation of upper airway obstruction, our data support the suggestion that gravity is the primary mechanism contributing to upper airway resistance during both the supine posture and sleep, since a larger effect was seen during nREM than during REM sleep.

The brief arousals caused either by apneas or hypopneas cause reduced or fragmented sleep. Overall we saw a significant reduction of ~70% in the number of respiratory-related arousals in these five subjects, accounting for virtually all of the reduction in the total number of all arousals during spaceflight. This suggests that contrary to our initial hypothesis, microgravity may improve sleep quality to some extent, especially for the subjects with positional sleep disordered breathing problems, obstructive sleep apnea, or upper airway resistance syndrome. For example, in the subject with the greatest AHI, the respiratory-related arousal index and arousal index decreased during spaceflight by 60% and 34%, respectively, with a concomitant increase in sleep efficiency.

Despite the improvement in sleep-disordered breathing, upon return to Earth the crew complained of significant fatigue. This likely resulted from a significantly shorter amount of sleep over virtually the entire flight (only ~6.1 hours per night inflight compared to almost eight hours pre-

flight), although clearly this was not from respiratory causes. It seems likely that the changes we observed during the post-flight period result from this fatigue, as opposed to some adaptive effects relating to the return to one-G.

CONCLUSION

Microgravity exposure greatly reduces the number of sleep-related apneas and hypopneas, significantly reduces snoring, and reduces the number of respiratory arousals in normal, healthy individuals. These changes were probably due to the elimination of the gravitationally induced changes in the upper airway anatomical structures. Indeed, the reduction in the total number of arousals in microgravity resulted almost entirely from the large reduction in respiratory-related arousals. From these data, we can infer that gravity plays a dominant role in the increase in upper airway resistance and obstruction that occurs after the transition to the supine posture and during all stages of sleep.

It has long been recognized that gravity might play a role in some forms of sleep-disordered breathing (especially obstructive sleep apnea (OSA)). However, these data provide the first direct evidence that the effect of gravity on the upper airways is dominant in causing obstruction. As such, the results may prove useful, increasing the emphasis on postural treatments for OSA. The simple expedient of decreasing the rearward gravitational effect on the upper airways by encouraging sleep in a semi-upright posture may prove more acceptable to some patients than other treatments, such as continuous positive airway pressure (CPAP).

Microgravity also resulted in a significant reduction in the hypoxic ventilatory response; however, the change that occurred was no different to that seen in the supine position. In contrast to the hypoxic response, there was no change in the response to inhaled CO_2 caused by either the supine position or by microgravity. Thus, the ventilatory control mechanisms in microgravity in these healthy subjects were much the same as they were in the supine position on the ground. The similarity is borne out by the similar numbers of central apneas and hypopneas observed in microgravity and on the ground, since central events result primarily from the ventilatory control system.

In conclusion, our data suggest that any sleep disruption caused by spaceflight or by microgravity has origins unrelated to the respiratory system.

Acknowledgements

The authors acknowledge the sustained and excellent support and cooperation of the crews of STS-90 and STS-95 and the many National Aeronautics and Space Administration and Lockheed Martin Engineering Services personnel who supported the missions. They particularly thank Jim Billups, Mel Buderer, Sherry Carter, Steve Cunningham, Marsha Dodds, Dennis Heher, Gerald Kendrick, Pat Kincade, Amy Landis, Angie Lee, Raoul Ludwig, Wim Martens, Suzanne McCollum,

Mary Murrell, Peter Nystrom, Carlo Peters, Carlos Reyes, Joseph Ronda, Steven Shea, and Timothy Snyder.

This work was supported by NASA contracts NAS9-18764 and NAS9-19434, and by National Institutes of Health Grant U-01-HL53208-01.

REFERENCES

A CLINICAL METHOD FOR ASSESSING THE VENTILATORY RESPONSE TO CARBON DIOXIDE. D.J.C. Read, in *Australas. Ann. Med.*, Vol. 16, pages 20–32; 1967.

A CLINICAL METHOD FOR ASSESSING THE VENTILATORY RESPONSE TO HYPOXIA. A.S. Rebuck and E.J.M. Campbell, *Am. R. Respir. Dis.*, Vol. 109, pages 345–350; 1974.

SLEEP APNEA SYNDROMES. K.P. Strohl et al., In *Sleep and Breathing*, N.A. Saunders and C.E. Sullivan, eds. Marcel Dekker, New Your; 1984.

PULMONARY FUNCTION IN SPACE. J.B. West, A.R. Elliott, H.J. Guy, G.K. Prisk, *J. Am. Med. Assoc.*, Vol. 277, No. 24, pages 1957–1961; 1997.

HYPERCAPNIC VENTILATORY RESPONSE IN HUMANS BEFORE, DURING, AND AFTER 23 DAYS OF LOW LEVEL CO_2 ESPOSURE. A.R. Elliott, G.K. Prisk, C. Schollmann, U. Hoffmann, *Aviat. Space Envir. Med.*, Vol. 69, No. 4, pages 391–396; 1998.

MICROGRAVITY AND THE LUNG. G. Kim Prisk, *J. Appl. Physiol.*, Vol. 89, pages 385–396, 2000a.

SUSTAINED MICROGRAVITY REDUCES THE HUMAN VENTILATORY RESPONSE TO HYPOXIA BUT NOT TO HYPERCAPNIA. G.K. Prisk, A.R. Elliott, J.B. West, *J .Appl. Physiol.*, Vol. 88, pages 1421–1430, 2000b.

MICROGRAVITY REDUCES SLEEP DISORDERED BREATHING IN HUMANS. A.R. Elliott, S.A. Shea, D.J. Dijk, J.K. Wyat, E. Riel, D.F. Neri, C.A. Czeisler, J.B. West, G.K. Prisk. *Am. J. Resp. Crit. Care Med.*, Vol. 164, pages 478–485, 2001.

The Effects of Spaceflight on the Rat Circadian Timing System

Authors

Charles A. Fuller, Dean M. Murakami,
Tana M. Hoban-Higgins, Patrick M. Fuller,
Edward L. Robinson, I-Hsiung Tang

Experiment Team

Principal Investigator: **Charles A. Fuller[1]**

Co-Investigators: **Dean M. Murakami[1]**
Tana M. Hoban-Higgins[1]

Technical Assistants: **Patrick M. Fuller,[1] I-Hsiung Tang,[1]**
Edward L. Robinson,[1]
Jeremy C. Fuller,[1] Steven Killian,[1]
Angelo De La Cruz,[3] Nick Vatistas,[1]
Andris Kanups,[2] Barbara Smith,[2]
Brian Roberts,[3]
L. Elisabeth Warren[1]

Engineers: **Jim Connolley,[3] Mike Skidmore,[3]**
John Hines,[3] Julie Schonfeld,[3]
Tom Vargese[3]

[1]University of California, Davis, USA
[2]Ohio State University College of Veterinary Medicine,
Columbus, USA
[3]NASA Ames Research Center, Moffett Field, USA

ABSTRACT

Two fundamental environmental influences that have shaped the evolution of life on Earth are gravity and the cyclic changes occurring over the 24-hour day. Light levels, temperature, and humidity fluctuate over the course of a day, and organisms have adapted to cope with these variations. The primary adaptation has been the evolution of a biological timing system. Previous studies have suggested that this system, named the circadian (*circa* ~ about; *dies* ~ a day) timing system (CTS), may be sensitive to changes in gravity. The NASA Neurolab spaceflight provided a unique opportunity to evaluate the effects of microgravity on the mammalian CTS. Our experiment tested the hypotheses that microgravity would affect the period, phasing, and light sensitivity of the CTS. Twenty-four Fisher 344 rats were exposed to 16 days of microgravity on the Neurolab STS-90 mission, and 24 Fisher 344 rats were also studied on Earth as one-G controls. Rats were equipped with biotelemetry transmitters to record body temperature (T_b) and heart rate (HR) continuously while the rats moved freely. In each group, 18 rats were exposed to a 24-hour light-dark (LD 12:12) cycle, and six rats were exposed to constant dim red-light (LL). The ability of light to induce a neuronal activity marker (*c-fos*) in the circadian pacemaker of the brain, the suprachiasmatic nucleus (SCN), was examined in rats studied on flight days two (FD2) and 14 (FD14), and postflight days two (R+1) and 14 (R+13). The flight rats in LD remained synchronized with the LD cycle. However, their T_b rhythm was markedly phase-delayed relative to the LD cycle. The LD flight rats also had a decreased T_b and a change in the waveform of the T_b rhythm compared to controls. Rats in LL exhibited free-running rhythms of T_b and HR; however, the periods were longer in microgravity. Circadian period returned to preflight values after landing. The internal phase angle between rhythms was different in flight than in one-G. Compared with control rats, the flight rats exhibited no change in HR. Finally, the LD FD2 flight rats demonstrated a reduced sensitivity to light as shown by significantly reduced *c-fos* expression in the SCN in comparison with controls. These findings constitute the first demonstration that microgravity affects the fundamental properties of the mammalian circadian timing system, specifically by influencing the clock's period, and its ability to maintain temporal organization and phase angle of synchronization to an external LD cycle.

INTRODUCTION

Two of the most fundamental environmental influences that have shaped the evolution of organisms on Earth are gravity and the cyclic changes occurring over the 24-hour day. The rotation of the Earth relative to the Sun creates the fluctuation in light and dark, temperature, and humidity over the course of a day. Organisms have adapted in many ways to cope with these cyclic variations in the environment. The primary adaptation has been the evolution of a biological timing system, which anticipates these daily fluctuations. This system has been named the circadian (*circa* ~ about; *dies* ~ a day) timing system (CTS). For example, although daily rhythms in sleep/wake, body temperature, hormonal activity, and performance have been measured in many organisms, they are not simply passive responses to changes in the environment, but rather are generated by a neural pacemaker, the circadian clock, located in the brain (Moore-Ede, 1982). In mammals, this neural pacemaker is called the suprachiasmatic nucleus (SCN). The SCN contains a genetic clock that affects our physiology to create the circadian rhythms of the body. In addition to possessing an endogenous clock, the SCN detects cues from the environment so that internal rhythms can keep time with the external day—a process called entrainment. The primary environmental time cue or zeitgeber (*zeit* ~ time; *geber* ~ giver) detected by mammals is light. All of the circadian rhythms of the body must be coordinated to peak or fall at certain times of the day. Therefore, there is a specific time when body temperature is high to meet the activities of the day, when the stomach is active to anticipate the arrival of food, or when the hormone melatonin is secreted to affect sleep. All of these physiological circadian rhythms must peak at their own specific time for the body and mind to operate efficiently. When these various circadian rhythms are no longer coordinated in time, body temperature may peak at bedtime or food may arrive when the stomach isn't ready. As a consequence, the mind and body do not operate well and difficulties arise in coping with stress. People experience this with jetlag, shift work, and as new parents. It takes time for circadian rhythms to readjust and regain temporal coordination. Until then, individuals will not be at their best. Controlled animal studies in our laboratory have shown the difficulty in coping with a stressor when circadian rhythms are not coordinated (Fuller, 1978).

The gravity of the Earth (one-G) is a constant environmental force that has shaped the anatomy and physiology of virtually every organism. Animal studies have shown that exposure to high gravity levels (hypergravity) reduces body mass, decreases food and water intake, increases metabolism, and decreases activity, body temperature, and performance. A very interesting finding was that animals exposed to twice normal gravity (two-G) exhibited a dramatic change in the amplitude and phase of the circadian rhythms of heart rate, body temperature, and activity, suggesting a disruption in rhythm coordination (Fuller, 1994). We were also interested in whether exposure to two-G might affect the function of the clock responsible for generating circadian rhythms, the SCN. We tested this question using two groups of rats. One group was exposed to two-G for two days, while a control group remained at one-G. At the appropriate time, all of the rats were exposed to light for two hours. Immediately after the light exposure, we used the biological marker *c-fos* to see if the light activated neurons in the SCN. Control rats that remained at one-G had a high number of *c-fos*-activated neurons within the SCN. However, rats at two-G did not have a significant number of *c-fos*-activated neurons in the SCN (Murakami, 1998). This study demonstrated that exposure to two-G could affect the normal function of the pacemaker generating circadian rhythms. Such a disruption in circadian rhythms and SCN function could help to explain the underlying causes of space adaptation syndrome, sleep disturbance, and reduced performance experienced by astronauts in space.

Often, to study the mammalian biological clock, a circadian scientist will house the subjects (animals or humans) in an environment with no time cues. This is referred to as "constant conditions" and can occur either in constant darkness or in constant light. Under constant conditions, the clock is no longer synchronized by environmental time cues, and thus is free to express its own internal period. The period of the clock differs slightly from individual to individual, but it is extremely stable within an individual. By using constant conditions in the Neurolab experiment, we were able to look at the internal or "free-running" period of the clock on the ground and in space. Any changes in the period of the clock during spaceflight could then be interpreted as an effect of gravity on clock function.

The Neurolab mission (STS-90) gave us the unique opportunity to study the effects of the microgravity of space on rat body temperature and heart rate and their respective circadian rhythms. For the first time, we were able to study the effect of microgravity on mammalian circadian rhythms in constant lighting conditions where there were no specific time cues. This allowed us to test the hypothesis that microgravity affects the circadian timing system, including the amplitude and period (i.e., timing) of the circadian pacemaker. In addition, we were able to test the hypothesis that microgravity affects the normal response of the SCN to light by measuring *c-fos*, a marker for neuronal activity. Revealing the physiological, circadian, and neural pacemaker responses to spaceflight may help us understand the process of adaptation to microgravity.

METHODS

Subjects and Housing – Forty-eight adult male Specific Pathogen-Free Fisher 344 rats, weighing 350–370 grams, were used in this experiment. Care of the rats met all National Institutes of Health standards outlined in the *Guide for the Care and Use of Laboratory Animals*, and was approved by both the University of California (UC) Davis and NASA Institutional Animal Care and Use Committees. During preflight and postflight recording periods, the rats were housed individually in standard vivarium cages. During the flight recording period, the flight and ground control groups were housed individually in the NASA Research Animal Holding Facility cages. Food and water were available ad libitum during the entire experimental protocol.

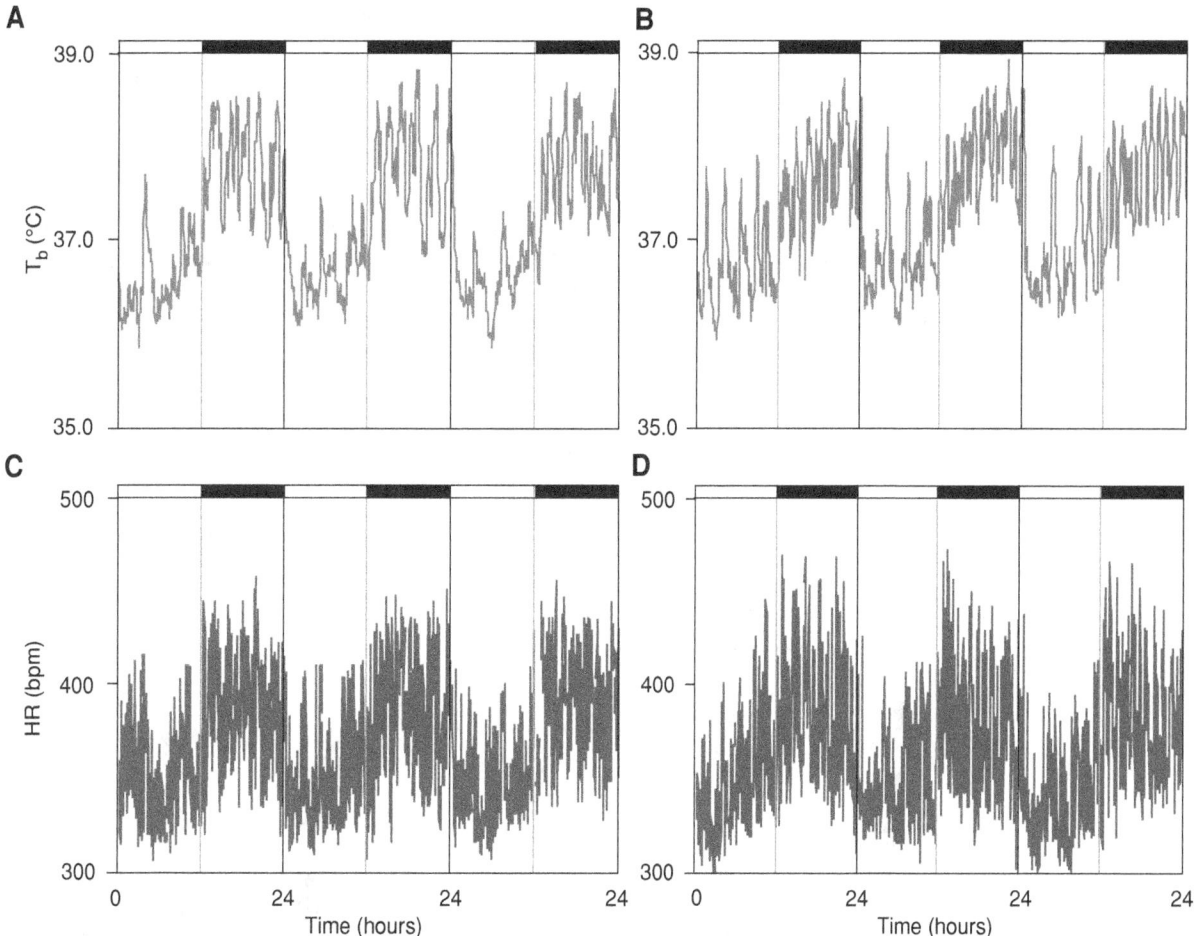

Figure 1. Graphs of three days of body temperature (T$_b$; °C) and heart rate (HR; beats/minute) data are plotted for a CTL rat (A & C) and an FLT rat (B & D). These data are from rats in a light-dark (LD) cycle during the flight portion of the experiment. The timing of the LD cycles is indicated by the light and dark bars at the top of each graph. Distinct circadian rhythms of T$_b$ and HR are evident in both rats. As is expected in a nocturnal animal, both T$_b$ and HR are higher during the dark, the rat's active period.

Experimental Protocol – Each rat had a biotelemetry unit placed in the abdomen for chronic recording of body temperature (T$_b$) and heart rate (HR). The rats were divided into two groups: a flight (FLT) group (n=24) and a ground control (CTL) group (n=24). Telemetry data were collected during a two-week preflight and a 16-day flight period. The LL CTL and FLT animals were also measured during a two-week postflight period. In both the FLT and CTL groups, 18 of the rats were housed in a 24-hour light-dark cycle (LD 12:12; 100:0 lux) and six were housed in constant dim red-light (LL; 30 lux). The constant dim red-light permitted monitoring of the rats by the crew during the flight. Brain tissue was collected on four different days during flight: on flight day two (FD2) and flight day 14 (FD14) and postflight on post-recovery day two (R+1) and post-recovery day 14 (R+13). Identical procedures and timelines were used for the ground CTL group. Prior to tissue collection, one-half of the rats in each group were exposed to a light pulse. A light pulse of 100 lux was delivered for 60 minutes at a time of day when light is known

to elicit robust *c-fos* expression in the SCN in one-G. All brain tissue was fixed in 4% paraformaldehyde. Sections of the SCN were then made and stained for *c-fos* expression using methods described below.

Biotelemetry – A custom telemetry receiver system (Neurolab Biotelemetry System) was developed by NASA Ames Research Center (Sensors 2000 Program) using commercially available components (Konigsberg Instruments, Inc., Pasadena, CA). This system allowed for heart rate and body temperature to be transmitted via radio waves to a recording system while the rats moved about freely (see technical report by Buckey and Fuller in this publication).

Under anesthesia, a sterilized biotelemetry transmitter was inserted into the abdomen. Electrocardiogram leads and an antenna were positioned under the skin. The rats recovered for two weeks before recording began.

c-fos Immunohistochemistry – The brains were fixed in 4% paraformaldehyde fixative and placed in 30% sucrose solution. Brains from each group were then frozen and sectioned coronally on a microtome in 50-μm sections. Sections were washed, treated with antibody and dye to reveal the presence of *c-fos*, and mounted on slides for examination under the microscope according to the procedure of Murakami et al., 1998. The results were quantified by counting the number of *c-fos* immunopositive cells in each SCN section.

Circadian and Statistical Analyses – Phase, mean, and amplitude of the T_b and HR rhythms were determined using a least-squares sine/cosine fit. The average daily mean, rhythm amplitude (calculated as the mean to maximum of the best-fit Fourier function) and acrophase (calculated as the time of the maximum of the best-fit Fourier function) of T_b and HR were

calculated for each group. The difference, in hours, between the acrophase of T_b and the onset of the light (Lon=hour 0) was used to compare the relative timing of the rhythms in an LD cycle. The relative timing of the acrophases of the two rhythms was used to examine the relationship between the T_b and HR rhythms in LL. Period (τ) was calculated by both Enright's Periodogram method and a linear/nonlinear least-squares cosine fit spectral analysis. Repeated-measures analysis of variance was used to compare preflight (one-G), flight (μG), and postflight (one-G) data. Specific mean comparisons were made using Tukey's HSD post-hoc test. An α of <0.05 was considered statistically significant. All data are summarized as mean±standard error.

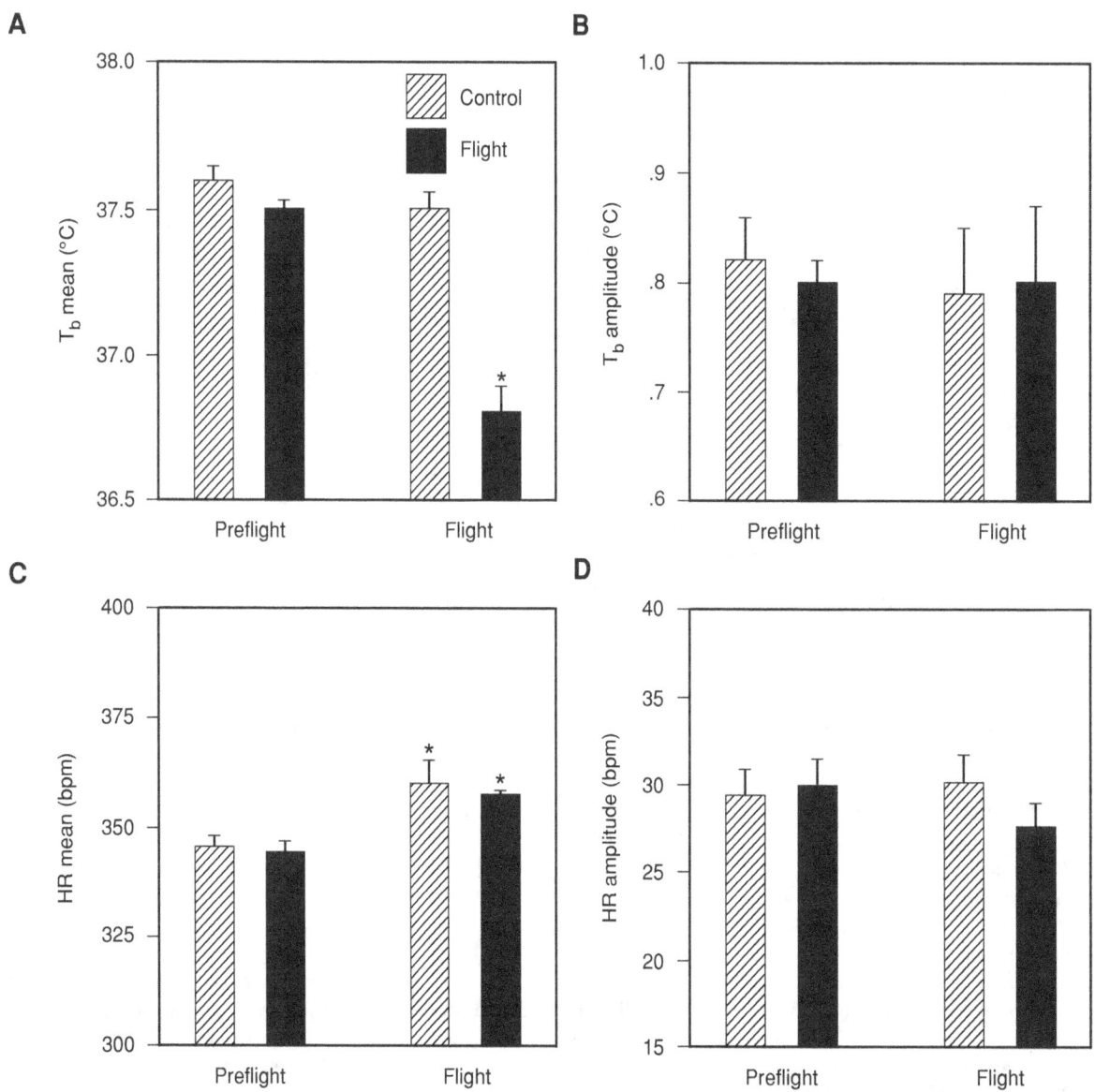

Figure 2. Daily mean (A) body temperature (T_b; °C) and (C) heart rate (HR; beats/minute) are presented as histograms along with average amplitude of the respective rhythms (B & D). These histograms are averages (+ SEM) from all LD CTL and FLT rats for the preflight and flight periods. In the FLT rats, the mean daily T_b was significantly lower during flight than preflight. Mean HR was significantly increased for both CTL and FLT rats during flight compared to preflight. There were no other significant differences between FLT and CTL or between experiment segments. (* = p < 0.05)

The Neurolab Spacelab Mission: Neuroscience Research in Space

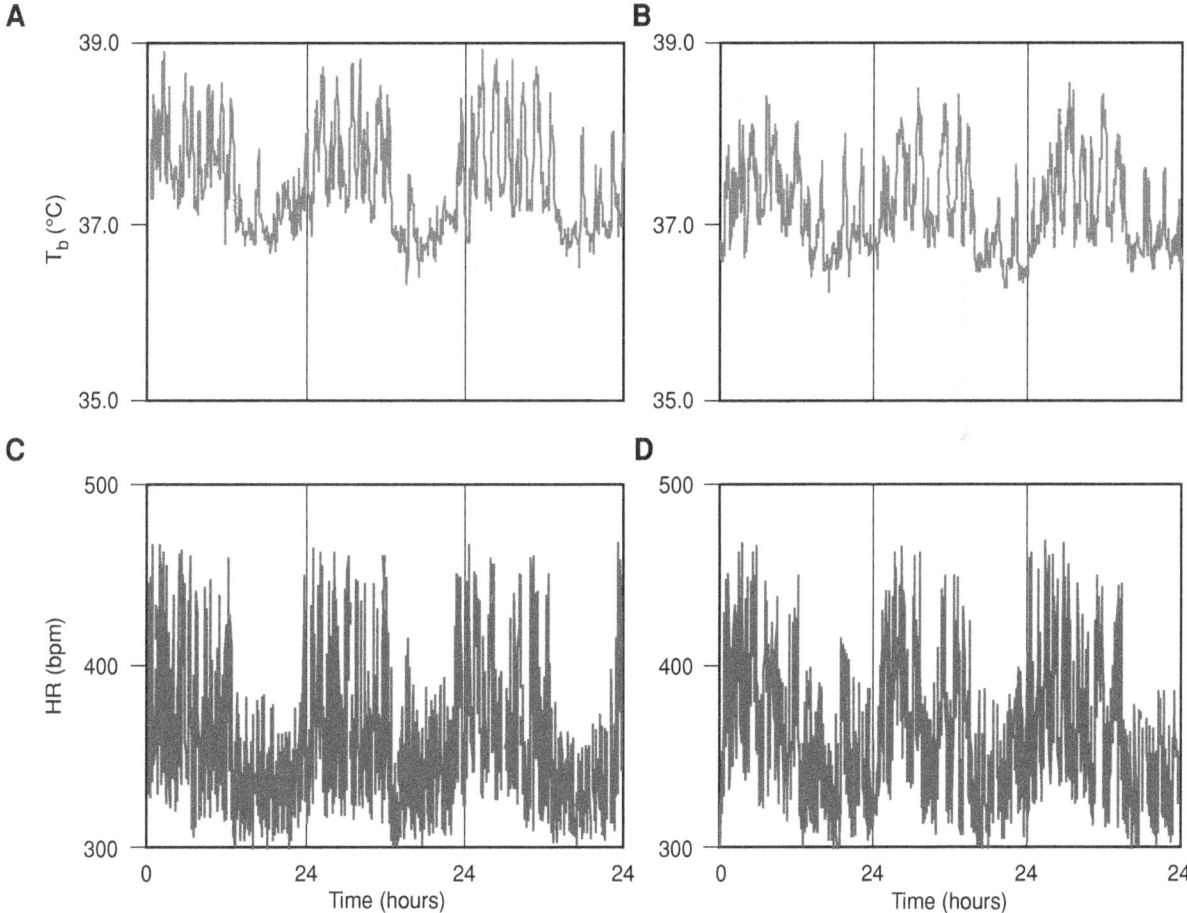

Figure 3. Graphs of three days of body temperature (T_b; °C) and heart rate (HR; beats/minute) are plotted for a CTL rat (A & C) and an FLT rat (B & D). These data are from rats in constant dim red-light (LL) during the flight portion of the experiment. Persistent circadian rhythms are evident in both variables in each rat.

RESULTS

LD Body Temperature (T_b) – Examples of T_b circadian rhythms recorded in LD during the flight period are shown in Figure 1. Data from a CTL rat are shown in Figure 1A and from a FLT rat in Figure 1B. Body temperature is higher during the dark, as expected in a nocturnal animal. Average mean T_b and circadian amplitude are presented as histograms in Figure 2. Mean daily T_b for the LD rats significantly decreased from preflight to flight (Figure 2A). CTL and FLT groups did not differ preflight. No changes were seen between mean daily T_b in CTL from preflight to flight. The mean daily circadian T_b amplitude (Figure 2B) did not differ significantly between preflight and flight or between FLT and CTL groups.

LD Heart Rate (HR) – HR data recorded from the same rats at the same time as the T_b data shown in Figures 1A and 1B are presented in Figures 1C (CTL) and 1D (FLT). HR is highest when the rats are awake and active, during the dark. This corresponds to the time of the highest T_b. The average mean and circadian amplitude are presented in Figures 2C and 2D. Mean daily HR of both CTL and FLT rats was significantly higher during flight. Since the response was similar in both groups, it is

likely due to the changes in the cage environment between preflight and flight. The mean daily amplitude of the HR circadian rhythm (Figure 2D) was not significantly different between preflight and flight or between the FLT and CTL groups.

LL Body Temperature (T_b) – Circadian rhythms of T_b and HR recorded under LL are presented in Figure 3. Data from a CTL rat are shown in Figure 3A and from a FLT rat in Figure 3B. Clear, persistent T_b circadian rhythms are evident in LL. The daily mean and circadian amplitude are graphed in Figure 4. As in the LD FLT rats, mean T_b was lower during flight; however, mean T_b did not differ significantly among preflight, flight, and postflight periods (Figure 4A). In contrast to the LD FLT rats, in LL the average T_b rhythm amplitude was significantly higher in microgravity compared to pre- and postflight (Figure 4B). In CTL rats, the mean circadian T_b rhythm did not change over time. Mean amplitude of the circadian T_b rhythm also did not differ between FLT and CTL in preflight or postflight measurements (Figure 4B).

LL Heart Rate (HR) – As with T_b, HR circadian rhythms persisted in LL in both CTL (Figure 3C) and FLT (Figure 3D) rats. As seen in LD, mean daily HR in both FLT and CTL groups exhibited a significant increase from preflight to flight

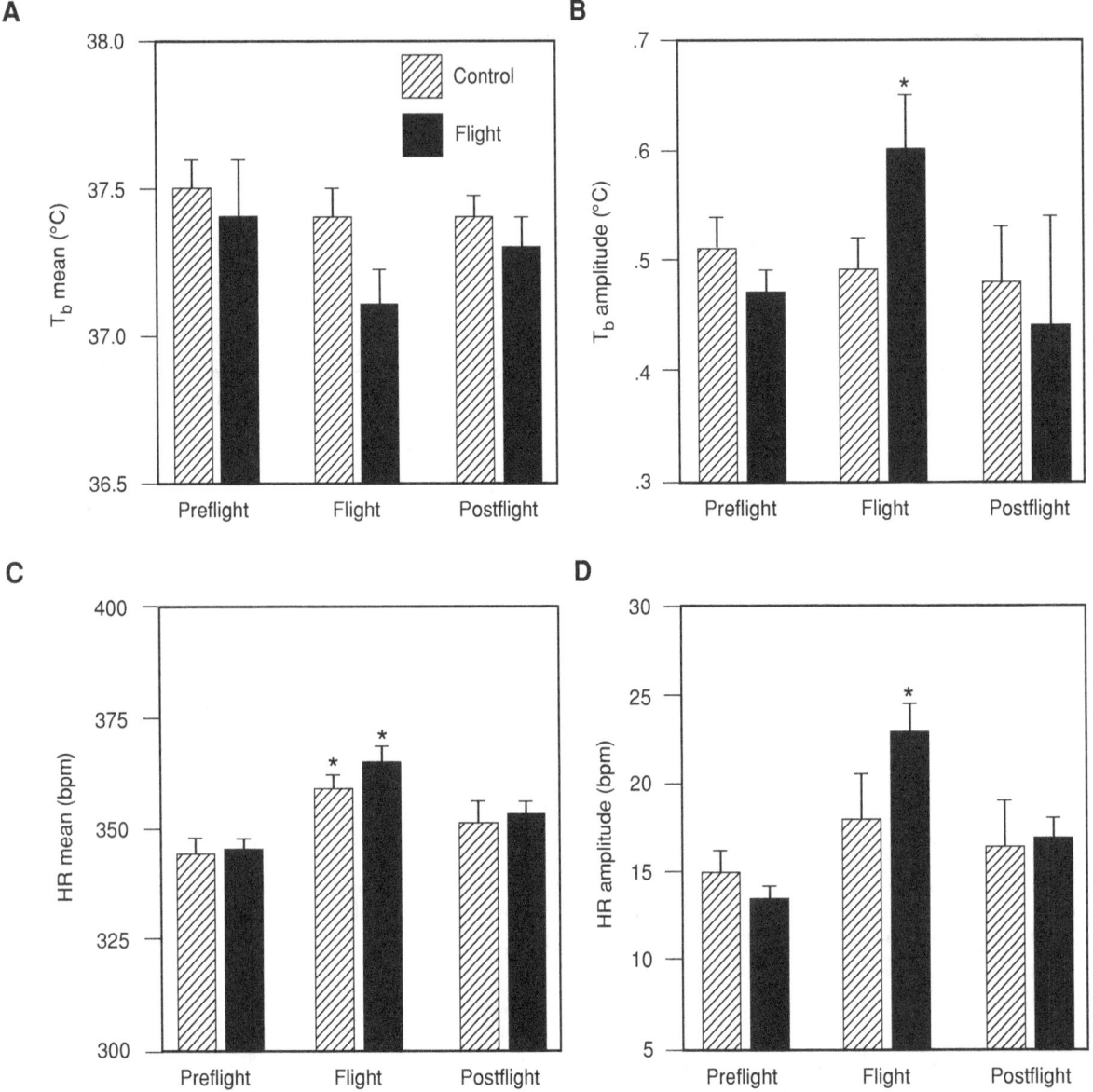

Figure 4. Daily mean (A) body temperature (T_b; °C) and (C) heart rate (HR; beats/minute) are presented as histograms along with the amplitudes of the respective rhythms (B & D). These data represent averages (+ SEM) from all LL CTL and FLT rats for the preflight, flight, and postflight periods. As was seen in the LD FLT group, there was a reduction in the average daily mean Tb during flight in the LL FLT animals, but it did not achieve significance. There was a significant increase in the amplitude of the T_b rhythm in the LL FLT rats during flight compared to pre- and postflight. As was seen in the LD rats, the LL CTL and FLT rats demonstrated a significant increase in mean HR during flight compared to pre- and postflight. There were no other significant differences between FLT and CTL groups or between experiment segments. (*=p<0.05)

(Figure 4C). The mean amplitude of the HR rhythm also increased significantly preflight to flight in the FLT rats (Figure 4D). The mean amplitude of the HR rhythm did not differ significantly between preflight and postflight. Daily mean HR and average amplitude of FLT rats were not significantly different from CTL during pre- and postflight (Figure 4D).

LD T_b-HR Phase – There was a significant phase delay in the timing of the T_b rhythm during flight in FLT rats compared to CTL rats during the flight period (Figure 5A) and compared

to preflight. The HR rhythm of the FLT group appeared to have a less-stable phase relationship with the LD cycle as compared to the CTL group; however, the difference was not significant.

LL T_b-HR Phase – In LL, the relative timing of the T_b and HR rhythms was altered during spaceflight compared to preflight and postflight as well as to CTL measurements (Figure 5B). As was seen in the LD FLT rats, there was a significant delay in the relative timing of the T_b rhythm, resulting in a

The Neurolab Spacelab Mission: Neuroscience Research in Space

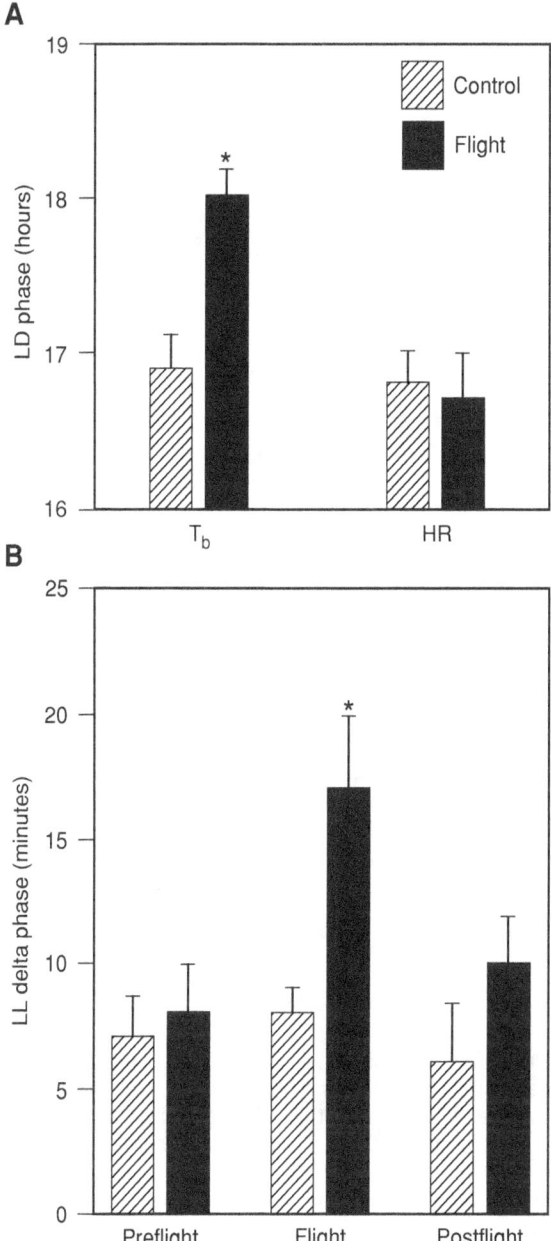

Figure 5. The relative timing of the body temperature (T_b) and heart rate (HR) rhythms during flight are presented as histograms. (A) shows the average (+SEM) time of acrophase (hours after light onset) of the T_b and HR rhythms for the LD CTL and LD FLT groups. The T_b rhythm was significantly delayed in the FLT group compared to the CTL group. However, there was no difference in the timing of the HR rhythm. Data from the LL CTL and LL FLT groups for the preflight, flight, and postflight segments are presented in (B). Since there was no LD cycle to use as a reference point, data are expressed as the average difference (+SEM) in time in minutes between the acrophase of the two rhythms. During flight, there was a significant delay in the relative timing of the T_b rhythm with respect to the HR rhythm in the FLT group compared to pre- and postflight and compared to the CTL group. (*=$p<0.05$)

longer time lag between the acrophases of the two rhythms. Preflight the relative timing of the two rhythms in the FLT and CTL LL groups was similar to that of the preflight FLT and CTL LD groups.

LL Circadian Period – Recording data under LL allowed us to calculate the period of the T_b and HR rhythms. The FLT LL group had a significant increase in the period (τ) of both the T_b (Figure 6A) and the HR (Figure 6B) rhythms in microgravity as compared to preflight and postflight measurements. In contrast, the LL CTL group did not show any significant changes in τ over time. Preflight and postflight τ was similar for both FLT and CTL groups.

LD SCN c-fos – To test the effects of microgravity on the sensitivity of the circadian pacemaker to light, we used a one-hour light pulse to stimulate production of *c-fos* (a marker for neuronal activity) expression in SCN neurons. We compared the number of *c-fos*-labeled neurons in the SCN of rats exposed to light with rats that were not exposed to a light pulse. FLT and CTL animals were exposed to a one-hour light pulse on FD2, FD14, R+1, and R+13. Since the production of *c-fos* in response to light differs over the day, all light pulses were given at the same time beginning at two hours after lights-out. Figure 7 shows a photomicrograph of the SCN of two CTL LD animals. The rat labeled Dark (Figure 7A) did not receive a light pulse, while the rat labeled Light (Figure 7B) did. The *c-fos* produced in response to the light pulse is visible in the darkly stained cells (Figure 7B; arrows). No such staining is visible in the Dark rat. This normal pattern of *c-fos* expression in light-pulsed rats was seen in all CTL rats. However, when compared with the CTL rats exposed to light, the FD2 FLT rats exposed to light had significantly lower levels of *c-fos* expression. A similar depression of *c-fos* expression, although not significant, was also observed in the R+1 FLT rats exposed to light. The FD14, R+1, and R+13 FLT rats exposed to light demonstrated a pattern of response that was closer to the response of the CTL rats. Although fewer *c-fos* immunopositive cells appeared in the FLT group compared to the CTL group, the difference was never significant.

DISCUSSION

This study demonstrated that rats exposed to spaceflight exhibit significant changes in circadian timing. The flight rats maintained in a normal LD cycle had T_b rhythms that were consistently delayed throughout the flight. In contrast, the HR rhythm maintained a relatively normal phase relative to the LD cycle during flight. As a consequence, the T_b and HR rhythms had an altered and less stable phase relationship during flight. This response has also been documented in monkeys (Fuller, 1996) and humans (Gundel, 1993). This loss of internal coordination of circadian rhythms is consistent with the condition of jetlag in humans. The present results in rats suggest that astronauts living in the space environment for extended periods of time, such as on the space station, could experience desynchronized circadian rhythms. Therefore, the

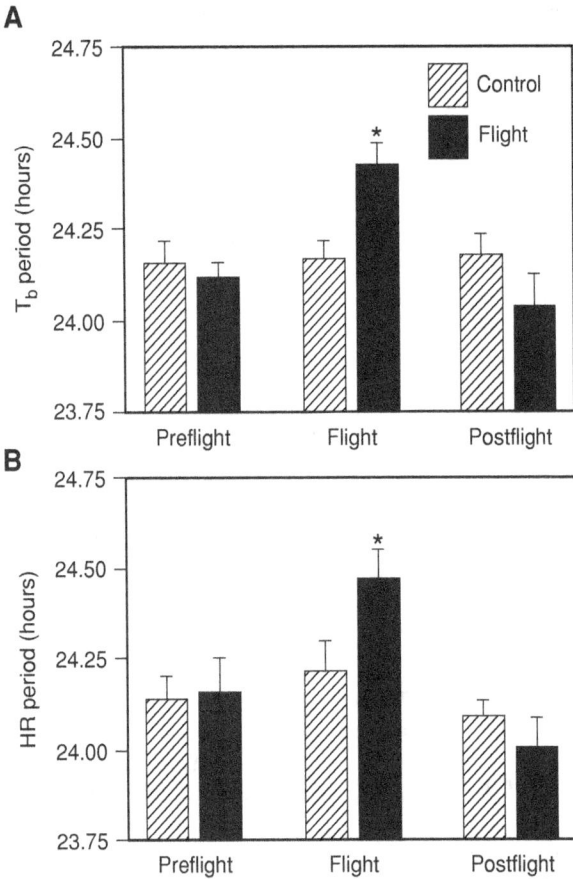

Figure 6. Average period (hours + SEM) of the (A) body temperature (T_b) and (B) heart rate (HR) rhythms of the FLT LL and CTL LL rats are presented as histograms for the preflight, flight, and postflight segments of the experiment. There was a significant increase in the period of both rhythms in the FLT group during flight compared to the preflight and postflight as well as compared to the CTL group. (*=p<0.05)

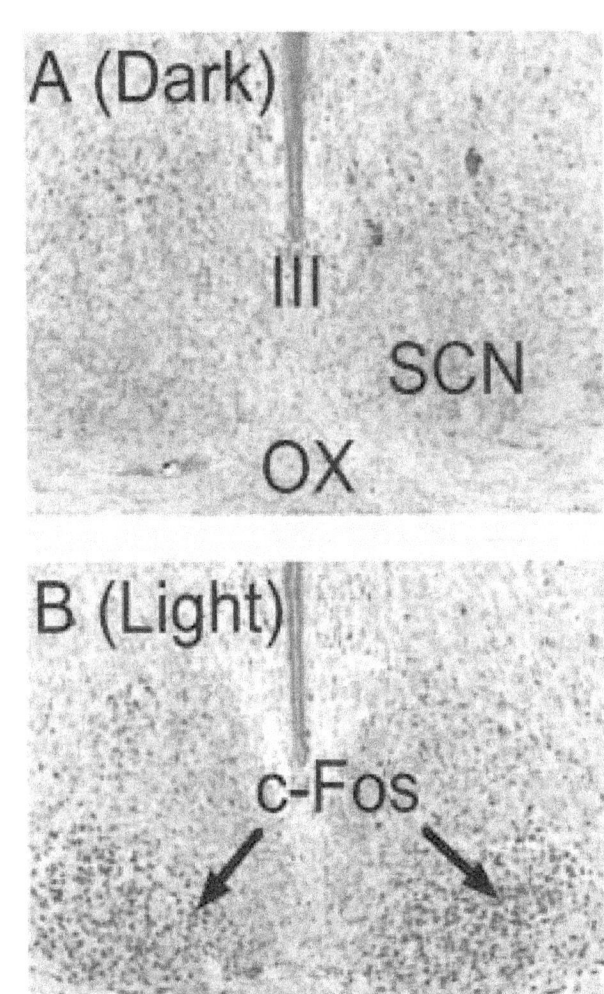

Figure 7. Samples of SCN from two CTL LD rats treated to reveal the presence of *c-fos*. A light pulse was presented to the rat in (B) at a time of day it would be expected to elicit *c-fos* expression within the SCN. The rat in (A) did not receive a light pulse, and thus acted as a control. As expected, without a light pulse (A), there was no expression of *c-fos*. In contrast, the *c-fos* expressed in response to the light pulse (B) is evident as the darkly stained cells (Arrows). (III = third ventricle; OX = optic chiasm)

astronauts may feel as if they are jet-lagged and have performance deficits even though they are maintained in a normal LD cycle.

This study also demonstrated that there was a significant decrease in the induction of *c-fos* in SCN neurons by a light pulse during the early part of the flight. This could be caused by a decrease in the sensitivity of either the eye or the SCN to light, or to a shift in the phase of the pacemaker such that light exposure occurred during a less-sensitive period. It will be important to understand which of these alternatives occurs during spaceflight since each can have a direct effect on how well light entrains circadian rhythms of humans in microgravity

conditions. When humans do not receive adequate light levels to entrain circadian rhythms, T_b rhythms can become out of phase with the other rhythms. Therefore, the evidence for reduced responsiveness of the SCN to light in the FLT rats provided by lower *c-fos* induction could explain the gradual change in phase of the T_b rhythm relative to both HR rhythms and the LD cycle. In humans, a syndrome that has been linked to possible circadian dysfunction is seasonal affective disorder or winter depression. An effective clinical treatment for seasonal affective disorder has been to expose patients to light of sufficient intensity. This light treatment not only causes remission of depression, but it also shifts the patient's circadian rhythms.

It will be important to understand if there is also a critical threshold of the light level required on the space station to minimize the risk of circadian dysfunction.

The flight rats maintained in LL exhibited a significant increase in the period of their circadian rhythms. It is highly likely that the period of the pacemaker (SCN) was affected by spaceflight since both T_b and HR circadian rhythms exhibited an identical change in period. In addition, the timing of the T_b rhythm became delayed in flight compared to the HR rhythm. Both of these responses (i.e., period and phase differences) returned to preflight values after the flight. It is not clear how spaceflight can induce such changes in the circadian pacemaker. Changes in the intensity of ambient light levels during LL can significantly alter the period of circadian rhythms (Moore-Ede, 1982). The reduced c-fos response in SCN neurons following a light pulse might suggest that a change in the light input to the pacemaker was responsible for the change in period. However, reduced photosensitivity of the pacemaker, analogous to a decrease in ambient light intensity, would be expected to shorten the period of the rhythms as shown in previous observations of intensity-period relationships.

The vestibular system's response to changes in gravity has been shown to cause a number of autonomic effects. It is possible that the vestibular system affects the SCN either directly or indirectly through autonomic nuclei. New animal models (knockouts, transgenic) have recently been developed to explore physiological changes related to vestibular system function. Our recent studies, using several new genetic mouse models, suggest that the vestibular system may mediate the effect of gravitational changes on circadian rhythms (Murakami, 1998; Fuller, 2000). Further studies employing these new genetic mouse models will increase our understanding of the neural mechanisms mediating the physiological responses and process of adaptation to space.

Acknowledgements

The authors wish to recognize and thank Dr. Frank Sulzman for his vision and dedication in bringing the Neurolab mission, NASA's contribution to the Decade of the Brain, to fruition. We would also like to extend our deepest appreciation to the dedicated personnel at the NASA Ames Research Center and the NASA Kennedy Space Center. We would like to thank the dedicated project team personnel. In particular, we are grateful to Angelo De La Cruz and the dedicated NASA engineering staff: Jim Connelly, Mike Skidmore, John Hines, Julie Schonfeld, and Tom Vargese. We would also like to extend our appreciation to NASA Headquarters for providing our team the opportunity to be a part of Neurolab. Thanks are also due to the wonderful staff team at the UC Davis Chronic Acceleration Research Unit and the many student team members including Jeremy Fuller, Steven Killian, and Gabriel Ramirez. In addition, we would like to acknowledge the skill of our professional veterinary surgical team, Drs. Andris Kanups, Barbara Smith, and Nick Vatistas. This paper is dedicated to the memory of Dr. Nick Vatistas, a scientist, surgeon, colleague, and friend—we miss him dearly. (Supported by NASA Grant NAG2-944 and NIH Grant HL53205.)

REFERENCES

THERMOREGULATION IS IMPAIRED IN AN ENVIRONMENT WITHOUT CIRCADIAN TIME CUES. C.A. Fuller, F.M. Sulzman and M.C. Moore-Ede. *Science,* Vol. 199, pages 794–6; 1978.

THE CLOCKS THAT TIME US: THE CIRCADIAN TIMING SYSTEM IN MAMMALS. M.C. Moore-Ede, F.M. Sulzman and C.A. Fuller. Harvard University Press, Cambridge, MA; 1982.

SLEEP AND CIRCADIAN RHYTHM DURING A SHORT SPACE MISSION. A. Gundel, V. Nalishiti, E. Reucher, M. Vejvoda and J. Zulley. *J. Clin. Invest.,* Vol. 71, pages 718–724; 1993.

INFLUENCE OF GRAVITY ON THE CIRCADIAN TIMING SYSTEM. C.A. Fuller, T.M. Hoban-Higgins, D.W. Griffin and D.M. Murakami. *Adv. Space Res.,* Vol. 14, pages 399–408; 1994.

PRIMATE CIRCADIAN RHYTHMS DURING SPACEFLIGHT: RESULTS FROM COSMOS 2044 AND 2229. C.A. Fuller, T.M. Hoban-Higgins, V.Y. Klimovitsky, D.W. Griffin and A.M. Alpatov. *J. Appl. Physiol.,* Vol. 81, pages 188–193; 1996.

CHRONIC 2G EXPOSURE AFFECTS C-FOS REACTIVITY TO A LIGHT PULSE WITHIN THE RAT SUPRACHIASMATIC NUCLEUS. D.M. Murakami, I-H. Tang, C.A. Fuller. *J. Grav. Physiol.,* Vol. 5. pages 71–78, 1998.

EFFECTS OF 2G ON CIRCADIAN RHYTHMS IN BRN 3.1 MICE. D.M. Murakami, L. Erkman, M.G. Rosenfeld and C.A. Fuller. *J. Grav. Physiol.,* Vol. 5, pages 107–108; 1998.

EFFECTS OF 2G ON HOMEOSTATIC AND CIRCADIAN REGULATION IN THE HET MOUSE. P.M. Fuller, T.A. Jones, S.M. Jones and C.A. Fuller. *Soc. Neurosci. Abstracts,* Vol. 26, page 207; 2000.

Technical Reports

To accomplish the research goals of the Neurolab mission, special techniques, procedures or equipment were sometimes needed. This section highlights a selection of the technical accomplishments of the Neurolab experiment teams. These procedures and techniques may be useful for those planning space research in the future.

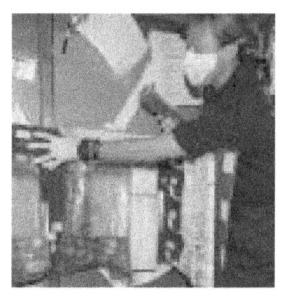

Measurement of Place Cell Firing in the Hippocampus in Space

Authors

James J. Knierim, Gina R. Poe,
Casey Stengel, Bruce L. McNaughton

Experiment Team

Principal Investigator: **Bruce L. McNaughton**[1]

Co-Investigators: **James J. Knierim,**[2]
Gina R. Poe[1]

Technical Assistants: **Kathy Dillon,**[1] **Shanda Roberts,**[1]
Veronica Fedor-Duys[1]

Engineers: **Casey Stengel,**[1] **Krzysztof Jagiello,**[1]
Vince Pawlowski[1]

[1]University of Arizona, Tuscon, USA
[2]University of Texas-Houston Medical School, Houston, USA

RATIONALE

The goal of our experiment was to understand how the part of the brain that provides information on location and navigation (the hippocampus) encodes that knowledge. One way of addressing this question is to record the activity of neurons in the hippocampus while an animal performs various behavioral tasks. This kind of electrophysiological recording typically requires electrically shielded labs and a variety of amplifiers and computers to acquire, process, and display the signals. For spaceflight use, however, the system must be small, reliable, and able to provide the crewmembers data on whether useful information is being recorded.

To meet these requirements, we designed a system that allowed the collection of complex neurophysiological measurements in space. The system was composed of an interface box, which connected to the electrodes measuring the brain signals and processed the multiplexed data, and a computer, which stored and displayed these signals. These two portable instruments contained all of the electronics necessary to perform in space what usually takes racks of equipment to perform in a laboratory on Earth. The system designed for spaceflight forms the basis of a state-of-the-art commercial system now available to perform multichannel neurophysiological recordings in the laboratory.

HARDWARE DESCRIPTION

The measurements from the brain cells are made with tetrodes—recording probes made from four pieces of very fine wire (0.0005 in) twisted together. To perform the experiment, data are acquired from 12 tetrodes and two single-channel reference electrodes. Since each tetrode has four separate recording tips, each experiment requires the acquisition, storage, and processing of 50 separate channels of electrophysiological data. Figure 1 shows the equipment arrangement used initially in our ground-based laboratory to record these signals from the hippocampus. This system uses multiple computers to process and display the data, amplifiers to filter and condition the signal, and audio monitors to listen to the activity of the electrodes.

Additional electronics are required to deliver electrical stimulation to the "reward centers" of the rat's brain, which serves as a pleasurable reinforcer to perform the behavioral tasks. Finally, a video tracking system is used to acquire, digitize, and store the position of the animal at a rate of 20 samples/second. In the laboratory, this configuration of numerous computers, monitors, switches, amplifiers, and electronics takes up considerable space. For use in the tight confines of the Space Shuttle, where the entire recording system must be stowable and quickly assembled/ disassembled, a more compact design of the hardware was essential.

Figure 2 shows the interface box that was designed to connect to the tetrodes. This system has the capability to record from two arrays of tetrodes simultaneously (96 channels) as well as 24 additional channels of brain EEG activity. The two main components of this interface box are a "digital control and data communications board," which uses digital logic circuits to acquire the simultaneous channels of data, convert them to data packets, and send these packets to the data computer, and a custom 12-bit, 5-MHz analog-to-digital converter with 160-channel input multiplexer board. This box also contains boards to amplify and filter the signals, to track and digitize the locations of colored light-emitting diodes on the animal's head to compute location at 60 samples/second, and to deliver electrical current to provide stimulation to the brain's reward centers. Figure 3 is the

computer system that stored and displayed the data from the interface box, generated an audio output of the brain signals, and served as the user interface. This computer, a modified Sun Sparcstation, took the raw data from the interface box and produced a computer display that showed the firing of cells in the hippocampus. These displays were used by the crewmembers on board to assess whether the experiment was running properly, and a video downlink enabled the investigator team on the ground to monitor the progress of the experiment as it occurred.

Figure 4 shows the system during ground testing before the mission. The entire rack of equipment and bank of computers

Figure 1. This is the equipment arrangement used initially in a ground-based laboratory to process signals from the hippocampus. This system uses multiple computers and a rack of amplifiers to process and display the data. For spaceflight use, however, a more compact system is essential and was the focus of hardware development for this experiment.

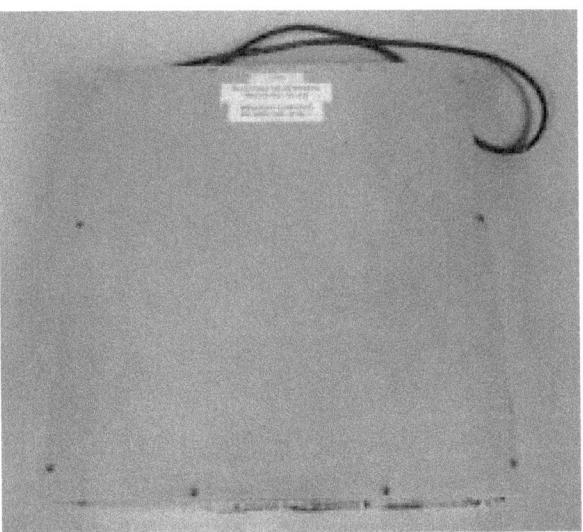

Figure 3. The computer system stored and displayed the data from the interface box. This computer is based on a Sun Sparcstation.

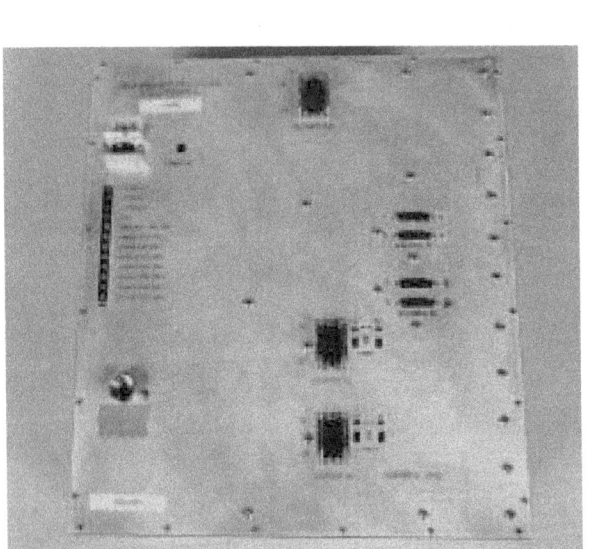

Figure 2. An interface box was designed to connect to the tetrodes. This box collects the very-low-level signals from the brain cells, amplifies and filters them, and packages them in a way so that they can be efficiently transported to the data computer. This box also contains circuitry for video tracking and for delivery of a low-level brain stimulation current.

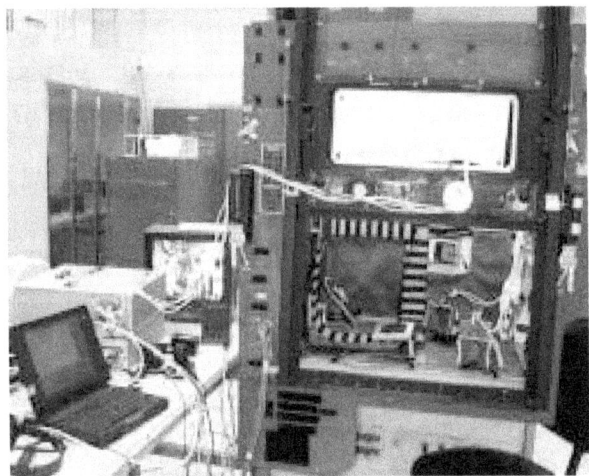

Figure 4. This shows the hardware arrangement during a crew training session before the Neurolab mission. The studies were conducted in the general-purpose workstation, which appears on the right. On the table are the interface and computer systems. A laptop computer served as the video monitor for the data computer.

previously used (Figure 1) have been replaced with the two boxes and the laptop computer on the table at left. The rat performed its task inside the general-purpose workstation (GPWS), and the neural signals were carried from the workstation to the interface box with white cables made of fine-wire conductors. Figure 5 shows a schematic of the arrangement of the entire system during the mission. The interface box and computer system were mounted above the GPWS and connected to the tetrodes inside the GPWS. A laptop computer provided experimental control and monitoring to the crewmembers. The system could be unstowed and assembled within minutes, and its performance was reliable in providing clean neural signals.

APPLICATION

The result of this development effort was a compact, robust system for collecting complex electrophysiological data. This system developed for spaceflight was further refined and is now commercially available (Neuralynx, Inc., Tucson, AZ). The system allows the simultaneous recording of 160 channels of electrophysiologial data in a compact design that is very resistant to electrical noise and other common problems in neurophysiological recordings. It runs in combination with a personal computer under the Windows operating system.

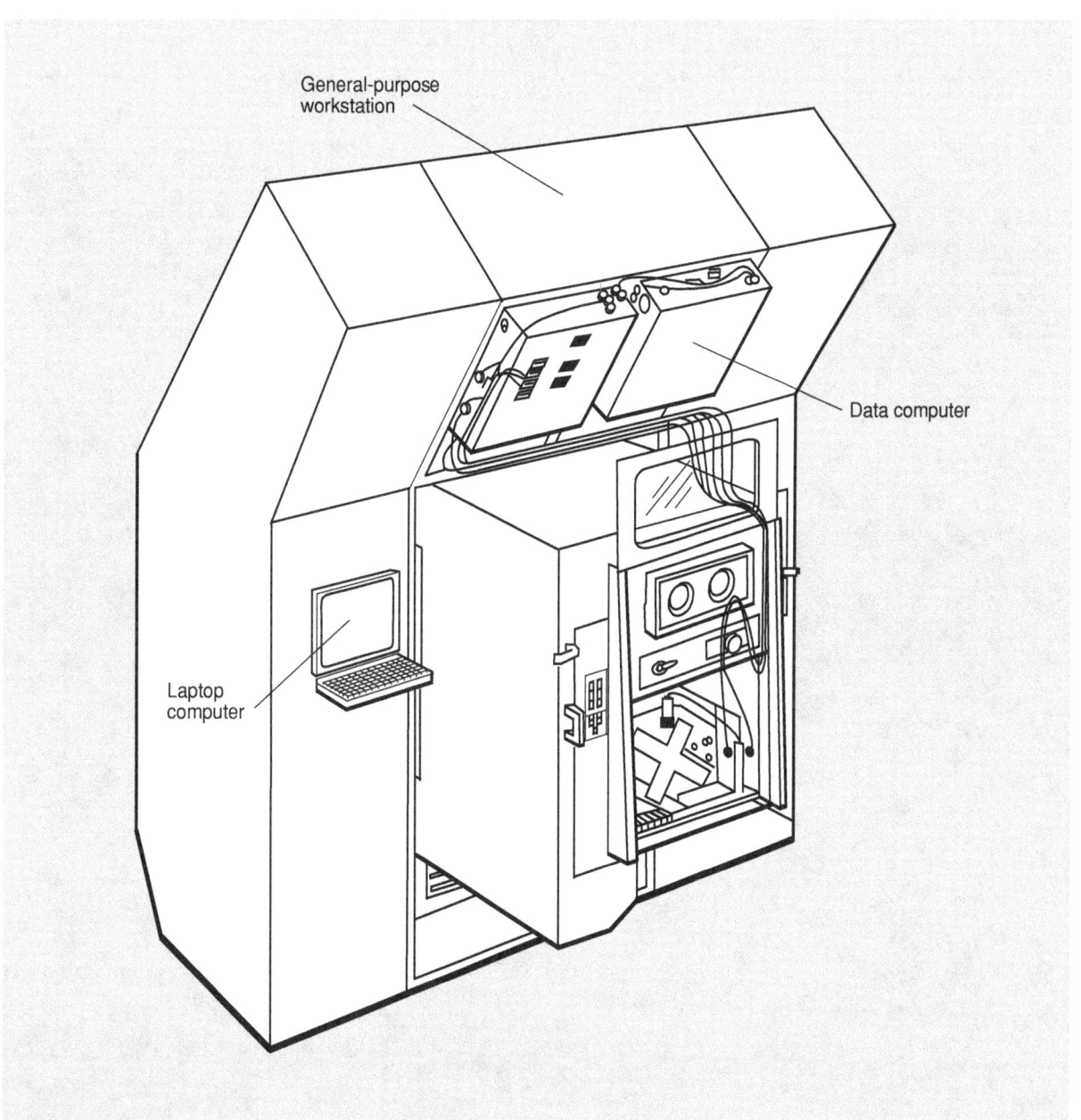

Figure 5. This schematic shows the ultimate arrangement of hardware that was used on the Neurolab mission.

Portable Sleep Monitoring System

Authors

Derk-Jan Dijk, Jay C. Buckey, Jr., David F. Neri,
James K. Wyatt, Joseph M. Ronda, Eymard Riel,
G. Kim Prisk, John B. West, Charles A. Czeisler

Experiment Team

Principal Investigators: **Charles A. Czeisler,[1]**
John B. West[2]

Co-Investigators: **Derk-Jan Dijk,[1] Ann R. Elliott,[2]**
James K. Wyatt,[1] G. Kim Prisk,[2]
David F. Neri,[1,3] Joseph M. Ronda,[1]
Rod J. Hughes[1]

Technical Assistants: **Eymard Riel,[1] Karen Smith[1]**

Engineers: **Janelle M. Fine,[2] Trevor K. Cooper,[2]**
Jennifer Jackson,[1] Alex McCollom,[1]
Wim L.J. Martens,[4] Carlo Peters[4]

[1]Harvard Medical School and Brigham and Women's Hospital,
Boston, USA
[2]University of California, San Diego, USA
[3]NASA Ames Research Center, Moffett Field, USA
[4]TEMEC Instruments BV, Kerkrade, The Netherlands

RATIONALE

The sleep studies on Neurolab required detailed measurements to assess sleep accurately. On Earth, these measurements would typically be done in a fully equipped sleep laboratory. The operational demands of the Neurolab mission (limited space, power, and time) required a system that could perform all the usual sleep laboratory measurements in a small, portable package. The sleep system assembled for Neurolab included a special suit for measuring respiratory movements, sleep headgear that provided electroencephalography and eye movement data, and a portable recording device that could process and save the information. This system demonstrated that high-quality polysomnographic sleep data could be collected in a very demanding operational environment, and illustrates how sleep could also be monitored routinely in homes and clinics on Earth.

HARDWARE DESCRIPTION

The sleep monitoring system that was developed for this spaceflight experiment consisted of a portable digital sleep recorder, a custom-fitted sleep cap, a respiratory inductance plethysmography body suit, a cable harness, an impedance meter, and a computerized signal-quality assessment system. This system was also used for all preflight and postflight recordings using the same procedures as those used inflight.

Sleep net

Sleep was recorded using a sensor array (e-Net Physiometrix, North Billerica, MA) that was placed on the head. The modified sleep net was an integrated set of components consisting of a reusable customized headpiece and disposable silver/silver chloride hydrogel biosensors (Hydrodot Biosensors, Physiometrix, N. Billerica, MA). The electrophysiological head and face electrode sites were integrated into an elastic lattice cap that was secured on the head by a chin and neck strap. Sleep nets were individually tailored to each subject to ensure proper fit and reproducibility of electrode site placement. The electrodes recorded brainwaves with an electroencephalogram (EEG), eye movements with an electro-oculogram (EOG, left, right outer canthus), and muscular activity around the chin with an electromyogram (EMG, submental). Electrodes were positioned in sockets within the sensor array placed according to the International 10-20 System: two reference electrodes behind the ear, one forehead ground electrode, two EOG electrodes, four EEG electrodes (C3, C4, O1, O2), and four chin EMG electrodes. After placing the electrodes in the sleep net, the function of each electrode was verified by impedance checking prior to each recording (maximum impedance 10 Kohms) The shielded wire leads on the outside of the sleep net were combined into a single connector that attached to the digital sleep recorder (DSR) via a single connector. Figure 1 shows the blue sleep net connected to the digital sleep recorder.

Digital Sleep Recorder

All sleep recordings were acquired on a modified Vitaport-2 DSR (Temec Instruments, Kerkrade, The Netherlands). The

Portions of the text in this report are from Dijk, 2001, used with permission.

DSR is a portable, modular battery-operated sleep recorder with a 12-bit digital-to-analog converter (DAC). For the experiments during the Neurolab mission, the DSR consisted of the recorder base with three modules. EEG, EOG, and EMG signals were recorded through the first module. Cardiorespiratory signals were recroded through the second module. The third module, an eight-channel DAC, was used to output acquired signals to other analytical devices. (For details, see Dijk, 2001, and Elliott, 2001.) The 24-channel Vitaport-2 with the DAC attached measured 4×9×15 cm. Data were stored on an 85-Mb Flash random access memory (RAM) card, which provided the capability of recording the Neurolab experiment signals for more than 10 hours using four standard AA batteries. Sixteen channels were used to record and store the data. In addition to the EEG, EOG, and EMG signals, other measurements included respiration via nasal/oral airflow (three-pronged thermistor adhered to upper lip; EdenTec Corporation, Eden Prairie, MN); rib cage and abdominal motion (respiratory inductance plethysmography); snoring (microphone attached to the neck at the level of the larynx); light (detector incorporated into microphone on throat); arterial oxygen saturation via pulse

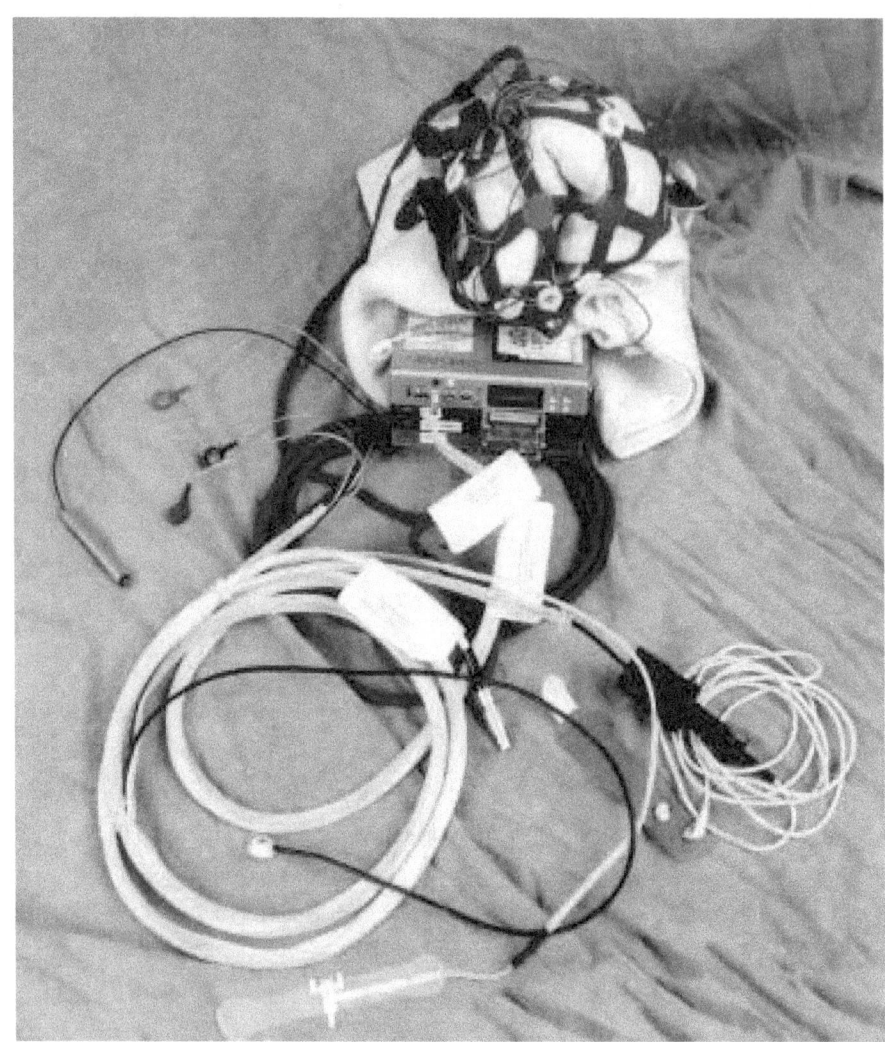

Figure 1. The blue sleep net is seen at the top of this figure resting on a white towel. Under the sleep net is the DSR. The DSR connects both to the sleep net and to the harness containing the connectors for the electrocardiogram, pulse oximeter, nasal thermistor, microphone, light sensor, respiratory inductance plethysmograph, and event marker.

oximetry (SaO$_2$: Ohmeda Flex-probe; Ohmeda Medical, Inc., Columbia, MD, adhered to left ring finger); heart rate via electrocardiogram (lead II, 256 Hz); and an event marker. The DSR and the harness that connected to it are shown in Figure 1.

EEGs were low-pass filtered at 70 Hz and high-pass filtered with a time constant of 0.33 second and sampled at 256 Hz. To optimize the bandwidth and to limit the size of the data file on the personal computer miniature communications interface adapter (PCMCIA) card, the Vitaport-2 carried out an on-line software moving averaging filtering with a cutoff frequency of 64 Hz, and the EEGs finally were stored at 128 Hz.

EOGs were low-pass filtered at 35 Hz and high-pass filtered with a time constant of one second and sampled at 128 or 256 Hz. An on-line software moving averaging filtering was applied with a cutoff frequency of 32 Hz, and the results were stored at 64 Hz.

EMG signals were low-pass filtered at 100 Hz and high-pass filtered with a time constant of 0.015 second and sampled at 128 or 256 Hz, and the results were stored at 128 Hz.

The data collected on the DSR were stored on PCMCIA 85 MB Flash RAM cards (SanDisk, Sunnyvale, CA). During the STS-90 mission, data were transferred to a microcomputer and downlinked from the Space Shuttle to Mission Control at Johnson Space Center, allowing inspection of the data by the investigators after each inflight sleep recording. Sensor impedances were checked using a NASA-customized impedance meter (inflight) or a GRASS EZM4 impedance meter (pre- and postflight). During and after instrumentation but before sleep, all physiological signals were displayed in real time, on the screen of a laptop, using an expert system for astronaut assistance (Callini, 2000). This allowed inspection of signal quality before the sleep recordings.

Figure 2. The respiratory inductance plethysmograph suit has wires for measuring both chest and abdominal respiratory movements incorporated into the top portion. The shorts have a pocket that holds the DSR. The various Velcro straps are used to route the harness.

Respiratory Inductance Plethysmograph Suit

The sensors needed to measure the movement of the rib cage and abdomen during breathing (called respiratory inductance plethysmographs) were integrated into a custom-fitted two-piece (vest plus shorts) Lycra body suit (Blackbottoms; Salt Lake City, UT). The suit is shown in Figure 2. The rib cage and abdominal wires for the inductance plethysmography measurements were sewn into the vest section of each suit, with the chest band at the level of the nipples and the abdominal band over the umbilicus. Because the spine lengthens in microgravity, the vest section had adjustable shoulder straps and was held in place by attachment to the shorts with integrated Velcro strips to ensure proper location of each band at all times. A single harness connected all leads from the subject's torso to the DSR.

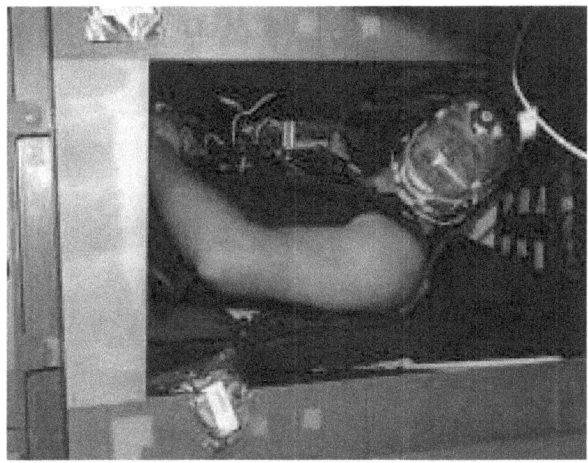

Figure 4. Payload specialist Jay Buckey is shown here, fully instrumented for sleep and resting in the sleep station in the middeck of the Space Shuttle *Columbia*.

Figure 3. Instrumentation: Payload Specialist Jim Pawelczyk applies hydrogel biosensors to Mission Specialist Rick Linnehan (foreground) while Payload Specialist Jay Buckey arranges straps to secure the sleep net on Mission Specialist Dave Williams (background).

Complete Ensemble

During the mission, the astronauts instrumented each other (Figure 3). The resulting complete ensemble is shown in Figure 4. In this picture, a crewmember is fully instrumented and ready for sleep in the sleep station on the Space Shuttle *Columbia*, which has a door slide for darkness and sound attenuation and a built-in ventilation system.

APPLICATION

The sleep ensemble designed for Neurolab could be used to measure sleep in a variety of settings, including patient homes, remote clinics, and hospitals without fully equipped sleep laboratories. It could also be used in studies of sleep in shift workers in the field—e.g. on oil rig workers—the crews of ships, or interns in hospitals.

REFERENCES

EFFECTIVENESS OF AN EXPERT SYSTEM FOR ASTRONAUT ASSISTANCE ON A SLEEP EXPERIMENT. G. Callini, S.M. Essig, D.M. Heher, and L.R. Young. *Aviat. Space Environ. Med.*, Vol. 71, pages 1023–1032; 2000.

MICROGRAVITY REDUCES SLEEP DISORDERED BREATHING IN HUMANS. A.R. Elliott, S.A. Shea, D-J Dijk, J.K. Wyatt, E. Riel, D.F. Neri, C.A. Czeisler, J.B. West, G.K. Prisk. *Am. J. Respir. Crit. Care Med.,* Vol. 164, pages 478–485; 2001.

SLEEP, PERFORMANCE, CIRCADIAN RHYTHMS, AND LIGHT-DARK CYCLES DURING TWO SPACE SHUTTLE FLIGHTS. D-J Dijk, D.F. Neri, J.K. Wyatt, J.M. Ronda, E. Riel, A.Ritz-De Cecco, R.J. Hughes, A.R. Elliott, G.K. Prisk, J.B. West, and C.A. Czeisler. *Am. J. Physiol. Regul. Integr. Comp. Physiol.*, Vol. 281, pages R1647–1664; 2001.

Neurolab Virtual Environment Generator

Author
Charles M. Oman

Experiment Team

Engineering Team: Charles M. Oman[1]
Jonathan Adams,[3] Terence Adams,[3]
Amjad Al-Hajas,[3] Clif Amberboy,[3]
Herbert Anderson,[3] Andrew Beall,[1]
Ron Bennett,[3] Floyd Booker,[2]
Huong Charles,[3] Jean Cheng,[3]
Dennis Grounds,[2] Jennifer Krug,[3]
Karen Lawrence,[3] Angelene Lee,[2]
Grace Lei,[3] Al Leyman,[3] Louise Li,[3]
Trent Mills (Project Engineer),[3]
David Muecke,[3] Kim Nguyen,[3]
Pradip Shah,[3] Felix Silvagnoli,[3]

[1]Massachusetts Institute of Technology, Cambridge, USA
[2]NASA Johnson Space Center, Houston, USA
[3]Lockheed Martin Engineering Services, Houston, USA

RATIONALE

A virtual environment generator (VEG) computer graphics workstation, along with associated subject interface devices and restraint system, was developed to support the sensory-motor and performance investigations aboard the STS-90 Neurolab mission. The VEG was a custom-designed, low-power-consumption, rack-mounted Windows NT-based 200-MHz PC graphics workstation equipped with a wide field-of-view (64 degrees×48 degrees) color, stereo liquid crystal head-mounted display. The system also included an electronics box, interfaces for Spacelab data, a keyboard, a flat panel display, a headset, microphones, and a thigh-mounted joystick. Software included a Session Manager graphical user interface, as well as Experiment and Archive Managers to render scenes, collect data, and analyze results for individual experiments. Other accessories included an optical head/hand tracker (not flown) and a subject restraint system (SRS). When wearing the SRS, the subject could connect the harness at each hip to a pair of floor-mounted 147 Newton (33-pound) constant force springs. This force would pull the subject toward the floor of the Spacelab and allow the crewmember to "stand" in weightlessness. Alternatively, the springs could be disconnected and the crewmember would float with only light restraint provided by cloth tethers. Science results obtained using the Neurolab VEG are described in the accompanying article by Oman, et al. The VEG served as a prototype for a higher performance virtual-reality system under development for the International Space Station.

OBJECTIVES

When the Neurolab mission was conceived in the early 1990s, virtual-reality (VR) technology had advanced to the point where NASA advisory committees realized it could be exploited in space neuroscience studies, human factors research, and even onboard crew training. One recommendation was that a flight-qualified VR system be developed for use on Spacelab missions. Progress in producing real-time computer images, tracking head movements, and generating both auditory and tactile stimuli allowed scientists to create completely artificial environments and virtual objects, and experiment with new types of "immersive" human computer interfaces. NASA working groups and

university workshops defined preliminary requirements for the system, based on research needs and current or soon-to-be-available technology. Proposals to use the system were solicited for Neurolab (NASA, 1993). Responsibility for the development was assigned to the Integrated Project Development Laboratory in the Space and Life Sciences Directorate at Johnson Space Center. Lockheed Martin Engineering Services (LMES) served as the contractor. Scientists from two sensory-motor and performance experiments tentatively selected for the mission served informally as advisors. The virtual environment generator (VEG) workstation also ultimately served as a prototype for the International Space Station Human Research Facility graphics workstation.

Figure 1. Crewmember preparing for VEG session on flight day four, holding head-mounted display shroud in right hand.

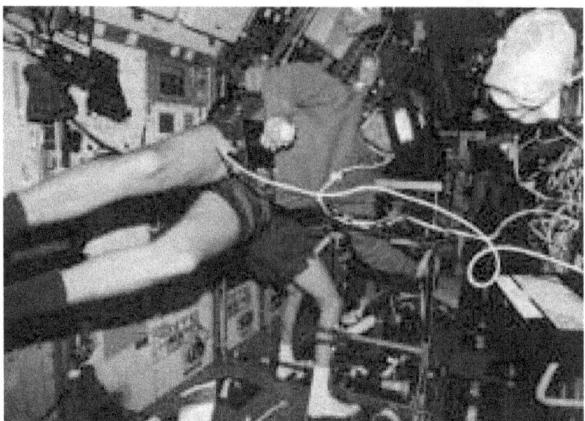

Figure 2. Crewmember performing E136 object recognition experiment in left shoulder-down condition on flight day four.

DESCRIPTION

Figure 1 shows a photograph of the VEG in use aboard Neurolab. To shorten development time, constrain costs, and promote design maturity, the VEG used commercial-off-the-shelf system components. Spacelab physical space constraints, and the requirement for 360-degree immersive display capability, precluded use of surround-screen displays (e.g., Cruz-Neira, 1993), so the VEG used the more compact and widely used head-mounted display (HMD)/head-tracker approach (Sutherland, 1968; Fisher, 1986). Scientists wanted the highest resolution, largest field-of-view, color stereo HMD available. Cost and safety concerns excluded head-mounted solutions using cathode ray tubes. Instead liquid crystal displays (LCDs) were preferable. A red-green-blue compatible color stereo LCD HMD display (Proview-80, Kaiser Electro-Optics, Inc., Carlsbad, CA) was selected, which provided 640×480 pixel monocular resolution and a relatively large 64 degrees×48 degrees monocular field-of-view (Figure 2). The HMD could be worn with spectacles. The interpupillary distance was adjustable. A removable black plastic shroud occluded outside vision. A pair of small fans was added to provide additional ventilation inside the shroud. The HMD connected to the VEG microcomputer via a small (25.4 cm×25.4 cm×7.6 cm) external electronics box. For head and hand tracking, several electromagnetic systems were evaluated, but none could meet VEG accuracy specifications when operating near Spacelab equipment racks. Instead, an optical tracking system (Innovision Systems, Inc., Warren MI) was chosen that used a pair of infrared charged-coupled device cameras with built-in infrared sources (Qualisys MCU240) clamped to overhead handrails to track small spherical retroreflective markers on the HMD and hand. Two-dimensional marker position data computed in the camera were processed by a dedicated Pentium processor mounted in a Spacelab Standard Interface Rack Tracker Electronics Drawer (49 cm wide, 18 cm high, 70 cm deep). Proprietary Innovision software computed HMD and hand position and orientation, and sent these data to the VEG workstation via an RS-422 computer link. Tracking volume was approximately one cubic meter.

A 500-watt Spacelab power constraint determined the choice of computer graphics workstation. Scientists had originally proposed the use of a high-end, Unix-based graphics workstation, but this typically used several kilowatts of power. However, graphics accelerator cards became available for PC games and workstations, which met VEG graphics performance requirements. This made it possible to use the spaceflight-qualified NASA Life Science Laboratory Equipment (LSLE) Microcomputer II for the VEG by adding dual Z-13 Open GL [graphics language] graphics accelerators (Intergraph Corporation, Huntsville, AL), each driving one eye of the HMD. This three-programmable controller interface (PCI) card system was able to render up to 1.2 million triangles/second and fill at rates to 38 million texels/second per eye. The VEG LSLE Microcomputer II (Figure 3) fit in a second Spacelab standard interface rack, and used a Zandac passive backplane (six industry standard architecture (ISA) slots, six PCI slots, one central processing unit (CPU) slot) and a 200-MHz Trenton Pentium Pro single board computer, with 512 K level 2 cache, and 256 MB of error-corrected coded RAM. The computer had an internal isolated power supply, was cooled by fans, and consumed less than 250 watts of 28-V power. Interfaces for SCSI2, PCI,

Figure 3. LSLE Microcomputer II - VEG configuration. Front view.

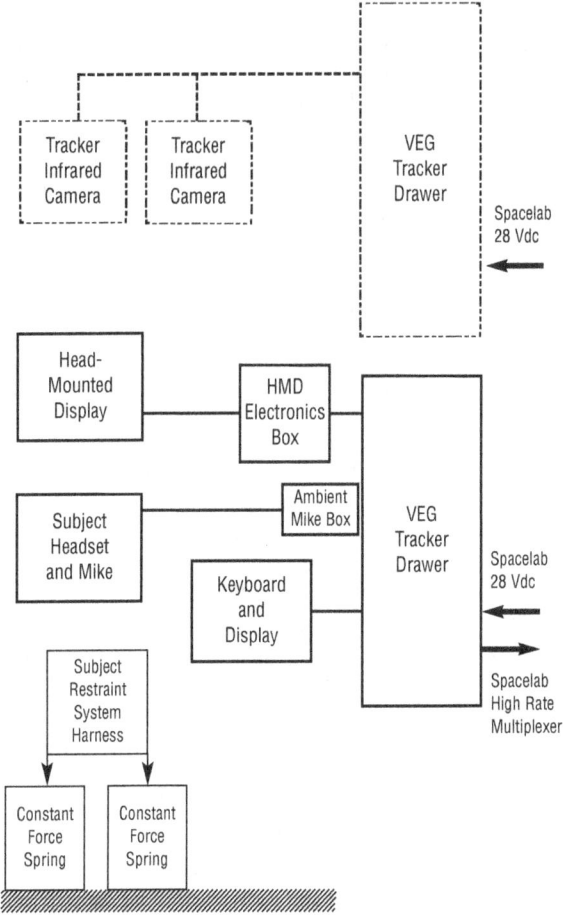

Figure 4. VEG system component hardware block diagram.

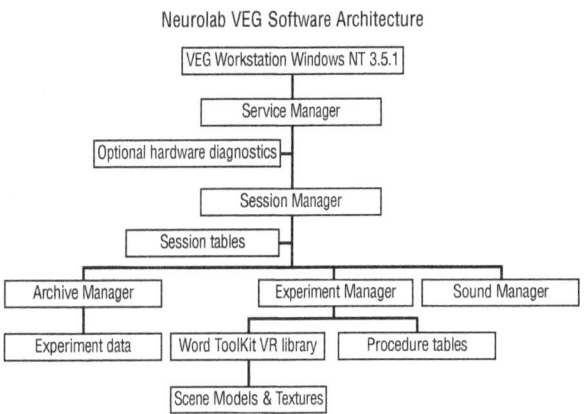

Figure 5. VEG software block diagram.

integrated drive electronics (IDE), personal computer miniature communications interface adapter (PCMCIA), RS-422, parallel, serial, and floppy disks were provided. Primary program and data storage was on three 2.1 GB IDE hard drives (Seagate ST32140A). Onboard physical archiving and backup boot capability was provided using four removable miniature PCMCIA hard drives (Viper 340 MB). Spacelab telemetry downlink was provided using a custom 64-Kbps, high-rate multiplexer board and initial rate integrating gyro-B (IRIG-B) timing generator board. A digital signal processing board with voice compression capability (Bitware Snaggletooth ISA digital signal processor (DSP) board with Bitsi, Digital Voice Systems, Inc., Westford, MA, 4 Kbps card) was provided to digitize subject voice comments dictated into a head-mounted microphone. Two options were provided for external user control and display at the system level: (1) a standard LSLE Microcomputer II keyboard and display could be clamped to Spacelab handrails and directly connected; or alternatively, (2) the user could connect an external portable laptop computer running remote user software (PC Anywhere, Symantec, Cupertino, CA). A thigh-mounted subject interface device (SID) was provided for use at the experiment level while the subject was wearing the HMD. This was a two-axis, four-button/hat-switch right-handed joystick

(Thrustmaster Flight Pro RS-232), equipped with Nomex/Velcro straps and thigh pad. To reduce directional sound cues, ambient sounds were recorded using a single microphone mounted near the VEG front panel, amplified, and fed back binaurally to a headset worn by the subject beneath the HMD. Voice comments made by the subject were separately recorded by a small boom microphone attached to the subject's headset. A VEG computer system hardware block diagram is shown in Figure 4.

Figure 5 shows an overview of the VEG software configuration. An important design goal was to make the VEG operable by a single crewmember, who served as both subject and operator. The VEG workstation operating system (OS) was Microsoft Windows NT v. 3.5.1. The system booted to a Service Manager routine, which checked the system configuration, including the joystick and HMD, and allowed the user to enter the Grenwich Mean Time and either start the VEG Session Manager or run diagnostic routines. The VEG Session Manager was a single screen, pull-down graphics user interface (Figure 6) written in Microsoft Visual Basic. Subjects

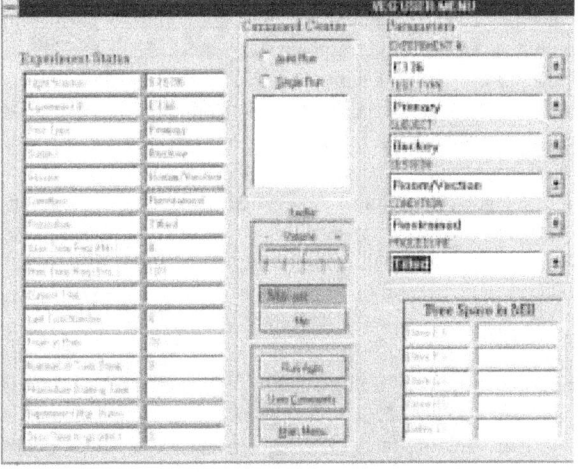

Figure 6. VEG session manager user menu.

employed pull-down menus to identify themselves. The Session Manger then suggested the experiment and particular procedure they should run during that session, based on a plan provided by the experimenters. If desired, subjects could run individual procedures out of the predefined order. The display showed the experiment selected, the estimated time required to complete the procedure, and the time remaining in the session. The percentage of data storage capacity remaining on each hard drive was also displayed. Subjects could enter experiment notes in a text window, and toggle on/off the subject microphone. When the subject was ready, he/she would launch Experiment Manger, Archive Manager, and Sound Manager processes and don the HMD. The Experiment Manager interpreted arguments passed by the Session Manager, determined which procedure to run and the parameters, activated joystick sampling, loaded and rendered the virtual scenes, detected user button pushes, and performed some experiment data analysis on the fly. The Experiment Manager was written by one of the Massachusetts Institute of Technology experimenters (ACB) in Microsoft Visual C v.4, using World Tool Kit, a library of VR C extensions (Sense-8 Corp., Mill Valley, CA). Experiment procedures and examples of the three-dimensional scenes rendered are described in the accompanying article by Oman, et al. (see science report by Oman et al. in this publication.) Virtual worlds used in the experiments typically had several hundred textured polygons. Scenes were updated at 15-30 Hz. The Experiment Manager compiled science data in both raw and processed form, and piped it to the Archive Manager process. The Sound Manager provided compressed voice into the Archive Manager. The Archive Manager merged science, sound, and engineering data into a single telemetry stream, backed it up on both IDE and PCMCIA hard drives, and sent it to the Spacelab high-rate multiplexer. Ground support computers in the Neurolab Science Center decommutated the telemetry stream, and displayed selected science and engineering data for the investigators. No command uplink capability was required.

The subject restraint system harness was a wide, blue Nomex-padded adjustable belt equipped with adjustable, padded shoulder straps. The harness had D-rings on both hips to attach cables from the constant force springs, and additional D-rings on both hips and the center of the back to attach the optional cloth tethers used when the subject was tested floating. The harness design was adapted from a NASA treadmill harness design used on Shuttle and Mir. It relatively comfortably distributed downward loads between the shoulders and pelvis.

When the experiment required downward subject restraint, the Subject Restraint System spring assemblies attached to the Spacelab deck using fast pins, 20 in. apart. The spring assemblies were constant-force spring reels (Vulcan Spring and Mfg. Company, Telford, PA) of the type conventionally employed as tool lifters. The subject floated between the springs, pulled the steel cable from each, attached it to the harness D-rings using a carabiner, and then stood up in place. With the springs extended into their working range, they each applied 147 N (33 lbs) of downward force to the harness. The spring force was calculated to vary only ±6.7 N over the 95th-percentile adult user.

When the experiment required free-floating conditions, the subject could attach cloth tethers between D-rings on the hip and deck, and run a third tether from the D-ring on the back of the harness up over overhead handrails and down to a point within easy reach. The back tether could then be tightened and secured so that the subject floated beyond normal reach of any surrounding surface.

Three complete VEG systems (flight, flight alternate, and training) were fabricated. In addition, two functionally equivalent computer systems and HMDs were built for development and use in pre- and postflight data collection activities.

DISCUSSION

Although the VEG optical tracker met its procurement specifications, the software proved unable to reliably identify the multiple targets mounted on the HMD or hand in three dimensions under simulated operational conditions. Six months prior to the mission, the tracker drawer and cameras were removed from the Neurolab payload at the investigators' request. A backup VEG computer was flown instead in the available space. The science impact was limited, but the "virtual ball catch" reserve experiment session was deleted from the timeline. The VEG was integrated in Neurolab Rack L3. Accessories were stowed in Rack L10 and overhead stowage drawers. Flight alternate and ground VEG systems were also used successfully by the E136 investigators in a total of seven baseline data collection sessions before and after the mission.

Experience in training and early simulations showed that using a separate laptop computer and remote control software to control the VEG was impractical due to data communication delays. An alternative method using a directly connected keyboard and display was used thereafter.

Neurolab was launched April 16, 1998, and flew for 16 days. The flight VEG was activated and used successfully as planned by all four subjects on flight days three and four. The crew discovered they could satisfactorily free-float in Spacelab without the use of the cloth tethers. During the flight, the crew commented that the HMD produced surprisingly compelling motion illusions, which were dramatically reduced when the SRS harness was worn. Investigators were able to assess samples of flight data in both real time and near real time during the mission. Flight alternate and ground VEG systems were also used successfully by the investigators in baseline data collection sessions before and after the mission.

From the science user's perspective, the VEG was a major advance over the mechanical rotating dome displays previously used in Spacelab vection research. The engineering team provided extraordinary support, especially in the face of all the late problems related to the tracker development. As flown, the complex VEG hardware and software performed well. There were two component failures. The HMD LCD florescent backlight failed in one eye. Fortunately, this occurred early in an extra session on flight day 16. Data on one subject were obtained, however. The manufacturer has since changed the backlight type. Also, the analog-to-digital converters built into several of the joysticks exhibited multi-valued behavior. The problem was circumvented in postflight analysis software. In terms of design lessons, the HMD ventilation fans proved ineffective and noisy. The crew disconnected them in flight. The 4.5-minute reboot time slowed VEG troubleshooting procedures. A small amount of science data was lost at the end of each procedure due to an interprocess data communication problem. The gain on the subject's microphone was preset too low, so most of the subject's contemporaneous voice comments were inaudible and never recovered. Voice recovery efforts were complicated by the use of non-industry-standard and proprietary hardware voice compression.

As flown, the Neurolab VEG lacked a head tracker, so strictly speaking it was not a fully immersive VR display system. However, the experiments achieved their major science goals, and the VEG served a second purpose as the prototype for NASA's VR suite aboard the International Space Station Human Research Facility. The new system will use a faster workstation, and will be equipped with higher performance graphics accelerators and a hybrid optical-inertial tracker. The first science users will be two Neurolab investigators, who formed an international team to develop VOILA, an experiment to study visuomotor orientation in long-duration astronauts [VOILA]. Other applications are foreseen in human factors research, robotics, and onboard crew training.

Acknowledgements

The VEG system was designed and built over a three-year period by a NASA/LMES engineering team that was responsible for the success of this development. After Neurolab, many VEG team members moved on to other assignments, and several have left LMES and NASA. The Principal Investigator (CMO) was an advisor to the engineering team and, at the editors' request, prepared this report to document the VEG design challenges and solutions and the team's significant contribution to the success of the mission. He takes responsibility for all errors and omissions. Preparation was supported by NASA Contract NAS9-19536,

REFERENCES

A HEAD-MOUNTED THREE-DIMENSIONAL DISPLAY. I. Sutherland. Proceeding of the Fall Joint Computer Conference, AFIPS, Arlington, VA., AFIPS Conference Proceedings, pages 757–764; 1968.

VIRTUAL ENVIRONMENT DISPLAY SYSTEM. S. S. Fisher, M. McGreevy, J. Humphries and W. Robinett. *Comput. Graphics,* 9–21; 1986.

ANNOUNCEMENT OF OPPORTUNITY, SPACELAB LIFE SCIENCES-4, NEUROLAB. NASA. Washington, DC, NASA; 1993.

SURROUND-SCREEN PROJECTION BASED VIRTUAL REALITY: THE DESIGN AND IMPLEMENTATION OF THE CAVE. C. Cruz-Neira and D. J. Sandin. *Comput. Graphics,* Vol. 27, pages 135–142; 1993.

Inflight Central Nervous System Tissue Fixation for Ultrastructural Studies

Experiment Team

Principal Investigator: **Gay R. Holstein**

Co-investigator: **Giorgio P. Martinelli**

Mount Sinai School of Medicine, New York City, USA

Authors

Gay R. Holstein, Giorgio P. Martinelli

RATIONALE

The cerebellum is a brain structure critical for coordination, timing muscle contractions during movement, and motor learning. Anatomical studies have shown that neurons in the cerebellum are structurally modifiable. These investigations have demonstrated changes in neuron shape, as well as in the three-dimensional structure of the specialized contacts between nerve cells (synapses) as a result of altered behavioral experiences. Physiological studies have provided some insight into the functional significance of those structural modifications in the performance of motor tasks. The objective of our Neurolab experiment was to evaluate the fine structure of the cerebellum at several time points during and following spaceflight in order to identify architectural alterations in the rat cerebellum that correlate with the adaptation to microgravity, and re-adaptation to Earth's gravity. This required the ability to preserve (fix) the tissue acquired in space so that it could be analyzed after landing.

A major technical challenge that faced our experiment was determining the optimal fixation strategy for the brain tissue collected during flight, while ensuring the safety of the crew and the Shuttle. Tissue fixation is necessary in order to preserve the integrity of the biological specimens intended for future study. The essence of good tissue fixation for anatomical studies is preservation of fine ultrastructural detail (the structure of the individual nerve cells and their connections). This detail is usually seen with electron microscopes. Most laboratories conducting similar studies of the central nervous system (CNS) inject chemical preservatives (aldehydes, most commonly paraformaldehyde and glutaraldehyde) through the blood vessels of the brain in a process called perfusion fixation. Although perfusion fixation is the method of choice for ultrastructural investigations, the method requires relatively large volumes of these toxic chemicals, which may escape into the closed Shuttle environment and harm the crew. Alternative procedures such as rapid specimen freezing, which can be used in some types of ground research, were not possible for the specific goals of this experiment because they disrupt brain ultrastructure. In addition, microwave fixation was not possible since microwaves may interfere with the Shuttle's operating systems. As a result, a series of ground studies was conducted to develop and validate an immersion fixation procedure for spaceflight, in which tissue specimens are harvested and placed into vials of liquid fixative that then diffuses into the tissue over time. This approach tends to provide poorer-quality tissue fixation than that obtained by perfusion fixation, since pathologic degradation of the specimen usually occurs during the diffusion-dependent time interval between euthanasia and actual specimen fixation. Nevertheless, this technique was attempted to avoid the possible release of dangerous chemicals into the Shuttle environment and to provide a method that could be used successfully in space.

259

PROCEDURE DESCRIPTION

The first ground study was designed to establish the optimal time period required for immersion fixation of the cerebellum. The specific objective of this experiment was comparing the ultrastructural tissue preservation at 18 days vs. 30 days of immersion fixation. Initially, 12 rats were sacrificed. The brains were removed, placed in vials containing ice-cold (4°C) paraformaldehyde (PF) fixative, and refrigerated. After 18 days, half the specimens were removed from their vials and sectioned as they were to be cut for the flight experiment. These slices of cerebellum were subsequently processed and then cut into 50-nm sections for electron microscopic analysis. The second set of brains remained in fixative for 30 days, and was then treated in the same way (see Table 1).

Twenty electron micrographs were taken of each brain. The number of holes in the tissue, the sizes of the holes, the total area, and the percent area of holes per electron micrograph and per brain were determined in coded specimens in order to apply an objective quantitative measure to evaluate the condition of each cerebellum. As indicated in Table 1, these data were used to rank order the best six conditions. In general, this quantitative assessment indicated that 5.59% of the total area of sampled brain was occupied by pathologic holes attributable to poor fixation in the 30-day immersion-fixed specimens, in comparison with 2.54% in the 18-day immersion-fixed tissue. These quantitative results were substantiated by qualitative observations of poor tissue preservation and pathologic degradation in the 30-day fixed samples not observed in the 18-day fixed tissue (see Figures 1 and 2). The results demonstrated clearly that 18 days of immersion fixation provided far better tissue quality than 30 days of fixation. They also reinforced the importance of rapid brain harvesting.

The second ground study was designed to determine the optimal fixation sequence for immunocytochemical studies of immersion-fixed cerebella. Immunocytochemistry is a method that can be used to identify specific molecules associated with brain cells. Often, tissue preservation requirements for

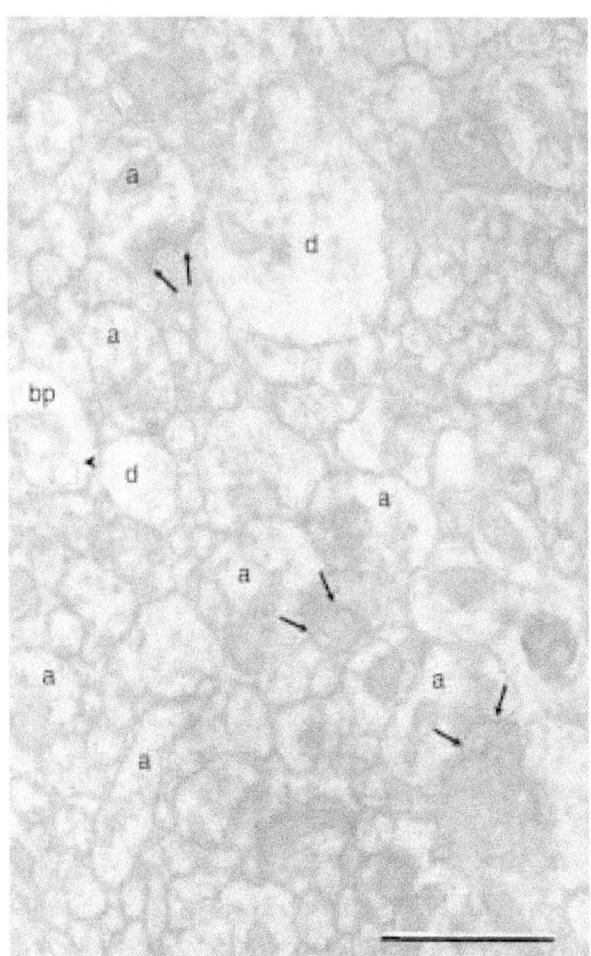

Figure 1. Electron micrograph from a rat cerebellum immersion-fixed in 4% paraformaldehyde for 18 days. The ultrastructure preservation is acceptable, with few bloated profiles (bp), torn membranes (arrowhead), or signs of pathologic degradation of the tissue. Well-preserved axon terminals (a) containing clusters of synaptic vesicles often formed synaptic contacts (arrows) with adjacent dendrites. Other large dendritic cross-sections (d) are also well fixed. Scale bar: 1 µm.

Table 1. The effect of fixative composition, duration of immersion fixation, and brain extraction time on ultrastructural tissue preservation.

Rat #	Fixative	Duration in fixative	Brain extraction time	Results
1	2% PF	30 days immersion	2.5 min.	Rejected.
2	2% PF	18 days immersion	2 min.	**Ranked #4**.
3	4% PF	30 days immersion	2 min.	Rejected.
4	2% PF	18 days immersion	2 min.	**Ranked #5**.
5	2% PF	30 days immersion	3.5 min.	Rejected.
6	4% PF	18 days immersion	1.5 min	**Ranked #1**.
7	2% PF	30 days immersion	1.5 min.	Rejected
8	2% PF	18 days immersion	1.5 min.	**Ranked #2**.
9	4% PF	30 days immersion	2 min.	Rejected.
10	2% PF	18 days immersion	2 min.	**Ranked #6**.
11	2% PF	30 days immersion	1.5 min.	Rejected.
12	4% PF	18 days immersion	2 min.	**Ranked #3**.

immunocytochemistry conducted in conjunction with electron microscopy are different from those for electron microscopy alone. Since one aim of the Neurolab experiment was identifying changes in cellular molecules during and following spaceflight, a technical study to compare ultrastructural tissue preservation and immunocytochemical staining obtained with different 18-day immersion fixation strategies was required. Initially, four sets of conditions were evaluated, as indicated in Table 2.

The brains were obtained using the same procedure as for the first experiment. After 18 days in the immersion solution, the specimens were removed from their vials, and processed for ultrastructural (electron microscopy) analysis or for immunocytochemistry to visualize the presence of GABA, a neurotransmitter used by many cells in the cerebellum. Thin

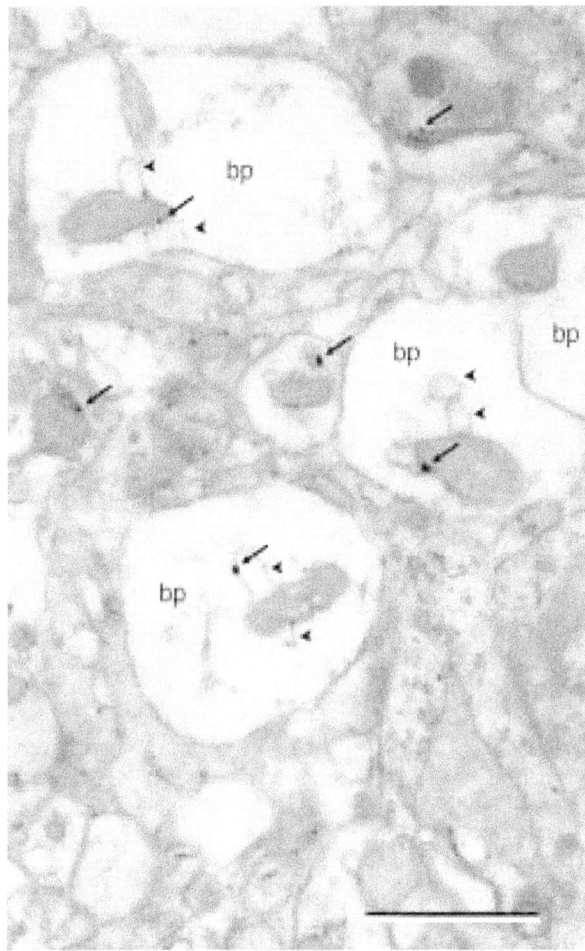

Figure 2. Electron micrograph from a rat cerebellum immersion-fixed in 4% paraformaldehyde for 30 days. The ultrastructure preservation is unacceptable, with many bloated profiles (bp), torn membranes (arrowheads), and signs of pathologic degradation of the tissue (arrows). Scale bar: 1 µm.

sections were evaluated quantitatively for tissue condition, and qualitatively for immunoreactivity in known GABAergic neurons. The quantitative assessment of tissue quality indicated that significantly better preservation of the brain tissue was obtained using the higher concentration of PF fixative for the extended exposure period. However, little immunocytochemical staining was observed. As a result, we concluded that (1) the tissue must remain in fixative (not in buffer) for the 18 days; (2) improved ultrastructure is obtained with 4%, rather than 2%, PF; but (3) a lower concentration of glutaraldehyde should be utilized in the fixative to improve immunocytochemical staining.

On the basis of these results, three additional fixation sequences were evaluated for use with each of two possible immunocytochemical staining methods (see Table 3). In particular, for three additional fixation protocols, half of the brain tissue was used for "pre-embedding" immunocytochemistry, and the other half for "post-embedding" studies. Using the former technique, the immunocytochemical procedure is conducted on brain slices before they are prepared for electron microscopy. Using the latter technique, the immunocytochemistry is performed directly on thin sections that have already been prepared and then cut for electron microscopic analysis. In general, the former technique provides larger sample sizes, whereas the latter approach carries the advantage of better tissue quality. Having evaluated these two methods with three additional fixation sequences, it was concluded that the optimal fixation conditions for the Neurolab experiment were: 4% PF with 0.1% glutaraldehyde for 45 minutes, then 4% PF for 18 days, all at 4°C. We further concluded that better tissue quality and staining were obtained when the immunocytochemistry portion of the study was conducted after the brain slices were prepared for electron microscopy ("post-embedding" immunostaining).

The last fixation experiment was performed in order to verify that the fixatives selected for the Neurolab experiment could withstand the storage requirements for the flight. The fixative solutions were to be prepared on the ground, two days prior to Shuttle launch. Since brains were to be collected for this experiment on flight days (FDs) two and 14, it was criti-

Table 2. The effect of fixation conditions on immunocytochemical tissue staining.

Rat #	Fixation conditions	Tissue treatment	Evaluation of staining and fixation
1	2% PF/0.35% glut for 45 min 2% PF for 18 days	ICC	Little ICC staining, mediocre tissue fixation.
2	2% PF/0.35% glut for 45 min 2% PF overnight PBS for 18 days	ICC	Little staining, worst fixation.
3	4% PF/0.35% glut for 45 min 4% PF for 18 days	ICC	Little ICC staining, good tissue fixation.
4	4% PF/0.35% glut for 45 min 4% PF overnight PBS for 18 days	ICC	Little staining, poor fixation.
5	2% PF/0.35% glut for 45 min 2% PF for 18 days	ICC	Little ICC staining, mediocre tissue fixation.
6	2% PF/0.35% glut for 45 min 2% PF overnight PBS for 18 days	ICC	Little staining, worst fixation.
7	4% PF/0.35% glut for 45 min 4% PF for 18 days	ICC	Little ICC staining, good tissue fixation.
8	4% PF/0.35% glut for 45 min 4% PF overnight PBS for 18 days	ICC	Little staining, poor fixation.
9	2% PF/0.35% glut for 45 min 2% PF for 18 days	EM-stereology	Ranked #2
10	2% PF/0.35% glut for 45 min 2% PF overnight PBS for 18 days	EM-stereology	Ranked #4
11	4% PF/0.35% glut for 45 min 4% PF for 18 days	EM-stereology	Ranked #1
12	4% PF/0.35% glut for 45 min 4% PF overnight PBS for 18 days	EM-stereology	Ranked #3

Abbreviations: electron microscopy (EM), glutaraldehyde (glut), immunocytochemistry (ICC), paraformaldehyde (PF), phosphate-buffered saline (PBS).

Table 3. The effect of glutaraldehyde concentration on immunocytochemical tissue staining.

Rat #	Fixation conditions	Tissue treatment	Staining and fixation evaluation
1	4% PF/0.05% glut for 45 min 4% PF for 18 days	Pre-ICC	Mediocre fixation, mediocre staining
2	4% PF/0.1% glut for 45 min 4% PF for 18 days	Pre-ICC	Acceptable fixation, mediocre staining
3	4% PF/0.25% glut for 45 min 4% PF for 18 days	Pre-ICC	Acceptable fixation, little staining
4	4% PF/0.05% glut for 45 min 4% PF for 18 days	Post-ICC	Mediocre fixation, acceptable staining
5	4% PF/0.1% glut for 45 min 4% PF for 18 days	Post-ICC	Acceptable fixation, acceptable staining
6	4% PF/0.25% glut for 45 min 4% PF for 18 days	Post-ICC	Acceptable fixation, little staining
7	4% PF/0.05% glut for 45 min 4% PF for 18 days	Pre-ICC	Mediocre fixation, mediocre staining
8	4% PF/0.1% glut for 45 min 4% PF for 18 days	Pre-ICC	Acceptable fixation, mediocre staining
9	4% PF/0.25% glut for 45 min 4% PF for 18 days	Pre-ICC	Acceptable fixation, little staining
10	4% PF/0.05% glut for 45 min 4% PF for 18 days	Post-ICC	Mediocre fixation, acceptable staining
11	4% PF/0.1% glut for 45 min 4% PF for 18 days	Post-ICC	Acceptable fixation, acceptable staining
12	4% PF/0.25% glut for 45 min 4% PF for 18 days	Post-ICC	Acceptable fixation, little staining

Abbreviations: glutaraldehyde (glut), paraformaldehyde (PF), pre-embedding immunocytochemistry (pre-ICC), post-embedding immunocytochemistry (post-ICC).

cal to determine whether the fixatives remained potent for the 16 days between solution preparation and tissue immersion for the FD14 brains, and then for the 18 additional days of immersion fixation. For this experiment, the ultrastructural tissue preservation and immunocytochemical staining for GABA were compared in brains fixed using the protocol developed in the previous experiment. One half of the brain was placed in freshly prepared fixative (4% PF with 0.1% glutaraldehyde in 0.1M phosphate buffer for 45 minutes, then 4% PF for 18 days). The other half-brain was placed in fixatives (same solutions) prepared 16 days before euthanasia. After 18 days of immersion fixation, the cerebella were cut into slices and then processed for electron microscopy. We found no differences in the ultrastructural tissue preservation of fresh vs. stored fixative.

APPLICATION

The fixation studies described above were verified in an integrated experiment verification test held at NASA/Ames Research Center (ARC). This test involved three experimental groups that were representative of the subject groups planned for the Neurolab Shuttle mission. They included a flight group, a vivarium group, and a hypergravity group. To simulate hypergravity at two times the Earth's gravitational force, this latter group was placed in the 24-foot centrifuge facility at ARC. The 24-foot centrifuge consists of a central vertical shaft spindle driven by a 25-horsepower motor. Attached to the top of the spindle, approximately six feet from the ground, are 10 radial arms. Each arm holds two enclosures 23.5 in. high×39.5 in. wide ×22 in. deep. The centrifuge was set at 19.99 RPM to create a two-G environment at the floor of the animal cages. As with the planned Shuttle mission, each experimental group above had four sets of subjects: FD2 (N=4), FD14 (N=9), recovery (R) day+1 (N 4), and R+13 (N=7). The cerebellum from each of these 72 rats was immersion-fixed according to the protocol developed previously, and then sectioned and processed for electron microscopy. Tissue analysis verified that the fixation protocol developed for flight could provide acceptable tissue quality for conducting the electron microscopic and immunocytochemical studies proposed for the Neurolab experiment. In fact, most of the brain tissue subsequently obtained from the Neurolab mission proved to be acceptable for ultrastructural study. Observations of cerebellar tissue from FD2 rats are included elsewhere in this volume, and illustrate the utility of this fixation approach.

Acknowledgements

The authors are deeply indebted to Dr. Louis Ostrach and Ms. Lisa Baer of NASA Ames Research Center, and Dr. Ewa Kukielka, Ms. Rosemary Lang, and Mr. E. D. MacDonald II for invaluable technical assistance. Supported by NIDCD grant DC02451 and NASA grant NAG2-946.

Biotelemetry for Studying Circadian Rhythms

Authors
Jay. C. Buckey, Jr., Charles A. Fuller

Experiment Team

Principal Investigator: **Charles A. Fuller**[1]

Co-investigator: **Dean M. Murakami**[1]
Tana M. Hoban-Higgins[1]

Technical Assistants: **Patrick M. Fuller**[1]
I-Hsiung Tang[1]
Edward L. Robinson[1]
Jeremy C. Fuller[1]
Steven Killian[1]
Angelo De La Cruz[3]
Nick Vatistas[1]
Andris Kanups[2]
Barbara Smith[2]
Brian Roberts[3]

Engineers: **Jim Connolley**[3], **Mike Skidmore**[3],
John Hines[3], **Julie Schonfeld**[3],
Tom Vargese[3]

[1]University of California, Davis, USA
[2]Ohio State University, College of Veterinary Medicine, Columbus, USA
[3]Ames Research Center, Moffett Field, USA

RATIONALE

Body temperature and heart rate change in synchrony with the 24-hour day. Several other physiologic parameters, including hormones and activity, also show daily rhythms. The day-to-day rhythms of these changes are called circadian rhythms. Circadian rhythms are altered by weightlessness; however, studying the nature of these changes in space requires robust, reliable, and small instrumentation. Continuous heart rate and body temperature measurements are particularly important, but a system requiring external electrodes and connecting wires would not be practical in space. One solution to this problem is to use biotelemetry. With biotelemetry, small sensors are placed within the body and the information is transmitted outside the body by radio waves. In this way, measurements can be made continuously, without restraining wires or external electrodes. A biotelemetry system was developed for use with the adult rats on Neurolab. With this system, continuous heart rate and body temperature data were collected during flight. This system demonstrated how physiologic data could be collected easily with minimal crew involvement.

HARDWARE DESCRIPTION

The system, which was based on a calibrated biotelemetry unit from Konigsberg Instruments (Konigsberg Instruments, Inc., Pasadena, CA), records body temperature and heart rate and transmits them outside the body using radio frequency (RF) waves. Prior to flight, rats were implanted with the transmitters. These rats were anesthetized with 3% isoflurane in pure medical-grade oxygen. Each rat's abdomen was shaved and

sterilized. A midline incision was made in the abdomen, and a sterilized biotelemetry transmitter was inserted into the peritoneum. A mid-thoracic antenna and two axillary leads for measurement of electrocardiogram (ECG) were positioned under the skin. The rats recovered for two weeks before baseline recording began for the selection of experimental groups.

Six of the Research Animal Holding Facility (RAHF) cages were modified to support the biotelemetry experiment. These cages are similar to other RAHF cages in that they have

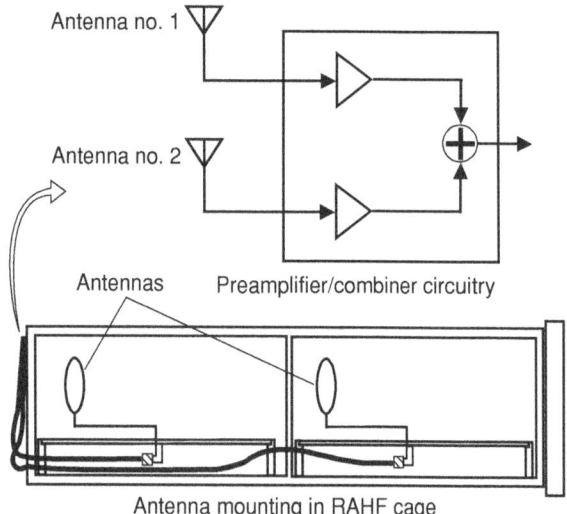

Figure 1. Six RAHF cages (bottom) were modified to support the biotelemetry measurements. Antennas were placed next to the cage living area; and a preamplifier, which was mounted on the back of the cage, integrated the information from the two antennas.

two compartments in which rats are singly housed with feeding, watering, and waste systems. Antennas—located on the inside of each RAHF cage compartment, opposite the watering lixits —received the telemetry information (Figure 1). The antennas were made up of a single-side, copper-clad personal computer board (PCB) with cables running along the top of the cage and out through the back of the cage. Located on the outside of the RAHF cage was a small, black, anodized aluminum box. This box held a PCB circuit that combined the signals from the two antennas and sent them to receivers located in the Neurolab biotelemetry chassis (NBC). Figure 2 provides an overview of the system.

The NBC, the main control unit for the biotelemetry system (Figure 3), was located on the right side of the RAHF, just below the main control panel for the RAHF system. The system required minimal crew intervention during flight.

Figure 4 shows the clear fluctuations in heart rate and body temperature recorded by the system.

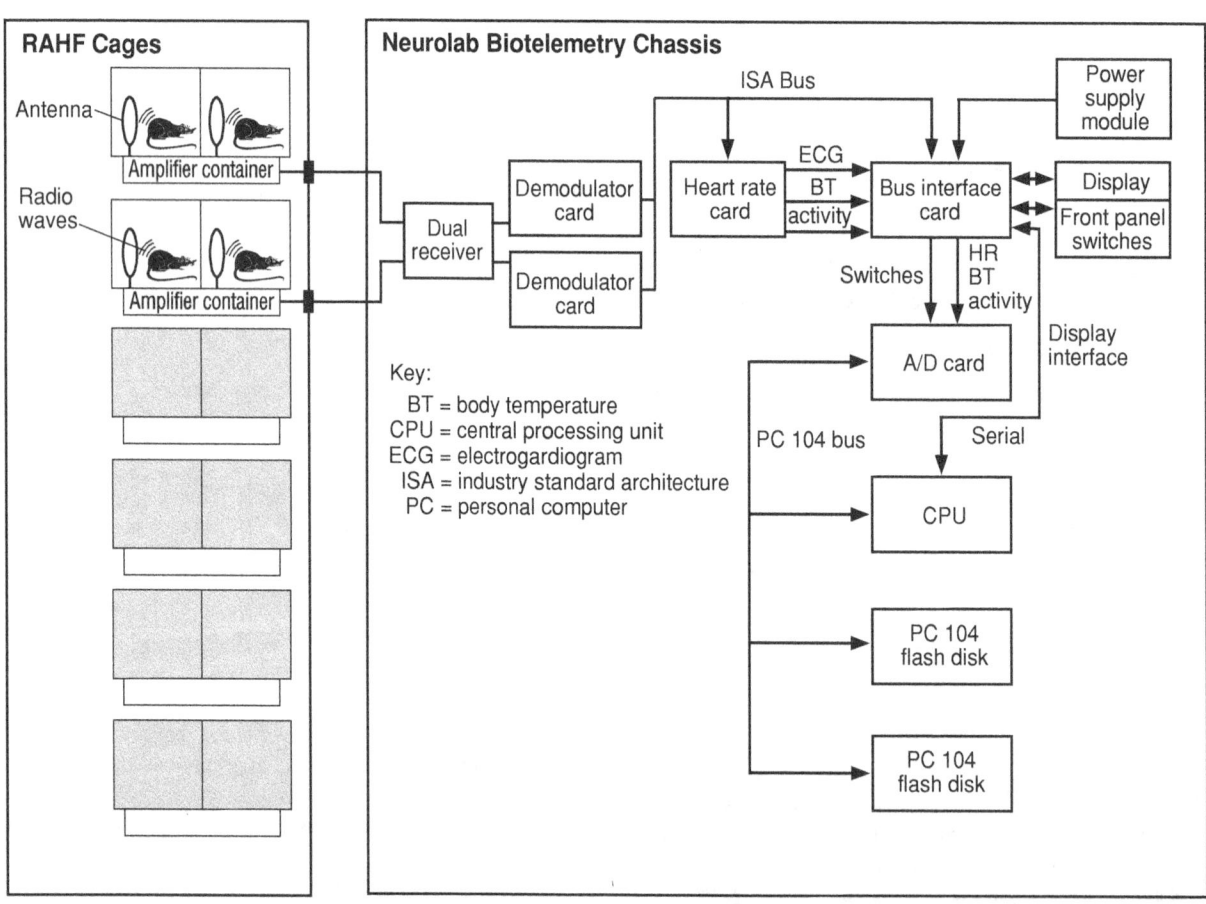

Figure 2. Data from the rat were transmitted to the antennas in the individual cages (shown as loops within the cage). These data were then send to the Neurolab biotelemetry system, which included the necessary electronics to process and store the heart rate and body temperature data for analysis after flight.

The Neurolab Spacelab Mission: Neuroscience Research in Space

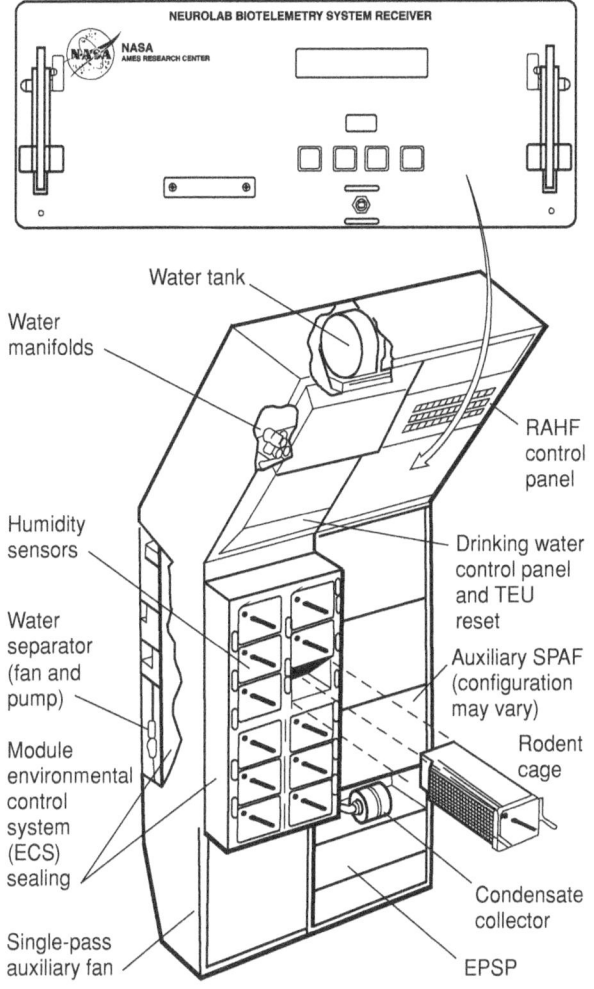

Key:
EPSP = experiment power switching panel
SPAF = single pass auxiliary fan
 TEU = thermal electric unit

Figure 3. Schematic diagram showing the RAHF and the Neurolab biotelemetry system. The system worked well and required minimal crew intervention.

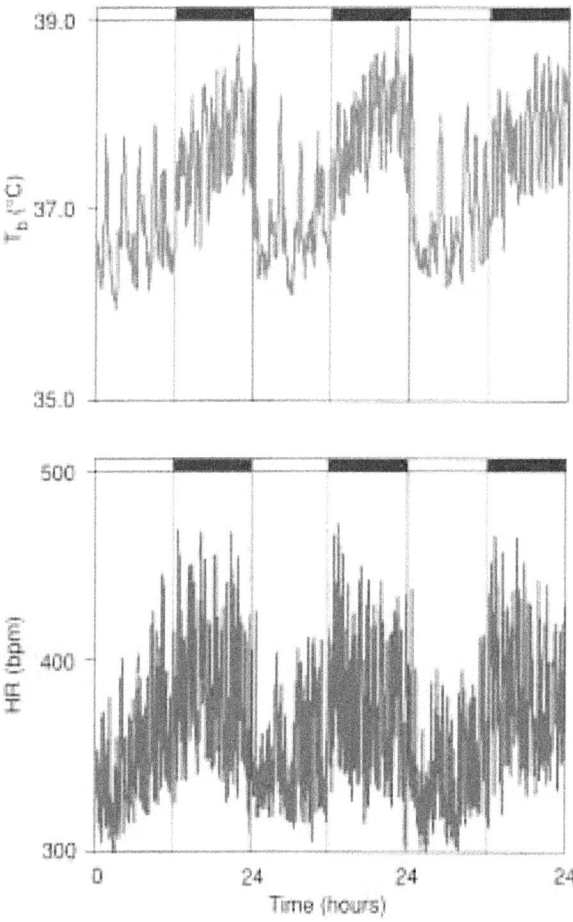

Figure 4. Variations in body temperature (top) and heart rate (bottom) recorded from flight rats on the Neurolab mission.

APPLICATION

Biotelemetry offers many advantages for spaceflight research. Since crew time to perform experiments is often at a premium on spaceflights, systems that can operate autonomously and still provide high-quality data are extremely useful. For example, on Neurolab, an ingestible, telemetry-based, body temperature sensor was used to monitor body temperature for human sleep experiments. Applications of biotelemetry to human medicine are likely to increase in the future.

Microneurography System for Spaceflight Use

Experiment Team

Principal Investigator: **David Robertson**

Co-Investigators: **Italo Biaggioni,
Andrew C. Ertl,
Rose Marie Robertson**

Vanderbilt University, Nashville, USA

Authors

Andrew C. Ertl, Jose Limardo, Rob Peterson,
James A. Pawelczyk, Amit Pathak,
Sachin Y. Paranjape, André Diedrich, Italo Biaggioni,
Rose Marie Robertson, David Robertson

RATIONALE

The autonomic (or automatic) nervous system helps maintain a relatively constant blood pressure under varied everyday conditions. Blood pressure regulating reflexes, "baroreflexes," control how fast and hard the heart contracts as well as the constriction or dilation of arterioles—the small muscular blood vessels that supply capillaries with blood. Arteriole constriction raises blood pressure and dilation lowers it, so the control of arterioles is a powerful way to regulate blood pressure. When blood pressure is high, nerve activity from the brain to the arterioles is low, allowing a relaxation of the arterioles. When blood pressure is low, increased nerve activity causes arterioles to constrict. This arteriolar control can be measured by detecting the nerve traffic to the arterioles using a specialized technique called microneurography. Microneurography involves placing a very small needle electrode in a nerve (such as the peroneal nerve near the knee) and recording signals from the appropriate nerve fibers.

Microneurography has been instrumental in revealing the mechanisms involved in disease states such as cyclosporine-induced high blood pressure, congestive heart failure, and low blood pressure with standing. Microneurography can also be used to study the adaptation to spaceflight. After exposure to microgravity, astronauts show varying degrees of a syndrome known as orthostatic intolerance. The symptoms of orthostatic intolerance are lightheadedness and dizziness that can lead to the feeling of an impending faint (pre-syncope) or actual fainting (syncope) with sufficiently prolonged upright posture. Severe orthostatic intolerance while landing the Space Shuttle, or during an emergency after landing, could present a significant risk to the crew and the spacecraft. As a result, understanding this problem fully is a priority for NASA life scientists, and was a focus of the autonomic nervous system studies on Neurolab.

In prior studies on orthostatic intolerance, information regarding the impact of microgravity on autonomic nervous system circulatory control had been largely inferred from indirect measurements obtained preflight and postflight. With microneurography, the responses of the autonomic nervous system can be measured directly, and this ability helps answer longstanding questions about the cardiovascular adaptation to weightlessness. To use this technique in space, however, a compact, robust system that met rigorous safety standards was needed.

Existing microneurography systems could not be used without undergoing modification. The University of Iowa's Department of Bioengineering produces a widely used research microneurography system. The unit was unsuitable for spaceflight, however, because it is packaged in a large, 19-in.-wide rack-mounted box and would take up too much space in a space vehicle. Also, the system was not designed to meet all of NASA's electrical safety requirements. In particular, current-limiters were needed in the preamplifier to prevent electrical shock if other protections failed. A portable, robust device that met all of NASA's requirements was needed to accomplish microneurography successfully on Neurolab. This report focuses on the development of the equipment and techniques used to measure sympathetic nerve traffic (microneurography) in humans during the Neurolab experiments.

267

HARDWARE DESCRIPTION

A microneurography system consists of three components: an electrode, a preamplifier, and an electronics system. The electrode is placed in the nerve to capture the nerve signal. A preamplifier takes the very low-level signal from the electrode and amplifies it so that it can be further used. An electronics system processes the signal so that it can be interpreted and used for analysis.

Electrodes

Typically, the recording electrode is placed in the peroneal nerve, which carries nerve traffic to blood vessels in the muscles of the calf. Figure 1 shows a magnetic resonance imaging cross-section of the area just below the knee. The peroneal nerve, which carries nerve traffic to the arterioles, is very close to the surface at this point. The recording electrode was positioned in this nerve. A second, reference electrode was also placed just under the skin over a bony area. The electrodes were very fine needles, like acupuncture needles, and so inflicted no pain when they were inserted.

The electrodes used were standard 200-mm microneurography electrodes that were obtained from Frederick Haer Inc. (Bowdoin, ME). The recording electrode was made from tungsten metal with a taper that were coated with epoxy resin so that the final 2–5 µm was exposed for recording. The reference electrode was made of uncoated tungsten. Each electrode was examined for defects prior to packaging and sterilization. Both electrodes were packaged and sterilized by ethylene oxide exposure. A picture of the electrode is shown in Figure 2.

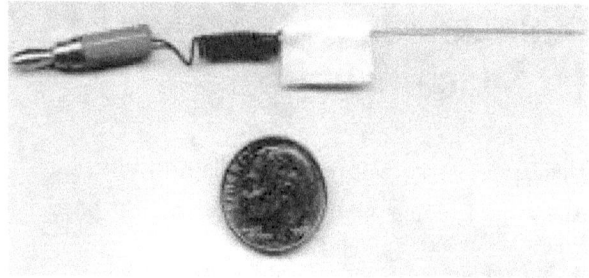

Figure 2. This figure shows the electrode (needle) used to make the measurements. The white flag on the top helped grasp the needle. The needle itself is approximately the size of an acupuncture needle.

Preamplifier

Wires from the electrodes were connected to the preamplifier, which was strapped to the thigh just above the knee using a Velcro strap. The preamplifier was housed in an 8.3 cm×7 cm×3.2 cm aluminum box. To prevent injection of current through the active and reference electrodes, two 100-µA current limiters were placed inline with the electrodes. These limiters ensured that electrical current could not be introduced into the nerve during a recording. The gain of the preamplifier was 100. To make clear, noise-free recording, the preamplifier needed to make good contact with the skin. To accomplish this, the bottom of the preamplifier was a grounding plate made of copper. An electrolyte gel was also used on the plate. To prevent oxidation of the preamplifier grounding plate, the surface was plated with a thin layer of gold.

The unit had no sharp corners, edges, or protrusions. The materials supplied were free of toxic offgassing materials, which is crucial in the enclosed habitable areas of spacecraft. Quick-release cables were provided at the connections between preamplifier and amplifier to permit rapid egress in case of an emergency when the measurements were being taken. The final configuration of the preamplifier and control box is shown in Figure 3.

Control Box

The control box provided signal processing and the controls for the operator to fine-tune the nerve signal for recording. The aluminum control box (16.5 cm×10.2 cm×5 cm) provided controls for gain, low pass frequency, high pass frequency, RC integration time, direct current (DC) offset, noise rejection, and volume. The raw nerve traffic signal from the preamplifier was frequency filtered, full wave rectified, and integrated through the control box to produce the final signal for recording and analysis. Butterworth filters with a flat range response were available on the system with filtered bandwidth of 300 Hz to 10,000 Hz. The resulting signal was amplified to reach a total gain of 70,000–160,000. The microneurography control box interfaced with the microcomputer on the Spacelab, which recorded the processed neurogram at 2,000 samples per second with 12-bit resolution.

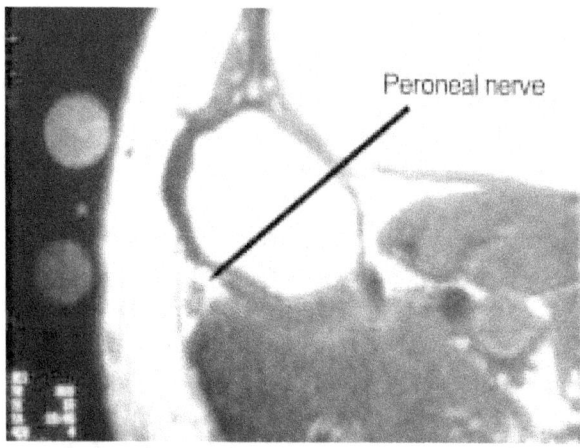

Figure 1. This image shows a cross-section of the leg near the knee. The dark areas are muscle, and the skin surface is to the left in the picture. The white areas under the skin are subcutaneous fat. The needle was advanced through the skin and subcutaneous fat into the peroneal nerve. When the proper nerve bundles were entered, a characteristic sound was heard on the headphones.

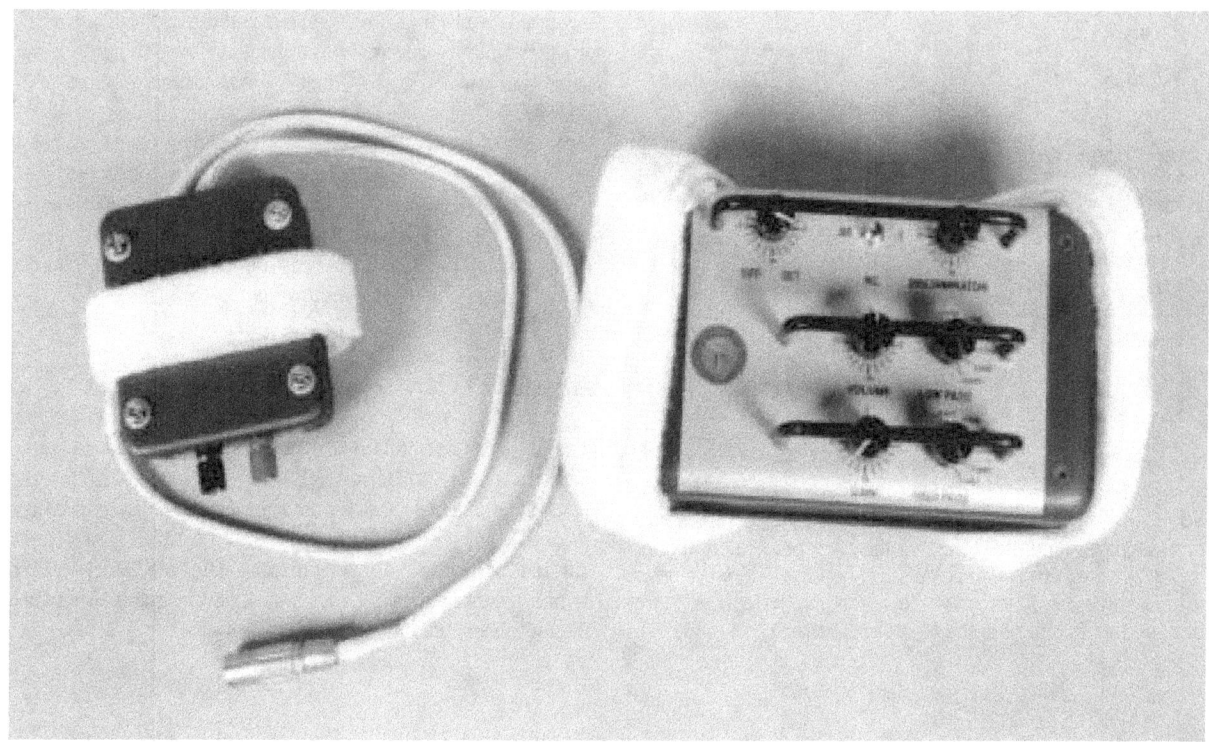

Figure 3. This image displays the final microneurography unit. The preamplifier is on the left and the control box on the right. The preamplifier was placed on the thigh above the knee using the white Velcro straps. The microneurography needle connected to the preamplifier at the black and red connectors. The preamplifier connected to the control box through a cable that included a quick-release fitting (not shown). The control box had the controls necessary to produce a high-quality microneurography signal. The Velcro straps allowed the control unit to be securely mounted where the operator could reach it easily.

Twenty-eight volts were supplied to the unit from the Spacelab rack-mounted power supply. The control box contained a DC/DC converter unit to convert the voltage to ±12Vdc to operate the subsystems in the unit.

During microneurography, the operator would listen to the signal from the unit to determine whether the electrode was properly placed. Headphones were obtained off the shelf (Sennheiser GmbH, Wedemark, Germany) to provide high-fidelity audio feedback from the amplifier to the operator. A cable extension was added to allow freedom of movement for the operator during the experiments. During flight, noise-canceling headphones (David Clark, Inc., Worcester, MA) were used.

Microneurography procedure

The microneurography procedure involved three steps: (1) locating the peroneal nerve, (2) finding the nerve traffic to the arterioles, and (3) recording the signal. To locate the nerve, the crew used a small electrical stimulator. When the tip of the stimulator was over the nerve, the foot would twitch. This location was then tagged with a marker. The skin was prepared with alcohol, and the operator donned sterile gloves. Also, the operator established an electrical connection to the subject using a grounding strap. In this way, stray electrical signals were not introduced into the microneurography signal. The operator

Figure 4. Payload specialist Jim Pawelczyk, Ph.D., is shown in this preflight photograph performing microneurography. The preamplifier is on the leg of the crewmember who is lying inside the silver lower body negative pressure device.

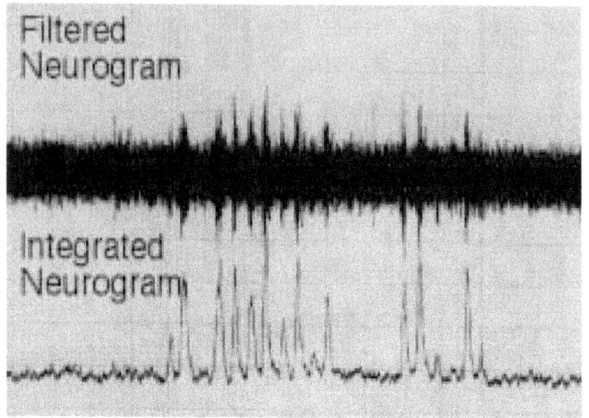

Figure 5. A sample microneurography tracing from the unit is shown. The top panel shows the bursts of nerve activity that occur when blood pressure is low. When these bursts are summed over time, they produce the clear peaks that appear on the bottom tracing. These bursts of nerve activity were used to determine the function of the autonomic nervous system before, during, and after the Neurolab mission.

advanced the electrode into the nerve and listened for the characteristic sound that occurred when the nerve bundles carrying the nerve traffic were entered. After that, adjustments were made to optimize the signal and to confirm that nerve traffic to the arterioles was being recorded. The amplified signal was relayed to a microcomputer. The microcomputer was used to collect data, display feedback, and control operations during the experimental protocols. Figure 4 shows the microneurography procedure being performed preflight in a mockup of the Spacelab. A sample microneurography tracing is shown in Figure 5.

APPLICATION

The ability of the microneurography system to measure sympathetic nerve traffic reliably in space demonstrates its potential for use in Earth-based applications. Its portability means that it can be used in a variety of settings for research studies and clinical assessments. The electrical safety improvements added for the NASA development mean that the unit meets the highest safety standards.

Chronic Recording of Regenerating VIIIth Nerve Axons with a Sieve Electrode

Authors

Stephen M. Highstein, Allen F. Mensinger,
David J. Anderson, Christopher J. Buchko,
Michael A. Johnson, David C. Martin,
Patrick A. Tresco, Robert B. Silver

Experiment Team

Principal Investigator: **Stephen M. Highstein[1]**

Co-Investigators: **Richard Boyle,[2] Allen F. Mensinger,[1] Kaoru Yoshida,[3] Shiro Usui,[4] Timothy Tricas[5]**

Engineer: **Anthony Intravaia,[2] David J. Anderson,[6] Christopher J. Buchko,[6,7] Michael A. Johnson,[6] David C. Martin,[6] Patrick A. Tresco,[8] Robert B. Silver[9]**

[1]Washington University School of Medicine, St. Louis, USA
[2]NASA, Ames Research Center, Moffett Field, USA
[3]University of Tskuba, Tskuba, Japan
[4]Toyohashi University of Technology, Toyohashi, Japan
[5]Florida Institute of Technology, Melbourne, USA
[6]University of Michigan, Ann Arbor, USA
[7]Guidant Corp., Santa Clara, USA
[8]University of Utah, Salt Lake City, USA
[9]Marine Biological Laboratory, Woods Hole, USA

RATIONALE

One of the major goals of the Decade of the Brain (the 1990's) was the implementation of chronic interfaces to the nervous system. Such interfaces could serve as conduits for signal transmission past injured nerves or be used to control prostheses. One approach to this interface has been the sieve electrode. In this design, a small piece of silicon, with multiple small holes placed through it, is inserted into a nerve. Nerve fibers grow through the holes of the sieve and make contact with an electrode there. In this way, nerve traffic in the nerve fibers can be measured directly and signals can be introduced into the nerve. More than a quarter century has elapsed since the concept of the sieve electrode as a neural interface was first introduced, yet numerous technical difficulties have prevented its routine use. Hardware complications for a silicon-based wafer electrode have included difficulties in the consistent micromachining of quality electrodes, interfacing the microprobes with manageable recording leads, and developing transdermal (through the skin) connectors for continuous data acquisition. On the biological side, channeling of nerve fibers through the sieve pores has proven to be problematic even in nerves that exhibit vigorous regeneration.

The present study reports consistent success and repeated sampling of multiple, single-regenerating axons via a transdermal connector in awake, unrestrained fish. Long-term, chronic recording has been achieved by implanting a sieve electrode in the path of regenerating vestibular nerve fibers from the otolithic organs of a fish, *Opsanus tau* (oyster toadfish). This approach to nerve recording is ideal for spaceflight use (see science report by Boyle et al. in this publication) and also holds excellent promise for several clinical applications.

TECHNIQUE DESCRIPTION

The University of Michigan Center for Neural Communication Technology fabricated sieve electrodes possessing nine recording sites within a porous matrix using micromachining techniques (Figures 1A, B, C). The electrode consists of a recording head with a 20-μm-thick silicon support rim surrounding a four-μm-thick internal diaphragm. The diaphragm is interspersed with iridium-coated active sites (5–20 μm diameter) and additional noncoated pores (3–10 μm diameter). The active sites are integrated via silicon leads into a seven-mm silicon ribbon cable that terminates in a rectangular bonding pad that was ultrasonically bonded to a flexible circuit board, and connected via a transdermal lead to a nine-pin connector (PI Medical, Joure, The Netherlands). The impedance of each active site (0.2 to 1.5 MΩ) was determined before implantation, and regularly tested during recording sessions.

Initial implants were conducted using unmodified electrodes. Later implants incorporated several postproduction additions. The first was to electrostatically coat the electrode head with a porous thin film of ProNectin-L (SLPL) (Protein Polymer Technologies, Inc., San Diego, CA). A semipermeable nerve guide tube (NGT) (450-μm diameter, 400-μm total length) fabricated from poly- (acrylonitrile-) vinyl chloride was then attached to the head and secured with Loctite 3341 medical adhesive. Finally, the entubulated, coated probes were dipped into a neural adhesive [0.4% solution of protamine sulfate (Sigma) and poly-d, l-lysine (Sigma)] just prior to implantation.

Toadfish were obtained from the Marine Biological Laboratory in Woods Hole, MA, and maintained at 15°C in flow-through aquariums. All fish care and experimental procedures conformed to American Physiological Society and institutional guidelines.

Fish were anesthetized in 0.001% MS-222 (Sigma) and injected with 0.1 ml of 2% pancuronium bromide (Sigma). The vestibular (VIIIth) nerve was transected with iris scissors at one of two locations: (1) the anterior ramus of the VIIIth nerve, which included portions of semicircular canal and utricular nerves; or (2) the anterior portion of the saccular nerve. The sieve electrode was lowered into the transection site with a micromanipulator. The transected ends of the nerve were approximated to both sides of the electrode, or teased into NGTs. The bonding pad of the sieve electrode was mounted with cyanoacrylate gel to the cranium. The remainder of the cranium was sealed in a similar fashion, and the incision was sutured.

After surgery, the fish were tested biweekly for neural activity. Unrestrained and unanesthetized fish were placed in a small aquarium. The nine active sites of the sieve electrode were recorded differentially, and neural activity was recorded via a Cambridge Electronic Design 1401 computer interface.

For morphological examination, the toadfish were deeply anesthetized in 0.01% MS-222, and the tissues were preserved using perfusion fixation. The implant was examined to determine whether the electrode head was encapsulated with neural tissue. If so, the electrode and surrounding tissue were dissected from the brain and placed overnight in 20% sucrose

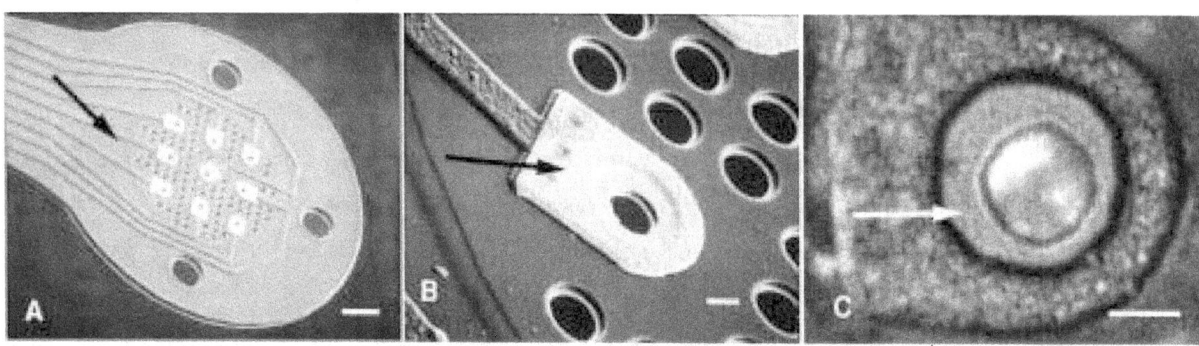

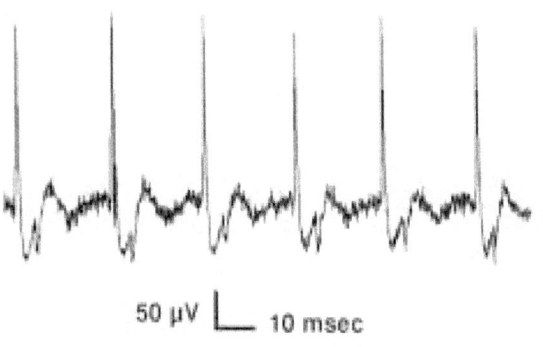

Figure 1. A. Scanning electron micrograph of the sieve microelectrode showing the nine active sites and the silicon leads (black arrows). The iridium-lined active sites were eight μm in diameter, and the unlined support pores were five to eight μm in diameter. B. Scanning electron micrograph of the eight-μm iridium-lined active site. C. Confocal micrograph of regenerating axons growing through an active site of an entubulated, SLPL probe (white arrow), 60 days following implantation. To help with visualization, the neural tissue has been reacted with a secondary tetramethylrhodamine antibody to neuron-specific enolase. D. Neural activity from the active site in Figure 1C 45 days following transection. Scale bars (μm): A=100; B=8; C= 5μV; D=5μV, 10 msec.

solution in phosphate buffer. The tissue/electrode block was then placed in an embedding medium and sectioned with a cryostat in 40-μm increments. The block was examined after each section to ascertain the presence of tissue contiguous with the electrode. If a gap in the tissue was detected before reaching within 200 μm of the electrode face, the implant was considered unsuccessful. If sectioning revealed a continuous line of tissue, the tissue/electrode block was sectioned to within 100-200 μm of the electrode face. The block was then reversed, and the process was repeated. The block was viewed with a Zeiss 510 laser scanning confocal microscope. Images were imported into Abode Photoshop and printed with a Kodak XLS printer.

Morphological examination demonstrated that 28 to 35 days was the minimum time needed for the axons to regenerate into or through the implant site, and that by the end of this period, the electrode head was routinely encased in neural tissue. The addition of SLPL increased the percentage of fibers that regenerated into the tubes, and established contact with the electrode face. Further success was attained with the addition of the nerve glue to the coated, entubulated electrodes.

A confocal micrograph taken from the distal face of an entubulated SLPL probe indicated that five of the nine active sites and approximately 50% of the nonlined pores are filled with neural material. Figure 1D shows a close-up of a 20-μm active site illustrating an axon traversing the pore (arrow). The neural recordings in Figure 1 were obtained from this site.

To date, seven implanted electrodes have yielded electrophysiological data. Three uncoated electrodes yielded spontaneous signals from single active sites following implantation in the anterior ramus of the VIIIth nerve. Four SLPL electrodes with NGTs, one in the saccular nerve and three in the utricular portion of the anterior ramus, yielded both spontaneous- and evoked-action potentials in response to mechanical stimulation. The earliest detection of neural activity occurred 29 days following nerve transection, with signals persisting for up to 42 days in individual fish.

Results demonstrate that axons will regenerate through a sieve electrode, and that chronic recordings are possible from these electrodes. Histological examination revealed that axons would grow on or through surfaces coated with extracellular matrix molecules such as the genetically engineered polymer SLPL. The polymer contains both GAGAGS crystalline silk-like sequences as well as several repeats of one of the identified laminin binding domains (IKVAV), making it a "super-sticky" laminin analogue. The filaments of polymer provided high surface area for cell attachment, and maintained the ability to transfer electrical signals between the nerves and the device. The NGTs channeled the nerves through the sieve, and the neural glue provided greater adhesion of the cut axons to the inside of the tube, greatly increasing the number of axons regenerating through the coated electrodes.

The present study was able to detect neural activity within 29 days of transection with a median delay of 42 days post-transection. This was significantly faster than the 49- to 175-day intervals reported in previous studies, emphasizing the advantage of transdermal leads for repeated sampling. Although morphological examination determined that up to seven active sites contained neural tissue, the maximum number of active sites that yielded neural activity was three in any one fish. It is assumed that electrical activity can only be recorded from sites in proximity to nodes of Ranvier. As the axons regenerate through the pores, the nodes of Ranvier would be distributed at different distances along the axons. Although the internodal distance is foreshortened in regenerating nerves, thereby increasing the chances of a node being near an active site, the minimum distance a node can be situated from an active site and produce viable recordings remains to be determined. The variability in internodal distance may explain why the number of recording sites remained less than the number of active sites containing neural material.

Experiments have demonstrated that regenerating axons will grow through the sieve electrode, and that neural recording is possible from unrestrained animals.. The eventual integration of the sieve electrode with a telemetry device will allow chronic recording from freely moving animals, and determine how the nervous system functions in a natural environment. This information should greatly enhance our knowledge of the neural mechanisms of behavior.

APPLICATION

The device holds excellent promise for several clinical applications. The ability of mammalian peripheral nerves to partially regenerate following injury suggests that the device could serve as a conduit for signal transmission past injured nerves. In systems that have limited natural regeneration, such as spinal cord, neurotrophins could be incorporated with the SLPL molecule to stimulate sufficient growth to have nerve sprouts enter the electrode and establish a neural interface for prosthesis control.

Acknowledgements

The authors would like to thank J. Hetke for help in design and electrode manufacture, M. O'Neill with fish maintenance, and R. Rabbitt and K.S. O'Shea for valuable discussions and advice. This work was supported by NASA Life Science and Klingenstein fellowships and NIH R21-RR12623 to AFM, NASA NAG-2-0945 and NSF-AO93-OLMSA-02 to SMH, the NIH NINDS Neural Prosthesis Program (Contract NS 5-2322) and NIH/NCRR P41 RR09754-04 and an NSF (MCB) grant to RBS.

Surgery and Recovery in Space

Authors

Jay C. Buckey, Jr., Dafydd R. Williams, Danny A. Riley

Experiment Team

Principal Investigator: **Danny A. Riley[1]**

Co-investigator: **Margaret T.T. Wong-Riley[1]**

Technical Assistants: **James Bain[1], Sandy Holtzman[1], Wendy Leibl[1], Angel Lekschas[1], Paul Reiser[1], Glenn Slocum[1], Kalpana Vijayan[1], Carol Elland[2]**

[1]Medical College of Wisconsin, Milwaukee, USA
[2]NASA Ames Research Center, Moffett Field, USA

RATIONALE

Nerves can have a profound effect on the skeletal muscles that they serve. To understand the development of antigravity muscles in weightlessness, it is important to study not only the muscles, but also their innervation. These nerves originate from motor neurons in the spinal cord. To study specific spinal neurons, they must be labeled to distinguish them from other nerve cells in the spinal cord. This is commonly done using retrograde labeling, which involves depositing a small amount of a label in the muscle of interest. The label is taken up by the motor nerve terminals and transported back up the nerve axons to the nerve cell body in the spinal cord. When tissue sections of the spinal cord are examined under a light microscope, labeled nerve cells serving the injected muscle are readily resolved.

Performing retrograde labeling, however, presents technical challenges. First, for this experiment, the muscles to be labeled are very small—a millimeter or two in diameter—so the label must be injected in microliter quantities using a fine needle. Second, microsurgery is required to expose the muscle for accurate injection. Third, proper anesthesia and recovery techniques are essential to ensure that the rats can tolerate the procedure easily and recover without difficulty. Fourth, in space this must be done within a workstation that keeps the work area about an arm's length away from the eyes.

For Neurolab, the procedures and techniques to perform recovery surgery were developed and used for the first time in space. Five, 21-day-old rats were anesthetized, and their soleus and extensor digitorum longus leg muscles were injected with label. The surgical incisions were closed with wound adhesive. All rats tolerated the procedure well, and they all recovered uneventfully. The wound sites were healing cleanly when examined two days after the surgery. The procedures worked well and demonstrate that, if needed in the future, challenging and delicate surgical procedures can be performed in space.

PROCEDURE DESCRIPTION

Anesthesia

Administering proper anesthesia for small rats can be difficult. Too much anesthesia can kill the rat, and inadequate anesthesia is unacceptable. The difference between a fatal and an acceptable dose may be just a few tenths of a milliliter. To develop the anesthesia protocol for Neurolab, extensive testing was performed preflight. A combination of acepromazine, xylazine, and ketamine (a Ketaset cocktail) was used for anesthesia. Different doses were tested to determine which dose provided the best anesthesia, while allowing for a full recovery. A nomogram chart was developed, based on preflight testing, that listed the dose to give based on the length of the rat's tail. Tail length was shown to correlate strongly with the rat's weight. Neurolab was not equipped to measure the rat's body mass in microgravity. The nomogram defined individualized doses appropriate for the rat's size. The anesthesia was administered into a forelimb muscle using an insulin-type syringe with a 25-gauge needle. Typical doses were 0.3–0.5 cc. Figure 1 shows the contents of an anesthesia kit. The syringes were prefilled, and there was Velcro attached to the syringe body and the syringe caps. The syringes were placed in bags with removable seals. In use, the card holding the syringes was removed from the plastic bag and attached

on the inside of the general-purpose workstation (GPWS) with Velcro. As needed, the syringes were removed individually from the card and replaced after injecting.

During the mission, the anesthesia protocol had to be modified slightly. The rats on the mission had not grown as large as rats of the same age had grown on the ground. In addition, inflight, several rats had unexpected foreshortening of their tails. This meant that tail length could no longer be used as an accurate predictor of the body weight. Preflight training, however, had provided experience with the use of the anesthetic and its effectiveness. The typical dose that had been used in preflight training was reduced by approximately 25%. This provided excellent anesthesia. One rat received an extra 0.1 cc dose because of a delay in the anesthetic effect.

Procedure

Figure 2 shows the GPWS during a preflight training session with the front door open. Before the procedure, the crewmember would unstow the necessary equipment from stowage areas in the Spacelab and place it into the GPWS. The equipment included the cork-topped worktable (Figure 3) and all of the necessary kits and surgical tools (small scissors, retractor, forceps). In addition, a clear plastic box with a screen lid was used as the area in which the rats could awaken safely from

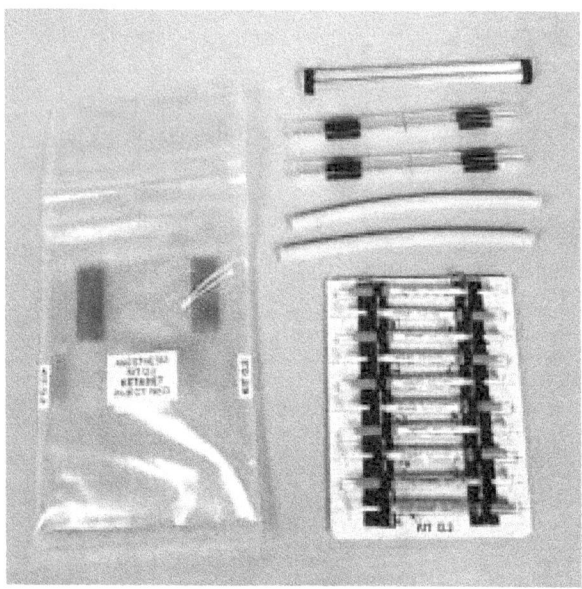

Figure 1. Anesthesia kit for the injection procedure. The individual anesthesia syringes were preloaded with anesthetic, and Velcro was placed both around the syringe and around the cap so they could be used easily in space without floating away. The syringes were mounted on a card that could be placed on the wall of the GPWS within easy reach of the operator. All kits that contained fluids were placed in sealed bags (shown empty and open here) so the fluids would be contained if a syringe broke and would not leak into the spacecraft cabin.

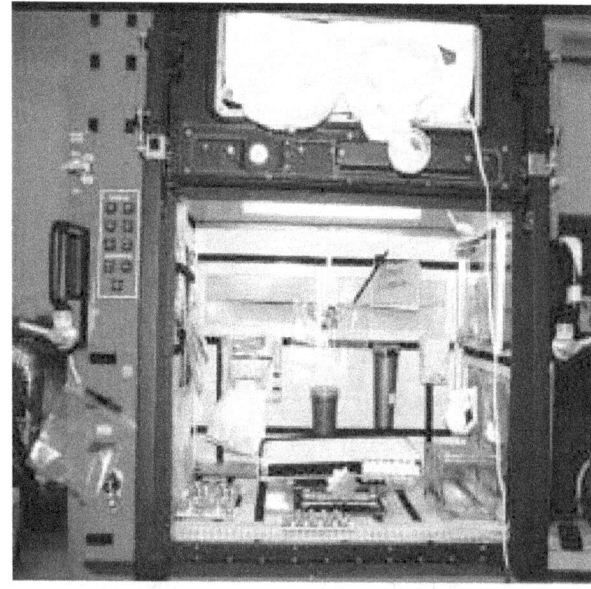

Figure 2. Interior of the GPWS. The workstation provided a containment environment that was isolated from the spacecraft cabin and allowed for procedures (e.g., surgery, use of fixatives, tissue collection) that would involve working with fluids and rodents. A table with a cork top could be placed within the GPWS and secured to the floor with Velcro. All of the other necessary materials could be attached to Velcro on the walls. In the bottom right of the workstation is the clear plastic box used to hold the rats.

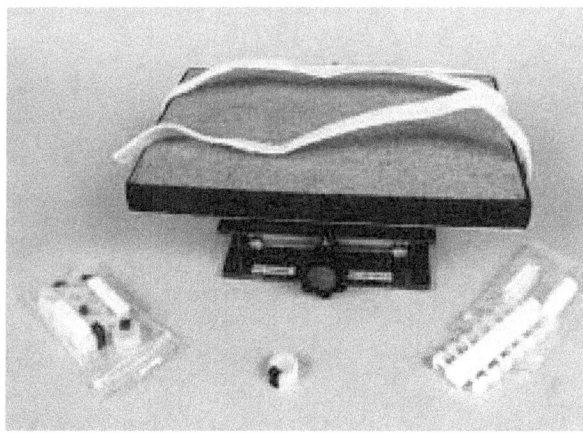

Figure 3. This worktable allowed for a variety of procedures to be performed in the GPWS. The table was covered with an absorbent pad, and equipment could be attached to the table either with pins into the cork or by Velcro straps that attached to the Velcro on the sides and bottom of the table. Height was adjustable.

Figure 4. Payload Specialist Jay Buckey is shown wearing magnifying surgical loupes that the crewmembers used while doing delicate work within the workstation.

surgery. Once the workstation was appropriately configured, the crewmember would don gloves, close the GPWS door, and insert his or her arms through the Tyvex gauntlets into the workstation.

Another crewmember would pass an animal cage through the side door of the GPWS, and the rats were transferred to a clear holding box in the workstation. To perform the procedure, anesthesia was administered to a rat and the crewmember would wait 5-10 minutes for it to take effect. Once the rat was sufficiently anesthetized and nonresponsive to foot pinch, the rat was secured to the worktable with a Velcro strap placed loosely around the abdomen. The leg was fixed in position with tape. Prior to beginning surgery, the crewmember cleansed the rat's leg with alcohol.

Since the worktable was approximately an arm's length from the crewmember's eyes and the area of interest was a couple of millimeters in size, magnifying surgical loupes were used (Figure 4). A small incision was made in the rat's leg, the muscles were exposed, and the label was injected. The incision was closed using a skin adhesive (Nexaband). After surgery, the rat was placed in a plastic box to recover. When

all of the rats were awake, they were transferred into a cage and rehoused in the Research Animal Holding Facility.

This procedure was performed on FD13 of the Neurolab mission, and it was uneventful. The anesthesia was adequate for all rats, and they all recovered. Examination of the leg wounds two days after surgery showed that the wounds were healing well. Postflight assessment verified that the label had been placed in the correct muscles but, due to technical problems with the label, it was not carried retrogradely to the spinal cord as intended. This procedure was the first time that survival surgery with all of its components (anesthesia, surgery, recovery) had been performed in space.

APPLICATION

This effort showed that the technical problems with performing delicate surgical procedures in space (securing tools, adequate anesthesia, clear field of view) could be solved. This information will be useful for future, longer missions, where surgery may need to be done in space.

Adaptation to Linear Acceleration in Space (ATLAS) Experiments: Equipment and Procedures

Authors

Bernard Cohen, Gilles Clément, Steven Moore, Ian Curthoys, Mingjia Dai, Izumi Koizuka, Takeshi Kubo, Theodore Raphan

Experiment Team

Principal Investigator: **Bernard Cohen[1]**
Gilles Clément[2]

Co-Investigators: **Steven Moore[1]**
Ian Curthoys[3]
Mingjia Dai[1]
Izumi Koizuka[4]
Takeshi Kubo[5]
Theodore Raphan[6]

[1]Mount Sinai School of Medicine, New York City, USA
[2]Centre National de la Recherche Scientifique, Paris, France
[3]University of Sydney, Sydney, Australia
[4]St. Marianna School of Medicine, Kawasaki, Japan
[5]Osaka University, Osaka, Japan
[6]Brooklyn College of the City University of New York, New York City, USA

RATIONALE

During orbital spaceflight, the acceleration of gravity is reduced from one-G on the Earth's surface to about 10^{-6}-G in orbit. Nevertheless, the head accelerations associated with changing direction and with turning are unaffected. Therefore, the body must selectively adapt to the relative absence of gravity. As a result of this reorganization, astronauts frequently exhibit difficulty with balance upon landing that could pose a serious problem if they had to function efficiently in a gravitational environment immediately after a long-duration spaceflight. They can also become disoriented when in motion. These deficits are as yet incompletely understood. The otoliths, located in the inner ear, sense head accelerations and generate reflexes to the eyes and postural muscles that maintain gaze and posture when moving in a gravitational environment. They are also directly involved in sensing the direction of gravity and contribute importantly to the sense of spatial orientation. It is likely that changes in otolith-mediated reflexes that occur as a result of adaptation to microgravity could be responsible for postflight problems with gaze and balance.

The purpose of two Neurolab experiments (see science reports by Moore et al. and Clément et al. in this publication) was to study how spatial orientation and reflexes originating in the otolith organs might be affected by adaptation to microgravity. Flight and ground-based instruments were developed for the Neurolab mission to study how orienting otolith-ocular reflexes and the perception of the spatial vertical are changed by adaptation during spaceflight. These instruments included a flight-rated centrifuge that delivered measured amounts of linear acceleration along different body axes in flight, a functionally equivalent centrifuge to do ground-based testing, and an apparatus that statically tilted the head and body with regard to gravity before and after flight. The static tilt apparatus allowed the teams to obtain data that could be compared with the results of centrifugation. Binocular, infrared video-oculography, a noninvasive technique, was utilized to measure eye movements in three dimensions while the astronauts were being centrifuged. Equipment was also developed that could deliver visual stimuli to calibrate the induced eye movements and to induce optokinetic nystagmus (OKN) and ocular pursuit during centrifugation and static tilt. This report describes the development and use of this equipment. This unique hardware performed flawlessly in flight, provided the first inflight "artificial gravity," and advanced the technology of eye movement recording.

HARDWARE DESCRIPTION

To perform the experiments on the mission successfully, the equipment had to be able to generate controlled accelerations to the inner ear, present moving visual patterns to the test subject, and record eye movements during the stimuli. To accomplish these goals, the following equipment were developed:

a. Ground and flight centrifuges,
b. Tilt chairs for postflight use,
c. A visual display,
d. An eye movement recording system, and
e. A videotape processing facility.

Centrifuges

One technique for producing accelerations to the inner ear in space is to rotate subjects about a distant axis in a centrifuge to generate centripetal linear acceleration. Until the present experiments, the response to this kind of centrifugation had not been formally studied during spaceflight. Two centrifuges were developed for these experiments. The flight model was built by the European Space Agency (ESA) (Figures 1-3), and a ground centrifuge based at Johnson Space Center (JSC) in Houston, TX, that was built by Neurokinetics Inc., Pittsburgh, PA, was used for pre- and postflight studies (Figure 4). Both centrifuges were functionally identical. Subjects were seated with the body vertical axis parallel to the axis of rotation at a radius of 0.5 m, and were oriented facing or back to the direction of motion (tangentially) either left-ear-out (LEO) (Figure 2) or right-ear-out (REO) (Figure 4). In this configuration, rotation at a constant velocity of 254 degrees/second and 180 degrees/second

provided one-G and 0.5-G, respectively, of centripetal acceleration along a line joining the ears (the subject's interaural axis). Accelerations at the onset and end of rotation were 26 degrees/second². Subjects could also be positioned lying-on-back (LOB) (Figure 3) so they lay supine along the arm of the centrifuge with their heads 0.65 m from the axis of rotation. In the LOB configuration, rotation at a constant velocity of 223 degrees/second and 158 degrees/second provided centripetal acceleration along the subject's body vertical axis with a magnitude of one-G and 0.5-G at the level of the interaural axis, respectively. Because the center of rotation was about an axis approximately through the navel, the centripetal acceleration was directed down (toward the feet) at the head and up (toward the head) at the feet; i.e., toward the center of rotation. Thus, there were potentially conflicting estimates of the direction of the accelerations from some of the tilt receptors in the body and the otoliths. A restraint system, consisting of a five-point harness—thigh, shoulder, and neck pads—and a knee strap held the body firmly in place (Figure 1). A custom-made facemask, consisting of a fiberglass front and back shell that was molded to the bony features of the skull, restrained the subject's head. The subject held handgrips, mounted on either side of the chair, that incorporated a push button used to calibrate the video system (see below) and an emergency rotation-abort switch. The subjects wore a set of headphones to provide a masking noise that eliminated external auditory cues.

Tilt Chairs

The centrifuge used for ground-based testing at JSC was also used as a static tilt chair (Figure 4). With subjects in the REO orientation, the centrifuge chair assembly could be tilted about an

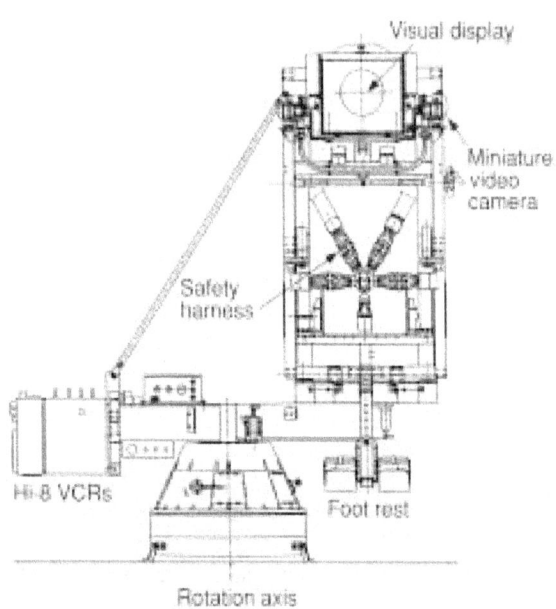

Figure 1. Schematic of the flight centrifuge built by the European Space Agency.

Figure 2. The flight centrifuge in the Spacelab module, which was located in the cargo bay of the Orbiter *Columbia*. A crewmember, who is a subject for inflight centrifugation, is oriented left-ear-out, facing the direction of motion. The centripetal linear acceleration is directed along the interaural axis, which lies on a line joining the ears (photograph courtesy of NASA).

Figure 3. Inflight centrifugation in the lying-on-back (LOB) orientation. During centrifugation, the centripetal linear acceleration was aligned with the subject's body vertical (Z) axis (photograph courtesy of NASA).

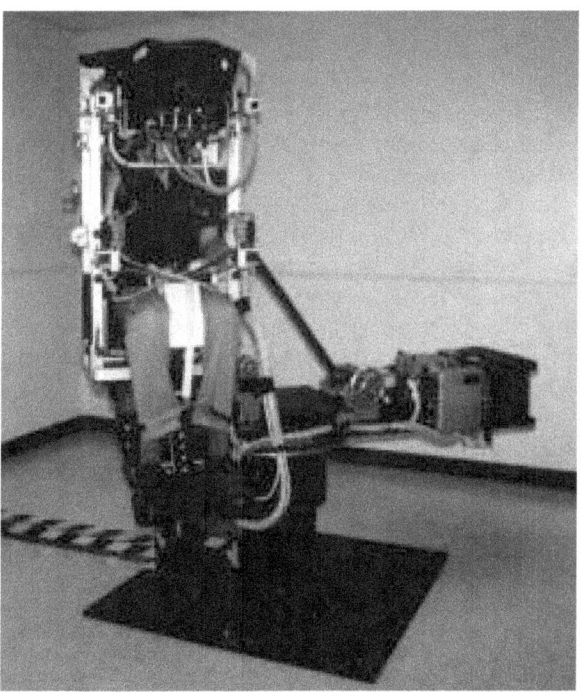

Figure 4. Ground centrifuge used for baseline data collection at Johnson Space Center (JSC). The crewmember subject is about to be centrifuged in the right-ear-out (REO) configuration.

axis located underneath the subjects' seat. As a result, subjects were tilted in roll at angles between zero degrees (upright) and 90 degrees left-ear-down. A digital inclinometer was used to set the angle of tilt. A second tilt chair, built by Leigh McGarvie and Nicholas Pasquale of the Mount Sinai Medical Center, was used to test the response of the astronauts to static tilt on the day of landing at the Kennedy Space Center (KSC), Cape Canaveral, FL (Figure 5). For this, subjects were seated in an automobile racing seat and firmly held in place by a five-point safety harness, adjustable padded shoulder and neck supports, and a fiberglass back shell to support the head. Subjects could be tilted in roll from the upright (zero degrees) in 15 degrees steps to 90 degrees left-ear-down about an axis behind their head. A pin mechanism locked the chair in place at each angle. A three-meter-diameter white hemisphere, centered at eye level, was positioned in front of the subjects to display points of light and optokinetic stimuli. A static five-point display, consisting of a center point and four eccentric points at ±10 degrees horizontal and vertical gaze angles, was used to calibrate eye movements.

Visual Display

A visual display, consisting of a 158×167-mm liquid crystal display screen and associated optics and electronics, was mounted in a box directly in front of the subject's face on the centrifuge chair (Figures 1-4). The visual display had a field of view of ±44 degrees horizontally and ±40 degrees vertically, and presented black dots on an amber background. This was used to display a sequence of dots at known gaze angles to enable calibration of the eye movement monitor. The display was also used to present optokinetic patterns (sequences of five-degree stripes that moved horizontally, vertically, and diagonally across the screen at 30 degrees/second) and smooth pursuit targets (small dots that moved horizontally

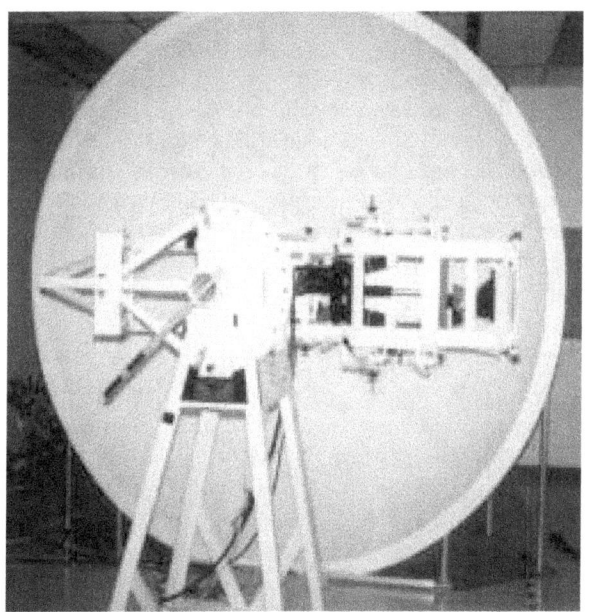

Figure 5. Static tilt chair, installed at KSC, that was used to test the astronauts prior to launch and on the day of landing. Here the chair is shown tilted 90 degrees in the roll plane. The white homisphere, which had a diameter of three meters, was used to project optokinetic patterns.

and vertically across the screen). Optokinetic patterns were presented for 20 seconds in each direction (up, right, down, 45 degrees left and down left), followed by 10 seconds of darkness. During smooth pursuit, the target was stationary at the center of the screen for a period ranging from 1.8 to 2.4 seconds, then stepped five degrees to one side. The target then moved at a constant speed (20 degrees/second) in the direction opposite to the step to a final position ranging from 25 degrees to 35 degrees. The duration for presentation of the target at zero degrees and the final position of target varied randomly to avoid anticipation by the subjects. The size of the step was adjusted so that the image of the target was on the fovea at the onset of pursuit. There were four trials in each direction (up, down, left, and right) in random order. The various visual stimuli were perceived at about the same viewing distance—i.e., about 60 cm—for all subjects. The pattern illumination was adjusted for each subject for best viewing of pursuit and calibration targets.

Eye Movement Recording

Binocular video recordings were obtained using two miniature NTSC video cameras that were mounted on the visual display unit (Figure 1), which provided video images at a frame rate of 30 Hz. Two rectangular banks of nine infrared (IR) light-emitting diodes (LEDs) (wavelength 950 nm), attached to each camera, were used to illuminate the subject's eye. The LEDs were not visible to the subject. Images of the subject's eyes were directed onto the charge-coupled device of the video camera via an IR beam-splitter mounted on the visual display unit, which was transparent to light in the visible range and allowed the subject a clear view of the visual display. The horizontal and vertical camera position and focus could be adjusted manually, by the operator, using a small video monitor as a guide, to obtain clearly focused images of the subject's eyes. Two custom-made Hi-8 video cassette recorders (VCRs), mounted on the opposite end of the rotator beam to the subject chair, were used to store the video images of the eyes. Several experimental parameters, such as centrifuge velocity and timing information, were also recorded onto the Hi-8 videotape along with the image of the eye. Only one eye was recorded in the tilt apparatus shown in Figure 5.

Tape Processing Facility

Following the Neurolab mission, ground and inflight Hi-8 videotapes were dubbed onto Betacam SP tapes for post-processing. The tape processing facility consisted of a Betacam VCR (Sony UVW-1800), an IBM-compatible personal computer (PC) fitted with custom-made video digitization and display hardware, and a video monitor. Processing of the video data was automated, with two seconds of images digitized and stored in the PC. The digitized video images were then processed field by field (where a single field consisted of either the odd or even lines from a complete video frame), providing a sampling rate of 60 Hz.

The coordinates of the pupil center in the image field were calculated using a partial ellipse fit based on the work of

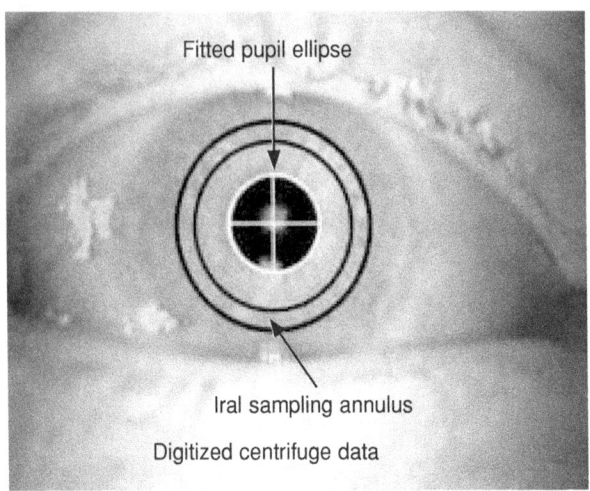

Figure 6. A single frame from the Hi-8 videotape illustrates the techniques used to measure 3D eye position. The center of the pupil was detected and fitted with an ellipse. An annulus, centered on the pupil, was used to sample the gray-level information of pixels within the iris. A circular cross-correlation was then used to measure rotation about the line of sight. A "bar code" at the bottom of the image (not shown) contained digitized data regarding centrifuge velocity, mission elapsed time, and other parameters important for describing the experiment.

Zhu et al. (Zhu, 1999). Location of the center of the pupil at 25 known gaze positions, obtained during calibration (see visual display above), formed the basis of a three-dimensional (3D) spherical model of the eye, which was used to determine horizontal and vertical eye position from the pupil center of subsequent images. Torsional eye position was obtained using the gray-level intensity information of the iris (iral signatures) obtained from a circular iral sampling annulus centered on the pupil (Figure 6). A reference iral signature was sampled from images acquired while the subject fixated on the center point of the calibration display. Pattern matching of the reference signature with iral signatures obtained from subsequent images provided a measure of torsional eye position (Moore, 1996). Improved accuracy of torsional computation was achieved using geometric algorithms that compensated for the eccentricity of the iral sampling annulus according to eye orientation (Moore, 1996). Eye position in head coordinates was represented as Euler angles in a Fick (yaw, pitch, roll) rotation sequence.

APPLICATION

There were a number of innovations in the research equipment developed for this flight. Firstly, the precisely controlled human centrifuge allowed study of the effects of sustained levels of 0.5-G and one-G of side-to-side and head-to-foot linear acceleration on otolith-induced orienting eye movements. The effects of "artificial gravity" on otolith-induced orienting eye responses in space had not been determined before. Secondly, this was the first use of binocular 3D video-based eye movement recordings during controlled vestibular stimulation in space. Finally, this was the first use of visual stimuli for study of visual-vestibular interactions during centrifugation, either on Earth or in space. The development of this technology has been invaluable for testing the otolith tilt-translation hypothesis (see science report by Clément in this publication) and has facilitated study of vestibulo-sympathetic reflexes (Kaufmann, 2002). It will also provide the means for evaluating the efficacy of artificial gravity as a countermeasure to deconditioning of vestibulo-ocular and vestibulo-sympathetic reflexes during long-term space travel. In clinical medicine, the acceleration provided by centrifugation in darkness with concomitant visual stimulation may be a useful test of vestibular function. There may also be value in using centrifugation in the rehabilitation of patients who have had the inner ear removed on one side.

REFERENCES

A Geometric Basis for Measurement of Three-Dimensional Eye Position Using Image Processing. S.T. Moore, T. Haslwanter, I.S. Curthoys, and S.T. Smith. *Vision Res.,* Vol. 36, pages 445–459; 1996.

Robust Pupil Center Detection Using a Curative Algorithm. D. Zhu, S.T. Moore, and T. Raphan. *Comput. Meth. Prog. Bio.,* Vol. 59, pages 145–157; 1999.

Vestibular Control of Sympathetic Activity; An Otolith-Sympathetic Reflex in Humans. H. Kaufmann, I. Biaggioni, A. Voustaniuk, A. Diedrich, F. Costa, R. Clarke, M. Gizzi, T. Raphan, and B. Cohen. *Exp. Brain Res.* Vol. 143, pages 463-469; 2002.

Visual-Motor Coordination Facility (VCF)

Authors

Otmar Bock, Barry Fowler, Darrin Gates,
John Lipitkas, Lutz Geisen, Alan Mortimer

Experiment Team

Principal Investigator: **Otmar Bock[1]**

Co-Investigator: **Barry Fowler[2]**

Co-Authors: **Otmar Bock,[1] Barry Fowler[2],
Darrin Gates[3], John Liptikas[2],
Lutz Geisen[1], Alan Mortimer[4]**

[1]German Sport University, Köln, Germany,
[2]York University, Toronto, Canada
[3]Bristol Aerospace Ltd., Winnipeg, Canada
[4]Canadian Space Agency, Ottawa, Canada

RATIONALE

Astronauts' anecdotal observations, surveys of video footage, and recent quantitative studies suggest that hand-eye coordination (visual-motor coordination) may be degraded during spaceflight. To better understand the fundamental role gravity plays in coordinating movement, and to determine human performance limitations on space missions, detailed, well-controlled studies on visual-motor coordination are needed. Careful recording of movements such as grasping, pointing, and tracking targets can reveal a great deal about the adaptations taking place during spaceflight.

Experiments during space missions differ dramatically from typical laboratory research, imposing very high demands on equipment and procedures. Since space and mass are both limited on missions, the hardware should be small and lightweight. Also, since developing space-qualified equipment is expensive, the hardware should be deployable for a number of different experimental objectives. Setup and operation of the equipment should be simple, since the crewmembers will have a variety of backgrounds, time is limited during space missions, and the principal investigator is not readily available to supervise operations.

To meet these constraints, we designed the Visual-motor Coordination Facility (VCF) according to the following criteria:

- fits into a single Spacelab locker
- easy to set up and stow
- no calibration or signal quality judgements needed
- turnkey operation
- largely automated nominal procedures
- largely automated error diagnostics

To accommodate a variety of experimental paradigms for present and future use, the VCF employs computer-generated visual objects as stimuli. The subject can then point at, grasp, or track the visual object presented. The size, location, and movement of the target object are freely programmable depending on the skill to be tested. The VCF registers the positions of one or two fingers of one hand and, using computer control, can allow or prevent visual feedback about the hand movements. A trigger switch is also available to measure reaction times or other responses.

The VCF is a compact, versatile, easy-to use device for the testing of visual-motor skills. In addition to space applications, the VCF could be useful for visual-motor skill testing on Earth, particularly in mobile, non-laboratory scenarios. Possible application areas include industrial, sports, clinical, and seniors' home environments.

HARDWARE DESCRIPTION

Figure 1 illustrates a Neurolab crewmember using the VCF. The crewmember is viewing an object on the computer screen. The VCF tracks the movements of the gloved hand.

The main functional component of the VCF is a laptop computer, oriented with its LCD facing up, as shown schematically in Figure 2. The computer presents visual targets, records hand position data and trigger switch events, stores these data to diskette and hard disk, and controls the peripherals.

Figure 2 shows that the subjects are located to the rear right of the VCF and observe the computer display through a reflective Lexan "mirror." Due to the orientation of this mirror, the virtual image of the computer display will appear in a vertical plane at eye level, 30 cm ahead. The mirror is rear-covered by a black shutter, which can be raised or lowered by a servomotor under computer control or by manual overdrive. When the shutter is lowered, the screen serves as a full mirror; when it is raised, it serves as a semitransparent mirror, allowing subjects to see both the computer display and bright objects straight ahead—equivalent to a head-up display.

The subjects wear a black Spandex glove on their dominant hand, with microlamps attached to the thumb and index fingertips. The lamps are connected to the VCF computer's parallel port by a plug on the corresponding side of the VCF. The computer can turn one or both lamps on, depending on the experimental paradigm. Mounted on the other side of the VCF is a molded handgrip with a built-in trigger switch at its top, which is connected to the VCF by a plug on its side. Subjects can conveniently grasp the handle with their nondominant hand to stabilize body posture, and at the same time depress the trigger switch with their thumb if the paradigm so requires. Left-handers will connect the glove to the left, and the handgrip to the right VCF plug, while right-handers will do the opposite. The computer checks this connection pattern, and thus determines whether the subject is left- or right-handed. This allows the device to modify target presentation depending on handedness. In our study, the display for right-handers was shifted towards the right edge of the VCF and vice versa for left-handers, which made it easier for subjects to "reach

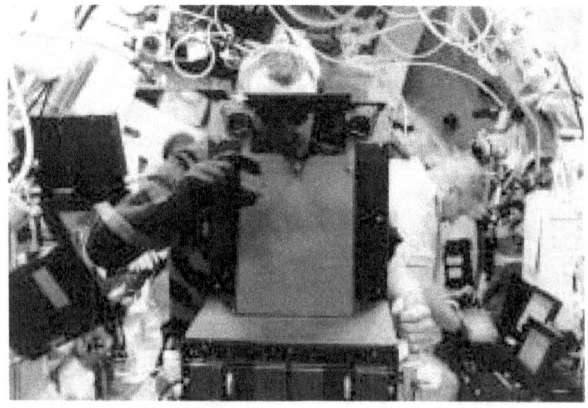

Figure 1. The VCF in use during the Neurolab mission.

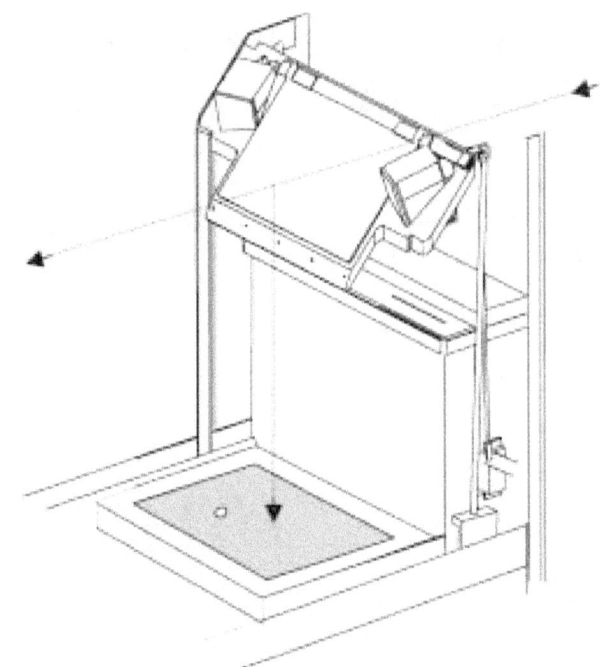

Figure 2. Schematic view of the main VCF components. The laptop computer screen is oriented horizontally (shaded area, with a white dot representing a visual target). The computer itself is vertical, with the diskette drive slot at the top. Located above the screen is the Lexan sheet with a back cover: if the back cover is lowered (as shown), the Lexan serves as full mirror; if the back cover is raised by rotating it about a hinge at the top, the Lexan becomes semitransparent. The two miniature cameras can be seen to the left and right of the back cover. Subjects look into the VCF from the rear right, and perceive the computer screen as being straight ahead in their frontal plane (dashed lines).

around" the VCF structure when responding to the presented targets.

Two miniature charge-coupled device (CCD) cameras on the left and right of the mirror record the subjects' workspace from different angles. The camera signals are multiplexed to an image processing board, where they are digitized in real time. A threshold and a center-of-gravity routine are then applied to determine the locations of bright objects in the field of view, which are stored as fingertip positions in camera coordinates. To prevent the cameras picking up light sources other than the fingertip lamps, their field of view is limited to the VCF workspace by a Lexan sheet covered by black cloth, which is located five cm beyond the virtual target plane. This arrangement reliably screens out undesired light sources. Occasionally, depending on the subjects' hand posture, lamps could be hidden from a camera's view (see Figure 3).

Figure 3 shows schematically the VCF, background sheet, handgrip, and approximate location of the dominant hand in the cameras' field of view. In this presentation, the subject looks into the viewing aperture of the VCF from the front left, with forehead and chin resting against a padded head-rest and

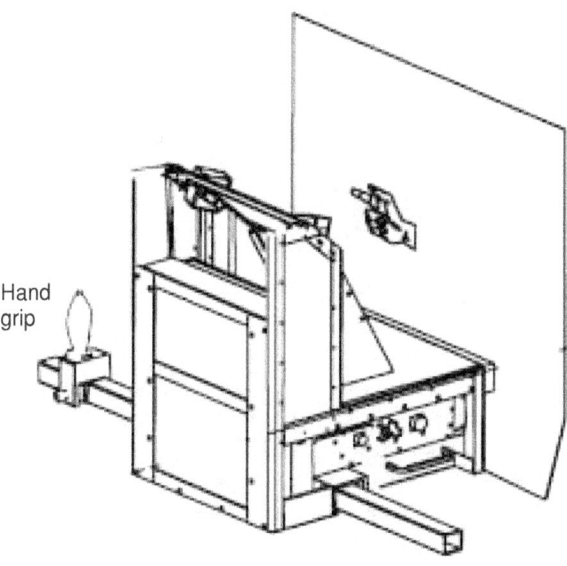

Hand grip

Figure 3. Schematic view of the VCF. The device is mounted on a square bar, along with the handgrip for the nondominant hand. Also shown is the Lexan sheet, which screens off external light sources, and the position of the subject's hand in the workspace.

chin-rest. The viewing aperture effectively blocks out visual and acoustic distractions from other activities around the subject.

Two clamps and a square recess in the VCF baseplate allow a secure attachment of the hardware. On Neurolab, the VCF was attached to a height-adjustable mounting structure, which could be collapsed to the Space Shuttle floor between sessions. The subject's body was secured to the equipment by the head-rest, chin-rest, handgrip, waist band, and footloops

SOFTWARE DESCRIPTION

The program implements a battery of visual-motor tests, described elsewhere in this book (see science report by Bock et al. in this publication). In a typical experimental session, the subject sets up the VCF by mounting it on a suitable support structure, and plugging in the glove, handgrip, and +28 Vdc power supply. After turning on the unit, user dialogs are displayed on the VCF computer screen in upside-down proportional graphical characters of 30×25 (h×w) pixels. These appear right side up to the subject who is viewing the virtual display area. Subjects communicate confirmatory decisions to the VCF by pressing the trigger switch, and non-confirmatory ones by withholding a trigger press (the keyboard cannot be used for this purpose, as it is not easily accessible). Subjects also provide inputs to the program by inserting diskettes that identify the subject's name, specify the order of experimental paradigms for the particular session, and indicate which part of the session has already been completed. The latter feature is particularly useful when the session must be interrupted and the VCF

turned off due to other inflight activities. Data collection can resume later where it left off.

After startup, the subjects typically see several messages, prompting them for a few diagnostic checks (e.g., "are the fingerlamps lit?" "is the shutter open?"). Subjects press the trigger switch to confirm successful diagnostics, and the software starts displaying targets and recording lamp positions for the first experimental paradigm. Rest breaks are programmed about every 20 seconds to prevent fatigue and to allow the recorded data to be saved to diskette and hard disk during these intervals. Achievement scores are displayed during the rest breaks for motivation purposes. Subjects can terminate the rest break anytime by depressing the trigger switch.

To achieve highly accurate readings of experiment elapsed time, the standard PC Timer 0 is set to one millisecond rather than 18.6-millisecond resolution. To provide the program with this enhanced time resolution, an interrupt procedure is introduced to handle the one-millisecond Timer 0 interrupt while ensuring normal function of the regular timer.

To ensure that even briefly presented (e.g., moving) targets are completely displayed on the computer screen, target presentation is synchronized with the screen refresh pulse. The actual target onset is recorded and stored in experiment elapsed time.

To record trigger switch presses asynchronously to target displays, the switch is connected to the ACQ input of LPT1. This arrangement, along with the proper programming, allows us to record trigger presses with an accuracy of one millisecond.

Repeatedly drawing the targets on the screen would present a large computational burden for the computer. To avoid this, targets are predrawn in all possible positions in background color. The software cycles through the color lookup table to make the target successively "appear" in different locations. This cuts the memory transfer to the graphics controller by a factor of 200 to 9000, depending on target size.

All recorded data are saved to the hard drive and, as a backup, to diskette. Finger position data are stored in camera coordinates along with a time stamp (i.e., t, x1, x2, y1, y2). Subsequent software converts these data into 3D Cartesian coordinates, with a position accuracy better than three mm, and a recording rate of 30 to 50 lamp positions per second. Additional data analysis software is available to extract kinematic parameters from the recorded data, such as reaction time, peak velocity, and movement amplitude.

The VCF software performs a number of automated diagnostic routines. For example, if one of the finger lamps is not visible by one of the cameras over several samples, the software will interrupt data recording, open the shutter so the subject can see the hand, and display an instruction to correct hand posture or lamp alignment. After rectifying the problem, the subject returns to data collection by depressing the trigger switch. Other diagnostic routines check for faulty diskettes and diskette drives, proper shutter position, and operator errors such as premature removal of the diskette. The experiment will automatically run off the hard drive if diskettes cannot be accessed; and conversely, the experiment can run off a bootable "hard-drive bypass diskette" if the hard drive should fail.

APPLICATIONS

The VCF is a compact, versatile, easy-to use device for the testing of visual-motor skills. It was well accepted by the Neurolab crew, and functioned flawlessly during the mission, allowing us to collect the data we hoped for. Data analysis revealed an intriguing pattern of performance changes in weightlessness.

Due to its properties, the VCF could be a useful device for visual-motor skill testing on Earth—particularly in mobile, non-laboratory scenarios. Possible application areas include industrial, sports, clinical, and seniors' home environments. We have started to evaluate the utility of a simplified VCF model in collaboration with two rehabilitation institutes in the city of Köln, Germany. Using different paradigms than on Neurolab, we are attempting to quantify the visual-motor deficits of patients with selected neurological disorders. Particular areas of interest are patients with problems in suppressing undesired "automated" movements, or those who need help in using advance information to prepare an upcoming movement. In addition, we are planning a study on the visual-motor performance of seniors, to establish the ability of elderly subjects to learn new motor skills.

The development effort for this experiment produced a robust and versatile system for testing visual-motor coordination, which can be used for important Earth-based applications.

Acknowledgements

VCF design, development and testing was funded by contracts of the Canadian Space Agency to Drs. Bock and Fowler, and of the German Space Agency (DLR) to Dr. Bock. The final VCF model was built and space-qualified by Bristol Aerospace, Ltd., under a contract by the Canadian Space Agency.

Perspectives From Neurolab

Two unique features of the Neurolab mission are discussed in this section. On Neurolab, the crew was closely involved in working with the research teams to shape the varied requirements of 26 experiments into a single set of workable procedures on an integrated timeline. Also, like many Earth-based laboratories, Neurolab had a significant animal research program, which presents challenges in space.

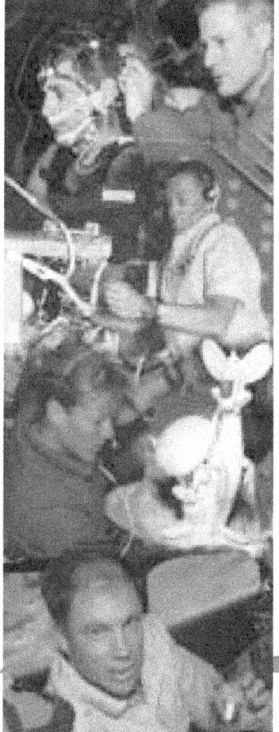

Crew Perspectives From Neurolab

Neurolab had a crew of nine, seven flight crew in orbit and two alternate payload specialists who served in the payload operations and control center during the flight. All the crewmembers were involved in shaping a complex scientific payload into a set of workable procedures that could be done in space. This process gave them a unique perspective on the mission as a whole.

STS-90 crewmembers (front row, left to right) Alternate Payload Specialist Alexander (Alex) W. Dunlap, Pilot Scott D. Altman, Commander Richard A. Searfoss, Alternate Payload Specialist Chiaki Mukai; (back row) Payload Specialist James (Jim) A. Pawelczyk, Mission Specialist Richard (Rick) M. Linnehan, Mission Specialist Kathryn (Kay) P. Hire, Mission Specialist Dafydd (Dave) Rhys Williams, Payload Specialist Jay C. Buckey, Jr.

Commander Rick Searfoss summarizes his impressions of the Neurolab mission in this way:

"Neurolab represents the absolute pinnacle in melding rigorous, detailed scientific investigation with the unique, unforgiving, and ever-present operational demands of human spaceflight. Building on the lessons of 15 previous Spacelab module missions and having very specialized and focused scientific objectives produced outstanding success with the most complex space research mission ever flown. Now, four years after I had the great privilege of leading the flight crew as we worked, trained, and flew the mission, I stand even more in awe of what the entire team—contractors, investigators, NASA life sciences and mission operations personnel, and flight crew—accomplished with this flight.

Early in my astronaut career, a very wise man taught me that the commander sets the tone and the essence of his job is to "work the matrix of relationships between all the crew members." Success in that endeavor invariably leads to mission success. I took that lesson to heart and tried to apply it every single day as launch drew closer and we finally flew. Further, with an undertaking as intricate as the Neurolab mission that philosophy has to be expanded to the "nth" power, where n is the number of different groups with very diverse backgrounds and perspectives, all of whom are crucial to a productive mission. I think virtually everyone on the whole team lived that philosophy and contributed to a marvelous sense of unity. After four years details fade, but what remains as strong as ever in my memory is the incredible human synergy of hundreds of dedicated people who made Neurolab work. And, from a very broad perspective, I really think that is what human space exploration is about, not the high-tech hardware, nor the fancy flying, nor even, really, the payloads, but the ability to pull together in the quest to do things never done before. And in that sense, there has never been a more superlative human space mission than Neurolab since we started this whole wonderful business of flying people to space 40 years ago."

Payload Commander Rick Linnehan echoes the importance of teamwork and good working relationships among the crew in the success of this complex mission:

"My most vivid memories of the Neurolab mission are the ones involving the crew and the interpersonal relationships during training, flight, and even postflight. We all got along and worked so well together, even though there were grueling training and inflight schedules to adhere to and the associated stress that was inevitable. I consider everyone on that flight a very good friend. STS-90 was one of the most important scientific missions that NASA has ever flown. The inflight experiments that we conducted, and were a part of, were first-class science, and the information that was gathered will significantly contribute to the understanding of many physiologic and disease processes that occur in Earth's populations. These

data will also contribute to future spaceflight initiatives involving long-duration flights and, someday, planetary exploration. I am proud to have been a member of the Neurolab team."

Pilot Scott Altman also emphasizes the cooperative nature of the flight. Although he was mainly responsible for the operating the Shuttle and Spacelab systems, he also participated in the scientific payload:

"I was proud to be a part of such a scientifically ambitious effort that coordinated so many investigators and experiments from around the world into one mission. The entire team seemed to work together seamlessly as we planned the experiments and trained for the flight. I really enjoyed participating in the ball catch experiment and supporting all the scientific work by trying to fulfill the photographic and television documentation requirements for each experiment. I also have the claim to being responsible for more living organisms than any other crewmember—over 1500 crickets! I can report that they did well in zero-G and there was no chirping keeping us awake nights.

Neurolab was an incredible experience—both in space and on the ground. I was truly honored to be a part of it."

Crew involvement was critical to Neurolab. Some members of the crew began training on the mission almost two years before the flight began. Throughout the preflight period the crew was involved in refining procedures, developing equipment, and testing protocols. During the flight, the crew served both as operators and as subjects for the experiments inflight and as communicators and problem solvers on the ground. The result of this effort was that the science that could be done on Neurolab was at the level of what could be done in a laboratory on the ground.

Alternate Payload Specialist Chiaki Mukai notes:

"It was truly a fortunate opportunity for me to be a part of the Neurolab mission—the last Spacelab mission conducted in the golden age of the Space Shuttle Program for science utilization. It was one of the most comprehensive science missions ever conducted with various difficult techniques and procedures. The mission included recording microneurograms, using the virtual environment generator head-mounted display, riding in the off-axis rotator, recording from neurons in the rodent hippocampus, performing several surgical techniques (including the survival surgery on the rodents), and so on. The mission demonstrated how capable the Spacelab was by carrying out those techniques and procedures in a same manner we do in the laboratories on the ground. I believe the mission served as a role model for a multiuser science facility such as International Space Station (ISS). I hope the ISS will be completed soon and become mature enough to support the same kinds of challenging experiments as the Neurolab mission did."

The crew of the Neurolab mission included:

Commander Richard A. Searfoss, Col, USAF, ret. Col. Searfoss became an astronaut in 1991. He holds a B.S. in aeronautical engineering with honors from the USAF Academy and an M.S. in aeronautics from the California Institute of Technology. He is also a Distinguished Graduate of the USAF Fighter Weapons School; was named Tactical Air Command F-111 Instructor Pilot of the Year in 1985; and has received the USAF Commendation, Meritorious Service, Defense Superior Service, and Defense Meritorious Service Medals.

Pilot Scott D. Altman, a commander in the U.S. Navy, became an astronaut candidate in December 1994. Commander Altman holds a B.S. degree in aeronautical and astronautical engineering from the University of Illinois and an M.S. in aeronautical engineering from the Naval Postgraduate School. His honors include Distinguished Graduate of the USN Test Pilot School, Association of Naval Aviation 1987 Awardee for Outstanding Achievement in Tactical Aviation, and Navy Achievement and Commendation Medals.

Mission Specialist Kathryn (Kay) P. Hire, was the flight engineer for STS-90. A 1981 graduate of the U.S. Naval Academy, she earned her Naval Flight Officer wings then flew worldwide oceanographic research missions on board specially configured P-3 aircraft. She also completed an M.S. in space technology at Florida Institute of Technology. In 1993, she became the first female in the U.S. assigned to a combat aircrew when she reported for duty with Patrol Squadron Sixty-Two (VP-62).

Mission Specialist Richard (Rick) M. Linnehan, a veterinarian, was the payload commander for Neurolab. Dr. Linnehan earned a B.S. degree in animal sciences from the University of New Hampshire and a D.V.M. from Ohio State University College of Veterinary Medicine. He completed an internship in zoo animal medicine and comparative pathology at the Baltimore Zoo and Johns Hopkins University. NASA selected him as an astronaut in 1992.

Mission Specialist Dafydd (Dave) Rhys Williams, a medical doctor and Canadian Space Agency (CSA) astronaut, was selected as an astronaut in 1995. Dr. Williams' credentials include an M.S. in physiology; an M.D. and a Master of Surgery from McGill University, Montreal; two residencies—one in family practice from the University of Ottawa and the other in emergency medicine from the University of Toronto—and a fellowship in emergency medicine from the Royal College of Physicians and Surgeons of Canada. He was the head of life sciences at NASA-Johnson Space Center from 1998–2002.

Payload Specialist Jay C. Buckey, Jr., a research associate professor of medicine at Dartmouth Medical School, received a B.S. in electrical engineering at Cornell University and an M.D. at Cornell University Medical College. Dr. Buckey completed his internship in internal medicine at The New York Hospital-Cornell Medical Center and his medicine residency at the Dartmouth-Hitchcock Medical Center. He began his space career as a NASA Space Biology Research Fellow at the University of Texas Southwestern Medical Center.

Payload Specialist James (Jim) A. Pawelczyk, an associate professor of physiology at Penn State University, holds B.A. degrees in biology and psychology from the University of Rochester, an M.S. in physiology from Penn State University, and a Ph.D. in biology from the University of North Texas. He completed a postdoctoral fellowship in cardiovascular neurophysiology at the University of Texas Southwestern Medical Center.

Alternate Payload Specialist Alexander (Alex) W. Dunlap, a veterinarian and medical doctor, holds a B.S. degree (cum laude) in zoology and animal science from the University of Arkansas, a D.V.M. from Louisiana State University School of Veterinary Medicine, and an M.D. from the University of Tennessee College of Medicine.

Alternate Payload Specialist Chiaki Mukai, an astronaut for the National Space Development Agency of Japan (NASDA), holds an M.D. and a Ph.D. degree in physiology from Keio University School of Medicine, Japan, and is board-certified as a cardiovascular surgeon by the Japan Surgical Society. A NASDA astronaut since 1985, she has flown in space on the STS-65 and STS-95 missions.

Animal Care on Neurolab

Authors
Jay C. Buckey, Jr., Richard M. Linnehan,
Alexander W. Dunlap

ABSTRACT

The Neurolab mission carried rats, mice, snails, crickets, and two kinds of fish into space. This was the largest variety of living things ever flown on a space mission. On the whole, the experimental animals were healthy during the flight, but there was a notable exception. Although all the mice, adult rats, and most of the young (14 days old at launch) rats did well, more than half of the youngest rats, eight days old at launch, died in the Research Animal Holding Facility inflight. Several factors contributed to this, including housing design and inadequate monitoring capability. An understanding of why this occurred is important for the planning of future space missions.

INTRODUCTION

The Neurolab mission had an ambitious animal research program. The mission carried rats, mice, two kinds of fish, snails, and crickets to accomplish the research objectives. Both adult rats and two groups of young rats (eight days old at launch and 14 days old at launch) were flown. Habitats were provided to sustain the animals. Two different types of animal habitats were available for rats and mice: the Research Animal Holding Facility (RAHF) and the animal enclosure modules (AEM). Oyster toadfish were flown in the vestibular function experimental unit (VFEU) and swordtail fish were housed in the closed ecological biology and aquatic system (CEBAS). The CEBAS also contained snails. Crickets were housed in an incubator (BOTEX) that included a centrifuge to provide artificial gravity for a control group of crickets. The fish, snails, and crickets were not removed from their habitats inflight and the hardware sustaining them required minimal crew intervention, so they will not be discussed in this report. Most of the crew activity was devoted to the rats and mice.

The crew included a veterinarian, Rick Linnehan, D.V.M, who provided and supervised care to the animals inflight, and a veterinarian as an alternate payload specialist, Alex Dunlap, D.V.M., who communicated with the crew during the flight. A veterinary kit was onboard that provided treatment capability for minor problems. Animal checks were incorporated into the timeline. Most experimental procedures were carried out in the general-purpose workstation (GPWS). The GPWS provided a contained space where animals could be removed from their cages and where a variety of experimental equipment could be deployed and used (anesthetics, fixatives, surgical instruments, infusion pumps, work platforms, etc.).

The adult rats, the 14-day-old (at launch) rats, and the mice all did well on the mission. The youngest rats (eight days old at launch), however, had significant problems. More than 50% of these rats died on the flight. An investigation into this revealed that inadequate housing and the inability to provide effective monitoring contributed to this high mortality. The mortality was probably not a unique physiological effect of weightlessness, since rats of this age had flown successfully prior to Neurolab (although in a different kind of housing). For the future, cage designs for spaceflight should incorporate surfaces that allow for easy three-dimensional navigation and reliable ways to monitor the animals.

HOUSING

Research Animal Holding Facility (RAHF)

The RAHF is a temperature- and humidity-controlled facility for housing rodents (Figure 1). Light levels and lighting duration also can be adjusted. The RAHF consists of 12 cages, each of which can house two adult rats, or one mother rat and eight young. The cages each have individual feeding alcoves and watering lixits to provide food and water, respectively. Waste is controlled by airflow that enters the cage on one side and moves the waste to a tray on the opposite side. The individual cage walls were aluminum, and not all surfaces had areas that the rats could grasp. A diagram of the RAHF is shown in Figure 1.

The food system on the RAHF includes a method to measure food consumption. By pulling on a measuring tape, a rough estimate of the amount of food bar consumed can be obtained. The water system also sends data to the ground on how many times water moves through the lixit. Each cage also can detect activity, based on a light sensor. When an animal moves by the sensor, the light beam is interrupted, and this provides a rough measure of activity. Experience on previous spaceflights, however, had shown that weightless objects floating in the cage also trigger the sensor and render the data useless. As a result, the activity data were not used on Neurolab.

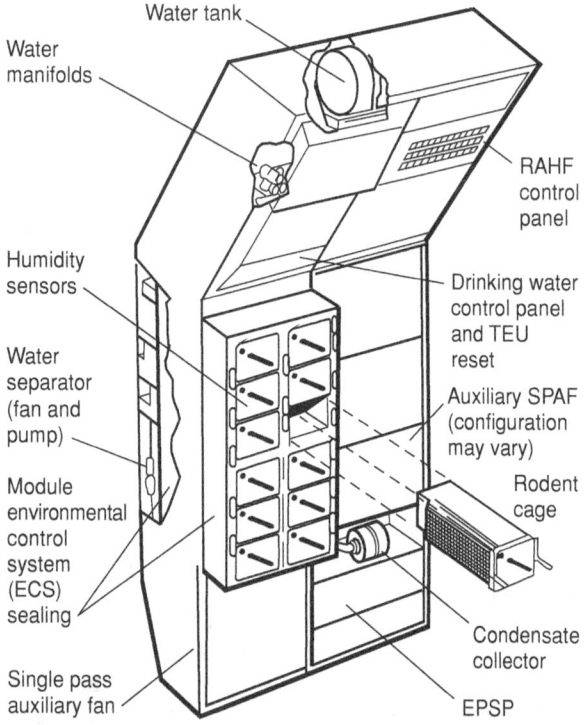

Figure 1. The Research Animal Holding Facility consisted of 12 cages that could house two adult rats or one mother rat and eight young rats. The various systems on the RAHF provided a temperature- and humidity-controlled environment. Water outlets in the cages (lixits) provided water to the rats when the touched the lixit. Food bars were dispensed through a mechanical system.

Each cage module also has a small window. From this window, however, the rear cage cannot be seen, and the view of the front cage is constrained.

Animal Enclosure Module (AEM)

The AEMs are animal habitats designed to fit in a middeck-sized locker (Figure 2). Instead of cages, the animals are housed as a group in an open area. Food is available at the center of the AEM, and water is provided through lixits. The cover of AEM is clear plastic, so the animals can be seen easily. The AEMs were not originally designed to allow for animals to be taken out during the flight. For Neurolab, a special attachment was built that allowed the crew to take animals from and return them to the AEM. A picture of the AEM is shown in Figure 2. On the NIH.R3 Space Shuttle mission, young rats (eight days old at launch) were flown successfully in an AEM.

Comparison of the two housing methods

The RAHF provided individual cages and tight control of environmental parameters. Also, individual food and water consumption could be measured. The RAHF had the disadvantage that the animals could not be easily seen. The AEMs provided group, rather than individual housing, and so did not provide individual measurements of food and water consumption. The rats, however, could be easily seen through the plastic lid. They could also move about easily on the wire surfaces that lined the AEM.

Treatment capabilities

The veterinary kit on board provided fluids for dehydration and antibiotics (enrofloxacin) in case of infection. Also, the kit had Nembutal for euthanasia.

Inflight loss of young (eight days old at launch) rats[1]

The most significant animal issue on the flight was the unexpected death inflight of more than 50% of the youngest rats (eight days old at launch). This occurred in the RAHF located in Spacelab rack 3 (RAHF3).

Twelve Sprague-Dawley female rats, each with a litter of eight young rats, were loaded into RAHF 3 cages 40 hours before launch. Integration of the cages into the RAHF was normal and data indicated that the RAHF was operating as planned. Lixit (water) counts demonstrated that the mothers were consuming water as expected, and this continued over the next two days while STS-90 remained on the launchpad awaiting launch. After launch while on orbit, telemetered data demonstrated normal water consumption and RAHF operation. During the prelaunch period, the Spacelab atmospheric carbon dioxide partial pressure gradually increased to 13 mmHg

[1]The authors would like to acknowledge the white paper written by Louis Ostrach at NASA-HQ, which also included information about this event that was used in this report.

Figure 2. The animal enclosure module provided group housing for the rats. Food was available on exposed food bars in the center of the module and water was available through lixits. The clear cover allowed for the rats to be easily seen.

(1.7%) and then dropped precipitously to ~2 mmHg after Spacelab activation. Based on limited data available in the literature, it is unlikely that these conditions would have affected either the mothers or the young rats.

On FD1, FD2, and FD6, the crew performed the scheduled feeder tape measurements and visual checks on RAHF3. The feeder tape measurements showed that food was being consumed; however, visual checks on the rats were problematic. Rat health could not be accurately assessed through the RAHF windows (this was a known problem from experience on the SLS-2 flight). The windows offered a partial view of the front cage; this view worsened over time as the inside of the windows became coated with particles. Also, since the rats often spent their time in the rear of the cages, usually they could not be seen.

The first planned activity involving the RAHF3 rats was scheduled for FD6. Due to changes in the timeline, however, testing on these animals was postponed and no RAHF3 cages were pulled on that day. The first time a cage was removed from RAHF3 was on FD8. At this time, it was found that two of the eight young rats in that cage had died. Since this was unanticipated, the remaining cages were pulled and an inventory of all the rats in RAHF3 was conducted. In 11 of the 12 cages, some young rats (ranging from one to six per cage) had died. Upon opening each cage, the crew rated whether the remaining rats appeared sick or healthy. Out of the 96 young rats that were launched, 38 had died and 19 appeared sick. The sick rats were given subcutaneous fluids and antibiotics (enrofloxacin). The fluid mix was 0.113 mg enrofloxacin/mL fluid. One mL of this solution was given to each sick rat. Despite this, 12 sick rats were clearly unable to survive and were euthanized. The young rats were redistributed to give the sick rats more opportunity to feed from a mother rat. The status of the rats and the work that had been done with them was put in a spreadsheet and sent to the ground. A private veterinary conference was held to discuss the findings and determine the best plan of action. At the end of FD8, there were 46 young rats remaining in RAHF3.

On each subsequent day, the crew pulled every cage in RAHF3, brought it to the workstation, and checked the health of all the rats. Some rats worsened and were euthanized. Most rats improved and those that appeared sick where given fluids and enrofloxacin. Some rats also received dilute Gatorade and were warmed with heated fluid bags. Spreadsheets were completed each evening to send to the ground to help with replanning. On FD9 there were 44 rats remaining, 40 on FD10, 39 on FD12 and 38 on FD13.

By FD13 the situation in RAHF3 had stabilized. Gel packs (water-containing gel) that the rodents could use to get water in addition to the lixits had been added to the cages. There was some evidence that the mothers were neglecting the young in some cages, and the rats were placed in cages where the other young rodents were doing well.

After landing, the most notable finding was that the rats and cages were soaking wet and the neonates were hypothermic. Two more young rats had died, and one had to be euthanized.

Issues and Recommendations

Ultimately, a group of the young rats did survive and a subset of them appeared healthy throughout. Overall, however, the mortality rate was extremely high, and this has to be taken into account when reviewing the results from the development experiments involving the youngest rats on Neurolab. The problems that occurred with the young rats highlight several issues that can be useful for future space research.

Housing

Young rats had flown in space prior to Neurolab and had done reasonably well. The NIH.R3 Shuttle flight test established a 90% survival in two litters of 10 young rats eight days old at launch. These animals, however, still weighed ~25% less than age-matched AEM ground controls and ~30% less than vivarium controls after landing. Also, this flight used AEMs for housing and not the RAHF. The RAHF cage, which does not have the wire surfaces that the AEM does, would not provide the rats with as many surfaces to grasp and use for navigation. Since in weightlessness the rat can float in three dimensions, all walls need to have surfaces that the rats can grasp. This is especially important because eight-day-old rats have their eyes closed and depend upon the mother for food. If they were to become detached from the mother and not have a way to navigate back, they could end up becoming dehydrated and malnourished. It is possible that this was one factor at work on the Neurolab mission. Also, a mission-length test with young rats using the exact RAHF cages was not performed prior to the mission (the RAHF had flown successfully several times prior to Neurolab). For future missions, the housing should be tested extensively to make sure that it could provide adequate support.

Monitoring capability

The problems on Neurolab might have been averted or minimized if the crew or ground had had some indication of problems within the cages. As it turned out, however, none of the monitoring systems were adequate to detect the problems within the cages. Lixit counts remained normal in all cages. The food bar measurements were consistent with previous experience. The windows were not adequate to allow the crew to see the problems in the cages, and the activity monitoring system had been shown prior to Neurolab to be unreliable. For future habitats, the ability to have reliable monitoring of animals is critical.

Veterinary care

The crew on Neurolab had the ability to remove cages and work with the animals in the GPWS. This capability was very important to salvage the experiments and stop the decline in the health of the rats in RAHF3. The veterinary kit that was initially proposed for the mission had minimal capabilities.

The crew was able to add extra items to the kit preflight, such as the fluids and antibiotics, which subsequently were used on the mission. This experience demonstrated the importance of having the crew trained in the appropriate actions to take, involved in decisions on kit contents, and having the necessary tools on board.

CONCLUSION

The Neurolab mission had a significant animal care problem in space. The problem was unanticipated, and seemed unlikely based on the testing that preceded the mission. The crisis highlighted the importance of detailed preflight hardware testing, and on the provision of dependable monitoring systems for animals. Also, the ability of the crew to treat the animals inflight was an important factor in stabilizing the situation. In the future, it is likely that young rats can be flown successfully if the hardware is robust, reliable monitoring is present, and the crew can intervene if a problem arises.

Appendices

Authors and Affiliations

Adams, Gregory R.	University of California, Irvine, USA
Adams, Jonathan	Lockheed Martin Engineering Services, Houston, USA
Adams, Terence	Lockheed Martin Engineering Services, Houston, USA
Agricola, Hans	University of Jena, Jena, Germany
Al-Hajas, Amjad	Lockheed Martin Engineering Services, Houston, USA
Amberboy, Clif	Lockheed Martin Engineering Services, Houston, USA
Ames, Jimey E.	Hunter Holmes McGuire Department of Veterans Affairs and Medical College of Virginia at Virginia Commonwealth University, Richmond, USA
Anderson, David J.	University of Michigan, Ann Arbor, USA
Anderson, Herbert	Lockheed Martin Engineering Services, Houston, USA
Angelini, Daniela	IRCCS Fondazione Santa Lucia, Rome, Italy
Bain, James	Medical College of Wisconsin, Milwaukee, USA
Baisch, Freidhelm J.*	DLR Institute of Aerospace Medicine, Köln, Germany
Baldwin, Kenneth M.*	University of California, Irvine, USA
Beall, Andrew C.	Massachusetts Institute of Technology, Cambridge, USA
Beck, E. J.	DLR Institute of Aerospace Medicine, Köln, Germany
Benavides, Luis	New York University School of Medicine, New York City, USA
Bennett, Ron	Lockheed Martin Engineering Services, Houston, USA
Berthoz, Alain*	Collège de France, Paris, France; The European Laboratory for the Physiology of Action, Paris, France, and Rome, Italy
Biaggioni, Italo	Vanderbilt University, Nashville, USA
Bideaux, Patrice	Université Montpellier II, Montpellier, France
Blanc, Emmanuelle	Université Montpellier II, Montpellier, France
Blomqvist, C. Gunnar*	University of Texas Southwestern Medical Center, Dallas, USA
Bock, Otmar*	German Sport University, Köln, Germany
Booker, Floyd	Lockheed Martin Engineering Services, Houston, USA
Boyle, Richard	Oregon Health Sciences University, Portland, USA; NASA, Ames Research Center, Moffett Field, USA
Buchko, Christopher J.	University of Michigan, Ann Arbor, USA; Guidant Corp., Santa Clara, USA
Buckey, Jr., Jay C.	Dartmouth Medical School, Hanover, USA
Centini, Claudia	Università di Pisa, Pisa, Italy
Charles, Huong	Lockheed Martin Engineering Services, Houston, USA
Cheng, Jean	Lockheed Martin Engineering Services, Houston, USA
Cirelli, Chiara	Università di Pisa, Pisa, Italy
Clément, Gilles*	Centre National de la Recherche Scientifique, Paris, France
Cohen, Bernard*	Mount Sinai School of Medicine, New York City, USA
Comfort, Deanna	York University, Toronto, Canada
Connolley, Jim	NASA Ames Research Center, Moffett Field, USA
Cooke, William H.	Michigan Technological University, Houghton, USA
Cooper, Trevor K.	University of California, San Diego, USA
Cox, James F.	Hunter Holmes McGuire Department of Veterans Affairs and Medical College of Virginia at Virginia Commonwealth University, Richmond, USA
Curthoys, Ian	University of Sydney, Sydney, Australia
Czeisler, Charles A.*	Harvard Medical School and Brigham and Women's Hospital, Boston, USA

---

*Principal Investigator

Affiliations of the authors are at the time of the STS-90 mission.

Dai, Mingjia	Mount Sinai School of Medicine, New York City, USA
d'Ascanio, Paola	Università di Pisa, Pisa, Italy
Dechesne, Claude J.	Université Montpellier II, Montpellier, France
DeFelipe, Javier	Instituto Cajal (CSIC), Madrid, Spain
De La Cruz, Angelo	NASA Ames Research Center, Moffett Field, USA
Demêmes, Danielle	Université Montpellier II, Montpellier, France
Denslow, Maria J.	Harvard Medical School and Beth Israel Deaconess Medical Center, Boston, USA
Diedrich, André	Vanderbilt University, Nashville, USA
Dijk, Derk-Jan	Harvard Medical School and Brigham and Women's Hospital, Boston, USA
Dillon, Kathy	University of Arizona, Tucson, USA
Drummer, C.	DLR Institute of Aerospace Medicine, Köln, Germany
Eckberg, Dwain L.*	Hunter Holmes McGuire Department of Veterans Affairs and Medical College of Virginia at Virginia Commonwealth University, Richmond, USA
Elland, Carol	NASA Ames Research Center, Moffett Field, USA
Elliott, Ann R.	University of California, San Diego, USA
Ertl, Andrew C.	Vanderbilt University, Nashville, USA
Fedor-Duys, Veronica	University of Arizona, Tucson, USA
Fine, Janelle M.	University of California, San Diego, USA
Förster, Susanne	University of Ulm, Ulm, Germany
Fowler, Barry	York University, Toronto, Canada
Fu, Qi	University of Texas Southwestern Medical Center, Dallas, USA
Fuller, Charles A.*	University of California, Davis, USA
Fuller, Jeremy C.	University of California, Davis, USA
Fuller, Patrick M.	University of California, Davis, USA
Gaffney, F. Andrew	Vanderbilt University, Nashville, USA
Gao, Wenyuan	University of Texas Health Science Center, San Antonio, USA
Garcia-Segura, Luis Miguel	Instituto Cajal (CSIC), Madrid, Spain
Gates, Darrin	Bristol Aerospace Ltd., Winnipeg, Canada
Gaven, Florence	Université Montpellier II, Montpellier, France
Geisen, Lutz	German Sport University, Köln, Germany
Gerzer, R.	DLR Institute of Aerospace Medicine, Köln, Germany
Giller, Cole A.	University of Texas Southwestern Medical Center, Dallas, USA
Gomez-Varelas, Nicole	NASA Ames Research Center, Moffett Field, USA
Grounds, Dennis	NASA Johnson Space Center, Houston, USA
Haddad, Fadia	University of California, Irvine, USA
Harbaugh, Heidi	NASA Ames Research Center, Moffett Field, USA
Harder, Justin	Harvard Medical School, Brigham and Women's Hospital, Boston, USA
Harding, Shannon	New York University School of Medicine, New York City, USA
Harrison, Jeffrey L.	University of Texas Health Science Center, San Antonio, USA
Hayes, Nancy L.	University of Medicine and Dentistry of New Jersey, Robert Wood Johnson Medical School, Piscataway, USA
Heer, Martina	DLR Institute of Aerospace Medicine, Köln, Germany
Highstein, Stephen M.*	Washington University School of Medicine, St. Louis, USA
Hines, John	NASA Ames Research Center, Moffett Field, USA
Hoban-Higgins, Tana M.	University of California, Davis, USA
Holstein, Gay R.*	Mount Sinai School of Medicine, New York City, USA
Holtzman, Sandy	Medical College of Wisconsin, Milwaukee, USA
Horn, Eberhard R.*	University of Ulm, Ulm, Germany
Howard, Ian P.	York University, Toronto, Canada
Hughes, Rod J.	Harvard Medical School and Brigham and Women's Hospital, Boston, USA
Intravaia, Anthony	NASA, Ames Research Center, Moffett Field, USA
Iwasaki, Kenichi	University of Texas Southwestern Medical Center, Dallas, USA
Jackson, Jennifer	Harvard Medical School and Brigham and Women's Hospital, Boston, USA
Jagiello, Krzysztof	University of Arizona, Tucson, USA

Jenkin, Heather L.	York University, Toronto, Canada
Johnson, Michael A.	University of Michigan, Ann Arbor, USA
Jüngling, Susanne	German Sport University, Köln, Germany
Kalb, Robert	Children's Hospital of Philadelphia, Philadelphia, USA
Kämper, Günter	University of Ulm, Ulm, Germany
Kaneko, Michiyo	Tokyo Kaseigakuin Junior College, Tokyo, Japan
Kanups, Andris	Ohio State University, College of Veterinary Medicine, Columbus, USA
Karemaker, J. M.	DLR Institute of Aerospace Medicine, Köln, Germany
Katahari, Kiyoaki	Fukushima Medical University School of Medicine, Fukushima, Japan
Katsuda, Shin-ichiro	Fukushima Medical University School of Medicine, Fukushima, Japan
Killian, Steven	University of California, Davis, USA
Kissell, Amy	University of California, San Diego, USA
Knierim, James J.	University of Arizona, Tucson, USA; and University of Texas-Houston Medical School, Houston, USA
Koizuka, Izumi	St. Marianna University School of Medicine, Kawasaki, Japan
Kosik, Kenneth S.*	Harvard Medical School, Brigham and Women's Hospital, Boston, USA
Krol-Sinclair, Jerzy	Harvard Medical School and Brigham and Women's Hospital, Boston, USA
Krug, Jennifer	Lockheed Martin Engineering Services, Houston, USA
Kubo, Takeshi	Osaka University, Osaka, Japan
Kuielka, Ewa	Mount Sinai School of Medicine, New York City, USA
Kuppinger, Marlene	University of Ulm, Ulm, Germany
Kuusela, Tom A.	University of Turku, Turku, Finland
Lacquaniti, Francesco	The European Laboratory for the Physiology of Action, Paris, France, and Rome, Italy; IRCCS Fondazione Santa Lucia, Rome, Italy; Università di Roma Tor Vergata, Rome, Italy
Lane, Lynda D.	Vanderbilt University, Nashville, USA
Lang, Rosemary	Mount Sinai School of Medicine, New York City, USA
Lawrence, Karen	Lockheed Martin Engineering Services, Houston, USA
Lee, Angelene	NASA Johnson Space Center, Houston, USA
Lei, Grace	Lockheed Martin Engineering Services, Houston, USA
Leibl, Wendy	Medical College of Wisconsin, Milwaukee, USA
Lekschas, Angel	Medical College of Wisconsin, Milwaukee, USA
Levine, Benjamin D.	University of Texas Southwestern Medical Center, Dallas, USA
Leyman, Al	Lockheed Martin Engineering Services, Houston, USA
Li, Louise	Lockheed Martin Engineering Services, Houston, USA
Limardo, Jose	Vanderbilt University, Nashville, USA
Linnehan, Richard M.	NASA Johnson Space Center, Houston, USA
Liptikas, John	York University, Toronto, Canada
Llinás, Rodolfo R.	New York University School of Medicine, New York City, USA
MacDonald II, E. Douglas	Mount Sinai School of Medicine, New York City, USA
Marchiel, Brian	Centre National d'Etudes Spatiales (CNES), Toulouse, France
Martens, Wim L. J.	TEMEC Instruments BV, Kerkrade, The Netherlands
Martin, David C.	University of Michigan, Ann Arbor, USA
Martinelli, Giorgio P.	Mount Sinai School of Medicine, New York City, USA
Matsumoto, Shigeji	Nippon Dental University, Tokyo, Japan
McCollom, Alex	Harvard Medical School and Brigham and Women's Hospital, Boston, USA
McIntyre, Joseph	The European Laboratory for the Physiology of Action, Paris, France, and Rome, Italy; Collège de France, Paris, France; IRCCS Fondazione Santa Lucia, Rome, Italy
McMahon, Diane	Centre National d'Etudes Spatiales (CNES), Toulouse, France
McNaughton, Bruce L.*	University of Arizona, Tucson, USA
Mensinger, Allen F.	Washington University School of Medicine, St. Louis, USA
Miller, Thomas	Medical College of Wisconsin, Milwaukee, USA
Mills, Trent	Lockheed Martin Engineering Services, Houston, USA
Miyake, Masao	The University of Tokyo, Tokyo, Japan
Miyamoto, Yukako	Fukushima Medical University School of Medicine, Fukushima, Japan
Moller, K.	DLR Institute of Aerospace Medicine, Köln, Germany

Moore, Steven	Mount Sinai School of Medicine, New York City, USA
Mortimer, Alan	Canadian Space Agency, Ottawa, Canada
Muecke, David	Lockheed Martin Engineering Services, Houston, USA
Mukai, Chiaki	National Space Development Agency of Japan, Ibaraki, Japan
Murakami, Dean M.	University of California, Davis, USA
Nagaoko, Syunji	Fujita Health University, Aichi, Japan
Nagayama, Tadanori	Fukushima Medical University School of Medicine, Fukushima, Japan
Natapoff, Alan	Massachusetts Institute of Technology, Cambridge, USA
Neri, David F.	Harvard Medical School and Brigham and Women's Hospital, Boston, USA; NASA Ames Research Center, Moffett Field, USA
Neubert, Jürgen	German Aerospace Establishment, Köln, Germany
Nguyen, Kim	Lockheed Martin Engineering Services, Houston, USA
Nowakowski, Richard S.*	University of Medicine and Dentistry of New Jersey, Robert Wood Johnson Medical School, Piscataway, USA
Oishi, Hirotaka	Fukushima Medical University School of Medicine, Fukushima, Japan
Okouchi, Toshiyasu	Fukushima Medical University School of Medicine, Fukushima, Japan
Oman, Charles M.*	Massachusetts Institute of Technology, Cambridge, USA
Paiva, Manuel	Université Libre de Bruxelles, Brussels, Belgium
Paranjape, Sachin Y.	Vanderbilt University, Nashville, USA
Parker, Kevin A.	University of Texas Health Science Center, San Antonio, USA
Pathak, Amit	Vanderbilt University, Nashville, USA
Pawelczyk, James A.	The Pennsylvania State University, University Park, USA
Pawlowski, Vince	University of Arizona, Tucson, USA
Peters, Carlo	TEMEC Instruments BV, Kerkrade, The Netherlands
Peterson, Rob	Vanderbilt University, Nashville, USA
Poe, Gina R.	University of Arizona, Tucson, USA
Pompeiano, Maria	Università di Pisa, Pisa, Italy
Pompeiano, Ottavio*	Università di Pisa, Pisa, Italy
Prisk, G. Kim	University of California, San Diego, USA
Raphan, Theodore	Brooklyn College, New York City, USA
Ray, Chet	University of Texas Southwestern Medical Center, Dallas, USA
Raymond, Jacqueline*	Université Montpellier II, Montpellier, France
Reiser, Paul	Medical College of Wisconsin, Milwaukee, USA
Riel, Eymard	Harvard Medical School and Brigham and Women's Hospital, Boston, USA
Riewe, Pascal	University of Ulm, Ulm, Germany
Riley, Danny A.*	Medical College of Wisconsin, Milwaukee, USA
Ritz-De Cecco, Angela	Harvard Medical School and Brigham and Women's Hospital, Boston, USA
Roberts, Brian	NASA Ames Research Center, Moffett Field, USA
Roberts, Shanda	University of Arizona, Tucson, USA
Robertson, David*	Vanderbilt University, Nashville, USA
Robertson, Rose Marie	Vanderbilt University, Nashville, USA
Robinson, Edward L.	University of California, Davis, USA
Ronda, Joseph M.	Harvard Medical School and Brigham and Women's Hospital, Boston, USA
Ross, Muriel D.*	NASA Ames Research Center, Moffett Field, USA
Schiavi, Richard	Vanderbilt University, Nashville, USA
Schonfeld, Julie	NASA Ames Research Center, Moffett Field, USA
Sebastian, Claudia	University of Ulm, Ulm, Germany
Shah, Pradip	Lockheed Martin Engineering Services, Houston, USA
Shimizu, Tsuyoshi*	Fukushima Medical University School of Medicine, Fukushima, Japan
Silvagnoli, Felix	Lockheed Martin Engineering Services, Houston, USA
Silver, Robert B.	Marine Biological Laboratory, Woods Hole, USA
Skidmore, Mike	NASA Ames Research Center, Moffett Field, USA
Slocum, Glenn	Medical College of Wisconsin, Milwaukee, USA
Smith, Barbara	Ohio State University College of Veterinary Medicine, Columbus, USA

Smith, Karen   Harvard Medical School and Brigham and Women's Hospital, Boston, USA
Smith, Theodore   Massachusetts Institute of Technology, Cambridge, USA
Stengel, Casey   University of Arizona, Tucson, USA
Steward, Oswald   Reeve-Irvine Research Center, University of California at Irvine, Irvine, USA
Sulica, Dan   New York University School of Medicine, New York City, USA

Tahvanainen, Kari U. O.   Kuopio University Hospital, Kuopio, Finland
Tang, I-Hsiung   University of California, Davis, USA
Temple, Meredith D.   National Institute of Biomedical Imaging and Bioengineering, Bethesda, USA
Tononi, Giulio   Università di Pisa, Pisa, Italy
Tresco, Patrick A.   University of Utah, Salt Lake City, USA
Tricas, Timothy   Florida Institute of Technology, Melbourne, USA

Usui, Shiro   Toyohashi University of Technology, Toyoshashi, Japan

Varelas, Joseph   NASA Ames Research Center, Moffett Field, USA
Vargese, Tom   NASA Ames Research Center, Moffett Field, USA
Vatistas, Nick   University of California, Davis, USA
Venet, Michel   Università di Roma Tor Vergata, Rome, Italy
Ventéo, Stéphanie   Université Montpellier II, Montpellier, France
Vijayan, Kalpana   Medical College of Wisconsin, Milwaukee, USA

Wago, Haruyuki   Fukushima Medical University School of Medicine, Fukushima, Japan
Waki, Hidefumi   Fukushima Medical University School of Medicine, Fukushima, Japan
Walton, Kerry D.*   New York University School of Medicine, New York City, USA
Warren, L. Elisabeth   University of California, Davis, USA
Wesseling, K. H.   DLR Institute of Aerospace Medicine, Köln, Germany
West, John B.*   University of California, San Diego, USA
Wiederhold, Michael L.*   University of Texas Health Science Center, San Antonio, USA
Williams, Dafydd R.   NASA Johnson Space Center, Houston, USA
Wong-Riley, Margaret T. T.   Medical College of Wisconsin, Milwaukee, USA
Wyatt, James. K.   Harvard Medical School and Brigham and Women's Hospital, Boston, USA

Yamasaki, Masao   Fukushima Medical University School of Medicine, Fukushima, Japan
Yoshida, Kaoru   University of Tskuba, Tskuba, Japan

Zacher, James E.   York University, Toronto, Canada
Zago, Myrka   The European Laboratory for the Physiology of Action, Paris, France, and Rome, Italy; IRCCS Fondazione Santa Lucia, Rome, Italy
Zhang, Rong   University of Texas Southwestern Medical Center, Dallas, USA
Zuckerman, Julie H.   University of Texas Southwestern Medical Center, Dallas, USA

Neurolab Team Picture

*The Neurolab mission international experiment team
of crewmembers, scientists, engineers, technicians, and managers.
Photo taken at the Fifth Neurolab Investigator's Working Group meeting
held August 21- 23, 1996, at the European Space Agency ESTEC facility,
Noordwijk, the Netherlands.*

Acronyms and Abbreviations

A

AEM	animal enclosure module
AGC	asynchronous ground control
AHI	apnea-hypopnea index
ALFE	astronaut lung function experiment
ALS	amyotrophic lateral sclerosis
AM	amplitude modulation
ANA	aortic nerve activity
ANOVA	analysis of variance
AP	area postrema
ARC	Ames Research Center
ASI	Agencia Spatiale Italiano
ATLAS	adaptation to linear acceleration in space

B

bpm	beats per minute
BT	body temperature

C

C_2H_2	acetylene
$CaCO_3$	calcium carbonate
CBF	cerebral blood flow
CCD	charge-coupled device
CCW	counterclockwise
CDT	Central Daylight Time
CEBAS	closed equilibrated biological aquatic system
CF	climbing fibers
CN	cranial nerve
CNES	Centre National d'Etudes Spatiales (French Space Agency)
CNS	central nervous system
CO_2	carbon dioxide
CPAP	continuous positive airway pressure
CPU	central processing unit
CRISP	CRickets In SPace
CSA	Canadian Space Agency
CTL	control
CTS	circadian timing system
CW	clockwise

D

DAC	digital-to-analog converter
DBP	diastolic blood pressure
DC	direct current
DLR	Deutschen Zentrum für Luft- und Raumfahrt (German Space Agency)
DSP	digital signal processor
DSR	digital sleep recorder

E

ECG	electrocardiogram
ECS	environmental control system
ECu	external cuneate nucleus
EDL	extensor digitorum longus
EEG	electroencephalogram
EGF	epidermal growth factor
EIT	electrical impedance tomography
EM	electron microscopy
EMG	electromyography; electromyogram
EOG	electro-oculogram
EPSP	experiment power switching panel
ESA	European Space Agency
ESS	Experimental Support Scientist
ESTEC	European Space Research and Technology Center

F

FC	foot contact
FD	flight day (e.g., FD2)
FL	foot lift
FLT	flight
fps	frames per second
FRA	fos-related antigen
FS	flocculent substance

G

GABA	gamma-aminobutyric acid
GH	growth hormone
GIA	gravitoinertial acceleration
glut	glutaraldehyde
GMT	Greenwich Mean Time
GPWS	general-purpose workstation

H

HCVR	hypercapnic ventilatory response
HD	head direction (cells)
HDT	head-down tilt
HMD	head-mounted display
HR	heart rate
HVR	hypoxic ventilatory response

I

IACUC	Institutional Animal Care and Use Committee
IANA	integrated aortic nerve activity
ICC	immunocytochemistry
IDE	integrated drive electronics
IEG	immediate early gene

IGF-I	insulin-like growth factor-I
IO	inferior olive
IOBe	subnucleus b
IODM	dorsomedial cell column of the inferior olive
IOK	dorsal cap of Kooy
IOM	medial inferior olive
IOPr	principal inferior olive
IR	infrared
ISA	industry standard architecture
ISS	International Space Station

J

JSC	Johnson Space Center

K

KSC	Kennedy Space Center

L

LA	linear acceleration
LBNP	lower body negative pressure
LC	locus coeruleus
LCD	light crystal display
LD	light-dark (cycle)
LED	left-ear-down
	light-emitting diode
LEO	left-ear-out
LL	low light (dim red light)
LMES	Lockheed Martin Engineering Services
LMS	Life and Microgravity Spacelab
LOB	lying-on-back
LP	light pulse
	linear position
LRt	lateral reticular nucleus
LRtPC	parvicellular part of the nucleus
LSLE	Life Science Laboratory Equipment
LTP	long-term potentiation
LVe	lateral vestibular (nuclei)

M

MAP	mean aortic pressure
MBP	mean blood pressure
MF	mossy fibers
MFB	medial forebrain bundle
MG	medial gastrocnemius
MHC	myosin heavy chain
MIT	Massachusetts Institute of Technology
mRNA	messenger RNA
MSNA	muscle sympathetic nerve activity
MV	microvilli
MVe	medial vestibular nuclei
MVN	medial vestibular nucleus
MVS	mean value synapses

N

N_2	nitrogen
NASA-HQ	NASA-Headquarters
NASDA	National Space Development Agency of Japan

NC	normal control
NGF 1B	nerve growth factor
NGT	nerve guide tube
NIDR	National Institute of Dental Research
NIH	National Institutes of Health
NLP	non-light pulse
nREM	non-rapid eye movement
NSF	National Space Foundation
NTS	nucleus of the tractus solitarius

O

O_2	oxygen
OCR	ocular counter-rolling
OI	orthostatic intolerance
OKN	optokinetic nystagmus
OS	operating system
OSA	obstructive sleep apnea
OTTR	otolith tilt-translation reinterpretation
OVAR	off-vertical axis rotation

P

P	postnatal day (e.g., P8)
PBS	phosphate-buffered saline
PC	personal computer
PCB	personal computer board
PCI	programmable controller interface
PCMCIA	personal computer miniature communications interface adapter
PCNA	proliferating cell nuclear antigen
PDGF-b	platelet-derived growth factor
PF	paraformaldehyde (fixative)
	postflight day (e.g., PF2)
P.L.P.	paraformaldehyde, l-lysine, sodium periodate
PRM	probed recall memory
PSI	position-sensitive interneuron
PTU	propylthiouracil
PVT	psychomotor vigilance test

R

R	recovery day (e.g., R+0)
RAHF	Research Animal Holding Facility
RAM	random access memory
RC	receptor cell
rCHR	roll-induced compensatory head response
REM	rapid eye movement (sleep)
REO	right-ear-out
RF	radio frequency
RR	respiratory rate
RT	reaction-time (task)

S

SaO_2	arterial oxygen saturation
SBP	systolic blood pressure
SC	supporting cell
SCN	suprachiasmatic nucleus
SD	standard deviation
SE	standard error

SEM	scanning electron microscope
SID	subject interface device
SLPL	ProNectin-L
SLS	space life sciences (e.g., SLS-2)
SMI	Sternberger Monoclonals Incorporated
SPAF	single pass auxiliary fan
SPF	specific pathogen free
SpVe	spinal vestibular nuclei
SRS	subject restraint system
ST	statoconia
SuVe	superior vestibular (nucleus)
SV	subjective vertical
sv	synaptic vesicles
SWS	slow wave sleep

T

3D	three-dimensional
T3	thyroid hormone
TA	tibialis anterior
Tb	body temperature
TC	time constant
TD	thyroid deficient
TDT	total dark time
TEM	transmission electron microscopy
TEU	thermal electric unit

TPR	total peripheral resistance
TST	total sleep time
TTC	time-to-contact

U

UC	University of California

V

VC	vacuole
VCF	Visual-motor Coordination Facility
VCR	vestibulocollic reflex
	video cassette recorder
VD	visual dependence
VDR	vitamin D3 receptor
VEG	virtual environment generator
VFEU	vestibular function experimental unit
VIV	vivarium
VN	vestibular nuclei
VOILA	visuomotor orientation in long-duration astronauts
VOR	vestibulo-ocular reflex
VR	virtual reality
VRI	visual reorientation illusion
VSR	vestibulospinal reflex

Index

Index

corkscrew 97
immature 95
surface 95, 97, 100, 101
ventroflexion 97, 98
Rodriguez, Shari 36
roll 135
cricket 138, 139
head 134
lateral 140
Ronda, Joseph 232
Ross laboratory 40
rotation(s) 2, 6, 70, 71
body 69, 80
clockwise (CW) 6, 189
constant-velocity 7, 13
counterclockwise (CCW) 6
direction 13, 14
head **4**
off-vertical axis (OVAR) 12
on-center 15
perception 6, 9
pitch 66
roll 66
Royal College of Physicians and Surgeons
of Canada 293
RS-422 254, 255
Rummel, J. 80
Russian Academy of Science 56
Ruyters, Gunter iii

S

saccule(s), sacculus 1, 12, 16, 24, 39, 42,
43, 45, 125, 143, 144, 145
macula 2, **3**
safety 83, 178, 212, 259
electrical 267, 270
standards 270
Saito, Mitsuro 184/185
SanDisk 250
Sandoz, Gwenn 17
SAS Institute, Inc. 97
Sato, Mutsumi 159
Sawyer, S. 80
Schmitt, Didier 36
Schonfeld, Julie 241
Schulirsch, M. 88
Scion Corporation 97
Scion Image 97
Seagate ST32140A 255
sea hare (*also* Aplypsia californica) 123,
124, 130
Searfoss, Richard A. 36, 291, 292, 293
semicircular canal(s) **3**, 12, 15, 46, 272
Sennheiser GmbH 269
sense(s) 1, 51, 52, 133
down 61
immersion 80
position 1, 36, 46, 51, 64, 74, 134
subjective 69
up 61
vision 46
Sense-8 Corp. 256

sensitivity 40, 43, 45, 46, 48, 49, 158, 207
afferent 49, 155
aortic nerve 153, 154
baroreflex 151, 154, 159
increased 48, 49
light 233
lower, reduced 139, 155, 233
nerve 47, 48
otolith 17
preflight 72
response 47, 48, 49
roll stimulus 141
sensor(s) 1, 40, 133, 152, 177, 188, 190,
193, 207, 215, 219, 224, 225, 251, 263
body temperature 265
light 296
otolith 49
pressure 91, 171, 176
regenerated 135
vision 133
Sensors 2000 Program 235
sensory deprivation 134, 143
sensory integration v, 51, 52, **53, 93**
sensory system(s) 96, 101, 134, 136, 143
gravity 134
non-gravity 134
septum 29
Shea, Steven 232
shoulder(s) 57, 59, 71, 256, 280
Sigma Chemicals Pyt Ltd. 46, 163, 272
signal(s) 28, 31, 35, 52, 59, 70, 101, 152,
171, 172, 177, 178, 179, 188, 190, 212,
245, 246, 250, 264, 267, 268, 269, 270,
271, 273
brain 171, 245, 246
brainstem **4**
cardiorespiratory 250
chemical 96, 102
EEG 250
electrical 59, 96, 102, 269, 273
EMG 59, 250
EOG 250
gravity 28, 34, 45, 46, 147
HD system 67
inner ear 1, 33, 36
microneurography 269
nerve 46, 268
neural 60, 246
neuronal 65
otolith 5, 6
position sense **53**
self-motion 46
sensory 34, 36, 56, 57, 84, 96
sympathetic 177, 198
very-low-level 226, 268
vestibular 28, 64
visual 33, 36, **53**
sinus
carotid 198
sigmoid **209**
straight **209**
sitting 151, 171, 176

Skidmore, Mike 241
skill(s) 55, 83, 84, 88, 96, 161, 285
human 83
manual 83, 84
visual-motor 285, 288
skin 57, 59, 61, 98, 189, 190, 197, 235,
263, 268, 269, 271
Skinner, N. 80
Sklare, Daniel iii
Skwersky, A. 80
Skylab 70, 219
sleep v, 29, 30, 207, 208, **209**, 211, 212,
213, 214, 216, 217, 218, 219, 220, 223,
224, 227, 230, 231, 234, 246, 249, 250,
251
deep (*also* delta) 34, 230
deprivation 28
disruption 223, 224, 230, 231
fragmented 231
hypnotic-induced 213
inflight 217
non-rapid eye movement (nREM) 214,
216, 228, 231
polysomnographic recording 214, 216,
217
rapid eye movement (REM) 29, 34,
211, 214, 216, 217, 219, 228, 231
regulation 29
sleep stage composition 213
slow wave (SWS) 34, 214, 216, 217
stages 214
structure 216, 217, 220
unmedicated 213
sleep apnea(s) 207, 223, 228, 231
central 224, 226, 231
sleep disturbance(s) 27, 211, 212, 216,
220, 234
countermeasure 207, 208, 211, 213
sleep duration
changes 215, 220
sleepiness 207, 212, 216
sleep loss 211, 219
cumulative 220
sleep medication(s) 207, 212, 216, 224
sleep net(s) 207, **209**, 214, 215, 249, 251
sleep-wake cycle 219
normal 28
regulation 27, 28
Smith, Barbara 241
Smith, Karen 220
Smith, Mike 184
Smith, Robin 220
snail(s) 91, 92, 123, 124, 125, 126, 127,
295
adult 125
developing 91, 124, 125
flight-reared 123, 126, 127, 128
freshwater 123
ground-reared 123, 126, 127
larval 123, 124, 125
pond (*also* Biomphalaria glabrata)
124, 125, 126, 130